T0155430

The Four Great Inventions
of Ancient China:
Their Origin, Development, Spread and Influence in the World

Pan Jixing

Translation Team

Translator: Zhu Yue

(Other translators sorted by alphabetization):

Bao Man	Hao Tugen	Hu Jian
Li Qing	Liu Sheng	Wu Xiaofang
Xia Beijie	Xu Liyue	Ye Lan
Zhu Linglin	Zhuang Xiaoling	

Edit Team

Editor: Yao Shuo

(Other editor sorted by alphabetization):

Han Jiwei	Hu Xueyin	Huang Chengqun
Li Panfeng	Yu Xiumei	

Preface

Science and technology in ancient China had long been well-developed. There were more than one hundred China's important discoveries and inventions which had great influence in the world. Therefore, during a long-term period China held the safe lead in the aspect of science in the world. Among them, papermaking, printing, gunpowder and compass should be the super and revolutionary inventions shaking the world since two millennia according to their influence on the social-historical course and the development of science and culture. The famous British scholar Francis Bacon (1561~1626) said that these mechanical inventions have changed the whole face and the state of things throughout the world, because the consequence of such inventions affected not some parts of an area but the whole world; their influence lasted not a short period of time but several hundred years. This could not been done by any empire, religion and great man in the history.

In the past, treatises on the history of papermaking, printing, gunpowder and the compass were published by Chinese and foreign scholars, but such kind of books were only limited to study one or at most two of them. The book systematically and profoundly studying the history of the four great inventions together has not appeared yet in China and abroad. In view of this, University of Science and Technology of China Press (中国科学技术大学出版社) planned to publish such a book in 1998, and asked me to write it. After effort of two years, its preliminary draft was finished in 2000, and through a series of revisions the final text was finished in 2002. It is entitled *Zhongguo Gudai Sida-faming: Yuanliu, Waichuan ji Shijie Yingxiang* (《中国古代四大发明——源流、外传及世界影响》, *The Four Great Inventions of Ancient China: Their Origin, Development, Spread and Influence in the World*).

In the process of research and writing, we try hard to use a comprehensive research method combining together the following six aspects: textual exploration of sources in sino-foreign languages, study of unearthed ancient materials, investigation of workshops preserving the traditional techniques, scientific explanation of ancient things with the knowledge of modern science and the comparison between the Chinese and foreign techniques. We try to absorb the newest results of research done in China and abroad and the newest materials of archaeological discoveries in order to reflex the newest academic level in the 21st century. Of course, to do such a complicated research relating to many disciplines of social and natural sciences is a new attempt for us, I would like to express my thanks to readers in China and abroad for their corrections and comments on this book. Let us summarize the important results of research in this book as follows.

According to the record of the *Hou Han Shu* (《后汉书》, *History of the Later Han Dynasty*, 445), paper was invented by an eunuch Cai Lun (蔡伦, 61~121 A. D.) in the year of 105, however, a series of archaeological discoveries since the 20th century in China and microscopic analysis of unearthed paper made before the time of Cai Lun proved that papermaking originated in the Western Han Dynasty (西汉, 206 B. C. ~24 A. D.) in the 2nd century B. C. The earliest paper was made of hemp fibres from waste rags. Although Cai Lun was not the inventor of paper, but he summed up the technical experience of papermaking in the Western Han, organized to make hemp paper of high quality and promoted the development of papermaking, so he also made contribution in the history of paper. Bark paper from paper mulberry bark was first made in Cai Lun's time. During the Six Dynasties (六朝) (from 3rd to 6th centuries), papermaking was further developed, the quality of hemp paper was improved, its output greatly increased, and hemp paper became the most important writing material in China from the 4th century. Mulberry bark paper, rattan bark paper, sizing and coating technique also appeared in the same time. Paper was also used for making umbrella, kite, paper cuts, artificial flower and others.

During the Tang Dynasty (唐) and Five Dynasties (五代) (from 6th to 10th centuries), in addition to the above mentioned raw materials, bamboo and bark of Daphne and Hibiscus trees were also used for making paper. More kinds of processed paper appeared, such as wax-coated paper, powder- and wax-coated paper, sprinkled gold-silver paper, water mark paper, embossed paper, coloured paper and gelatinized paper. Apart from writing paper and art paper, paper was also used for making clothing, armour, lantern, playing cards, visiting cards, commercial documents and paper money burnt for funeral ("fire-paper"). Another achievement in this period was to produce large-size paper of more than 3 m long for publishing the list of successful candidates in the imperial examination.

China first used paper as writing material in a large-scale in the world and already entered the golden age of hand-written copy during the 4th to 6th centuries. However, with increasing works people had to expend a lot of time and labour for copying everyday. The further development of religion (especially Buddhism) and the social culture showed that hand-written copy could not satisfy the necessity from the society. This promoted the appearance of printing technique which finally replaced hand-writing labour with the mechanical replication method. Newly found textual records and archaeological findings proved that such replacement happened in the Sui Dynasty (隋) (6th century), and in the early Tang Dynasty (7th century), wood-block printing, the earliest form of printing, obtained its early development.

Printing first originated among the people and was related to their religious belief and activities, so the extant earliest prints are Buddhist sutras. In the Five Dynasties (907~960), the government started to print the *Jiu Jing* (《九经》, *Nine Kinds of the Confucianist Classics*), namely, this art entered the elegant hall and was accepted by every stratum in the society. After block printing was developed, copper-plate printing appeared in the Tang Dynasty from the 8th century, it was usually made by casting and used to print single page or short work. In the Northern Song Dynasty (北宋, 960~1126), printing was all-roundly developed. Apart from block printing and copper-plate printing, movable type printing was invented. The earliest type was wooden type which was directly derived from wood block. Between the 10th and 11th centuries, wooden types were used to print contracts and bills, the so called *yanyin* (盐引) and *chayin* (茶引) (licence for sale of salt and tea). Earthenware or pottery type printing was invented during 1041 to 1048. Non-metal-type printing then was transmitted to the minority nationality areas *Xixia* (西夏) and Uygur in northwest during the 12th to 13th centuries.

Movable metal-type printing appeared in the Northern Song Dynasty (11th century) on the basis of copper-plate and non-metal-type printing. Early metal type made by casting of bronze was used for paper money printing by the Song government. Many specimens of plates and printed paper money issued in the Song Dynasty, the Jin Dynasty (金, 1115~1234) and the Yuan Dynasty (元) during the 12th to 14th centuries were unearthed in China since the 20th century. Apart from bronze type tin-alloy type was also founded for book printing in the Southern Song Dynasty (南宋, from 12th to 13th centuries) for lack of copper. Metal-type printing was greatly developed in the Ming Dynasty (明) and the

Qing Dynasty（清）(the 14th to 19th centuries).

The *banhua*（版画）or printed pictures has existed with block printing from the beginning to the end. On early printed pictures the colours were filled in artificially. Then multi-colour mono-block printing appeared in the Northern Song Dynasty (11th century), multicolour multi-block printing or *douban*（饾版，assembled blocks）printing started no later than the Yuan Dynasty (13th century), and was greatly developed in the late Ming Dynasty (16th century). The technological processes of block printing, non-metal-type printing and metal-type printing are described in detail in this book.

The manufacture of gunpowder presupposes the practical use and purification of saltpetre. Saltpetre was used as a medicine to be drunk between the Qin Dynasty（秦）and the Han Dynasty, so it must have been purified. Natural saltpetre was easily confused with *puxiao*（朴硝，sodium sulphate，Na_2SO_4）in outward appearance, so the alchemist Tao Hongjing（陶弘景，456～536) put forward a method for identifying saltpetre by its flame colour (violet) in 500. Saltpetre was carefully studied by Chinese alchemists and pharmaceutists. When alchemists did their experiments in the late Tang Dynasty (9th century), they mixed saltpetre with sulphur, realgar and honey together and let them burn. As a result, deflagration happened, gunpowder mixture was thus discovered. Military gunpowder developed during the late Tang Dynasty to the Five Dynasties (the first half of the 10th century) was used in the battlefield in the early Song Dynasty (the second half of the 10th century).

The *Wujing Zongyao*（《武经总要》，*Collection of the Most Important Military Techniques*) written by Zeng Gongliang（曾公亮，999～1078) described three kinds of the earliest recipes of military gunpowder. Its average contents were: saltpetre (60.7%), sulphur (30.9%) and charcoal (8.4%). Because it was made in the paste form for safety, it could only be used as explosive and could not be used as propellant powder, so early firearms were made of gunpowder package launched by bow, catapult and by hand. During the end of the 11th century to the early 12th century, solid gunpowder of high nitrate was successfully made and used for making fireworks and firecrackers including the flying fire of reaction device. In this case, the function of gunpowder was fully played.

The war between the Song troops and the Jin troops promoted the improvement of firearms. In the early Southern Song Dynasty (1128), metal bombard of bottle shape appeared, in 1132, the *huoqiang* (火枪, fire lance or flame thrower) made of large bamboo tube was used for city defence. The two kinds of firearms became the ancestors of all tube-type firearms in the later time. As a hybridization product of early bombard and fire lance, the *tuhuoqiang* (突火枪, flame-spurting lance) was used in 1259. It was made of a bamboo tube and shot forth a kind of bullet (stone or metal ball) and flame together. Dr. Joseph Needham (1900~1995) coined a special name for it — erupter. In 1161, the Song soldiers launched the *pili pao* (霹雳砲, thunderbolt missile or rocket-propelled bomb) and *huojian* (火箭, rocket) to attack the Jin troops. The manufacture of solid powder of high nitrate and the new types of firearms led to the second technical revolution in the history of gunpowder and firearms and had a deep influence on the future in China and abroad.

During the 12th to the 13th centuries, bomb and grenade of hard shell, the *huochong* (火铳, hand-gun) and cannon were developed. Multi-rocket launcher, two-stage rocket and two-stage reciprocating rocket were made and used in the Ming Dynasty (the 14th to 16th centuries). At the same time, mine, time bomb and submarine mine were also developed. It is worthy to point out that an adventurer of the early Ming Dynasty, Wan Hoo (万虎, 1450~1500) made a bold attempt of rocket-flight with the help of 47 large rockets and sacrificed himself for this. In addition to historical description, we also discuss the combustion theory of gunpowder, bring to light the working principles and internal construction of various firearms, and make a series of technical restoration of them.

The compass was a product of long-term magnetic research in ancient China. As early as the Warring States period (战国, 475 B.C. ~221 B.C.), the Chinese not only discovered lodestone's property of attracting iron, but also simultaneously discovered its property of directivity and polarity, and used natural lodestone to make the earliest instrument for determining the directions, the *sinanyi* (司南仪, south-pointer). The south-pointer was further developed in the Han Dynasty, on its board the 24 directions were marked, which became the foundation of dial of the later compass. The early south-pointer consisted of two parts: a square earth plate of diviner's board and lodestone of spoon shape. Although the south-pointer could point the directions, but its sensitivity of polarity was not good, because the shape of spoon was not reasonable and the friction resistance between lodestone and bronze board was not avoidable. Therefore, a lot of efforts for changing its shape and operation method have been made since the 4th century.

Owing to the development of geomancy and foreign trade, it was necessary to do accurate survey of directions and look for an effective navigation method, which promoted the improvement of south-pointer. During the 4th to 9th centuries, the following improvements were achieved: (1) natural spoon-shaped lodestone was replaced by the artificially magnetized iron needle or fish-shaped iron leaf, hence the technical transition from the spoon to the needle was finished; (2) the square dial of the south-pointer was changed into a round dial board of 24 or 12 directions, hence the shape of dial board cast off the influence of diviner's board; (3) the magnetized needle was hung with silk thread on the round dial board, and it could point the south; (4) the needle was put into a small "water pool" on the round dial board, and it also could point the south. However, the "suspending needle" usually flickered owing to the influence of the surrounding airstream. The "floating needle" seemed to be the better. The two kinds of south-pointing needles were actually different from the south-pointer, and the floating needle was alike to the wet compass.

The earliest form of the magnetic compass, the *kanyu luopan* (堪舆罗盘, geomancer's compass), had already been used by geomancers in the late Tang Dynasty (the 9th century), because the *Guanshi Dili Zhimeng* (《管氏地理指蒙》, *Master Guan's Geomantic Instructor*) compiled in the 9th century recorded that the magnetic needle did not point the due south-north but south by west or north by east. Therefore, it talked about the magnetic declination which could not be found with the *sinan* or south-pointer used before the Tang Dynasty. The origin of the compass should thus be traced back to 9th century China. In the Northern Song Dynasty (11th century), there were more records and descriptions on the wet compass: the *zhinanzhen* (指南针, south-pointing needle) and *zhinanyu* (指南鱼, south-pointing fish). The newest archaeological discovery shows that the dry-pivoted compass was made and used in the Southern Song Dynasty (12th century). We have made new technical restoration of the various wet and dry compasses and talked about the application of the compass in navigation from the Song Dynasty to the Qing Dynasty (from 11th to 18th centuries).

A great quantity of historical facts proved that the Old World including Asia, Africa and Europe should be regarded as a whole, its various areas usually kept direct or indirect mutual exchange in the aspects of economy, culture and science and personnel through different channels in the past, a bridge of mutual communication among peoples in different cultural areas was thus built in the

East and the West. The famous Silk Road on land and on the sea has been the important channel for such exchange since the Han Dynasty, and China's four great inventions were just transmitted westward to other countries along the Silk Road. The Chinese people never monopolized such inventions, but shared them as a gift with the whole world, and thus made a great contribution to the mankind.

According to our research, China's papermaking technique was first introduced to Korea (4th to 5th centuries), Japan (early 5th century) and Vietnam (3rd century), then to India (the second half of the 7th century). Chinese paper was exported to Myanmar, Thailand, Cambodia, and Indonesia as early as the Tang, Song and Yuan Dynasties, but such countries established their own paper mills in the 13th century, in the Philippines even later (15th century). China's papermaking technique found its way westward from the Tang Dynasty. In 751, in the battle of Talas (now Dzhambul, Kazakhstan) between the Tang army and Arabian troops, some Chinese soldiers having papermaking background were captured in Central Asia, they were asked to teach this technique, so the first paper mill was built with the Chinese method in Samarkand, then in Baghdad (794) and Damascus (10th century) in the Arabian world. The Arabs also built new paper mills in Cairo, Egypt (900) and Fez, Morocco (1100) in North Africa. Arabian paper was greatly exported to European countries in the medieval age, and through Arabia papermaking art was afterwards introduced into Europe.

The first European country to set up paper mills was Spain in 1150 in Xatrva, then the Spanish themselves had a new paper mill in Vidalon in 1157. In Southern Europe, the first paper mills were built in Montefano (1276) and Bologna (1293) in Italy. France started making paper in Troyes from 1340, Germany introduced the art from Italy and built earliest paper mill in Nuremberg in 1390. Germany locating in Central Europe hence became the transfer station of transmiting papermaking art to other European countries: Basel, Switzerland (1433), Vienna, Austria (1498), Dordrecht, Holland (1586), Crakow, Poland (1491) and Moscow, Russia (1576). England began to produce paper from 1495 in Hertford. Till the end of the 17th century, various European countries had already built paper mills. In the American New World, the first paper mills were built in Mexico (1575) by the Spanish and in Philadelphia, USA (1690) by Englishmen. In Canada, the early paper mill was built in Andreas (1803) with the help of the Americans. In the early 19th century, China's papermaking art had gone its long spread route of thousand years and ten thousand kilometres throughout the world, paper became a commonly used material in the international society.

Wood-block printing technique was first introduced into Japan from China during 764 to 770, but movable wooden-type and metal-type printing were introduced into Japan in the 16th century through Korea. Korea developed block printing with the technique from China at the end of the 10th century, and introduced China's wooden-type and metal-type technique at the late 14th century earlier than Japan. Vietnam introduced block printing art in the 13th century and issued paper money in the late 14th century using movable type technique from China. The Philippines started block printing and metal-type printing in the 16th century owing to the effort of the oversea Chinese Gong Rong (龚容, 1538~1603), a native of Fujian (福建), in Manila. Block printing found its way to Thailand in the 16th century. However, India did not develop printing before the coming of the Europeans, and the Arabian empire too.

After the Mongol overthrew the Arabian domination in Central Asia and West Asia, printing began to develop there. The Mongol Il-Khanate (1250~1356) issued paper money in Tabriz, Persia, in 1292 with the Chinese way, and thus introduced printing technique from China. Persian scholar Rashid al-Din (1247~1318) described the course of issuing paper money and China's printing technique in his work written in Persian in 1311. Although the issuance of paper money failed in Persia, printing technique was successfully used in West Asia, and was transmitted from there to Egypt, North Africa, during 1300 to 1350 under the domination of the Mameluke Dynasty (1250~1517). This was proved by many unearthed printed matters in Arabian language in Egypt since the 19th century. The Jews living between Persia and Egypt also learned the art of printing and printed religious works in Hebrew at the end of the 14th century.

During the 12th to 13th centuries, medieval Europe introduced papermaking art through Arabia, but did not introduce China's printing art in time, various reading matters still depended on handwriting. Block printing was transmitted into Europe from China during the Renaissance period (1350 to 1400). Germany, Italy and Netherlands (now Holland) seemed to be the early European countries to develop printing. The extant earliest printed matters are religious pictures and playing cards introduced from China by the Mongol army. Later, printed books appeared. Because block printing was not quite suitable to European alphabetical languages, so China's movable-type printing was soon transmitted there. The Italian technician Pamfilo Castaldi (1398~1490) tried to print books with wooden type of large size in Feltre in 1426, the Dutch Laurens Janszoon (1395~1465) did the same work in Haarlem in 1440 and printed Latin grammer and other books with large wooden types. But the Europeans met technical difficulty when they tried to make wooden type of small size, because such type could not be used for lack of enough strength.

In view of the above-mentioned facts, the German Prokop Waldfoghel (1367~1444) once living in Prague of the Holy Roman Empire developed the so-called "*ars scribandi artificialiter*" (art for writing artificially) in Avignon during 1441 to 1446. The key of his art lied in casting of metal type for printing. Ten years later, another German Johannes Gutenberg (1400~1468) successfully cast metal type and printed the Latin edition of *The Bible*, indulgence and other books in Mainz from 1450. Since then, his technique was soon transmitted to other parts of Germany and even whole Europe: Rome, Italy (1465), Basel, Switzerland (1468) , Vienna, Austria (1469), Paris, France (1470), Valencia, Spain (1475), Brugges, Belgium (1475), London, England (1476), Delft, Holland (1477), and Moscow, Russia (1476). Metal-type printing was also introduced to American New World: Boston (1638) and Philadelphia (1690) of the United States.

The comparative research between the Chinese and Gutenberg's technique shows that both sides used the same working principles and fundamental technical processes, even the outward appearance of metal type was also the same, but there was a time difference of 300 to 400 years between them. Of course, Gutenberg developed metal-type printing suiting measures to his local conditions. For instance, he made oil-based ink and a special press of screw device for impression, Because European paper has been thicker than Chinese paper, it could not be printed with the brush by hand, the special measures thus must be taken, which should be regarded as his invention. Metal-type printing as a whole was invented in China earlier than Europe by 3rd to 4th centuries, so Gutenberg was not the inventor of typography, but his contribution to the development of printing must be recognized. Many ancient and modern western scholars thought that Gutenberg must have heard of the idea that metal-alloy could be cast into type used for book printing directly or indirectly from travellers going along the Silk Road from China. In this book, we investigated the Chinese background of Gutenberg's technical activities.

During the 12th to 13th centuries, the Mongol army carried various firearms to go on doubling westward expeditions in Moslem world and Europe. As a result, China's gunpowder and firearm techniques were transmited to Arabian Area during 1250 to 1260 and to Europe during 1260 to 1270. Before this, Chinese saltpetre as a medicine and fireworks as children's toys might have been brought or exported to the West by travellers and merchants. The extant ancient manuscripts on military topic written in Arabian and Latin languages during the 13th to 14th centuries talked about the purification of saltpetre, "flying fire" (firecrackers), fireworks, rocket, "fire-hots" (bomb), grenade, hand-gun, fire-lance, bombard, also described the manufacture of gunpowder and different recipes. In this period, some parts of Arabian area, such as Persia, Syria, Iraq in West Asia and Egypt in North Africa, had already made gunpowder and firearms.

In Europe, the British scholar Roger Bacon (1214~1292) and the German bishop Albert Magnus (1200~1280) were the first to introduce the knowledge of gunpowder, which came from Arabian sources, in their works. Since then, European technicians did a series of experiments to model on Chinese firearms in the 13th century. From the first half of the 14th century, gunpowder and practical firearms were made and used in European countries: Italy (1326~1331), Germany (1330~1340), France and England (1340~1347), Poland and Russia (1342~1348) and Spain (1371). Early European firearms still looked like Chinese ones, but from the second half of the 14th century some kinds of improved firearms appeared in Europe, for example, breech-loading cannon was cast of iron and introduced to China by the Portuguese in the 16th century, the Chinese called it *folangji* (佛朗机, Frankish culverin). After the Renaissance, European firearms were greatly improved and gradually surpassed those of China. In this case, it was once thought that gunpowder was invented by a German called "Bacthold Schwartz" during the 13th to 14th centuries, but through further research it was found that there was not such a person in the history, the view was groundless.

As to the spread of China's gunpowder technique in Asian countries, it was transmitted to Korea in its Koryo Dynasty during 1270 to 1280 under the domination of the Yuan. Mongol rulers allotted large quantity of gunpowder and firearms to Koryo to equip Korean army with the new weapons every year. In the early Ming, the Chinese Emperor Taizu (明太祖) ordered to allot a lot of saltpetre, sulphur, rockets, hand-guns and other firearms to Koryo for defending the attack from *wokou* (倭寇, Japanese pirates) in 1374, saltpetre and sulphur were then compounded into gunpowder in Koryo according to the Chinese recipes. In 1377, a special office of firearms was established in Korea under the control by Koryo Kings. Vietnam introduced gunpowder technique from China during 1250 to 1280, this technique found its way to India after 1400, the Indian developed a kind of large rocket and striked greatly at the British-French aggressors in the 18th century. Thailand, Cambodia and Indonesia introduced gunpowder technique in the 14th century. Although Japan was attacked by firearms of the Mongol troops in the 13th century, but Japan introduced gunpowder technique in the 16th century, later than other Asian countries.

The history of the spread of China's compass in foreign countries remains to be further studied. Accumulated materials are insufficient, so here we can only provide a preliminary research. The textual record shows that the earliest use of the compass in Arabian area appeared in 1232, a Persian author Muhammad al-Awfi (1202 ~ 1257) in his *Jami al-hikāyāt* (*Collection of Anecdotes*) talked about Arabian sailors finding their way by means of a fish-shaped piece of iron rubbed with a magnet. It should be identical with the *zhinanyu* (south-pointing fish) or the wet compass invented in China. In fact, Arabian sailors beginning to use the compass should be earlier than the year of 1232 by at least 50 years. Through Arabia, China's wet compass was introduced to Europe in 1180 or so. An Englishman Alexander Neckam (1157 ~ 1217) described it in 1190. England, France and Italy used the wet compass earlier than other European countries. The dry compass was first recorded by a French scientist Pierre de Maricourt (1224~1279) in 1250, later than that of China by more than 100 years. China's compass was also transmitted to other Asian countries: India (13th century), Korea (15th century) and Japan (17th century).

In short, till the 17th century, China's four great inventions were already transmitted to many important Asian, European, African and American countries in the world and exerted a tremendous influence on the development of the mankind civilization, arose a series of changes which finally changed the whole face of the world. Firstly, the appearance of paper in the East and the West immediately led to a revolution in the history of the development of scripts carrier and soon replaced other ancient writing materials. Although ancient books written on wooden-bamboo slips, papyrus, palm leaves, silk, parchment and so on were gradually lost, their paper transcripts could be spread from generation to generation. Thus, paper had a great exploit in preservation of historical-cultural heritage. The popularization of paper transcripts spurred the rapid spread of knowledge, information, research results and scholars' new viewpoints, hence promoted the great development of education and culture including literature, art, history, philosophy, religions and science, made the society more progressive than before. We illustrated this circumstance with examples of China, Arabian empire and Europe in the medieval age.

However, people had to spend a lot of time and labour on writing everyday. With further development of the culture, hand-written copy had already not suited the necessity from the society. In this case, printing was invented in China. It provided a great quantity of books with mechanical replication method, so people were liberated from hand-writing labour. The social function of printing was fully embodied in the Song Dynasty, almost all books written before and after the Song Dynasty in various fields were printed and widely popular in the society. Unprecedented academic prosperity appeared and the Chinese-style Renaissance arrived in advance. Printing also promoted the development of the Neo-Confucianism and the imperial examination system in the Song Dynasty. Neo-Confucianism, as an official philosophy in the feudal society, its ideological influence lasted in China for more than one thousand years. The imperial examination system ensured the admission of highly educated men to government service on the basis of knowledge and ability rather than class origin. Scholars came from common families thus had a chance to be selected as officials through examination at various levels, so the quality of officials were improved and the feudal rule was consolidated. The Chinese examination system was then introduced to the West, and became the basis of Europe-America civil service system in the modern time. In the Song Dynasty, China first issued paper money in the world. The new financial system strongly propelled the commodity economy forward. After this system was modelled in the West, the capitalist financial order was at last established.

After the Europeans introduced China's printing technique, its social effect expressed in Europe during the Renaissance period even more intensely than China and other East Asian countries. Printing promoted the development of education in Europe, soon afterwards educated readers and scholars creating the spiritual wealth doubled and redoubled. Printed books provided for progressive scholars a new place to publish their own views on the nature and society according to their independent research, thus the monopoly on knowledge and the truth by the Church and Pope were broken through. The medieval world outlook and scientific system upheld by the Church were overthrown by new facts observed in this period. Humanist writings of anti-feudalism exposed and criticized the dark aspect of the Vatican and feudal system, aroused the masses to struggle with the feudal forces. Protestants printed new editions of *The Bible*, reexplained the Christian doctrine and put forward their propositions of religious reform. At the same time, writers upheld to publish literary works in national languages and thus waked up the national consciousness in various countries. The Latin was no more the sole official written language, because the standard Italian, English, French and German used by writers could equally play the part of tool of language and literature for philosophical discussion and literary creation. European countries from then started to have their own national culture. In the process of forming national languages, printing played the role of midwife. The result of all the above-mentioned led to the emergence of the Renaissance, Scientific Revolution and Reformation Movements in Europe.

The use of gunpowder and firearms led to the revolution in the history of war and weapons in the world, and hence also to the transformation of the mode of operations, offensive and defensive tactics and the composition of the army. The range, destructive and antipersonnel force of firearms greatly surpassed any kind of cold weapons, solid castles and the knight group armed with heavy armour on which the feudal forces depended to maintain their rule could not keep out the attack from firearms, so the newly rising bourgeoisie used the resources of metals they controlled to cast powerful bombards or cannon and launched attack to feudal forces. The first salvoes of the 14th-century bombards spelled the death-knell of the castle, and hence of the Western military aristocratic feudalism, then gunpowder and firearms were used by the bourgeoisie to oversea aggrandizement and armed plunder of the wealth from colonies and dependencies to fulfil the primitive accumulation of the capital. Meanwhile, the Western countries also dumped their industrial goods to these areas and brought back gold, silver and raw materials for reproduction, so the world market was formed. On the other hand, the peaceful use of gunpowder also promoted the exploitation of the mine, canal-cutting, bridge and railway building. Therefore, gunpowder as a social productive force made an important contribution to the exploitation of resources and economical construction.

Gunpowder did not only destroy the medieval feudal economy and political rule, but also smashed the medieval outworn ideology. It played a midwife role for emergence of new science and technology. So the British chemist John Mayow (1640 ~ 1679) said: "Nitre (saltpetre) has made as much noise in philosophy as it has in war. " To Europeans' surprise that the combustion of gunpowder could be accomplished even in vacuum without the help of the air or wind, this could not be explained by European theory of four elements nor phlogiston theory. Through efforts of generations the theory of oxygen was established, which led to the chemical revolution in the 18th century. The next, European technicians found that the flight locus of shell always mercilessly departed from the orbit determined by Aristotle (i. e., parabola rather than acute-angle line), the ballistics research finally led to the establishment of dynamics, a new discipline of mechanics. Gunpowder was also thought to be a new kind of energy source, and cannon — a device converting heat source into mechanical work. This thought arouse European scientists to exploit a new type of power machines which afterwards led to the appearance of early steam engine and laid the initial basis for the future Industrial Revolution. Therefore, gunpowder and firearms exerted a good influence on the development of modern science and technology too.

The appearance of the compass opened a new chapter in the history of determination of directions by the mankind. It was made on the basis of magnetic principle and could work in 24 hours everyday in fair weather or foul on land or on the sea. What the Chinese first fitted their oceangoing ships with the magnetic compass led to a revolution in the history of navigation in the world. This made the Chinese fleet possible to leave far from coastal waters and sail to oceans in order to open new trade routes on the sea since the early Song Dynasty (10th century), the oceangoing voyage reached a high tide by Zheng He's (郑和) fleet in the early Ming Dynasty (the first half of the 15th century). After the Europeans introduced China's magnetic compass and the Chinese oceangoing ship-building technique, they began to leave from the Mediterranean Sea and offshore seas around the continent in the late 15th century. They continuously sailed into the Atlantic Ocean, Indian Ocean and Pacific Ocean and finally discovered the American New World. They also opened the new sea-route rounding the southern end of Africa (the Cape of Good Hope) and sailing into Asia from Europe. Without the help of the compass, the Europeans could not sail so far and hence set up colonies and new markets in Africa, America and Asia along newly found sea routes. The bourgeoisie expended the scope of their activities to the whole world, they all got rich from such activities, their influence on state affairs thus became more and more great. In related European countries, the capitalist political and economical system was gradually established according to the necessity of the bourgeoisie.

The use of the compass in navigation directly led to the great geographical discovery. Geographical knowledge explosively increased, which promoted the development of quantitative cartography. A new globe was given to us by the navigators in this period. Navigators also discovered magnetic declination in vast areas, which led to the development of a new discipline — magnetics. Although the Chinese had been familiar with it in ancient time, it was really new to the Europeans, and they positively have done much work in this field. Navigation and astronomy obtained more achievements to which a great attention was paid by people. Magnetics just served as a bridge between the two ones. During the age of great navigations, the compass and magnetics deeply struck root in the hearts of the people that the magnetic concept was used to explain the celestial motion and its orbit by astronomers, as the German Johannes Kepler (1571~1630) and the Englishman William Gilbert (1544~1603) did. Their work provided the basis for Isaac Newton's (1642~1727) synthesization. Secondly, the compass should be regarded as the oldest instrument of magnetic and electrical science. It was also the ancestor of all instruments with dial and pointer readings, without which it was difficult to imagine how the modern science could be developed. The ancients have never thought of that the turn of a small magnetic needle brought the world so many changes.

To sum up the above-mentioned, we can see that China's papermaking, printing, gunpowder and compass spread in Europe provided the necessary material premise for the rise of modern science and the formation of modern society. Europe was fortunate to be the place where modern science was born, but the edifice of modern science could not be built up only on the basis of ancient Greek science and the few scattered materials left from medieval Europe. In fact, more than 50% of basic inventions and discoveries forming the basis of modern science came from China. The Chinese technical contribution composed mainly of the four great inventions played a decisive role in the process of transformation from the ancient Greek science into the mechanical physical modern science. Science developed in various cultural areas in the East and the West like rivers which joined together and flowed into the sea of modern science, so every nation and cultural area had made respective contribution to the formation of modern science. However, Europe was the entrance to the sea, because Europe had the social condition suitable to absorb the technical wisdom of all mankind and develop the new science at that time. It should not be denied that the traditional Chinese science was a mighty current which flowed into the sea of modern science. The view that the development of modern science was only the European's affair and had nothing to do with other parts of the world is certainly wrong. Such kind of Eurocentrism view had been abandoned by honest European and American scholars of our time.

The Renaissance, Scientific Revolution and Reformation happened in Europe had gone through the baptism of the fire of anti-feudal struggle. They were the three forms of expression of the same social movement. Its final subsequence led to the collapse of the feudalist system and the rise of the capitalist system. This was then a progressive social phenomena and conformed to the law of the historical development. On the other hand, medieval China, the birthplace of the four great inventions, had once given out the brilliant light of science, but at last could not develop the modern science and modern society, conversely, lagged behind the West since the middle Ming Dynasty (16th century). To analyze why Europe and China were so different is outside the range of this book. After wasting the time of more than 400 years, the Chinese nation have been awoke and risen abruptly nowadays. They decided to develop the country with science and education and strive to hold a certain position in the advanced science and technology in the world. The Chinese people are full of confidence to make a new great contribution to the mankind in the 21st century.

Pan Jixing
July 2nd, 2002

Contents

Chapter 1　Invention, Development and Spread of China's Papermaking Technique

1.1　Origin of Papermaking in Light of Excavations of Paper Made in the Western Han Dynasty

1.1.1　Reasons for Paper Invention in China

The earliest paper was the rough paper made of hemp or linen or rags. The hemp cloth was woven with cannabis sativa and boehmeria nivea of Chinese origin. Other countries or regions had flax and jute for weaving, and rags could also be made for paper. But why was papermaking invented in China, not somewhere else? This question needs to be discussed. The factors contributing to people's papermaking are the social needs of the new writing materials, which in turn are related to socio-economic, cultural, educational and technical backgrounds. In the time of several centuries B. C. , there were only a few materials used for writing in the world, namely, bamboo, silk, sedge, sheepskin, pattra leaves and bark. At that time, Greece, Persia, Egypt, Roman Empire, India and other countries were still in the slave society with frequent wars and constantly changing maps, which was a fragmented period of turmoil. Slave owners or ruling groups only plundered wealth, land and struggle for hegemony, caring nothing about cultural and educational development. The classical writing material was enough to meet the social needs and no demand was made on new materials for writing.

Papermaking, as a new product derived from cultural and educational development, could not be created in the slave society. Only in the more developed feudal societies, where the social progress, social productive forces, culture and education were more advanced, could the invention of paper be promoted. China entered the feudal society as early as the 5th century B. C. , the Warring States period, earlier than the other countries. At that time for the rest of the world, the slave monarches launched wars. In 525 B. C. , Egypt was conquered by Persia, which was then conquered in 330 B. C. by the Greek Kingdom of Macedonia; in 146 B. C. , Roman Empire conquered Greece. India, in 323 B. C. was ravaged by Greece Alexandria, and it was not until when Ashoka reigned (273 B. C. ~ 232 B. C.) that India became a unified slavery country. After Ashoka's death, India was split again. Some of the ancient civilized countries were subjugated one by one because of foreign conquests, their cultures were destructed and even their original characters could not be retained.

In contrast, thanks to Qinshihuang, the first Emperor of the Qin Dynasty, China established a unified feudal empire in 221 B. C. and unified the Chinese writing characters. The Western Han Dynasty followed, which maintained the system. When the time came to Emperor Wen of Han (179 B. C. ~157 B. C.) and Emperor Jing of Han (156 B. C. ~141 B. C.), the rule of peace came into being with the all-round social prosperity. Emperor Wu of Han (140 B. C. ~ 87 B. C.) in the reign still retained this momentum. The state unity, the development of socio-economy, culture and education in addition to science and technology required a lot of writing materials. However, people felt that silk was costly and bamboo bulky, adding to inconvenience. Therefore, they went to explore new material cheap and simple for substitution, and paper came out at this time. Besides, in terms of all the world's classical writing materials, only silk is related to the word "paper", which has been proved by the Chinese character "纸" (*Zhi*, paper) containing a radical with 纟 in the historical work *Chronica* (732~1201). Paper and silk have something in common: smooth ink absorption with light and soft quality, being foldable, easy to cut, and made of pure fiber. The original paper appeared in a way as a substitute for silk. Silks in the manufacturing process would undergo *piaoxu*, floccule (floc) removal. The silk fiber placed on the bamboo mat in the water was hit and the crushed silk down on the mat was removed after the sun drying to form a paper-like sheet, which was then abandoned and not used. This process implies the original action of beating for papermaking, and it is easy to inspire people to produce a technical association of papermaking.

Collected in Beijing Gugong (The Forbidden City) Museum there is a painting album entitled *Cansang Luosi Zhichou Tushuo* (《蚕桑络丝织绸图说》, *Silkworm Mulberry Silk-Weaving Map Album*) by Wu Jiayou (about 1818~1893) in the Qing Dynasty in the 19th century under the rule of Emperor Guangxu (1891). The picture (Figure 1) shows that on the river there are two women, with bamboo sticks in hand, are squatting, hitting the silk down, which helps us know the ancient floccule removal operation. In ancient times, before silk-making, it was necessary to choose the cocoon and the single cocoon with square round shape was regarded as good cocoon, which could be used to make silk of high quality. Double cocoon and sick cocoon were regarded as the secondary and used for *Xu*, the wadding, for materials to keep warm. Before the silk floss was made, the secondary cocoon would be boiled in grass ash water to make it degumming, which

Figure 1   Ancient Chinese Cocoon Processing Chart, Taken from *Wu Jiayou's Painting Book* (1891)

would be then flooded in water and opened and rinsed with washing and beating at the same time. The secondary cocoon, along with the remains after boiling and reeling would be mashed, dried and twisted into thread. The records of these processes could be found a lot in the ancient books, as was said in *Xiaoyao You* (《逍遥游》, *Carefree Travel*) in *Zhuangzi* (《庄子》, *The Book of Zhuang Zi*, about 290 B.C.). People in the Song Dynasty (858 B.C. ~476 B.C.), good at making drugs to prevent hand tear, had been working on cocoon processing for generations, and this kind of work was mostly done by women.

The pure silk fiber would be obtained after the raw material was boiled in plant and grass ash water, flooded in water for sericin removal. Then the fiber bundle was soaked in water for swelling, made disperse by mechanical beat. These two steps were necessary for papermaking. Through beating and rinsing, the silk slags fell on the mat, which would then be dried and taken down, and a thin sheet was formed. It suggested that the fibers dispersed in the water could be collected or stored through the porous bamboo mats, and the textile approach was not absolutely necessary to form a silk-like sheet. If the silk fiber was replaced by the hemp fiber, the original paper could be formed. Artisans in the Qin and Han Dynasties made substitutes for silk just because of their thoughts of replacing materials.

The key is "to substitute hemp floc for silk one". Only when the raw materials replacement is completed can the new writing paper be made according to floc removal principle and technical operation. As we all know, China as the origin of sericulture and silk weaving technology, has long been known both in the East and the West. Pliny the Elder (23~79), Roman Empire scholar, mentioned in *Bowu Zhi* (*Naturalis Historia*, 73) that the Seres were good at making silks, which were transported to Rome from the eastern end to the western end of the earth. The Seres were Chinese people. In 1958, a number of silk products were unearthed from the Neolithic Ruins at Qianshanyang in Wuxing of Zhejiang Province. The year of the Ruins is 2750 ± 100 B.C.. According to the annual tree ring correction, the year is 5260 ± 135 B.C.. In 1980, in Yangshao Cultural Relics at Yangzhuang in Zhengding County of Hebei Province, two earthen silkworm pupas were unearthed, and identified by the experts as domestic silkworm chrysalis. At the same time the unearthed were 70 pieces of tools for silk making. The site is 5400 years ago. The archaeological findings prove that it was five thousand years ago that China raised silkworm for silk making. The Shang Dynasty oracle also has some text information about it. Records show that the history of Chinese people's operation of cocoon processing for floc collection is the longest, which cannot be compared by any other countries.

As early as the Neolithic Age, hemp and boehmeria nivea fibers were used as raw materials for weaving in China. For the unearthed Yangshao Cultural Relics, which were 6000 to 7000 years ago, at the bottom of the pottery, cloth pattern was often found. Johann Gunnar Anderson (1874~1960) thought that it was hemp cloth. In 1958, the unearthed linen fabric from Qianshanyang Ruins in Wuxing of Zhejiang Province was identified as boehmeria nivea cloth. The year is 2050 ± 100 B.C.. This shows that the north planted hemp while the south, boehmeria nivea. The history of their spinning and weaving was as old as that of the sericulture. In order to extract the fibers from the phloem of the hemp, it was necessary to do the retting treatment of the raw material, making it soft and white before weaving. For retting, records could be seen very early, as was said in *Chenfeng* of *Guofeng* in *Shi Jing* (about the 5th century B.C.), "In the pool of the east gate, the hemp retting can be done ... the boehmeria nivea retting can be done". "After softening, the fiber can be made for clothes", as Zheng Xuan noted.

After pool retting and purification, the fiber obtained would be as soft as silk. To achieve this effect, from the technical analysis, in addition to pool retting, lime water or grass ash water cooking (steaming and boiling) processing is to be done before textile fiber is obtained, which is collectively known as "retting". Since ancient times the retting, the technique legacy of Chinese ancestors, was made up of two processes. The retting method of hemp and boehmeria nivea was as follows. The first step was to cut the plant off after the hemp was grown up, remove its leaves and roots, and tie it into a bundle for the pool retting (fermentation) for a period of time of one or two weeks. Through the biochemical action of the mold, the pectin in the phloem and other soluble impurities could be removed, and the material could be made swelling and soft. Normally it took a period of one or two weeks. The second step was to beat the stalk and peel the bast. Since it was quite tough and the fiber was not yet extracted, it was necessary to make the bast into thin thread, which was mixed in lime for several days. People would use lime water to boil it and then put it into the bamboo basket, rinse it with water before spreading it on the mat for immersion and drying. The purpose of this treatment was to further remove the impurities by chemical means so as to obtain the pure hemp fiber, making it soft and white as silk, which was ready for thread weaving[1]. The good hemp cloth could only be achieved after the two procedures of the raw materials processing.

From the above, we can see that before the invention of paper, Chinese has accumulated the technical experience of the purification of hemp or boehmeria nivea fiber, and the retting process is just the preparatory stage of the papermaking process. Chinese ancient unique cocoon processing technology offers clue for papermaking technique. Since the Western Han Dynasty, the unprecedented unification of the country with social stability, economic and cultural prosperity, science and technology development gave rise to the growing demand for writing materials. China was the only major country that uses silk for writing. Due to high cost of this material, people were eager to explore cheap fiber raw materials made of silk substitutes. All of these factors combined led to the successful papermaking with adoption of relative principle and technology. In the 2nd century B. C. , only China had the social, economic, cultural and technical conditions needed for papermaking, and thus became the origin of papermaking. Other countries did not have these comprehensive conditions or only partial conditions, not enough to complete the invention of paper.

1.1.2　Paper Origination in the Western Han Dynasty

1. 1. 2. 1　Western Han Dynasty Hemp Paper Unearthed from 1930s to 1950s

Although the origin country of papermaking is China, there are different

---

[1]　王祯. 农书[M]. 缪启愉, 译注. 上海: 上海古籍出版社, 1994.

arguments about its invention time and inventors. In summary, there are two different opinions. For the first opinion, Zhang Yi and Fan Ye are the representatives, the latter argued that the paper was invented in 105 by Cai Lun, an officer in the Eastern Han Dynasty[1]. Zhang Yi believed that Cai Lun used plant raw materials to make paper, known as the modern paper. But the silk piece is made of silky fiber by textile method, and it is not paper, nor the ancient paper. The paper is made of the plant fiber by copy making. Before Cai Lun, there was history record of paper uses in the Han Dynasty and his contemporaries did not think that paper was invented by Cai Lun. The second opinion, as represented by Zhang Huai (686~758) in the Tang Dynasty and Shi Shengzu (1204~1278) in the Song Dynasty, held that the Western Han Dynasty (the 2nd century B. C.) had paper, and that Cai Lun was not the inventor of paper, but its improver. Their views are in line with the point of view of historical development, and seem correct in principle, which are still necessary to be confirmed by the archaeological evidence.

According to the criterion that practice is the only standard to test the truth, whether the ancient records are correct or not, depends on whether the test of archaeological practice can withstand the inspection. Theoretically, papermaking is a kind of collective labor rather than a job to be completed by a single person. It is impossible for one person to invent it all of a sudden. Owing to the low social status of the artisans in the feudal society, who rarely were written into the annals of history, the artisans' technological creation was often made in the name of certain powerful and noble people. The brush was considered to be invented by General Meng Tian in the Qin Dynasty; the paper, official Cai Lun in the Eastern Han ... so on and so forth. But in 1954 in *Chu Mu* (墓, tomb) of Changsha in Hunan Province, the brush of Warring States period was found, and in recent years the ancient paper before Cai Lun was unearthed. In *Qian Zi Wen* (*Thousand Character Classic*), it is said that "Lun Paper Tian Brush" (Cai Lun invented paper and Meng Tian invented brush) saying needs to be corrected according to archaeological findings. Today, few people believe that Meng Tian invented the brush, but still some people believe that Cai Lun invented the paper.

Since the 20th century, thanks to the development of archaeological excavations in China, a lot of material data, which the people from the

---

[1] 范晔. 后汉书:蔡伦传[M]. 上海:上海古籍出版社,1986.

Northern and Southern Dynasties to the Tang and Song Dynasties had never seen, were accessible for the researches on the issues of paper origin. The material data could be used to judge the right and wrong concerning the origins of papermaking between the two academic views of the Millennium dispute. In 1933, the archaeologist Dr. Huang Wenbi (1893 ～ 1966) discovered a piece of ancient paper

Figure 2   Western Han Dynasty
Hemp Paper (49 B. C.)

made by the Western Han Dynasty (Figure 2) in the archaeological excavations of the Fengsui Ting Pavilion site of the Han Dynasty in Luobunao'er (罗布淖尔, *Lop Nor*), Xinjiang (Xinjiang Uygur Autonomous Region). About the unearthed paper, he wrote in *Archaeological Record of Luobunao'er*:

> Hemp paper: *hemp, white, square, thin, incomplete, bout 4. 0 cm in length, 10. 0 cm in width. It is very rough and uneven* ...[1]

Finally, Mr. Huang concluded that according to what was discovered, undoubtedly, in the Western Han Dynasty, there was paper for writing. However, the paper was rough in the Western Han Dynasty, which was refined in Cai Lun's time.

In the autumn of 1942, archaeologists Mr. Lao and Mr. Shi in Gansu Province (now in Inner Mongolia) discovered a piece of written paper of the Han Dynasty, which was plant fiber paper after test (Figure 3). From the words written on the paper, it could be judged that the paper was the official document discussing weapon transportation.

Figure 3   Paper Unearthed in
Inner Mongolia in 1942

In May 1957, at the Baqiao Bricks and Tiles Plant Site in Xi'an City of Shaanxi Province, bronze mirror and bronze sword were found and the next day people from provincial museum were sent to investigate it and nearly a hundred

---

[1]   黄文弼. 罗布淖尔考古记[M]. 北平,1948.

pieces of artifacts were collected. Under the linen cloth there were layers of paper pieces, known as the Baqiao paper (Figure 4).

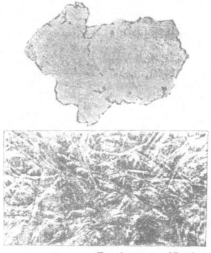

Ten times magnification

Figure 4　Western Han Dynasty Hemp Paper (140 B. C. ～87 B. C.) unearthed in Baqiao, Xi'an City (The upper picture is provided by Cheng Xuehua while the lower, Pan Jixing)

1. 1. 2. 2　Western Han Dynasty Hemp Paper Unearthed from 1970s to 1990s

In 1973, the Changcheng Archaeological Team of Gansu Province, with a planned scientific excavation at Jianshuijinguan Military Post Site of the Han Dynasty at the east bank of the Ejina River of Gansu Province, unearthed and sorted out the calendar inscribed wooden slip, silk piece, linen cloth, writing brush, ink stone, linen paper and other objects. Since the Han Dynasty, there were large-scale military activities, which continued till the end of the Western Han Dynasty.

Figure 5　Western Han Dynasty Hemp Paper unearthed in 1973

The unearthed were two pieces of ancient paper. Paper No. 1 (the original number EJT1:11) was from the residential area, white in color, 21 cm × 19 cm in size after flattening, thin and even in quality (Figure 5). The inscribed wooden slips of the same level of earth were mostly of Zhao and Xuan Periods, the earliest of which dated back to the 2nd Year under Emperor Xuan (52 B. C.). Paper No. 2 (the original number EJT30:3) was from the east side of the bedroom, dark yellow in color, 11. 5 cm × 9 cm in size, rather rough in quality. The earth layer belongs to Emperor Ping in Jianping Year (6 B. C.). Because Jinguan paper was from the Western Han Garrison Site, the archaeological community believes that the site is clear, every section of the site is clearly defined, and the cultural relic accumulation shows the earth layer relationship. In addition, the excavation has been planned by the professional archaeological team, therefore the cultural relics dating is scientific.

In December of 1978, in Zhongyan Village, Taibai Township, Fufeng County of Shaanxi Province, in the construction of water conservancy, the Han Dynasty architectural sites were found. Under the tile accumulation layer, in the circular pit was a cellar pot filled with more than 90 pieces of cultural relics including bronzeware, money, etc. For the bronzeware, there were pieces of decorative lacquerware stuffed with ancient paper, 6. 8 cm × 7. 2 cm in size for the largest piece after flattening, white in color, pliable, and with rusty green spots. After the experts' identification, the large pottery was thought to be the object sometime before or after the rule of Emperor Xuan (73 B. C. ~ 49 B. C.); the copper coin, between the Emperor Xuan and the Emperor Ping; the up-limit age of the paper, between the Emperor Xuan to Emperor Ping, no later than the Emperor Ping. As to archaeological excavation of Zhongyan paper, the earth layer is clear with a complete set of cellar artifacts of the Western Han Dynasty without any external disturbance, and original state is still maintained.

In October 1979, the Gansu Changcheng Joint Investigation Team made a large-scale excavation of the garrison site of the Western Han Dynasty in the Majuanwan area, 95 km northwest of Dunhuang in the ancient Silk Road. Here the garrison began from the Emperor Wu of Han Dynasty, prospered in time of Emperor Xuan and was abandoned in the year of 21. The ruins are well preserved and 337 kinds of cultural relics were unearthed, including silk fabrics, five baht money, ironware, copper arrow, fire maker, seal, wooden ruler, writing brush and ink stone, linen paper and 1,217 pieces of inscribed wooden slips. For the unearthed paper, Paper Ⅰ (the original number T12: 47), yellow in color, 32 cm × 20 cm in size with clear four sides, is the most complete piece of Chinese paper found so far. The wooden slip from the same earth level belongs to the period from 65 B. C. to 50 B. C.. Paper Ⅱ (the original number T10: 06) and paper Ⅲ (the original number T9: 26) amounted to a total of 4 pieces with fine texture. But with the animal fecal pollution, the paper was earthy yellow, and the calendar inscribed wooden slip from the same excavation unit was between the periods from 32 B.C. to 5 A.D.. Paper Ⅳ (the original number T9: 25) was white with a uniform texture; paper Ⅴ (the original number T12: 18), 2 pieces, with good texture, was of the period under Emperor Wang Mang during 8 and 23.

From June to September 1986, during the excavation of a large area of 11,000 m² in the outskirts of Tianshui at Fangmatan tombs of the Warring States, and the Qin and Han Dynasties, in the fifth Han Tomb, Gansu archaeologists found cultural relics such as pottery, lacquerware, woodware and paper. The paper (M5:5) placed on the chest of the dead had maps drawn on itself, with the remaining size 5.6 cm × 2.8 cm❶. The paper was yellow with stained surface, on which were painted fine black lines mountains, rivers, roads, etc. The painting method is similar to the 1973's unearthed silk map from No.3 Han Tomb (168 B.C.) in Mawangdui of Changsha City. ❷ The burial structure of Fangmatan No.5 Han Tomb was reminiscent of Qin Tombs, and the funerary objects were similar to those from the early Han Tombs in Shaanxi Province and Yunmeng, Hubei Province. Therefore, the archaeologists regarded the Tombs as the period of Wen and Jin (179 B.C. ~141 B.C.). This is by far the earliest paper map (Figure 6).

Figure 6   Western Han Dynasty Map Paper (176 B.C. ~140 B.C.), unearthed in 1986

From October 1990 to December 1992, the archaeologists of the Gansu Provincial Archaeological Institute carried out large-scale excavations at the site of the Han Dynasty in the Gobi Desert in the vicinity of the sweet water wells 64 km northeast of Dunhuang. Only in the range of 22,500 m², they found 35,000 pieces of wooden slips and 460 pieces of paper, in addition to pottery, lacquerware, woodware and silk, totaling 70,000 pieces. The earth layer of cultural relics was very clear, and the paper in the

❶   何双全.甘肃天水放马滩秦汉墓群的发掘[J].文物,1989 (2):1-11,31.
❷   马王堆汉墓帛书整理小组.长沙马王堆三号汉墓出土地图的整理[J].文物,1975 (2): 35-48.

soil was accompanied by calendar inscribed wooden slips, which contributed to time determination.

In summary, in 1933, 1942, 1957, 1973, 1978, 1979, 1986 and 1990 since the 20th century, China, in different locations such as in Xinjiang, Gansu, Shaanxi and other provinces, autonomous regions and districts, unearthed eight times the ancient paper produced before Cai Lun's time, from Emperor Wen, Jing to Xin Mang of Han Dynasty. Almost all the paper produced during the rules of all the Emperors in the Western Han Dynasty could be seen. In addition to blank paper, there were pieces of paper with painted maps and written words. Paper with no words on were still writable. All these prove that China invented paper as early as the 2nd century B. C. instead of Cai Lun's time, who was once thought to be the paper inventor in the 2nd century A. D..

1. 1. 2. 3   Documentary Records of Chinese Paper Before Cai Lun's Time

In addition to archaeological discoveries, there are some documentary records of Chinese paper before Cai Lun's time for reference. In the quotations of Tang and Song Dynasties, people could see the citations from *Sanfu Jiushi* or *Sanfu Gushi* (old stories) in the 4th century by Jin people on the paper uses in Emperor Wu's time of the Han Dynasty.

In the Han Dynasty, silk, wooden slips and paper were used simultaneously. In the Western Han Dynasty, on the wooden slips were found the earliest documentary records. On the wooden slip of the 1st century B. C. , the simplified Chinese character "纸" (*Zhi*, paper) appeared, which showed that the character was used in the beginning of the Western Han Dynasty (the 2nd century B. C.), indicating the new writing material such as Fangmatan paper.

Waiqi Zhuan in *Han Shu* (《汉书》, *History of the Western Han Dynasty*) records that in the 3rd year of Hongjia (18 B. C.), Emperor Cheng of Han Dynasty appointed Zhao Feiyan as the queen. For many years, she and her sister Zhao *Zhaoyi* could bear no child. On the other hand, in 12 B. C. , Cao Weineng, the Emperor's concubine, gave birth to the Emperor's son, which aroused hatred out of jealousy from Zhao *Zhaoyi*. After 10 days of the child's birth, the mother was put into the jail. The prison aide was ordered to wrap two poisonous pills with small thin paper and put it into a small box. Cao Weineng was forced to take the pills and poisoned to death. This kind of paper used in 12 B. C. could be used for writing. In the past people guessed it was "flake (floc) paper", which lacked evidence, for there was no historical record of "flake paper". Here it is to be the hemp paper, thinner than Majuanwan paper Ⅲ, made in the period of Emperor Cheng.

From the Western Han Dynasty, Emperor Xuan to Wang Mang period, there was some accumulation of hemp paper documents. The Eastern Han Dynasty Emperor Liu Xiu timely accepted the wealth more precious than gold and shipped it to the Luoyang Palace. Those Western Han Dynasty books and archives written on the silk, bamboo and hemp paper on 2,000 vehicles were transported into the Luoyang Palace in November 25th, as the first collection of books. The reason why the Emperor Guangwu valued these books was that he would take all these as a reference in the new Dynasty to develop rules, regulations and decrees as well as culture and education.

Emperor Guangwu liked to use paper as writing materials. He set up *Shangshutai* when he started to take his place, which was related to paper. After improvement, *Shangshutai* became the national government central institution, which was also called *Zhongtai* or *Taige*, being in charge of the seal and allocation of paper, brushes, ink and other items. In addition, according to *Baiguan Zhi* in *Hou Han Shu*, special official position was set up to look after the Emperor's paper, brush and ink and other treasures. In the Eastern Han Dynasty since the founding of the country (25 A.D.), the express provision was that the paper was the writing material for the Emperor and the government. There were special officials in charge of the management of paper, getting the pensions. It is worth noting that all this happened over 30 years before Cai Lun's time, because *Hou Han Shu* clearly states that this is what Emperor Guangwu ordered at the beginning of the founding of the country.

After Emperor Guangwu died, the Emperor's son Liu Zhuang ascended the throne, who is Emperor Ming (57~75). Emperor Ming and Zhang (76~88) inherited the cause of Emperor Guangwu, rectifying the officials, reducing the taxes, modifying the water conservancy and developing the culture and education. Over sixty years the early Eastern Han Dynasty seemed to be the reminiscent of the prosperity of the early Western Han Dynasty. Emperor Zhang in the year of 76 ordered Jia Kui to compile the teaching material based on *Chunqiu Zuoshi Zhuan* (《春秋左氏传》, *The Annals of History of Spring and Autumn*) and to teach "*Chunqiu Zuoshi Zhuan*" to 20 talents[1]. The 20 talents were given two sets of books written separately on bamboo slips and paper,

---

❶　范晔.后汉书:贾逵传[M].上海:上海古籍出版社,1986.

showing that the study was paid much attention to. At the beginning of the Eastern Han Dynasty, Confucian books had their written-paper version.

For the above mentioned Western Han Dynasty paper found through the archaeological excavations from 1933 to 1990, there are officially published excavation reports written by archaeologists. The reports have illustrated the unearthed paper preservation and dating basis and published the paper photos. Except the Baqiao paper, which is not the finding of the intentional excavation, the rest of the paper was discovered after the planned excavation of the Han Dynasty sites and tombs through scientific methods by archaeologists. The dating is done according to the site formation, burial structure, calendar inscribed wooden slip, copper coin, etc. , combined with the literature research and comparison with the paper of the known age from other tombs. A consensus in history has been reached for dating, artifacts and archaeological circles, leaving no doubt.

Documents and archaeological discoveries have shown that before Cai Lun's time, there was paper. Zhang Yi, Fan Ye and their followers, in fact, also were aware of the presence of paper before Cai Lun's time, nevertheless, they found that this paper could only be silk, and this would be a mistake in understanding. Whether the paper before Cai Lun's time was silk or not could be answered through the microscopic analysis of the unearthed ancient paper.

1. 1. 2. 4   Analysis of Unearthed Western Han Dynasty Paper

The age of the previously unearthed Western Han Dynasty paper has been determined by archaeologists using archaeological methods. In order to solve the issue of origin of papermaking, it is necessary to combine archaeological excavations with the analysis and testing of unearthed ancient paper. The past controversy over paper history has long been not resolved because of the absence of necessary conditions to achieve this combination. Besides, the results of the laboratory analysis are to be determined based on the technical definition of paper so as to judge whether the test sample is paper. The work can lead to scientific identification of the discovered ancient paper by archaeologists, providing the testimony for the paper argument. The test of the unearthed ancient paper began in China in the 1940s, but only for individual samples. The deep research on the unearthed paper was not carried out until the early 1960s.

The author was honored, from 1964, together with the archaeologists, to make a systematic analysis of the test on the unearthed Western Han Dynasty paper during 1957 to 1990. Working together with other scientific workers, we used the method as follows: first we used the relevant instruments for the macroscopic detection of the paper to obtain the average thickness, whiteness, basis weight, tightness and other technical data, using high magnifying glass to observe the appearance of paper characteristics; then we spun off the fiber and made the film after dyeing for microscopic detection. The fiber condition was revealed by high power optical microscope or scanning electron microscope. The maximum and minimum length, width, average length and width of the fiber as well as beating degree were measured. The paper fiber variety was determined and the fiber microscopic images were made. For each sample, the secret numbering was made in advance, the detectors would only know the number instead of the detection object, and only the record instrument display results were recorded. For every test, the test was at least done twice by different people in turn, and monitored by other people.

At the same time, with the same procedures, the parallel comparison and contrast test was carried out with the known, identified (by experts) fiber samples as cannabis sativa, boehmeria nivea, linen, jute, mulberry bark as well as those of known age including ancient hemp paper, bark paper and modern paper samples. The same secret numbering of samples was done so that the detectors recognized only the number instead of the object. For the above important tests, we did them in five professional research institutes at home and abroad and the test results had to undergo correction processing. These measures were to avoid subjective factors to make the sample testing as objective as possible.

From the high-power microscopic view of the vertical cross section, the horizontal cross section and the fiber assembly, we have seen single horizontal axial dislocations, but fewer double central axial dislocations in the fibroblasts. The horizontal transverse crevasse occurs regularly and equidistantly, with a few unidirectional fissures between horizontal and transverse fissures and with a few irregular fissures on the axis shift (Figure 7-A and 7-B), and the longitudinal crevasse was more and

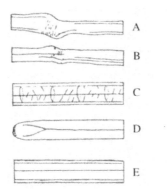

Figure 7 Microscopic Analysis of Longitudinal Section of the Western Han Dynasty Paper Hemp Fiber Cell, by Pan Jixing (1964)

more obvious. The punch holes are scarce and mixed with the fissures. In the change of the width of the fibroblasts, except for the feature of axial shift, the central width is uniform and almost unchanged, but at both ends it begins to be sharp, blunt (Figure 7-D), which shows another feature❶.

On the other hand, the fiber horizontal cross section is triangular (Figure 8-A to 8-F), and there are polygons, too (Figure 8-G to 8-P). The cell walls are thicker and have a hierarchical structure with a number of small fibers in each layer (Figure 8-A, 8-G, 8-J). The cell shape is diverse, but mostly elliptical (Figure 8-A~C, G~I, O~P). In the polygonal transverse incision, there is a protrusion outside the periphery, forming a flat view of the visible axis. All of these features are consistent with the characteristics of Sanko's annual herbaceous cannabis (cannabis sativa) fibers❷. The average fiber width is 18 μm~26 μm (1 μm = 10^{-3} mm), which also conforms to the variation of the hemp width (7 μm~32 μm). It is also found that some of the fiber characteristics are the same with the known boehmeria nivea for observation and contrast. Therefore, it can be said that the ancient paper raw materials are cannabis sativa and boehmeria nivea fiber. Not only the Baqiao paper tested in detail, but also the rest of the Western Han Dynasty paper are of the same material

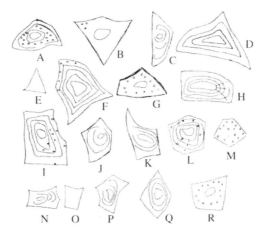

Figure 8　Microscopic Analysis of Cross Section of the Western
Han Dynasty Paper Hemp Fiber Cell, by Pan Jixing (1964)

---

❶　SINDALL R W. Paper Technology: An Elementary Manual on the Manufacture, Physical Qualities and Chemical Constituents of Paper and Papermaking Fibres[M]. 3rd ed. London: Griffin, 1920.

❷　潘吉星. 中国造纸技术史稿[M]. 北京: 文物出版社, 1979.

feature. Cannabis sativa and boehmeria nivea have been produced in China since ancient times.

For a variety of paper samples reviewed by the experts from China Textile Science Research Institute, no sample was found containing silk fiber, only hemp fiber was found. We also noted that the sample paper had little unbroken twist head, the same fiber of the paper, indicating that the paper was made from broken hemp cloth. Through microscopic observation, it could be seen that the hemp fiber purity was high with fewer miscellaneous cells, which was different from raw fiber with a lot of impurities, indicating that the raw material had undergone purification. The paper-like fiber was also different from the general hemp fiber, because it still contained impurities before weaving. The comparison through observation under the microscope between the Western Han Dynasty hemp paper fiber and the Tang Dynasty hemp fiber showed that the paper fiber purity in the Western Han Dynasty had little difference with its later counterpart. The non-fiber particles under the microscope were mostly sand particles, which was brought along with the unearthed paper pieces. The whiteness index showed that the whiteness of Baqiao paper and some Majuanwan paper was lower (25%), and the other whiteness of the Western Han Dynasty paper was 40%~45%. This indicated that the colored impurities in the raw material had been removed. The increase of the fiber purity and whiteness depends on the water cooking of the raw material rags with the grass ash, without which whiteness is difficult to improve, as shown in our simulation experiment. Low whiteness of paper is due to low cooking efficiency[1].

The average fiber length of the Western Han Dynasty paper was 0.9 mm~ 2.2 mm, with the longest being (1%) 10 mm. Those whose fiber was 0.5 mm~ 1.9 mm accounted for more than 50%. The average length of the unprocessed raw cannabis sativa fiber was 15 mm~25 mm, and boehmeria nivea raw fiber, 120 mm~180 mm. Through contrast, there was a difference from scores of times to hundreds of times, indicating that the fiber used in the paper was cut off, or had undergone shortening process. It was impossible for it to become short just because of natural rot and we did not observe this phenomenon. Although the paper has the unbroken hemp heads, they have been cut short,

---

[1]　潘吉星,苗俊英,张金英,等.对四次出土西汉纸的综合分析化验[Z].天津:天津造纸研究所,1981.

and under the microscope the incision could be seen. For all the samples of Western Han Dynasty paper under observation, no fiber bundle or rope head longer than 15 mm could be seen. Even if there were short fiber bundles or hemp heads, there were only a few.

Under the magnifying glass and microscope, the observation over the entire surface of the unearthed Western Han Dynasty paper shows that, whether it is positive or negative surface, the basic structure constitution displays the dispersion of individual fibers with vertical and horizontal non-directional intertwining (Figure 9 ~ 12). This is an important micro-physical structural feature of the plant fiber paper, which is fundamentally different from silk, sedge and tapa. Although for some individual parts it can be seen that the fiber is in the same direction, but this is only part of the phenomenon, which can not represent the whole. The average thickness of each piece of paper is 0.1 mm to 2.9 mm, normally 0.2 mm to 0.25 mm. There is a certain degree of mechanical strength. When the fiber is peeling off from the sample, certain force is necessary, and force is also necessary to tear the sample into pieces. Its tightness is 0.28 g/cm^3 to 0.33 g/cm^3, while the tightness of the modern Shaanxi Fengxiang hand-made hemp paper is 0.44 g/cm^3. This shows that the unearthed Western Han Dynasty paper is in line with the technical requirements of the handmade paper, and it is not the so-called natural accumulation of fiber, which can not have an average thickness of 0.2 mm~0.25 mm, nor is it possible to have 0.28 g/cm^3~0.33 g/cm^3 tightness and mechanical strength[1].

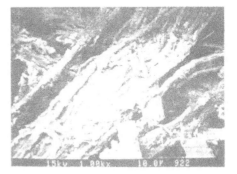

Figure 9   Photo of Western Han Baqiao
Paper Fiber Under Scanning Electron
Microscope×100, by Pan Jixing (1988)

Figure 10   Photo of Baqiao Paper Fiber
of Western Han Under Scanning Electron
Microscope×100, by Pan Jixing (1988)

---

[1]   潘吉星. 从考古发现看造纸术起源[J]. 中国造纸,1985,4(2): 56-59.

Under the microscope the observation shows that the Western Han Dynasty paper, Baqiao paper fiber cells did not encounter the strong mechanical damage, and generally the fibrillation or fine fibrosis was not high, but still in the sample, there existed crushed, broomized fiber (Figure 10). For the rest of the Western Han Dynasty paper fiber, there was a clear phenomenon of broomization. For Jinguan paper and Majuanwan paper, 40% of the total fiber was of broomization with a beating degree of about 50° SR (Figure 11 and Figure 12). From the technical point of view, the reason is that the hemp fiber raw material is cut short in advance, then cooked with grass ash water to become soft, and at last mechanically pounded with strong force for broomization. Otherwise, it is impossible for this phenomenon to occur. Baqiao paper in the manufacturing process undergoes cutting and pounding, but the cooking is insufficient, therefore both the whiteness and broomization degree are low.

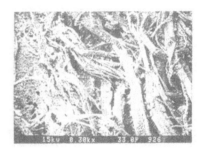

Figure 11   Photo of Jinguan Paper of Western Han Dynasty Under Scanning Electron Microscope×300, by Pan Jixing (1988)

Figure 12   Photo of Majuanwan Paper Fiber of Western Han Dynasty Under Scanning Electron Microscope×100, by Pan Jixing (1988)

Sometimes people are accustomed to using handmade paper cord to assess the Western Han Dynasty paper, and it seems that the unapparent cord means no copying. This is a misunderstanding. In fact, both in early and later times, papermaking could be done by using two tools. First, the woven mould was used for papermaking and there was apparent or unapparent weaving pattern instead of the cord or curtain pattern. Second, a cord pattern mould was used, then the paper was of the cord pattern. Jinguan paper had a clear texture, which was made from the woven mould. Zhongyan paper, Majuanwan paper, Xuanquan paper had the cord pattern, which was made from the cord mould. We found that among 16 pieces of Baqiao paper numbering 57~540, there were three or four pieces with rough cord pattern, each of which was 2 mm, showing the proof for papermaking of this process. The mould pattern of some paper could not seen clearly possibly because it was too thin or too thick or because of calendering, which did not mean no papermaking in the copying way.

The Western Han Dynasty paper samples were tested repeatedly one by one by means of the contemporary testing measure and a series of technical data were obtained. After the above phenomena were observed, these data and phenomena were synthesized. From the perspective of papermaking technology, the following test conclusions were reached. All the ancient Western Han Dynasty paper used the broken hemp cloth as the raw material, and after the basic processes of chopping→washing→grass ash water cooking→washing→mechanical pounding→woven or cord mould copying→drying, finally it was made. Thus from the perspectives including the appearance, physical structure, technical specifications and performance, it was concluded that it was the real plant fiber paper. Because of the rough or fine processing, the texture quality differed, and each had different uses, but all of them belonged to the category of paper[1]. Other domestic experts[2][3], after the test, came basically to the same conclusion.

1. 1. 2. 5   Early Papermaking Technology

From the 2nd century B. C. , for nearly a thousand years, the hemp paper had been the main paper kind. According to the laboratory analysis of the unearthed Western Han Dynasty paper, the basic judgments have been made on its manufacturing technology. As there are rare ancient books on the hemp paper manufacturing technology, to understand the detailed technical processes and tools and equipment to restore its technology, it is essential to do the field investigation on the technology of the manual hemp paper production. Because the hemp paper technology is developed from ancient times, it certainly contains some of the basic elements of the ancient times. To understand something present helps to understand its past. Therefore, in 1965 we knew its manufacturing process from Shaanxi Fengxiang Manual Hemp Paper Mill, which is as follows:

(1) Immersion wetting the broken hemp cloth → (2) Chopping → (3) Grinding→(4) Washing→(5) Mixing the hemp with lime water for grinding →(6) Stacking with lime slurry for a period of time→(7) Cooking the hemp

[1]   潘吉星. 中国科学技术史:造纸与印刷卷[M]. 北京:科学出版社,1998.

[2]   刘仁庆,胡玉熹. 中国古纸的初步研究[J]. 文物,1976 (5):74-79.

[3]   许鸣岐. 中国古代造纸起源史研究[M]. 上海:上海交通大学出版社,1991.

with the lime slurry→(8) Washing→(9) Further grinding→(10) Washing→ (11) Processing with the pulp→(12) Taking out the paper→(13) Pressing for dehydrating→(14) Drying paper→(15) Releasing paper→(16) Packaging.

The above procedure involves 16 processes. Though, for the early hemp paper, there might not be so many processes, some steps are essential, such as chopping, pounding, copy making and so on. If the raw material rags, hemp rope size and shape were not well cut in advance, it is difficult to make any treatment later. However, the dispersion of fiber could not be achieved only through chopping. The hemp material must be mashed before the pulp making. The early paper at least needs to undergo the following steps to be made:

(1) Immersion wetting of the broken hemp cloth → (2) Chopping → (3) Washing→(4) Cooking the hemp with the plant ash water→(5) Washing→ (6) Pounding→(7) Washing→(8) Processing with the pulp→(9) Copy making the paper→(10) Drying paper→(11) Releasing paper→(12) Packaging.

The above technological process includes 12 working procedures, based on which we did a simulation experiment in 1965 in Shaanxi Fengxiang Hemp Paper Mill. In order to make it close to the actual production situation, we used 20 kg ~ 25 kg broken hemp cloth, fag end, and operated with the actual production tools and equipment. The paper made was similar to the unearthed Western Han Dynasty paper, indicating that the production processes were the same. Paper made with simpler or more complex process was rougher or finer than the Western Han Dynasty paper, which either did not enter the paper category, or was close to the later paper. Therefore, the above process with 12 procedures should be the best way to reflect the early paper manufacturing process[1].

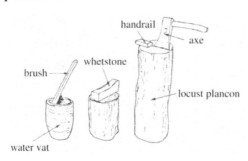

Figure 13　Hemp Cutting Tool,
drawn by Pan Jixing (1979)

(1) The mechanical pretreatment of raw materials (immersion wetting, chopping, washing) was as follows. After weighing, the hemp cloth, fag end, etc., were put into the basket, soaked in water, and the dust and mud were washed away. Then with the sharp axe the wetted hemp was cut into small pieces (Figure 13), and

❶　潘吉星. 从模拟实验看汉代造麻纸技术[J]. 文物,1977 (1)：51-58.

the metal, wood chips, feathers, leather and other debris were removed. After cutting, they were put into the basket and washed in the river.

(2) The chemical treatment of the raw materials (being soaked in grass ash water, cooking, washing) was as follows. The plant and grass were burnt into ash and put into the bamboo basket. With hot water immersion and filtration, the grass ash water with weak alkaline was made (Figure 14). The chopped hemp rag was soaked in grass ash water and then put into the cooking pot, and some lime water could be poured into, too. The cooking pot filled with water was made of iron (Figure 15). The grid was put on the pot, on which the open barrel was placed.

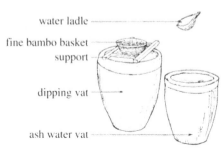

water ladle

fine bambo basket

support

dipping vat

ash water vat

Figure 14   Grass Ash Water Immersion Equipment, painted by Pan Jixing (1979)

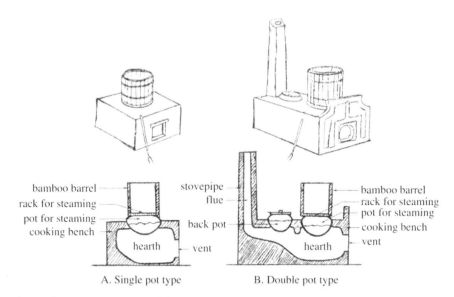

bamboo barrel

rack for steaming

pot for steaming

cooking bench

hearth

stovepipe

flue

back pot

vent

bamboo barrel

rack for steaming

pot for steaming

cooking bench

vent

hearth

A. Single pot type   B. Double pot type

Figure 15   Han Dynasty Cooking Pot for Papermaking, painted by Pan Jixing (1979)

The hemp was put into the bucket, and at the top mouth of the wooden bucket the grass ash and sacks were placed. The firewood was burnt for cooking to remove the color and impurities for fiber purification and softening for pounding. After cooking, the hemp was taken out and put into baskets, washed in river. The black liquor in pot was abandoned. There were a single pot and a double pot, and the latter also had another small pot. The afterheat was used to burn hot water. There was once a Han Dynasty pot unearthed.

(3) Hemp material pounding procedure was as follows. After cooking, the material was made soft and white, which was then put into the mortar for pounding (Figure 16) till the fiber became fine and thin. The purpose of pounding was to shorten the fiber and disperse it into fine ones for copying paper with tight texture. Since the fiber dispersion would affect the quality of paper, the procedure was time-consuming and laborious. The mashed hemp was to be rinsed in the river so that the remaining dust, dirt and other particles would be removed.

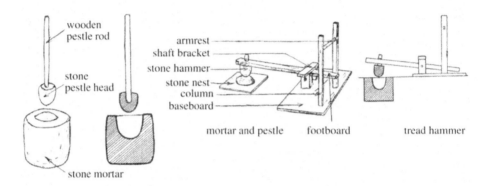

Figure 16   Hemp Pounding Equipment, painted by Pan Jixing (1979)

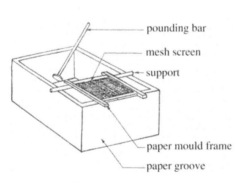

Figure 17   Paper Copying Slurry, painted by Pan Jixing (1979)

(4) The procedure of the processing with pulp and taking out paper is as follows. After pounding and washing, the hemp material that looks like white flocculence, is put into the rectangular wooden groove (Figure 17). The clean well water or spring water is added to make the suspension liquid, the pulp with appropriate consistency. Then through fully stirring the stick, the fiber is floating in the water. The pulp is not good if it is too thick or too thin. A spoonful of pulp can be taken out and slowly poured back into the groove, where the fibers are to be connected with each other in an appropriate manner. A piece of paper can also be temporarily taken out to see whether the thickness is appropriate, then the paper material or water can be added again. The paper

mould for the simulation experiment was designed temporarily, made into a rectangular wooden frame, and then the cauda equina or the bamboo curtain was fixed on the frame (Figure 18). With both hands holding the paper mould obliquely inserted into the slurry (not too deep), sway back and forth. Then lift it to drip water, and a wet piece of paper was made. Its thickness depends on the consistency of the slurry and the way of copying. This process is carried out by experienced workers.

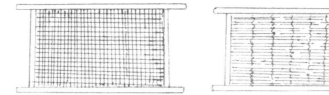

Figure 18   Two Kinds of Paper Cord in Han Dynasty, painted by Pan Jixing (1979)

(5) Drying paper and releasing paper procedure is as follows. After the wet paper is taking shape, water is still kept and the strength is not enough. It must be dried and dehydrated. With the above fixed paper mould for copying, if there is no dehydration, the paper can not be peeled off or released. So it should be dried outdoors in the sunlight. After the natural drying it can be released. In this way, the paper surface is not necessarily smooth, and it is to be calendered with fine stone before it becomes the finished paper (Figure 19). However, if it is used for packaging, no calendaring is needed.

Figure 19   Paper Calendering Operation, cited from Lin Yijun (1983)

The early hemp papermaking process is generally composed of these procedures. However, due to rough or fine operations, the paper texture is also different. But from the very beginning, the paper is produced through the combination of mechanical and chemical processing of the raw materials. This technique of the Western Han hemp paper has been used for the production of the bark paper and bamboo paper, which has been followed by other countries. The use of the grass ash water as a reagent for chemical treatment dates back for a long time. In early Qin time, it has been used for silk processing and laundry, and it is a matter of course when it is used for making paper. The unearthed Han paper has the cord (curtain) pattern, indicating that it is copying made with bamboo, and also some paper shows ribbing. The size of the paper cord depends on the size of the paper to be made. Generally its straight height is 24 cm~25 cm, and its horizontal length 35 cm~55 cm. With a fixed paper mould, only one copy of a piece of paper is made for one time, so many moulds are needed for preparation. Later, with the application of the mobile cord lathe, the paper cord tool and the frame can either be separated or combined, and one cord lathe can be used to copy and make tens of millions sheets of paper. Now the process of the Han Dynasty hemp papermaking is drawn as follows (Figure 20).

Figure 20   Han Dynasty Hemp Papermaking Process Flow Chart,
designed by Pan Jixing, painted by Zhang Xiaoyou (1979)
1 & 3. Washing material   2. Cutting material   4. Burning grass ash water
5. Cooking   6. Chopping   7. Slotting   8. Copy making   9. Drying paper, releasing paper

1.1.3　Cai Lun's Historical Role

1.1.3.1　Doubt on Statement of Cai Lun's Invention of Paper

Through the archaeological excavation and the ancient paper test, it is judged that paper originates from the Western Han Dynasty of the 2nd century B.C., earlier than the Eastern Han Dynasty in the early 2nd century, indicating that China has a long history of this invention. It is also the necessary affirmation on the creative labor of the nameless ancestors in the Western Han Dynasty. Therefore, the millennium dispute over origins of paper since the Tang and Song dynasties should be put to an end. The question to be discussed is how to reevaluate the historical role of Cai Lun.

Those who denied that the Western Han Dynasty had paper focused their attention on the earlier Baqiao paper and said that the year judgement was only a personal opinion of an archaeologist. There was mainly the fiber orientation arrangement without broomization phenomenon, "and therefore it was not paper" but "fiber accumulation". In fact, the judgment of this paper to be the artifact of the Western Han Dynasty is made collectively by the archaeological expert group of Shaanxi Province, not by any individual. From 1964 to 1988, more than 20 scientists from more than 10 scientific research institutions at home and abroad have tested the paper for up to 10 times, and all agree that it is hemp paper instead of a mess of wasted hemp. Besides, Jinguan paper, Zhongyan paper, Majuanwan paper, Fangmatan paper are all typical plant fiber paper[1]. About Baqiao paper, Japanese experts wrote as follows:

> *As to the world's earliest paper unearthed in Xi'an Baqiao, scholars have different views. Some scholars think it is rough hemp paper ... other scholars hold it is something of hemp fiber packing ... I think Baqiao paper is paper, at least I think it is what the word "Zhi" indicates when the specific character is first created.*[2]

The reason is that when this photomicrograph was observed, those who did not think it was paper held that the majority of the fibers were long and arranged in a parallel direction, but we could conclude that the whole fibers were disordered, and that the crushed fibers could be seen. On the other hand, Loulan paper was really the paper for writing. Through observation of the

---

[1]　久米康生. 出土紙が証言する前漢造紙[Z]. 東京: 百万塔, 1988(70): 1-5.

[2]　増田勝彦. 橋紙の化験結果に関する討論[Z]. 東京: 楼蘭文書紙と紙の歴史, 1988.

photomicrograph, the non-broken fiber and intact hemp fiber could also be seen. On the paper more than three hundred years later than the Baqiao paper (Wei Jin time), the non-paper part was still seen, which showed that the early paper processing was of low level, so it seemed that it was a matter of course to see the non-paper part.

The above argument is the most objective statement that the scientific practitioners have made over the test results of the ancient paper, and according to the study of the majority of Chinese and foreign scholars, for the previously unearthed ancient Western Han Dynasty paper, it can be said that there is no room for debate. Especially because of the discoveries of the Fangmatan map paper and the Western Han Xuanquan paper with Chinese characters, even without the test, it can be judged that they fall into the category of the early writing paper. Fangmatan paper was earlier than Baqiao paper, but the quality was better (Figure 6). Japanese scholars had comments on those who rejected the statement that the Western Han Dynasty has had paper. Since the statement that the Western Han Dynasty has had paper was announced in China, it was said that the statement lacked grounds, which was to maintain that Cai Lun was the inventor of paper. As a result of emotional issue, the statement that paper originated from the Western Han Dynasty was opposed, and so it was said that "Baqiao paper statement could not be finalized"[1].

Some people made use of our laboratory data analysis, and said that the technical specifications of the unearthed paper before Cai Lun were already close to that of the modern handmade paper, so they wondered whether the Western Han Dynasty people were able to create such paper. Our answer is that we should never underestimate the wisdom of the ancient craftsmen and their superb skills. As long as their fine craftsmanship is concerned, the fine artifacts often make modern people acclaim them as the peak of perfection, such as the man-made fish basin and light transmitting mirror, which are difficult for a modern physicist to design out. Moreover, they could not understand the manufacturing principle until they racked their brains to extreme. The origin of papermaking is a serious academic issue. Putting forward or refuting a certain point of view must be based on the evidence instead of the feelings, or the fiction and suppositions can not substitute for the arguments, and the administrative means cannot be used to suppress different views. In today's

---

[1]　中山茂.市民のための科学論[M].東京：社会評論社,1984.

society where science as well as academic democracy is flourishing, though the statement that the Western Han Dynasty has had paper is criticized by some people, the statement is still supported by more scholars both at home and abroad.

Scientific tests have proved that all the Western Han Dynasty writing paper before Cai Lun is of hemp or linen fiber, which is not essentially different from paper after Cai Lun's time. Archaeologists have made the scientific year time judgement over every piece of paper sample according to archaeological research methods. The test analyses have also proved that the paper before Cai Lun's time does not contain silk fiber, and of course is essentially different from the silk cloth. This has overturned Zhang Yi, Fan Ye and others' wrong assertion that the paper before Cai Lun's time is of silk and only Cai Lun creates the time of plant fiber paper.

1. 1. 3. 2   Cai Lun's Contribution to Papermaking

After the hemp paper was made in the Western Han Dynasty, the papermaking had experienced continuous improvement and then entered the stage of new development in the middle period of the Eastern Han Dynasty. After he acceded to the throne, Emperor Liu Zhao (89~105) in Yongyuan 14 years (102) appointed Deng Sui (81 ~ 121) as the empress, who was granddaughter of Deng Yu (2~58), the Eastern Han Dynasty's founding hero. She had been reading the classics and history when she was very young, and preferred to use paper. In early Yongyuan Year (89 ~ 100), the area of papermaking in China had been expanded, and the contributions to the palace should be good paper and ink.

When Empress Deng was appointed, it was in the late years of Emperor He. In December of the first year of Yuanxing (105), the Emperor died, and the infant Liu Long ascended the throne, respectfully called Empress Deng the Empress Dowager who was authorized to listen to politics. But Emperor Liu Long died before he was one year old, so the Liu Qin's son of the former Emperor's uncle was appointed Emperor An (107~125). The court was still held by the Empress Dowager, whose reign was 20 years and whose system lasted for all her life. Over 200 years after China had developed its paper, Cai Lun took to the stage of history. The book *Hou Han Shu* has the record of his life.

After Cai Lun entered the palace in 75 A. D. , he was the official of the court. From 76 A. D. to 86 A. D. , he started to serve as the door assistant official, who worked for the imperial court communication. In Emperor

Zhang's reign, the Empress Dou, who was not able to give birth to a child, incited Cai Lun to accuse Song *Guiren* who had a son and forced her to commit suicide. When the Empress Dou held court, she, in year 88, appointed Cai Lun who had the previous job done as the *Zhongchangshi* (the higher rank official) to look into politics, which was the beginning of the related official pre-governance in history. In year 89, Emperor He acceded to the throne, and Cai Lun maintained his position. Because Empress Deng loved paper, Cai Lun in year 105 looked over the papermaking, produced the paper and contributed to the palace. After Emperor He died, Empress Deng held office, who still put Cai Lun in the important position, appointing him the *Longtinghou* (龙亭侯, even higher official) in the year of 114, taking charge of 300 families which were in present Yangxian county of Shaanxi Province. In addition, Cai Lun was awarded as the *Changletaipu* (长乐太仆), whose power was at its peak. In the year of 121, Empress Deng died, and Emperor An took over the court. Because Cai Lun once had framed a case against his mother Song *Guiren* and forced her to death, the Emperor summoned for trial of Cai Lun, who was so self-aware that he committed suicide. The court eliminated his position and removed his fiefdom. He ended up in a miserable death for getting involved in the palace fight.

Since 91 A. D. when Cai Lun was an official of *Shangfangling* (尚方令), he did something good for the development of technology. *Shangfangling* was an official title or position, in charge of good weapon manufacturing for the Emperor including the sword of the Emperor. There were unearthed sword, stirrup and crossbow with inscriptions in Cai Lun's time, showing superior craftsmanship. When he served the positions of *Zhongchangshi* (中常侍) and *Shangfangling*, Cai Lun also went to the paper mill to investigate papermaking technology, selecting the skilled craftsman and gathering them in the Shangfang workshop to make paper for the imperial uses. In addition, he suggested that the court promote the use of paper. The Eastern Han paper also was once unearthed, such as the two pieces of paper with words made by Stein (1862~1943), which was excavated in 1901, in Luobunao'er (Lop Nor), Xinjiang. In 1959, a crumpled sheet of paper with dark powder was found in the tomb of the Eastern Han couple in Xinjiang, which might be used for women's eyebrows❶.

From 1990 to 1992, in the vicinity of Dunhuang, from the unearthed

---

❶ 李遇春. 新疆民丰县北大沙漠中古遗址区东汉合葬墓清理简报[J]. 文物,1960(6):6-12.

Xuanquan paper site of the Han Dynasty, the Eastern Han paper was found in addition to the Western Han Dynasty paper. The paper was yet to be analyzed, and according to the documentary records, the raw material used for hemp paper was not only the torn linen cloth as well as fag end, but also the old fishing net. Zhang Hua (232~300) in *Bowu Zhi* (《博物志》, *Natural History*, about 290)wrote, "Cai Lun, from Guiyang, began to tamper with the fishing net to make paper", which might have the supporting evidence. It is difficult to make paper from the fishing net. According to our simulated experiments, it would be necessary to strengthen the cooking and ramming processes if fishing net were used for papermaking.

The biggest contribution of Cai Lun was that he presided over the development of paper made of wood bast ("tree bark") fibers. In the Wei Kingdom, Three Kingdoms time, Dr. Dong Ba (200~275) in *Dahan Yufu Zhi* (《大汉舆服志》, Historical Record of the Han Dynasty) said: "Dongjing (Luoyang) has 'Cai Hou paper', namely Cai Lun paper. Paper made of old hemp was named the hemp paper; paper made of tree bark, paper; paper made of old fish net, net paper." This is an important record. Gu is Chu, the broussonetia papyrifera (mulberry leaf tree), whose phloem contains quality paper fiber. This tree has been born in China, north or south, since ancient times. In ancient time, the mulberry bark fiber was used for cloth weaving material. In 1907, Stein in Xinjiang unearthed yellow cloth made of very fine weaving, which was tested to "contain moraceae plants bark fibers, possibly mulberry tree". ❶

From the Western Han Dynasty to the early stage of the Eastern Han Dynasty (the 2nd century B. C. to the 1st century A. D.) when hemp paper was being made, in the process of collecting raw material of rags, inadvertently the rag made of mulberry bark fiber was mixed with the sackcloth, and thus the hemp paper with little mulberry fiber was created, which was of better quality. The broussonetia papyrifera containing many fibers, featured low cost while high quality, could be either planted or grow wildly. Cai Lun turned the unconscious process of making paper with mulberry fiber into a conscious process of papermaking, thus completing a technological breakthrough. The production of the paper should be seen as an invention, because it is more

---

❶　SERINDIA S A. Detailed Report of Exploration in Central Asia and Westernmost China[M]. Oxford: Clarenden, 1921.

complicated for the addition of some new technology and tools to use the mulberry fiber as raw material. The accomplishment of the invention was attributed to Cai Lun, who first presided over the fabrication of paper made of mulberry bark in 105 A. D.. This led to the production of a series of bark paper, such as mulberry paper, cane paper and daphne (bark) paper, which expanded the source of raw materials for paper. The paper became national paper in China in the 7th century.

In the first year of Yuan Xing under the reign of Emperor He (105), Cai Lun presented the hemp paper made in Shangfang to the court, and also proposed the paper promotion, which was approved. Ever since then uses of paper were popularized. Although before his time, there had already been paper, his contribution to papermaking was not to be neglected, which could be concluded as follows: Firstly, he summarized the technology experience from the Western Han Dynasty, the early Eastern Han Dynasty to that of his own time in hemp paper production, and organized the production of high quality paper. He then used old fish net as the raw material, so that he expanded the raw material range of hemp paper, thus improving the hemp paper technology. Secondly, he proposed to make paper from the mulberry bark, completing the technological breakthrough in invention of wood bast fiber paper, so he opened up new raw materials for papermaking, and led to the emergence of new paper series. Thirdly, he proposed to promote the production of papermaking at home, which was supported by the court, and the paper factories were therefore set up everywhere, which promoted the rapid development of the paper industry. In a word, Cai Lun is an innovator of papermaking technology, serving as a link between the past and the future.

To prove papermaking originated in the Western Han Dynasty, and to appraise Cai Lun of the Eastern Han Dynasty appropriately are to reflect the historical fact, and historical truth does not mean to work against Cai Lun, which does not make China invention eclipsed. Instead, the reflection is agreed on by scholars both at home and abroad. In recent years, publications of the United States, Britain, France, Germany and Japan have noted the current situation of domestic paper history disputes in China, and most of them agree with the statement that there was paper in the Western Han Dynasty.

Professor Qian Cunxun of the University of Chicago once wrote:

In recent years, a number of ancient paper since the 2nd century B. C. found around China proves that paper is invented and developed in the Han Dynasty ... That the Western Han Dynasty has had the plant fiber paper has been proved by other unearthed ancient paper in the archaeological excavation ... That there was paper before Cai Lun's time is not necessarily inconsistent with Cai Lun's achievements recorded in the history. He may well be the innovator of new materials in papermaking. ❶

Professor Sou Neiqing (Yabuuchi, 1906~2000), from Kyoto University of Japan enumerated the discovery of ancient Chinese paper in recent years, writing as follows:

Chinese papermaking originated from the early Han Dynasty, which is very clear. According to a 1965 microscopy, this type of paper was made from cannabis sativa. Thus, Cai Lun was not the first person to make paper, rather, he was the papermaking reformer, who first used the tree bast, the hemp, the cloth rag and the fish net to improve the papermaking technique. ❷

Professor Wilhelm Sandermann quoted from *Cai Lun Zhuan* (《蔡伦传》, *Cai Lun's Biography*) in *Hou Han Shu* the documentary records of paper uses before Cai Lun's time and the Western Han Dynasty paper unearthed in recent years, concluding that papermaking was invented in the Western Han Dynasty in the 2nd century B. C.. It was not Cai Lun who invented paper. ❸ France orientalist Dai Ren (Jean Pierre Drege) has been studying for nearly 30 years on the different opinions about the origin of the papermaking, who thinks that "the conclusion is that Cai Lun may not be the inventor of paper, but he improved the papermaking technique in about the year of 105". These comments from overseas scholars show that the excavations of the Western Han Dynasty paper have not led to confusion in the international community, and there is no doubt that it is in China that paper is invented.

1.1.3.3 Wrong Statements of Cai Lun's Invention of Paper

Although Cai Lun has contributed to the papermaking technique, he is not the inventor of the papermaking. Then how and when did the claim that paper

---

❶ NEEDHAM J. Science and Civilization in China: 5[M]. Cambridge: Cambridge University Press, 1985.

❷ 薮内清. 科学史からみた中国科学文明[M]. 东京: NHKブックス社, 1982.

❸ SANDERMANN W. Die Kulturgeschichte des Papiers[M]. Berlin: Springer Verlag, 1988.

was invented by Cai Lun came into being in history? The issue is to be analyzed.

More than three hundred years after Cai Lun's death, historian Fan Ye of the Northern and Southern Dynasties amended some section in *Cai Lun Biography* in *Dongguan Han Ji* (《东观汉记》, *A Book of Historical Records*), which changed from "being in change of Shangfang papermaking" into "making paper with tree bark, hemp, rags and fishing net". Therefore, it deviated from the original appraisal of Cai Lun in the Eastern Han historical record, mixing the invention of the Western Han people with the new contribution of Cai Lun and attributing everything to Cai Lun, which was not appropriate. Fan Ye's so doing was influenced by the thought of Zhang Yi in Wei Kingdom time, the Three Kingdoms.

In *Gujin Zigu* (《古今字诂》, *A Book of Word Study*), Zhang Ye divided paper into ancient paper and today's paper, believing that the ancient paper made before Cai Lun's time was of silk fabric used for writing, and also known as Fan paper. The so-called today's paper was made by Cai Lun from rags and also called *Zhi* (帋). But as to what he said, there was neither documentary evidence, nor physical evidence. Instead, he was playing with the words. During the Spring and Autumn period, the Warring States period and the Han Wei period, the silk fabric used for writing was generally called silk or silk cloth, which was never called paper or Fan paper. However, the fabrication of paper from hemp rags did not start from Cai Lun. In the Western Han Dynasty before Cai Lun's time, there had been paper made from hemp. Paper and silk are different without confusion of their names. It is unnecessary to name in such a way calling ancient paper or today's paper (present paper), which would be perplexing. Paper and silk are different things. Moreover, there is no confusion of names.

Zhang Yi's explanation of the word "paper" is not correct. When Fan Ye wrote in *Cai Lun Zhuan* (《蔡伦传》, *Cai Lun's Biography*) in *Hou Han Shu* in 445, he indiscriminately cited Zhang Yi's point of view, and said the ancients "used silk and called it paper", which mistook the definition of paper, confusing silk with paper. He also said that because of the high cost of silk, Cai Lun intentionally made paper with hemp material, so Cai Lun invented the plant fiber paper. Scholars of the Tang and Song Dynasties argued over Fan's statement, because they had read from the books of Han and Jin people including *Dongguan Han Ji*, *Fengfu Tongyi* and *Hou Han Ji*, etc. and saw the clear records of paper using before Cai Lun's time. All these early historical books did not say that paper was invented by Cai Lun. The present edition of *Dongguan Han Ji* derives from *Hou Han Shu*, which has deviated its original and is not well grounded. The later edited version of book should not be regarded as the same as the original one and it is to be checked or proofread with other ancient versions of books. Fan Ye, the author of the book *Hou Han Shu*, did not deal well with the historical materials left by his predecessors when he was writing the biography of Cai Lun. On the one hand, in *Baiguan Zhi* (《百官志》, *A Book of Officials' Records*), *Denghuanghou Zhuan* (《邓皇后传》, *Empress Deng's Biography*) and *Jia Kui Zhuan* (《贾逵传》, *Jia Kui's Biography*), he enumerates the records of paper using before Cai Lun's time; on the other hand, in *Cai Lun Zhuan* (《蔡伦传》, *Cai Lun's Biography*), he believed that it was Cai Lun who invented paper, therefore he fell into the trap of self-contradiction and could not be justified.

When the Emperor's son Li Xian (654~684) explained with notes in *Cai Lun Zhuan* in *Hou Han Shu*, he added some contents concerning Fan Ye's statement. It is not reliable to argue that Cai Lun's invention of paper is based on the historical record. Fan Ye's *Hou Han Shu* has its own uniqueness and features, but there is mistaken information in *Cai Lun Zhuan*.

It seems that *Cai Lun Zhuan* in *Dongguan Han Ji* written by people of the Eastern Han Dynasty is the earliest original records of Cai Lun. The book as the national history of the Eastern Han Dynasty is authoritative, however it does not mention that paper is invented by Cai Lun. When writing *Cai Lun's Biography*, Fan Ye was transcribing some erroneous statements concerning papermaking from *Gujin Zigu* by the Wei people and drew the erroneous conclusion that paper was invented by Cai Lun, giving rise to a long-term dispute over papermaking origin, which continued to modern times. In present view of this, *Cai Lun Zhuan* written by Fan Ye is somewhat biased, for it has neither the original records nor authoritativeness, sharing the same status as Tang and Song people holding dissent views, which cannot be judged as the highest standards for the right and the wrong. Only archaeological investigations can serve as the role of judgement. The archaeological evidence has proved the statement by the people of the Tang and Song Dynasties that paper was invented before Cai Lun's time is correct.

1.2 Papermaking Technology in the Wei, Jin, Northern and Southern Dynasties

1.2.1 Improvement and Popularization of Hemp Paper

The Eastern Han and Western Han Dynasties were the foundation period for papermaking industry and the main production was hemp paper. It was a period when paper and wooden or bamboo slips were used at the same time. The Wei, Jin, Northern and Southern Dynasties (the 3rd century to the 6th century) were the development stage of papermaking. In this stage, except the two large areas including Huanghe River and Yangtze River valley unified by the Western Jin (265~316), for the rest of the period, the whole country was in the state of secession of various regimes of different nationalities. Unlike the split western situations after the fall of the Roman Empire, China's divisions were mainly political, and the regimes still maintained a common culture, writing characters and mainstream ideas, and the integration of various ethnics and exchange of economy and culture still existed. The image of China as a whole was still clearly visible, and there were unified factors in the divisions. China did not experience the process of radical division as in the western world, and therefore the development of science and technology as usual was not affected, especially in the field of papermaking technology. Both the north and south regions needed paper regardless of the reigns or regimes, which further promoted the paper development.

The Wei, Jin, Northern and Southern Dynasties papermaking was a direct inheritance of hemp paper technology development from the Western Han and Eastern Han Dynasties of China. Moreover, there was obvious improvement on the technology as well. During this period, the hemp paper was still the main paper kind, but systematic test analyses of the unearthed paper of this period showed that the hemp paper whiteness had increased, the paper surface was smoother, the fibers were closer and the fiber bundles were fewer. Besides, the paper curtain patterns were generally visible. The beating or refining was significantly improved, and some Jin paper was up to 70° SR. The Northern and Southern Dynasty paper thickness was as thin as 0.1 mm~0.15 mm, while the Han Dynasty paper was generally 0.2 mm~0.3 mm. From the perspective of technical analysis, at this time, the paper was copied and fabricated through the foldable curtain bed paper cutting machine, which was more advanced and powerful compared to the fixed papermaking machine, similar to what is used at the traditional handmade paper mill at present. Its superiority was that it was able to copy out the thin yet uniform paper. In addition, it could make millions of paper copies without the need for changing operation to another machine, which was an invention of epoch-making significance. With this advanced machine for papermaking, a high degree of hemp beating was required so that the intensification of cooking and chopping was facilitated.

For the use of the foldable curtain bed papermaking machine, the working procedure of wet paper pressing was also added to the papermaking process and the paper drying mode was also changed at the same time. In the Han Dynasty, with a fixed paper machine, the paper copied would be directly dried in the sun, therefore more paper machines were needed and it was more time-consuming. With the flexible machine, after the paper was copied, the bamboo curtain was removed and the paper was moved from the curtain onto the wooden board and piled up layer by layer. After the water squeezing, the semi-dry paper was taken off, and with a brush the paper was transferred to the wooden board or the wall coated with fine lime surface for drying, which could be dried quickly. The single-sided smooth paper was therefore made. The paper surface sticking to the board or wall was smooth, called the front side, could be used for writing without calendaring or polishing, and the other side of the paper was called the back side. According to the observation and test over a large amount of paper from the Wei, Jin, Northern and Southern Dynasties, there were visible front and back sides, which proved that the above judgment was correct. The paper copying machine was made up of three parts including the wooden bed, the curtain and the side pillar (Figure 21). The curtain was usually made of fine bamboo stick (Figure 22). It was a thinner bamboo stick if there were nine or more curtain strips per centimeter (9 strips per cm ~ 15 strips per cm). A thicker bamboo stick

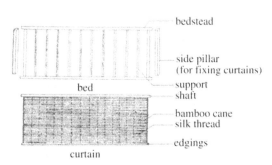

Figure 21   Flexible Curtain Bed Paper Mould, painted by Pan Jixing (1979)

contained five to seven curtain strips (mostly five). The northern non-bamboo region used stalks of achnatherum or hemerocallis fulva to make curtain (Figure 23).

silk thread —   — bamboo cane

Figure 22　Schematic Diagram of Curtain　　　Figure 23　Operation Chart of Curtain

Making, painted by Pan Jixing (1979)　　　Making, painted by Pan Jixing (1998)

According to the physical measurement of a great deal of paper from the Wei, Jin, Northern and Southern Dynasties, the length and width range of changes are listed as follows:

Table 1　Paper Size in Wei, Jin, Northern and Southern Dynasties
and Paper Copying Machine Dimensions

Time	Wei and Jin Dynasties		Northern and Southern Dynasties	
Type	Size A (small paper)	Size B (big paper)	Size A (small paper)	Size B (big paper)
Height (cm)	23.5~24.0	26~27	24.0~24.5	25.5~26.5
Length (cm)	40.7~44.5	42~52	36.3~55.0	54.7~55.0

From the size of the paper, the dimension of the paper copying curtain could be seen. In general, the range had been increased compared to the Han Dynasty paper. According to the need, there were divisions of big paper and small paper. In February 1972 in Tulufan City (Turpan) of Xinjiang, the unearthed Former Qin Dynasty (350~394) tombs had the writ of Jian Yuan 20 years (384), which had been written on a piece of original paper without cutting. After measurement, it was of the size of 23.4 cm×35.5 cm, similar to today's one page of "China TV News" or "Beijing Evening News", which was the small paper. The big paper should be as big as one page of today's *Guangming Daily* or *The New York Times*. Wang Xizhi, the Jin Dynasty calligrapher, was able to write a lot of words on this kind of paper. 1500 years ago, it was an uneasy job for a person to make such a large piece of paper with the curtain machine at that time. The people's praises of the paper offered notes or annotations to our test over the ancient paper. For example, Fu Xian, man of letters in the Western Jin Dynasty (239～294) in *Zhi Fu*❶ (《纸赋》, *Praise for Paper*) wrote a poetic essay, meaning of which is as follows:

> *Low-level writ becomes high-level one just because of different ways of writing.*
>
> *Classic books system at any time changes, so does the materials of writing.*
>
> *As rope knots were replaced by bone text, bamboo or wooden slips are finally replaced by paper for writing.*
>
> *Despite its raw material of rags, good paper is white, pure, fine, well cut while inexpensive for writing.*
>
> *On the paper shine brilliant words, and men of letters are fond of the paper's look as well as its writing.*
>
> *Foldable and flexible, the paper is convenient for collecting in addition to writing.*
>
> *When one is alone faraway and missing families and friends, the feelings can be sent across a distance with thousands of miles just by a paper letter of writing.*

Zhi Fu was once translated into English❷. In the Southern Dynasty, Emperor Xiaoyi (508~554) also praised the paper in a poem, as follows:

> *Squarely cut, the paper is as white as snow and frost.*
> *Brilliantly made, the paper keeps feeling and record.*

In the Han and Wei Dynasties, both silk and paper were the writing materials simultaneously, and paper became popular as a new material. In the Jin (the 4th century) time after the production of a large number of square, white and smooth paper, people did not have to use expensive silk and heavy bamboo slips for writing. Instead, they got used to writing with paper. As a result, paper became the dominant writing material. The ruler made it clear that the court should only use paper instead of the bamboo slips. Since the Eastern Jin Dynasty (317～420), the paper was used to copy-write *Sishu* including Confucian classics, history, philosophy and literature works; public

---

❶　傅咸.纸赋[M]//严可均.全上古三代秦汉三国六朝文：全晋文.北京：中华书局,1958.

❷　HUNTER D. The Literature of Papermaking[M]. Ohio：Mountain House, 1925.

and private documents; contracts and religious classics. The copy writing was prevailing at that time. The books were made into scrolls, with a fixed writing format. With each piece of paper writing 400 to 500 words, every piece was glued together so as to form a long roll. Scrollable and stretchable, the roll could be scrolled with the roller axis, convenient for use. The book collection, private or public, augmented considerably. In the Southern Dynasty (431), the inner government library had a collection of 64,000 volumes of books. Emperor Liangyuan in Jiangling had a collection of book with 70,000 volumes. In the reign time of Emperor Liangwu (502 ~ 549), "within the Four Seas (nationwide) every household has a collection of books of literature and history". The proliferation of paper books greatly promoted the development of social culture, education, science and technology, and also made religious prosperity. After testing, the writing books of this period found in Dunhuang stone chamber and the underground paper books unearthed revealed that most of them were accomplished with the hemp paper of higher quality.

The improvement of the paper quality and the increase of the paper dimension promoted calligraphy and painting art in the Wei, Jin, Northern and Southern Dynasties so much that they entered into a new phase, causing the change of Chinese characters. Writing on the hard bamboo slips, the brush pen was constrained by the space and quality of the writing material, so it was difficult to give the writing the full play and the writing speed could not be accelerated. In addition, with about ten to twenty words written, it was necessary to change the slip. On the opposite, on the big smooth paper, white and ink absorbent, the brush could dance or move freely like a dragon or a snake, and the calligraphy charm could be fully revealed. *Xiaozhuan* and *Lishu* (official script) were the main style in the Han Dynasty. Starting from the Qin Dynasty, *Xiaozhuan* are strokes of the even and uniform thickness, while *Lishu* shows the change of thickness, adding to the artistic play. The Wei, Jin, Northern and Southern Dynasties witnessed the prevalence of different scripts including *Kaishu*, *Lishu* and *Xingshu*. The writing speed was significantly accelerated and the strokes were more fluent and unrestrained, which reflected the writing material transition from the bamboo slips to paper.

Avadana, *Zhufo Yaoji Jing* (*Buddhas Scriptures Collection*), and *Sanguo Zhi* (*The History of the Three Kingdoms*) of the Eastern Jin Dynasty (Figure 24) represent the popular calligraphy art in this period. In the Jin Dynasty, thanks to the widespread use of paper, famous calligraphers like Wang Xizhi and Wang

Xianzhi appeared. The two calligraphers mainly wrote in *Kaishu* and *Xingshu* styles.

Similarly, painting on paper could reveal a better artistic effect. Gu Kaizhi（345～406）, the Jin Dynasty painter, and others took the lead to paint with brush on paper the birds, animals, vegetation, insect and fish, which were not yet handed down. Only Zhang Yanyuan's work about "Famous painting of past dynasties" (about 874) had them recorded. In 1964 in Tulufan (Turpan) of Xinjiang was unearthed a handwork of folk artists, the Eastern Jin Dynasty colored painting of the landlord life, with a full length of 106.5 cm, and 47 cm in height. It was a connection of six pieces of hemp

Figure 24   *Sanguo Zhi* of Eastern
Jin Dynasty unearthed in Xinjiang
Uygur Autonomous Region,
from *Cultural Relics*, 1972(8)

paper by test. This is the earliest paper drawing (Figure 25), reflecting the new fashion of painting by people at that time.

Figure 25   Eastern Jin Dynasty Colored Folk Painting unearthed in 1964 in Tulufan,
from *Unearthed Cultural Relics of Xinjiang*（1975）

1.2.2   Development of New Materials and New Uses of Paper

In the Han Dynasty, rags were mainly used to make hemp paper. In the Eastern Han Dynasty, Cai Lun summarized the hemp paper technology experience and created the papermaking technique from wild bark fiber, which opened up a new source of raw materials. However, the papermaking process was more complex than that of the hemp paper, and it was not easy to make the white paper, therefore it failed to promote. After the Jin Dynasty, due to the development of papermaking technology, bark paper production gradually increased and raw materials range further expanded. Su Yijian (958~998), in the Song Dynasty, wrote in Volume Ⅳ of *Wenfang Sipu* (《文房四谱》, *Several Records of Study*) that paper, based on some documentary, was made of mulberry bark❶.

Zhang Hua, the Western Jin scholar, wrote letters to his friend Lei Huan using mulberry paper. It could be inferred that in the Western Jin Dynasty paper was made of mulberry fiber. Since ancient times, China developed sericulture and raised silkworm with mulberry leaves and mulberry trees could be seen everywhere. The tender skins of the branches were peeled off to make paper while the growth of mulberry was not affected. In addition, other new uses increased at the same time. In 1901, Julius von Wiesner (1853~1913), the Austrian botanist, tested official document paper of the $3^{rd} \sim 5^{th}$ century paper of the Wei and Jin periods unearthed at Luobunao'er (Lop Nor) of Xinjiang, and found that there was mulberry paper. In 1972, there were unearthed pieces of writing paper in Tomb 196 at Astana in Turpan of Xinjiang, which were also mulberry paper after test. 42.6 cm in length, the paper is thin and white, with fine fiber and textile of concord❷. We also occasionally tested the paper made of mixed pulp of hemp and bark fibers. This type of paper took advantage of different fiber raw materials and reduced the production costs, which was a new breakthrough in raw materials blending. Therefore, it is of a great technical and economic significance.

If the Eastern Han Dynasty mulberry paper originated from the Yellow River Basin in the central region (now Luoyang in Henan Province), then till the Wei, Jin, Northern and Southern Dynasties, it spread rapidly to the vast southern region in Yangtze River Basin, and then shifted to the Yuejiang River delta in South China (now Guangdong) area, reaching north Vietnam. In the north it extended to Beijing.

Broussonetia papyrifera is moraceae woody. Its leaves are similar to

---

❶　苏易简. 文房四谱:纸谱. 丛书集成本[M]. 北京:商务印书馆,1960.

❷　潘吉星. 新疆出土古纸研究[J]. 文物,1973 (10):52-60.

mulberry leaves. Since the Han Dynasty, its bast fiber was used for clothes weaving fiber as well as papermaking. In the Wei and Jin Dynasties, the yield and producing areas of mulberry paper were further expanded. In the Northern and Southern Dynasties, broussonetia papyrifera were more cultivated, especially for papermaking. Jia Sixie (473~545), the Wei Dynasty agronomist, wrote a special chapter in *Qimin Yaoshu* (《齐民要术》, *Ancient China's Agriculture Encyclopedia*) on this tree planting. His book reflected the agricultural production status of the middle and lower reaches of the Yellow River in the 6th century.

According to the writing, in the 6th century, there were "paper mulberry shops", which were specialized in the cultivation of trees for making paper, as well as the middlemen who bought mulberry bark in the city. "So long as the mulberry can make paper," Jia Sixie pointed out, "there is a lot of profit." In 1973, at Thousand Buddha Cave of Dunhuang was unearthed the Northern Wei Dynasty (454) scripture which was written on the mulberry bark paper. In 1972, from the tombs of Astana of Xinjiang were unearthed three pieces of writing paper dated back to the 4th year under Jianchang (558) and 16th year under Yanchang (576), which were mulberry paper and bark paper.

From the beginning of the Jin Dynasty, at Shanxi River area, up to Cao E River upstream, south of Sheng County of Zhejiang, the wild rattan fiber was used to make paper. This area was rich in vines, and Shanxi River water was clear and suitable for papermaking. In history, the famous rattan paper was originated here. Later in other vine production areas, rattan paper was also made. Fan Ning, then the governor, ordered his subordinate to use rattan paper instead of the local paper for writing. Hence the rattan paper was considered to be good paper and it was promoted by the official Fan Ning. The "local paper" could not be understood as the toilet paper, but the paper of poor quality made locally at that time.

The Jin Dynasty paper workers from the practice discovered a scientific rule: all the textile fiber used for weaving could be used to make paper. This rule was realized from the experience of hemp papermaking in the Han Dynasty. The application of this rule led to the creation of substitutes of hemp paper like mulberry paper, rattan paper and other paper series. China never experienced raw material supply shortage although paper consumption increased gradually. However, for a long time in Europe, only hemp paper was made and used, therefore in the 18th century the raw material crisis broke out. As to the bast paper, thanks to its high quality with fine and flexible fiber, the paper was suitable for high-level uses, especially for calligraphy and painting. Besides, it was also good for printing.

With the increase in paper production, the use of paper was further extended to other aspects of daily life to replace expensive materials like silk, meeting the needs of writing materials in society. In the Wei, Jin, Northern and Southern Dynasties, there appeared many paper products as paper umbrella, paper kite, paper flowers, paper cutting, origami and toilet paper. The umbrella, which was widely used by people on rainy days, had appeared since the Northern Wei Dynasty (386~543). The surface of the umbrella was originally silk, which was not waterproof. When the paper brushed with Chinese wood oil was used for the umbrella surface, it was not only cheap, but also rain-proof. In the past, the carriage of the literati and officialdom had a cover, and in the Northern Wei Dynasty, the cover was made of thin bamboo strips covered with oiled paper, which was umbrella. The invention of the umbrella was also the indication of the birth of the waterproof paper.

Emperor Gao Yang (529~559), in view of the popular uses of umbrellas from the upper down to the lower levels, established a hierarchy, i. e., the Emperor used red, yellow umbrellas while the ordinary people used blue umbrellas. Paper umbrella was introduced by people from the northern minority areas into the central plains, and after the Tang and Song Dynasties, it was used over the whole nation (Figure 26), and the umbrella color classification system was also passed down. Later the umbrella spread to the rest of the world.

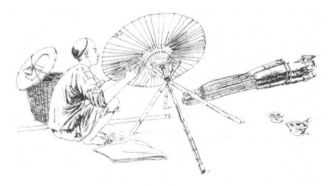

Figure 26　Umbrella Invented in China, from Colored Painting of Qing Dynasty

In ancient times, the kite, called *Zhiyuan*, was made of bamboo strips covered with silk. Paper kites were recorded in the Northern and Southern Dynasties. Gao Yang, the Northern Qi ruler, built a high wooden platform outside the palace of Yecheng (now Santai villages in Linzhang, Henan Province). It is said that after the overthrow of the Eastern Wei (534~550), the imperial descendants of the Eastern Wei Dynasty were crushed by the Emperor, and the Emperor's son together with the prisoner rose to the Jinfeng (golden phenix) platform and jumped. However, after gliding through the air for some time, the Emperor's son landed safely near the palace. It had to be a big kite, and he became the first person in the world flying in the air by the kite, even though the experiment was risky. A thousand years later, the Englishman Baden F. Smyth (1860~1937) successfully accomplished the flight experiment of flying kites in 1894. So the paper kite flying by the Emperor's son was of historic significance in aviation history❶.

Since the kite was able to lift off (Figure 27) and fly in the air, it was not a simple toy, but had scientific research value and other practical value. According to the history record, while Emperor Liangwu fought the traitor Hou Jing (503 ~ 502), Hou Jing won several cities of Liang, and at today's Nan Jing, both

Figure 27   Kite, from Wu Jiayou (1890)

sides fought fiercely. The Liang force, besieged in the city, let out the kite and asked for help outside the city❷. So the kite could also be used for military purposes.

Paper is able to be used for artists as the materials of paper-cut, which embodies a strong national art style. Its origin could be traced back to the Wei and Jin Dynasties, and it approached to its maturity in the Northern and Southern Dynasties. Color paper could be cut into geometric patterns, human figures, birds and animals, flowers and grasses, etc. With the style of elegant appearance, they could be placed on doors, windows or walls, and sometimes they could be used in specific festivals. From October to November 1959, at Asitana (Astana), Tulufan (Turpan) of Xinjiang, from the 4th century to the

---

❶   NEEDHAM J. Science and Civilization in China: 4[M]. Cambridge: Cambridge University Press, 1965.

❷   司马光. 资治通鉴[M]. 上海: 上海古籍出版社, 1987.

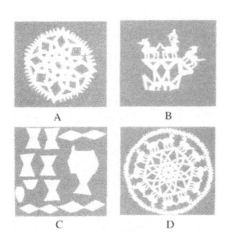

Figure 28　Papercut of Sixth Century
of Gaochang unearthed in 1959
in Xinjiang Uygur Autonomous Region,
from *Cultural Relics*, 1959(6)

A. Octagonal flower　B. Remnants of deer pairs
C. Diamond papercut　D. Restoration of B

7th century tombs, the papercut was unearthed. In the tomb of 306 (59TAM306), there were diamond-shaped and waist-shaped papercuts. There were another two pieces of round papercut, one was of the blue color, cut into a geometric pattern of octagonal flowers. With the outer serrated circle, the middle of the papercut was of the diamond-shaped and triangular pattern, belonging to the radiation-type papercut (Figure 28-A)[1]. The other papercut was an incomplete yellow piece (Figure 28-D). After recovery according to the paper cutting principle, it was a pattern of symmetrical deer, the periphery being a tooth circle, and the middle the hexagon. For each side, there were two deer standing back to back with their tails connected. Totally there were 12 deer. The writ of the tomb was written in Zhanghe 11 years (541)[2].

In 1959 from the Astana No. 303 tomb excavation, there was also unearthed yellow paper cut into a circular pattern, being the papercut with pair monkey flower pattern. Its age was Heping (peace) first year (551), Gaochang State (531~640) year number.

The Northwest Gaochang paper-cut art is undoubtedly imported from the Central Plains. Li Shangyin (813~858), the Tang Dynasty poet, once wrote about the papercut in his poem.

There were also man-made paper flowers related to paper cutting. The paper was dyed into different colors, and then by means of cutting, kneading, folding and other treatment, was made into branches, leaves, flowers, finally the bouquet of flowers, for interior decoration.

When Sun Fang (Sun Qizhuang) was appointed official in Changsha (now

❶　新疆博物馆.新疆吐鲁番阿斯塔那北区墓葬发掘简报[J].文物.1959(6):13-21.
❷　陈竞.从新疆古剪纸探中国民间剪纸的渊源[M]//陈竞.中国民间剪纸艺术研究.北京:
　　北京工艺美术出版社,1992.

Changsha City, Hunan Province), there was a story of the boy with paper flowers written in *Xisi Ming* (*West Temple Inscription*), which occurred in his tenure of office in Changsha. It could be seen that in the Jin Dynasty there had been paper flowers, which came into being at the same time or later than the paper cuttings. In this way, it also could be inferred that there should be a very early origin for origami, i.e., to fold the square paper or rectangular paper into a variety of origami art, and some processes of origami were included in the process of papermaking.

Another wide use of paper in daily life was for toilet paper, including stool cleaning and women's menstrual paper. This was essential for the health care of the masses to prevent illness. Because the paper was clean, soft, hygroscopic, one-time use, its use was better than any other material. For the records of toilet paper, there were rare historical materials, but people of the Wei, Jin, Northern and Southern Dynasties had used it. As for the packaging of paper, people of the Han Dynasty had used it, and the consumption was also quite large.

1.2.3   Paper Sizing (Gluing), Coating and Dyeing Technique

In order to improve the performance of paper, Jin people used sizing or gluing technique. The early sizing agent was a plant starch paste mixed into the pulp, and the copied paper sizing could be realized to avoid water bleeding in writing or painting. For the record in Europe, it was in *Papyrus Sive Ars Conficiendas Papyri*, in 1693 by the French J. Imberdi❶. European sizing paper was not earlier than the 13th century. In 1886, the 9th ~ 10th century Arabic paper was tested and sizing treatment was found, which was earlier than Europe. However, the Arabian papermaking technology was introduced from China.

In 1900, Aurel Stein (1862~1943) explored some of the Wei and Jin wood slips and Tang Dynasty paper books in Xinjiang's Hetian and Niya archeological excavation. The earliest of them was of the year 768, over 100 years earlier than Arabian paper. After test, it was the sizing paper. After that, in Gansu Province there were unearthed documents of the year 450 and the sizing

---

❶ HUNTER D. Papermaking: The History and Technique of an Ancient Craft[M]. 2nd ed. New York: Dover, 1978.

technique was further traced back to 450 A. D. ❶. The author in 1964 tested the paper of the book *Lücang Chufen* (412) in Beijing Library and found that the paper pulp contained the starch paste, so we continued to trace the year of sizing technology back to the period of the Eastern Jin Sixteen States time (304 ~439)❷. In 1973, we tested the unearthed paper from Xinjiang and noticed the hemp paper had been treated with sizing and calendaring. In fact, in the second half of the third century, the sizing application was there.

After the paper was treated with starch, although the writing effect of the paper was improved, it also had disadvantages as being brittle and of poor insect resistance. In order to overcome these shortcomings, surface coating technology was developed, which was an invention of paper processing technique. The method was to apply the white mineral powder to the paper table before calendaring. From the perspective of technological development, this is an improvement on surface sizing, in which mineral powder is substituted for starch.

The white mineral powder used for surface coating is usually chalk which is mainly calcium carbonate. Then there is the plaster ($CaSO_4 \cdot 2H_2O$), and lime and china clay ($Al_2O_3 \cdot 2SiO_2 \cdot 2H_2O$). The procedure is as follows: Firstly, grind the ingredients into fine powder, sift, and mix it with water into a milky suspension. Stir and remove the floating impurities. Secondly, cook the starch or glue in water, mix it with white powder suspension and stir again. Thirdly, dip in it evenly with broad brush comprising a row of pen-shaped brushes and coat it over surface of paper. Then dry. At last, do calendaring or polishing. If it is the single surface coating, the amount of white powder weighs 27% of paper; double surface, 30% of paper weight. The excessively thick coating would decrease the paper folding endurance while the paper weight would be unnecessarily increased. The paper used is of good quality, so the paper is sold at a higher price and mostly used as superior paper for superior purpose as writing.

Another paper processing is color dyeing, which would add to its aesthetic appearance, improve its performance, and expand its uses. The color paper began in the Han Dynasty, and Liu Xi of Eastern Han (66~141) explained the word "Huang" as the colored paper. After the Wei, Jin, Northern and Southern Dynasties, this technique was inherited and carried forward. The most

---

❶　中村长一. 纸のサイズ[M]. 大阪：北尾书局，1961.

❷　潘吉星. 新疆出土古纸研究[J]. 文物，1973（10）：52-60.

popular yellow paper was called Huang paper. The procedure of using this kind of paper to write a book before making a roll, was called "Zhuang Huang" (decoration and finishing). Most of the written classics in Dunhuang stone chambers are of this kind of yellow paper, which is light yellow in color and bitter in taste. This type of paper has the following effects: (1) moth proofing; (2) if there is a clerical error, the orpiment (As_2S_3) can be used for correction; (3) being formal. There is a sense of solemnity and the color of yellow means sacred. There were two ways to dye in the Jin Dynasty. First, writing was before dyeing; second, dyeing was before writing. The second is preferable. *Liu Bian Zhuan* (《刘汴传》, *Liu Bian's Biography*) in Volume 36 of *Jin Shu* (635) recorded that when Liu Bian came to Luoyang, he entered school and had a test. When the examiner ordered him to write, what he said meant that the paper was to be dyed yellow before writing.

After scrutiny over the unearthed ancient paper, it could be seen that most of the paper was dyed before writing. The yellow paper was mainly used for official government documents, Confucian classics and religious scriptures. The dye material used was from golden cypress, amur cork tree. The dry endothelium (inner skin) of deciduous trees was used. It was yellow, bitter, and aromatic. The dry endothelium (inner skin) was from the leaves of the family tree. The usual one was Huangbai (Phellodendrom amurense). Usually the phellodendrom amurense and phellodendrom sachalinense were used. In the Spring and Autumn period, the old tree over ten years was chosen. The old skin from the outside was removed whereas the inner skin was taken, half dried in the sunlight, flattened, cleaned, dried and ground into fine powder. The chemical analysis showed that phellodendri skin contained alkaloids and the main ingredient was berberine (C20H1905N), which was yellow, bitter and soluble in water. It was a yellow plant dye and a moth proofing agent, which could be used as medicine[1]. According to Jia Sixie, the later Wei man, who wrote a special section on dyeing and painting in *Qimin Yaoshu* described as follows:

> *For all the Huan paper, the whiteness was eliminated (the dye is coated to cover the white ground), and the color should not be too deep, otherwise it would become dark after a long time. People soak the phellodendrom sachalinense in water and cook. The dregs were thrown away and the liquid was used. To ensure the purity of the*

---

[1]   南京药学院.药材学[M].北京:人民卫生出版社,1960.

liquid, the repeated cooking processing was necessary, normally three times of boiling, filtering, and pressing. ❶

In 1964, we used hemp paper and mulberry bark paper to do the paper dyeing experiments. It was very effective to water cook phellodendrom sachalinense and do the dyeing. The color depth depended on the concentration of dye liquid and dye content, and color too deep was not suggested. As Jia Sixie put it, the longer the paper was stored, the yellower it would be. Thus it could be seen that the yellow paper in the Dunhuang stone chamber which was used to write classics in the Wei, Jin, Northern and Southern Dynasties, was originally dyed in light yellow but became deeper in color after a long time.

In addition to yellow paper, other kinds of colored paper were also produced in this period, which included paper with five different colors as blue, red, light blue, green and pink. Zhao (319~351), the time after the Sixteen States, the ruler Shi Hu (295~349) also ordered people to make five-colored paper in the capital city in 333. The dye used was mainly plant dyestuff. Red dyeing was done with red flower or hematoxylin, blue with indigo (blue), a mix of red, yellow and blue dye. By the way, kinds of secondary colors were obtained such as green, purple and orange. Purple dyeing could also be achieved with radix arnebiae seu lithospermi (purple grass).

1.3 Papermaking Technology in the Sui and Tang Dynasties and Five Dynasties

1.3.1 Development of Bark Paper and Rise of Bamboo Paper

After the Northern and Southern Dynasties, the Northern Zhou Dynasty (557~581) had made a unified effort to possess the Yellow River Valley and lower reaches of Yangtze River. In 581, Prime Minister Yang Jian seized political power of the Northern Zhou Dynasty and established the Sui Dynasty (581 ~ 618), further ending division between the Northern and Southern Dynasties and reunited China. After the Sui Dynasty, Li Yuan (565~635) and his son Li Shimin (599~649) established Tang Dynasty (618~649) during which unification was consolidated and social economic and cultural development was

---

❶　贾思勰. 齐民要术[M]. 北京:农业出版社,1961.

pushed to the peak. In this period, especially in the enlightened reign of outstanding Emperor Taizong (626 ~ 649) — Li Shimin witnessed famous "Prosperity of Zhenguan" (an era of peace and prosperity) in history, and followed by "Prosperity of Kaiyuan" governed by Emperor Xuanzong, Li Longji (685~762). However, the regime of the Tang Dynasty collapsed after the An-Shi Rebellion in 755, and local decentralization in the late Tang Dynasty led to a short repetition of feudal division in the Five Dynasties and Ten States Period. During 379 years of the Sui and Tang Dynasties, 326 years were a unified period.

During the Sui-Tang period, due to its political unification, the economy of Yellow River Valley shifted from recovery to flourishing stage on the basis of national integration. The economic development of Yangtze River Valley during the Northern and Southern Dynasties had approached to that of Yellow River Valley and continued to rise. Thus in this period, the two major valleys realized ultimate economic combination forming their distinct advantages, and the Grand Canal was one of bonds of this combination. This period witnessed a comprehensive and balanced development in agriculture, handicraft, and science and technology, as well as humanities and religion. The Sui-Tang period was a flourishing age in Chinese history and began the "Chinese Renaissance" preceding Europe for nearly a thousand years. British historian Herbert George Wells (1866~1946) deemed, "in the early Tang Dynasty, Chinese gracefulness, courtesy, cultural vitality and power made a sharp contrast with corruption, turmoil and division of the western world". Therefore, China possessed a leading position in the whole world. Under such circumstance, papermaking technology entered into a stage of rapid development.

From the perspective of the history of Chinese papermaking technology, the expansion of material sources was a sign of technological progress because the introduction of a new material tended to be accompanied by a set of new craft. The raw materials used in the Sui and Tang and Five Dynasties included hemp plants, paper mulberry bark, mulberry bark, rattan bark, daphne bark, hibiscus bark and bamboo, a few more varieties than those in the Wei and Jin Dynasties, and the Northern and Southern Dynasties. The emergence of bamboo was bound to develop a set of new technological processes due to its complexity in manufacture. The manufacture of hemp paper, as the main type of paper, was at its peak stage. Nevertheless, the output of bask paper increased remarkably compared to that of the previous dynasty and it competed for the leading position with hemp paper in quality. In addition to making paper from a single raw material, it was a new technological trend to make paper from wild plant fiber and the quantity of mixed-material paper was more than before. It could be traced back to this period that wastepaper was reused in the slot to make recycled paper. At this time, there were various methods for development and utilization of natural resources, which ensured a continual prosperity of papermaking in handicraft sector in national economy.

In the Tang Dynasty, besides rag, wild hemp fiber was also used as raw material to make paper. Volume 6 of *Compendium of Calligraphy* (847), written by Zhang Yanyuan, and Volume 2 of *Paper of Historical Calligraphy* (about 790), written by Dou Ji (760~820) both recorded: in the Tang Dynasty (713~741), Xiao Cheng (683~751) used wild hemp of west mountain and ancient paper mulberry in Guozhou (now Lingbao City of Henan Province) to make five-color stripe paper. Using wild hemp to make paper demanded a few more processes than using rags, however, the cost of whole production was greatly reduced as the raw material was available easily and cheap. China has abundant wild hemp, such as Corchoropsis tomentosa and Apocynum venetum and so on, all of which can be used to make paper. Emerging from the Jin Dynasty, rattan paper reached its peak in the Tang Dynasty, and its places of origin expanded from Shengxian County to the other places. *Geography Records of the New Book of Tang History* recorded that rattan paper was produced in Wuzhou (now Jinghua City of Zhejiang Province) and Yuhang District of Hangzhou City. Volume 26 of *Treaties on Prefectures and Counties in Yuan-ho Era* (841), written by Li Jifu, recorded that rattan paper was served in Wuzhou in Yuanhe Period (806~820) and in Xinzhou (now Shangrao City of Zhejiang Province).

Owing to the high quality of rattan paper, Emperor in the Tang Dynasty used it in the palace and imperial storehouse as senior document paper. *Hanlin Zhi* (819), written by Li Zhao (791~830) in the Tang Dynasty, pointed: "white rattan paper was used in awarding, enlisting, claiming, punishing and instructing while jute paper was used in comforting brigade ... and blue rattan paper was used in the worship at the temper." Drinking tea pervaded in the Tang Dynasty, Volume 2 *Usage of Instruments* in *The Book of Tea* (about 765), written by Lu Yu (733~804), pointed that first-class tea should be packed by white and thick rattan paper and then baked so as to avoid the change of taste. Many poets like Bai Juyi, Gu Kuang and so on liked writing poems in rattan paper. With sharp increase of the use of rattan paper and because of its limited quantity of raw material, Shu Yuanyu (about 765) came by the side of the Shanxi River only to find that rattan plants originally winding along the riverside for 450 miles were cut off, thereby *Sorrow of Ancient Rattan Plant by the Shanxi River* was written. It was a historical tragedy that the rattan paper decreased gradually and even became extinct since the Tang Dynasty.

Contrary to the rattan paper, Chupi paper (paper made from mulberry bark) had dramatically increasing output because paper mulberry was widespread, fast-growing and gained prosperous momentum artificially and naturally, thus becoming advanced calligraphy and painting paper. Volume 5 of *The Avatamsaka Sutra* (702), written by Buddhist Fa Zang (643~712) in the early Tang Dynasty recorded that Buddhist De Yuan, who built a clean vacant garden, planted various paper mulberries, flowers, and weeds and bathed when entering the garden and irrigated with perfumed water. Three years later, paper mulberries grew up and the fragrance pervaded the whole garden ... Then the bark was peeled off and dipped into perfumed water, and cleaned up, thus becoming the raw material of papermaking. After that, *The Avatamsaka Sutra* was exactly written in such paper and it also recorded that in the Yonghui Period (650~655), Buddhist Xiu De (585~680) in Dingzhou (now Dingxian County of Hebei Province), planted paper mulberries, flowers, and herbs in a clean garden for three years, irrigating it with perfumed water and then tidied them up to make paper. After that, he recruited famous scribe called Wang Gong in Weizhou who wrote *The Avatamsaka Sutra* in this paper. It showed that Buddhists not only planted paper mulberry, but also made paper for writing *The Avatamsaka Sutra* by themselves. They irrigated it with perfumed water in the garden, which indicated solemnity.

The Heart Sutra (600), stored in National Library in Beijing, was written in yellow Chupi paper by Emperor Yang in the Sui Dynasty, and each piece of paper was 25.5 cm × 53.2 cm with good quality. So was Taoism classic *The Supreme Secret*, which was written in the Kaiyuan Period (718) in yellow Chupi paper with wax covered on the surface, namely yellow wax letter paper. *Camp Roster in West Zhou* (716), written in the Kaiyuan Period, unearthed in Ashtar, Xinjiang, was in white Chupi paper. In the Tang Dynasty, many people wrote Buddhist scriptures in purple Chupi paper. Out of more than 40 kinds of Buddhist scriptures of the Sui and Tang Dynasties which we tested, around 25 percent of them were written in Chupi paper, hence accounting for a high proportion. Fiber of paper mulberry bark, thin and bright, could make close-knit and smooth tissue which looked like silk, thus getting its reputation as "Fragrant Silk Paper". Literates in the Tang Dynasty preferred Chupi paper, and literate giant Han Yu (768~824) called it "Chu Sir" (respectful name of the paper). Xue Ji (649~713), calligrapher in the Tang Dynasty, addressed Chupi paper respectfully as "the Duke of Chu" (the highest rank of ancient paper), just like artillery addressed by descendants as "the General". Early in the Tang Dynasty, Chupi paper had been traditional Chinese paper due to its rising social status. The character "Chu" was exactly the synonym of character "paper", and "Chu Mo" was also meant "paper and ink". Chupi paper was in supreme position among ancient paper, hence people in the Tang Dynasty thrived to develop it.

Mulberry paper was much more than in previous generations. Dunhuang sarcophagus discovered *The Lotus Sutra* (early in 7th century), written in yellow mulberry paper in Sui-Tang Period, and each paper was 26.7 cm × 43.5 cm with thin stripe veins. Unearthed in Astana in Xinjiang, household registration book in the Tang Dynasty, which bore the seal print of "The Seal of Dunhuang County", was made by using white mulberry paper. Lending Contract of Bai Huailuo (670) in the Zongzhang Period, and Field Distribution Record of Ning Hecai were also written on supreme white mulberry paper. In the Linde Period in the Tang Dynasty, Lending Contract of Teacher Bu (665) was transcribed in mixed-material paper made from hemp fiber and mulberry bark fiber. Held in the Palace Museum, Penta-Bull map, painted by Han Huang (723~787) in the Tang Dynasty, was a color painting in mulberry paper (Figure 29).

Picture 29 *Penta-Bull Map*, painted in Mulberry Paper by Han Huang,
artist in Tang Dynasty, held in the Palace Museum

Daphne bark paper and hibiscus bark paper were new types introduced in the Tang Dynasty. *Record of Marvels in South of the Five Ridges* (about 890), written by Liu Xun (860~920), said:

> *There were many Zhanxiang trees (namely Aquilaria malaccensis) in Luozhou County in Guangguan District, much like Juliu (a kind of tree) in shape with white and lush flowers and orange peel-like leaves, so they were used to make paper. The bark was white with fish-like veins in the surface, and widely used in Luozhou County, Yining County, and Xinhui County. The paper was inferior to Chupi paper as it could be broken and rotted by water easily.*

Terroir of South of the Five Ridges, written by Duan Gonglu (840~895), pointed: "there were many Aquilaria agallocha in Luozhou County, and much like Ju Liu (a kind of tree) in shape, whose bark could be used to make paper. Locals called such paper as 'Xiangpi paper'." Luozhou County, Yining County, and Xinhui County were in the South of the Five Ridges in the Tang Dynasty (now Guangdong Province). According to the passage quoted from *Newly Revised Canon of Materia Medica* (659), written by Su Jing in the Tang Dynasty, in *Compendium of Materia Medica* (1596): "Aquilaria agallocha resembled Ju Liu in shape with cyan bark and its leaves were like that of tangerine. Aquilaria agallocha was not withered in winter and produced white and round flowers in summer". Guiliu (an ancient tree), alternative name Juliu, deciduous tree classified as Juglandaceae, namely Pterocarya stenoptera, whose leaves grew alternatively with black and gray bark. Zhanxiang, classified as Thymelaeaceae, namely Aquilaria agallocha, was aiphyllium and its leaves grew alternatively with white flowers and taupe bark. Found in Guangdong Province, Guangxi Province, and Fujian Province, Aquilaria agallocha secrets resins for perfumery and its bast fiber could be used to make paper.

Aquilaria agallocha was classified as Aquilaria sinensis, namely local Aquilaria agallocha, and grew in the South of Five Ridges, which could be used to make perfumery and paper. It had yellow and green flowers, which showed the only difference in shape characteristic between Zhanxiang described by people in the Tang Dynasty, thus such paper should be named as Aquilaria agallocha bark paper. People in the Tang Dynasty also used other plant fiber of Thymelaeaceae to make paper. In the early 20th century, such paper was unearthed in Xinjiang and its raw material may be Daphne papyracea. Plants of

Thymelaeaceae were widely distributed in China and had many species. Once mastering the technical experience of making paper by bast fiber, workers in the Tang Dynasty would search and test various raw materials in the wild, therefore, choosing plants of Thymelaeaceae was natural. Among paper used for transcribing buddhism scriptures in Dunhuang sarcophagus, Daphne papyracea bark paper was included. Such paper coated with starch in the surface was 0. 13 mm~0. 16 mm in thickness, with thick stripe veins and each was 0. 25 cm in width.

In the process of making paper from wild plants, people in the Tang Dynasty even took the bark of Hibiscus manihot, classified as Malvaceae, as raw material, thus famous Xuetao paper came into being. *Exploitation of the Works of the Nature* (1637), written by Song Yingxing of the Ming Dynasty, recorded: Xuetao paper in Sichuan province took the bark of Hibiscus manihot as material, which was boiled in water with powder of Hibiscus flowers. The advantage of the paper did not lie in its material and quality, but beautiful color and spread very far. In particular, among five kinds of document paper with Emperor's reign marks unearthed in Xinjiang from the Dali period to the Zhengyuan period in the Tang Dynasty, there was mix-material fiber paper made from rag, mulberry bark and laurel bark. Laurie is aiphyllium, classified as Lauraceae, with alternate leaves, yellow flowers, and black brown barks, slightly similar to the Aquilaria agallocha in appearance. Its fruits have perfumed oil and distributed in the south of Yangtze River, hence the paper unearthed in Xinjiang must come from inland.

From this view, the raw material of making bark paper in the Tang Dynasty included major plants of Moraceae Gaudich, Thymelaeaceae, Malvaceae, Leguminosae, Lauraceae and Menispermaceae, far more than those in the Northern and Southern Dynasties. In view of 300 years of development of bark paper in the Tang Dynasty, it could clarify the manufacturing process of papermaking, of which each step was indispensable (Figure 30):

(1) Cutting down→(2) Peeling the bark→(3) Retting and degumming→ (4) Striping the cyan peel→(5) Washing→(6) Slurry ash water→(7) Digesting →(8) Poaching→(9) Removing residual cyan peel→(10) Cutting up→ (11) Pounding with pestle→(12) Washing→(13) Making slurry→(14) Picking up paper→(15) Squeezing water→(16) Stoving→(17) Taking off paper→ (18) Tidying and packaging.

Figure 30   Flowchart of Papermaking in Tang Dynasty,
designed by Pan Jixing, painted by Zhang Xiaoyou (1979)

1. Cutting down   2. Peeling   3. Bundling   4. Retting   5. Digesting in clean water
6. Striping the cyan peel   7. Slurry ash water   8. Digesting   9. Washing materials
10. Pounding materials   11. Making slurry   12. Squeezing   13. Drying in the sun
14. Packaging   15. Freighting

The raw materials were wild plants fibers containing 8.84% ~ 9.46% of pectin, 8.94% ~ 14.32% of lignin and other detrimental impurities, with cyan peel surrounding the phloem. Only by removing all the above substances, can paper be made. Therefore, after the bark is peeled off, it should be soaked in pond and digested at times, thereby removing such substances like pectin, and hemicellulose can loosen cortex for easy removal. It took more time to remove cyan peel, which could be beaten and ground by pestle and stone muller, and then washed in water along with treading, finally leaving the whiter bark material. The bark materials are tied into bundles, soaked in lime water, stacked for some time and then stewed in the digesters. At the same time, plant ash water is poured into bark materials from the top while the lignin and pigment change into black liquid with softer fibers and then washed. When the remaining cyan peel is removed and washed once more, the white bark is piled up into bundles, cut into pieces, pounded with pestle and washed in sack. After that, put it into paper slot, add water into it, stir it into pulp and then add rice-water. The concrete processes like picking up paper, squeezing moisture, drying in the sun and uncovering paper, are shown in the Figure 30.

Another great contribution in papermaking history made by paper workers in the Tang Dynasty was the invention of bamboo paper. As recorded in the article "Fine Paper in Xuzhu District" in the *National History in Tang Dynasty* (829), written by Hanlin academician Li Zhao (791~830):

> There are many kinds of paper, including Yanteng paper (paper made by the plant Yanteng) and Taijian paper (famous paper in ancient China) in the Kingdom of Yue (now in the surrounding area of Yangzhou City), Mamian paper in the Kingdom of Shu (now in the Sichuan Province) ... Liuhe paper in Yangzhou City and bamboo paper in Shaozhou (now in Shaoguan City in Guangdong Province).

Shaozhou was located in the South of the Five Ridges, which is now known as Shaoguan City in the north of Guangdong Province. This region has been rich in bamboo since ancient times, especially Phyllostach edulis, classified as Gramineae. *National History of Tang Dynasty*, finished in about Year 3 of the Dahe Period of Emperor Wenzong (829) in the Tang Dynasty, indicated that bamboo paper had sprung up in the southern areas rich in bamboo since the early 9th century. In *Record of Marvels in South of the Five Ridges* (895), written by Duan Gonglu (840~895) of the Tang Dynasty, the author pointed out that bark paper, made of Aquilaria agallocha in Luozhou of Guangdong Province, was inferior to mulberry paper and bamboo paper. Cui Guitu, scholar in the late Tang Dynasty (early 10th century), who made notes for *Record of Marvels in South of the Five Ridges*, pointed out that bamboo paper sprang in Guizhou (now Chun'an in Zhejiang Province), thus Zhejiang Province produced bamboo paper as well. Volume 3 of *Miscellany of Anecdotes* (926), written by Feng Zi of the late Tang Dynasty and the Five Dynasties, recorded:

> When Jiang Cheng was ten years old, his father was troubled for
> no paper. Hence he burned chaff and boiled and stewed bamboo to
> make paper for his father. Jiang Cheng's nickname was Hong'er,
> thus locals named the paper as Hong'er paper.

Plant ash is provided by burning chaff, and here digesting bamboo refers to manufacturing of bamboo paper. Different from hemp paper and bark paper, bamboo paper does not take bark fiber as raw material, but stem fiber. In other words, the entire bamboo pole is used to make paper, which initiates the emergence of wood pulp paper in Europe because the wooden pole is also the raw material of such paper. It is complicated to make paper by using bamboo pulp. China is a country rich in bamboo productions, thus making bamboo become a new source of raw material. It was a highly developed technique of papermaking in the Tang Period, which achieved breakthroughs and laid foundation for the development of bamboo paper in the Song Dynasty. In 1875, Englishman Thomas Routledge firstly succeeded in making paper from bamboo in the west, over a thousand years later than in China.

1.3.2 Expansion of Places of Origin and Uses of Paper

During the Sui and Tang Dynasties, because of the unification of nation and economic cultural exchanges between various regions, Han nationality and other ethnic groups reached their peak, and water and land communication networks connected regions closely, thus making papermaking technique spring all over the country, expanding papermaking areas and increasing papermaking workers including high level intellectuals who also became amateur papermakers. According to *Treaties on Prefectures and Counties in Yuan-ho Era* (841), *Geography Records of the New Book of Tang History* (1061), *Tongdian* — *The General Food Canon* (written by Du You, in 812), *Six Codes in Tang Dynasty* (written by Li Linfu, 739), *National History in Tang Dynasty* (written by Li Zhao, about 829), and archaeological materials uncovered in latest 50 years, papermaking places in this period included 36 cities as Chang'an (west capital city in ancient China), Luoyang (east capital city in ancient China), Xuchang, Fengxiang, Youzhou (now Beijing), Puzhou (Yongji City of Shanxi Province), Lanzhou, Shazhou (Dunhuang City of Gansu Province), Suzhou (Jiuquan City of Gansu Province), Laizhou (Huang County of Shandong Province), Xizhou (Turpan City in Xinjiang Uygur Autonomous Region),

Changzhou (Wujin City of Jiangsu Province), Jiangning (Nanjing City), Yangzhou (Liuhe City of Jiangsu Province), Hengzhou (Hengyang City of Hunan Province), Junzhou (Jun County of Hubei Province), Jingzhou (Jiangling City of Hubei Province), Luozhou (Lianjiang City of Guangdong Province), Shaozhou (Shaoguan City of Guangdong Province), Guangzhou, Yizhou (Chengdu City of Sichuan Province), Hangzhou, Yuezhou (Shaoxing City of Zhejiang Province), Wuzhou (Jinhua City of Zhejiang Province), Quzhou (Qu County of Zhejiang Province), Yanxian (Sheng County of Zhejiang Province), Muzhou (Jinhua City of Zhejiang Province), Xuanzhou (Xuancheng City of Anhui Province), Shezhou (She County of Anhui Province), Chizhou (Guichi District of Anhui Province), Jiangzhou (Jiujiang City of Jiangxi Province), Xinzhou (Shangrao City of Jiangxi Province), Fuzhou (Linchuan City of Jiangxi Province), Luoxie (Lhasa City of Tibet Autonomous Region), Changle (Fuzhou City of Fujian Province), Quanzhou and so on.

The above mentioned paper production places belong to 17 provinces, municipalities and autonomous regions respectively as Shaanxi Province, Henan Province, Beijing, Shanxi Province, Gansu Province, Shandong Province, Xinjiang Autonomous Region, Jiangsu Province, Jiangxi Province, Anhui Province, Jiangxi Province, Tibet Autonomous Region, and Fujian Province. The incomplete statistics generally reflects the distribution of paper producing areas, among which were important producing areas like Zhejiang Province, Jiangsu Province, Jiangxi Province, Anhui Province, and Sichuan Province, indicating shift of economic center from Yellow River Valley to Yangtze River Valley. Among the five provinces, there were governmental large-scale paper mills which were meticulously in the work and spared much effort at all costs, as well as medium and small ones in the folk. In addition to hemp paper, there were assorted varieties such as rattan paper, bamboo paper, and paper made from mixed-materials, as well as converted paper. The paper produced in different places was served internally and was also transported as commodity by land and sea to other counties in the east and the west, thus leading to the outspread of papermaking technology.

In the Sui, Tang and the Five Dynasties, paper was mainly served as cultural paper for transcribing books and documents. Although the total output of paper can hardly be estimated, it can be valued from several instances. Volume 35 of *Huiyao in Tang Dynasty* (a historical book that described the evolution of the various ordinances in the Tang Dynasty, 961), written by Wang Bo, recorded: in Year 3 of Dazhong Period (849), 365 volumes of books were copied in Jixian Academy in need of 12,000 pieces of hemp paper. *The Confucian Classics of Old History of Tang Dynasty* (945), written by Liu Xu, and the Volume 9 of *Six Codes in Tang Dynasty* (739), written by Li Linfu recorded: during the Kaiyuan Period (713 ~ 741), collection of Confucian classics, all together 4 books of 126,000 volumes, collected in the imperial storehouse, were transcribed in Jixian Academy, which demanded 41.42 million pieces of hemp paper. In addition to original version, transcripts and storage ones were needed as well. Jixian Academy alone consumed a large amount of paper, let alone paper consumption across the country. There were three or four thousand volumes of various works in a library cave in Dunhuang sarcophagus, most of which were the paper printings of the Sui, Tang and Five Dynasties. The public and private collections of other religious and cultural centers would be hundreds of times more, thus certainly making China enjoy the highest paper production in the world, exceeding the total output of all other countries.

The paper used by calligraphers and painters had higher requirements, and the immortal works increased compared to previous dynasties. Chinese character was changed from regular script combining with official script to regular script since the Sui Dynasty, which became fixed style of calligraphy. Regular script could be divided into three types including formal script, running script, and cursive script. There were inherited masterpieces of calligraphers, for instance, *Preface to the Orchid Pavilion Collection* (Figure 31), facsimiled by Feng Chengsu (650 ~ 710) in 707, *Self Book* (780) and *Manuscript of Memorizing Nephew* (Figure 32), created by Yan Zhenqing (709~785) in 758, *A Poem of Zhang Haohao*, created by Du Mu (803~853) and *The Regimen of Fairy*, by Yang Ningshi (873~954) and so on. Famous paintings in record were *Born of Gautama Buddha*, created by Wu Daozi (685~758), and *Penta Bull Map*, by Han Huang (723~787) and so on. Colored bird and flower painting of the Tang Dynasty, unearthed in Turpan of Xinjiang in 1969, was 201 cm in height and 141 cm in length with wide breadth linked by several pieces of paper. The surface was coated with white powder and calendered. The above mentioned works were mostly written in hemp paper, few in mulberry paper. There were more eminent works of calligraphy and painting in the literature, such as *The Change of Fahua*, painted by Zhan Ziqian (about 550~604) in the

Sui Dynasty, but it was not passed down.

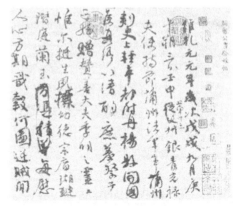

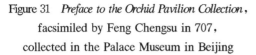

Figure 31　*Preface to the Orchid Pavilion Collection*, facsimiled by Feng Chengsu in 707, collected in the Palace Museum in Beijing

Figure 32　*Manuscript of Memorizing Nephew*, written by Yan Zhenqing in 758, selected from *The Calligraphy* 1978(2)

In addition to transcribing books and documents, blocking printing industry, upsurged in the Tang Dynasty, was another industry consuming large quantities of paper. The art of printing was initiated in folk and stimulated by religious activities attended by mass followers. In the Sui and Tang Dynasties, Buddhism developed rapidly and all sutras were required to be read repeatedly by followers under the enlightenment of Buddha. The transcribed mantras and scriptures were bound to be provided in Buddhist pagodas and temples so that they could be blessed by the Buddha for having lives with good luck and peace. Many followers practiced the teachings by buying Buddhist scriptures from the copyists who specially transcribed Buddhist scripture. Since transcribing scriptures took much time and they got into short supply, copying Buddhist scriptures by block printing came into being, which greatly relieved the problem. Thus in the early period, most prints were mantras and Buddhist scriptures. *Lotus Sutra*, printed and published in the early period of the Tang Dynasty (690~699), was unearthed in Turpan of Xinjiang in 1903. *Everbright Dharani Sutra*, printed and published in the Year 2 of Chang'an Period of Emperor Wuzhou (702). In 1966, it was discovered in the stupa of the Buddhist temple in Qingzhou of Korea. These two works were both published in Luoyang.

In the Tang Dynasty, there were also some non-religious prints published besides Buddhist prints, such as Chinese dictionary, almanac, and literature and so on. It was a period of time when transcribing books coexisted with printing ones because mechanical reproduction technology failed to replace handwork immediately. But after the Five Dynasties (10th century), due to nine Confucian classics which required to inscribe by government for intellectuals, block prints further developed and printing scriptures took the place of transcribing ones gradually. The combination of paper and printing technique provided a large number of cheap books and pushed ahead the development of cultural education and science and technology as well, while the art of printing stimulated the flourish of papermaking. The art of printing would be discussed in detail in the next chapter.

For the expanding sources of raw material and increasing production, paper in the Tang Dynasty was much cheaper than that of previous generations. Stein found an account book around the 8th century in a temple of Dunhuang, which recorded that the paper took *tie* as unit and each *tie* sold for different prices as 35 *wen* (price unit), 50 *wen* or 60 *wen*, one *tie* of paper selling 50 *wen* could make over 19 lanterns and each brush pen sold for 15 *wen*. In the first year of Jianzhong Period (780), a certain amount of silk was sold for 3,000 *wen*. In addition to umbrella, kite, paper flower and paper cutting which already existed in previous generations, people in the Tang Dynasty used cheap paper to make various daily necessities to replace expensive cloth, silk, and metal materials, such as paper fan, paper cap, visiting card, and paper lantern, paper bet net, paper cloth, paper screen, paper armor, paper money used for religions and ceremonial burial, window paper and so on.

Volume 2 of *Miscellany of Anecdotes*, written by Feng Zhi, recorded: "when Yang Yan was in charge of drafting stating announcement, peach-colored paper coated with frozen oil was used to paste windows of outside palace, and the paper window seemed much brighter." "From the bright paper window, we know it has been late; from the warm quit, we feel it has been spring" (纸窗明觉晚，布被暖知春), written by Bai Juyi, was selected in the *Late Bedding Poetry* in the Volume 17 of *Changqing Collection* (823). In the early stage, screen was comparatively heavy because it was made of plank, with a base underneath, and painted on the surface. Afterward, the screen framed with wood and decorated with silk pane appeared. The pane made of paper was not merely convenient for painting, but also for moving especially in folding shape. The poet Bai Juyi was against the tide because he just took white paper as pane of the screen, which was called "white screen", to pursue good virtue. He wrote in his poem *White Screen*: "why was white screen not inscribed and decorated? Keeping white screen clean and pure, I would gain myself nobility and integrity. The screen was framed with wood and decorated with white paper. Why should I leave such an elegant room?" Paper pane and paper window were cheap, which could be repaired easily.

The Book of Criticism, written by Lu Changyuan (709~757) of the Tang Dynasty, recorded: "during the Dali Period (766~779), there was an ascetic monk who never wore silk fabrics but paper clothes, so he was called paper-

clothing Zen master." Paper cloth, colored or not, was made of thick crumpled Chupi paper by sewing, which was against cold, breathable and light. During the year from 1964 to 1965, paper crown (Figure 33) of the Tang Dynasty was discovered in Astana tomb in Turpan of Xinjiang. *Life of Xu Shang*, Volume 113 of *New Book of Tang Dynasty*, pointed that Xu Shang (847~894) employed armors which were made of paper for thousands of soldiers when he led troops

Figure 33　Paper Crown of Tang Dynasty, unearthed in Turpan, selected from *Cultural Relic*, 1973(10)

fighting with Turkish army in the Xuanzong Period, and then bows and arrows hardly pierced. Body armors, made of multi-layer cardboard, light and convenient, could overcome firmness by gentleness and escape arrows, thus adopted by future generations.

At present, visiting cards for social use around the world are invented by China. In the poem *Appreciation to Le Tian Again*, Volume 23 of *Changqing Collection* (823) by Yuan Zhen (779~831) of the Tang Dynasty, there is a line:"最笑进来黄叔度，自投名刺战陂湖", the main idea of which is that we laugh at Huang Shudu most at present, for he recommended himself by using visiting card in Beihu. 名刺 (Ming Ci), namely visiting card, was generally two or three inches in width, having two colors such as red and white. Huang Shudu was a celebrity in the Eastern Han Dynasty called Huang Xian (75~122), and Beihu was water storage project constructed by Ma Zhen, prefecture chief in Kuaiji district in the Eastern Han Dynasty. The poem written by Yuan Zhen was to mock people at that time who tended to show off by using visiting card as the celebrity Huang Xian did, as if the small pond of Gusu (now Suzhou city) was likened to the Beihu in the Eastern Han Dynasty. The visiting card was introduced to Japan in the Tang Dynasty, and was called めいし (meishi) until now, which had the same name as that in the Tang Dynasty.

At present, the playing cards with characters and patterns, which are popular at home and abroad, come from China as well, and it was called Yezi Xi (card game in ancient China) in the Tang Dynasty. Volume 2 of *Collection of Notes of Du Yang*, written by Su E (850 ~ 930) of the Tang Dynasty, recorded: "In the Year 9 of the Xiantong Period (868), princess Tong Chang married, whose mansion was in Guanghua of Chang'an. Her father gave her coins of 50 million *guan* (equal to 2.5 billion *yuan*). One day, the whole family Wei gathered in the mansion, and they all fancied card game — Yezi Xi." This incident happened in 869. *Record of Xianding*, Volume 136 of *Extensive Records of the Taiping Era* (978), said: "Tang Litai, governor of Hezhou, played the card game with performing woman — Ye Maolian, thus the card game got the name 'Yezi Xi'." It can be seen that people at the lowest level played card game as well. *Yezi Xi*, Volume 33 of *Reading Notes in Spare Time* (1750), written by Zhao Yi (1727~1814) of the Qing Dynasty recorded: "Card game appeared in the Tang Dynasty. Figures in *Outlaws of the Marsh* were decorated in the playing card to inherit the game, but the name of which was changed." Playing card was made of cardboard in rectangle shape with hand-drawn patterns, which, later on, was printed and then added with colors by hand.

Ancient funeral was quite ceremonious so that metal money and utilities were buried in the tomb at the same time. After the Sui-Tang Period, great change took place in the funeral style so that metal money was changed into paper money, and carriages, horses and servants, made of paper, were cremated or buried in the tomb. Although such change led to high consumption of paper, compared to the burial of copper cash, it still saved much money. *Paper Money*, Volume 6 of *Notes of Experience of the Feng* (about 787), written by Feng Yan (726~790) of the Tang Dynasty, recorded: "at present, people has made mountainous paper money with delicate decoration for funeral. According to the ancient burial ceremony, precious jade and silk fabrics were buried after the event. Metal money was valued in previous generations, thus replaced by paper money." Volume 48 of *Pearls in the Garden of the Dharma* (668), written by Buddhist Dao Shi, said: "if white paper money was cut, it could be used as silver by ghost, and if yellow paper money was cut, it could be used as gold by ghost." In 1964, tomb No. 64 (TAM:34) of Astana in Xinjiang unearthed paper money for burial ceremony. Inferior paper was used to make paper money for burning, which was called fire paper. It was superstitious and wasteful to burn a multitude of paper money, and only the country rich in paper could have such an activity.

It is interesting that Chu money (money made from Chupi paper) offered to the dead became the alternative of copper cash in the field of economy in the real world, namely "flying cash", used in the Tang Dynasty. Such paper money was the prelude influencing the whole economic life of human being. *Shi Huo Zhi* in *Xin Tang Shu* (《新唐书·食货志》, "Farming, Cloth and Currency" in *The New Book of Tang*) recorded:

> *In Xianzong Period, copper coin was precious and copper ware was prohibited as well. Businessmen deposited money in Jinzouyuan (Beijing liaison offices of provinces), private banks and representative offices of military government and Jiedushi (regional military governor), so as to go anywhere without any burden. After the ticket was checked right, the copper cash would be returned back, so it was called "flying cash". However, this measure was latter forbidden by Jingzhaoyin.*

In other words, outside businessmen could deposit copper cash in liaison offices of provinces, private banks or representative offices of military government and *Jiedushi* to exchange a ticket if they did not want to take money back after selling the goods in Chang'an. The ticket was marked with the name, native place, amount of the saver and sealed by both parties. It was divided into two forms, one of which was given to the saver and the other was delivered to a definite province. When the businessman arrived, he could draw money by ticket if checked right. As a result, such measure of depositing money was called "flying cash" or "easy exchange", namely the bill of exchange in the contemporary world, because it greatly facilitated convenience of businessmen. Certainly, everyone was willing to pay for it due to its convenience. However, government loaned money from businessmen (actually plunder), leading to the credit crisis and *Jingzhaoyin* (administrator of Chang'an) forbade "flying cash" which began in the Year 4 of Yuanhe period of Emperor Xianzong (809) and initiated paper money in the Northern Song Dynasty.

1.3.3 Advances in Papermaking and Processing Technology

In the Tang Dynasty, paper and paper articles had been necessities of Emperors and numerous ordinary families. At the same time, papermaking technique and processing technology of paper both achieved new progress. The test of paper in this period indicated that fiber dispersibility of hemp paper and bark paper was remarkably improved with rare fiber bundles, tightly knotted fibers, and smooth surface. The thickness was generally between 0. 05 mm to 0. 14 mm, and thicker paper was between 0. 15 mm to 0. 16 mm. It is rare to find thicker paper than the above mentioned, while the thickness of paper in the Wei and Jin Dynasties and the Northern and Southern Dynasties was between 0. 15 mm to 0. 2 mm. Hence paper in the Sui and Tang Dynasties was thinner than that of previous generations, and Europeans could not even make such thin paper of the Tang Dynasty. Making thick paper was easy while making thin paper was much more difficult because pulp concentration was low and fiber was required to be highly dispersed. A series of technologies should be guaranteed that fine and tight papermaking curtain could be woven, paper was made in skilled techniques and wet paper should be uncovered without breakage.

Actual tests of veins from several hundreds of paper in the past dynasties proved that the texture can be divided into four grades: (1) rough line, each line was 0. 2 cm in thickness; (2) median line, each line was 0. 15 cm in thickness; (3) fine line, each line was 0.1 cm; (4) superfine line, each line was 0. 05 cm. Waving rough lines was easy, so was papermaking and vice versa. Paper with median lines and few with fine lines appeared since the Sui and Tang Dynasties, while most paper with rough lines was made before the Tang Dynasty, thus indicating that pulping technology entered a new stage after the Tang Dynasty. Although papermaking curtains were woven by bamboo canes, which were finer than previous generations, they could produce paper with larger format. Taking paper for transcribing Buddhism scriptures as an example, paper of the Tang Dynasty was about 30 cm in height, generally 36 cm~55 cm in length, and separately 76 cm~86 cm in length. *The Heart Sutra*, written in the late Tang Dynasty, was 94 cm in length.

In the late Tang Dynasty, giant Pi paper (as long as a bolt of silk), as the pioneering creation in papermaking technology, could also be made. *Record of Qingyi* (about 950), written by Tao Gu (903~970) recorded: "my late father had held several hundred pieces of paper, which was as long as a bolt of silk, smooth and tight, thick and white, called *Poyangbai* (name of the paper)." Tao Huan, Tao Gu's father, was in the Zhaozong Period of the Tang Dynasty (889~904), once was appointed as prefectural governor and his collected paper was made at that time. A bolt of paper was at least one piece (3 meters), and such large-format paper could be made in the Tang Dynasty. *Wenfang Sibao*

(《文房四宝》, *Four Books of Stationery*), written by Su Jianyi (958~996), recorded:

> *Li Bian, who founded the Southern Tang Dynasty, gave out a piece of Huifu paper on the day on which results were released for writing down the names of persons who passed the examination. Such paper was 3 meters long, one feet wide, and was as thick as a few layers of silk fabrics. Once the paper was released, people congratulated on each other and had great expectations from their children. Su Yijian went to the imperial storehouse of the Southern Tang Dynasty in Nanjing, surprisingly to find that a thousand pieces of Huifu paper was still deposited in this old building.*

"The last Emperor of the Southern Tang Dynasty" here referred to Li Bian, who founded the Southern Tang Dynasty (937~975), governing several places as Xuanzhou, Shezhou, Changzhou, Xingzhou, Chizhou and Fuzhou, which all were papermaking areas. In such places, paper used for announcement, one feet in width and three meters in length, could be made on the basis of technology in the late Tang Dynasty. In the Northern Song Dynasty, Su Yijian went to the imperial storehouse of the Southern Tang Dynasty in Nanjing, surprisingly to find that a thousand pieces of Huifu paper was still deposited in this old building, thus indicating a large-scale production of such paper ever before. It required many people to hold the papermaking curtain to make paper at the same time, and cooperate in harmony (Figure 34).

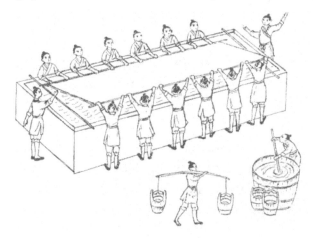

Figure 34　Diagram of Giant Bark Paper Manufactured by Many People
in Tang Dynasty, painted by Pan Jixing (1998)

The raw material of such paper was paper mulberry bark, and it was natural to be thicker because its format was three-meter long.

In the Tang Dynasty, making large-scale thin paper needed fine bamboo curtain, but papermaking process was finished only when the semi-dry paper was uncovered with no breakage and transferred to the board or wall space for baking. For this purpose, papermaking workers of the Tang Dynasty infiltrated mucilage of plants into pulp to increase lubricity of wet paper. That was so called paper potion or smoothing water by later generations. *Miscellany of Anecdote in Guixin* (about 1290), written by Zhou Mi (1232~1298), recorded:

> *The leaves of Abelmoschus manihot has to be made into paper potion used in manufacture of paper, or paper would be too sticky to be uncovered. If there were no Abelmoschus manihot, Carambola rattan (namely Actinidia chinensis), leaves of Hibiscus syriacus Linn, and wild grape could be alternatives, then the paper would be easy to be uncovered.* ❶

Zhou Mi pointed the key of uncovering paper without breakage. In the Dynasties of the Wei, Jin, Northern and Southern Dynasties, starch paste was put into pulp when hemp paper was made because it not only had the effect of sizing, but also had the function of lubrication. Simulated experiment proved that only when mucilage was put into pulp could paper be uncovered without breakage when bark thin paper was being made. In 1901, when testing document paper of the Tang Dynasty, unearthed in Xinjiang, Wiesner found emplastic substance from lichen, namely mucilage of plants. The function of mucilage was to improve suspension of fabric in pulp and increase lubricity of wet paper for easy separation. The usual mucilage was extracted from branches and leaves of Abelmoschus manihot which was classified as Malvaceae, and Actinidia chinensis which was classified as Actinidiaceae, and the latter was called Carambola rattan in ancient China. As the tradition of making paper, using mucilage of plants was hereafter inherited by successive dynasties and also followed by Japan and North Korea.

In the Tang Dynasty, paper for cultural use was also divided into two types as unprocessed paper and processed paper. The former referred to the paper which was dried after being picked up from paper channel with no processing

---

❶   周密. 癸辛杂识[M]. 上海:博古斋影印本,1922.

and treatment, while the latter referred to the paper which underwent processes and treatments on the basis of unprocessed paper, or was made by adding reagent into its pulp. In a large sense, processed paper included paper which underwent treatments as sizing, coating, waxing, padding and dyeing and so on. In a narrow sense, it referred to dyeing and sizing paper. Generally speaking, unprocessed and processed paper were both used for writing, but processed paper was much more suitable for writing regular script in small characters, color paintings, official documents, rescript and collection of books in the imperial storehouse. Thus each sector of imperial storehouse had several processed paper workers and decoration craftsmen. Sizing was one of the methods to make unprocessed paper into processed paper. After the Tang Dynasty, animal glue added alums replaced starch which was used in the previous dynasties and it was better than starch. *Essay of Ten Paper* (1100), written by Mi Fu of the Song Dynasty, recorded: "Yellow paper was made by adding glue into paper pulp, and imperial edicts in the Tang Dynasty were mostly written in yellow paper."

A variety of colored paper produced in the Sui and Tang Dynasties greatly exceeded that of previous generations. Apart from yellow paper, widely used, there were also many extant objects. Add yellow wax on the yellow paper, and yellow wax paper came into being, and if white wax was added on the white paper, namely white wax paper. In the Tang Dynasty, such pieces of paper were called Yinghuang paper or Yingbai paper. Volume 3 of *Famous Paintings in History*, written by Zhang Yanyuan, recorded: "as Zhao Guogong (first-class duke in ancient China), Li Jifu (758~814) was said to mount calligraphy works and paintings by Yinghuang paper, but the books made of Yinghuang paper were all terrible." Such paper was yellow and semitransparent, tougher and smoother than ordinary paper, thus called Yinghuang paper. It was not suitable for mounting paintings, but for writing or making rubbings from inscriptions, thus Prime Minister Li Jifu misused yellow paper. Volume 5 of *Record of Official Travel* (1233), written by Zhang Shinan, pointed: "Yinghuang paper was put on the hot iron and added with yellow wax evenly." According to the paper used for transcribing Buddhism scriptures, unearthed in Dunhuang sarcophagus, only few people of the Tang Dynasty used it, such as the work *Lotus Sutra* • *Nineteenth Virtues of Mage*, and so on. Majority of such paper was made in the early Tang Dynasty and the Middle Tang Dynasty, and few were seen in the late Tang Dynasty and the Five Dynasties. *Acta absurdum implevit Qieyun*, scroll book of the Tang Dynasty, collected in the Palace Museum in Beijing, was written in white hemp paper, calendered with double wax, thus was the sample of Yingbai paper. Paper in other colors could be waxed as well, and its another character was waterproof.

In the Tang Dynasty, paper was coated with white mineral powder by glue to the surface of paper, then waxed and calendered. As an innovation, such paper had both characters of powder letter paper (a kind of paper for document) and wax letter paper, so it was called powder wax paper and classified as double coated paper. *History of Book*, written by Mi Fu, recorded: "*Ode to Deadwood*, written by Chu Suiliang who was head of the secretariat in the Tang Dynasty, was a book of extension in powder wax paper. *Thousand Character Classic* was written and rubbed by monk Zhi Yong (560～620) in powder wax paper. There was a piece of hemp paper in the book, which was authentic version." Paper workers in the Tang Dynasty also learned the decorative arts of painters and weavers to decorate red and blue paper with gold and silver foils and powders, which were called gold flower paper and silver flower paper respectively. Taking gold flower paper as an example, gold was beaten into foils and then cut into small pieces of different shapes. Then glue was applied onto the colored paper, which was then covered with gold slice and leveled. It was also feasible to mix the glue with the dye liquor which was brushed on the paper, and to sprinkle gold pieces on the paper. The paper used was thicker bark paper that could be sprinkled with gold powders on one or both sides, and finally for writing. Such paper was mostly used by imperial storehouse and the rich families for it was expensive. *Hanlin Zhi* (819), written by Li Zhao (791~830) of the Tang Dynasty, recorded:

> All letters of appointment of political leaders were written in five-colored paper with gold flowers and silk. Letters for Zan Pu (governor of Tibet Autonomous Region) of the Tubo period were written in five-colored gummed paper decorated with gold flowers, put in a silver-locked sandalwood lacquer box and sent. However, letters sent to Ke Han (governor of Xinjiang) of the Huihu period, and to governors of Xinluo and Bohai were written in gold flower paper decorated with silk, and put in an inferior silver-locked sandalwood lacquer box.

Five-colored paper with gold flowers and silk meant that five-colored paper was decorated with pieces of gold and the four sides of paper were encrusted with woven fabrics with patterns to reinforce. Such paper was used by government to write appointment letters of generals and ministers of the county and letters to the rulers of Tibet, Xinjiang, and Bohai area, and the King of Xinluo, which, then with affixed imperial seal, were sent in a silver-locked sandalwood lacquer box.

Sometimes, people in the Tang Dynasty used particular processed paper, such as Yunlan paper designed by Duan Chengshi (803~863) in Jiujiang for letter paper. Yunlan paper, 9 inches (29.7 cm) in length with light blue cloud veins, was sent to his friend Wen Tingyun (812~870) in Xiangyang by Duan Chengshi. Then Duan Chengshi gave 50 pieces of such paper away to Wen Tingyun because he really liked the paper. The concrete preparation was to use light blue dye liquor to dye white bark paper into blue paper which was pound into muddy shape to make pulp. In the process of manufacture, the blue pulp should be poured in the proper part of wet paper, and the curtain should be swayed horizontally and gently in order that the pulp flowed on the wet paper forming waved-cloud veins, and then Yunlan paper came into being after being dried.

Xue Tao (literary name was Hong Du, 768~831), poetess in the Tang Dynasty, from an early age, lived in Chengdu for her father Xue Yun was an official here. During the Yuanhe period (806~820), she designed a red letter paper for writing the eight-line poem by the side of the Huanhua brook, a tributary of the Minjiang River in the suburb, and song together with 28 poets Bai Juyi, Yuan Zhen, Du Mu, Liu Yuxi and so on. As a result, such letter paper was famous around the literary world and was called Xuetao Letter Paper. According to the record of *Exploitation of the Works of Nature* (1637), such paper was made by using hibiscus bark as raw material and dyed by hibiscus juice. In the poem *Sending Cui Jue to Xichuan*, written by Li Shangyin (813~858), it said: "By the side of Huanhua brook, letter paper was in peach blossom color which was exquisite to write poems intoning jade hook." Here Huanhua Letter Paper referred to the Xuetao Letter Paper.

Embossing paper, water-marks paper and colored embossing paper were also made in the Sui and Tang Dynasties and Five Dynasties, which could reveal patterns besides water-mark lines when facing the light, thus making paper have a kind of indistinct beauty. Embossing paper was made by pressing board with carved convex reversible textures or patterns strongly on paper, and employing block printing principle but no ink. It could be called no-ink printing, and named as arch printing in the Ming Dynasty. Water-marks paper showcased patterns naturally after the curtain was waved, which was woven by silk thread of varying thickness or horsetail to make convex patterns, textures or characters. *The Record of Discrimination of Various Books*, written by Yang Shen (1542), recorded: "in the Tang Dynasty, there was a kind of paper called Juan paper, also named as Yanbo paper (namely water-marks paper), the texture of which was like ripples." "Yanbo", namely many ripples and Yanbo paper, was water-marks paper, often used as letter paper or paper for calligraphy and painting.

Volume 2 of *National History in Tang Dynasty* (829), written by Li Zhao, recorded that "Yuzi paper", one kind of embossing paper, was produced in Sichuan Province in the Tang Dynasty. When talking about such paper, *Four Books of Stationery* (986), written by Su Yijian, pointed: "it was using delicate fabric coated with adhesive which makes the textures obvious." After the textures were calendered, Yuzi paper, also called Luo paper, came into being. *Qingyi Lu* (《清异录》, *Qingyi Record*) (about 950), written by Tao Gu (903~970) of the Five Dynasties, said:

> The nephew of Yao Yi was skilled at making five-colored letter paper, smooth and delicate. Calendering boards were carved with different patterns such as landscape, vegetation, flower and fruit, phoenix and lion, insect and fish, the god of longevity, the eight immortals, and inscriptions on ancient bronze objects, which was called embossing booklet.

Yao Yi (866~940), literary name Wan Zhen, was imperial scholar in Chang'an, occupied an important position of the Later Liang Dynasty, the Later Tang Dynasty, and the Later Jin Dynasty. During 934 to 936, his son Yao Wei and brother-in-law made colorful embossing paper in the mansion of Kaifeng. They firstly invited painter to paint rough sketches of landscape, vegetation, phoenix and lion, insect and fish, the god of longevity, the eight immortals, and inscriptions on ancient bronze objects, which were then carved on agilawood carving board by type cutter. All patterns would be transferred to the five-colored paper when it was pressed on the carving board. In Europe, the earliest water-marks paper appeared in Italy in 1282 and even the earliest embossing paper made in Britain in 1796 was several hundred of years later than in China.

In the Five Dynasties and the Southern Tang Dynasty, Chengxintang paper (named after an ancient imperial palace) was produced eventually, which was famous in history. Chengxintang was a palace name of the ruler of the Southern Tang Dynasty in Jinling (now Nanjing). Li Yu (937~978), the last Emperor of the Southern Tang Dynasty (961~970), ordered to make such royal paper for calligraphy and painting, which was spread from storeroom to scholar-bureaucrats in the early Northern Dynasty. Many poets like Liu Chang (literary name Zi Fu, 1019~1068), Mei Xiaochen (literary name Shengyu, 1002~1060), Ouyang Xiu (literary name Yong Shu, 1007~1072) and Song Minqiu (literary name Ci Dao, 1019~1079) praised the paper in their poems. Cai Xiang (1012~1067) deemed that Chengxintang paper ranked first around the world. As depicted in the poems of the Song Dynasty, such paper was made in Shezhou (now She County of Anhui Province), Chizhou (now Guichi District of Anhui Province). It took mulberry bark as raw material and was manufactured out of pulp which was added with ice water collected in the twelfth lunar month. The paper, white and thick, with exquisite and scrupulous working of paper workers, featured fine shining fabrics, calendering on both sides, and smooth surface like jade plate, which was firm and easy to write on and had good ink acceptability. Emperor Li Yu liked to make paintings and poems using this paper and awarded officials. It had large output and even thousands of pieces remained till the Northern Song Dynasty, and each piece of paper enjoyed a value of over a hundred pieces of gold. Chengxintang paper was copied by successive dynasties since the Song Dynasty, served as top grade and mainly used by Emperors. To sum up, in terms of papermaking techniques, the Sui and Tang Dynasties and the Five Dynasties were periods during which there were numerous innovations, reaching an unprecedented technological peak.

1.4 Papermaking Technology in the Song and Yuan Dynasties, and the Ming and Qing Dynasties

1.4.1 Papermaking Technology in the Song and Yuan Dynasties

During 300 years of the Sui and Tang Dynasties, papermaking and paper processing technologies basically formed a complete technical system. The paper production sites were throughout the south and the north of China, and the use of paper extended to various aspects, laying a solid foundation for the development of papermaking for future generations. Going through short division of the Five Dynasties and Ten States, central China was soon united by the Northern Song Dynasty (960~1127). There were few ethnic minority regimes in the northern frontier, namely the Liao Dynasty (907~1125), established by the Khitan nationality, the Xixia Dynasty (1038~1227), established by the Dangxiang nationality, and the Jin Dynasty (1115~1234), established by the Jurchen (1115~1234). In 1125, the Jin destroyed the Liao and also destroyed the Northern Song Dynasty in 1126, thus ruler of the Song Dynasty moved to the south to establish the Southern Song Dynasty (1127~1279). In 1260, the Mongols founded Mongolian Nation, then vanquished the Jin and the Southern Song Dynasty, and founded the Yuan Dynasty, which achieved unprecedented unification. The Song and the Yuan Dynasties (960~1368), lasting for 408 years, inherited the achievements of papermaking technology from the Tang Dynasty in an all-round way and had further improvement.

The hemp paper, developing for a millennium since the Han and Tang Dynasties, began to decline from the Song Dynasty and was only used in ethnic minority areas in the north. This was mainly because cheaper bamboo paper and bark paper with superior quality achieved great development. In the 9th century of the late Tang Dynasty, production technology of bamboo paper springing up in Guangdong and Zhejiang areas soon spread to various places. Up to the Northern Song Dynasty, the places of origin and production of bamboo paper increased dramatically. The Yangtze River Valley and vast areas of the south of the Yangtze River were rich in various types of bamboo, at least 50 species were suitable for making bamboo paper, and bamboo fiber cells accounted for 60% to 70% of the total cell area, which provided a cheap and abundant source of fiber, making rapid rise of bamboo paper. Paper made from bamboo was not as tenacious as hemp paper and bark paper, and the early bamboo paper was faint yellow, but people liked to use it because of its low price. Volume 4 of *Four Books of Stationery*, written by Su Yijian (958~996) of the Northern Song Dynasty, recorded: "Now in Zhejiang area, fresh bamboo was used to make paper to write secret letter, and nobody dares to open it casually as it was fragile."

Kuiji Chih (1202), written by Shi Su (1147~1213) in the Jiatai Period of the Song Dynasty, recorded, after Su Shi (literary name Dongpo, 1036~1101) returned from Hainan Island to which he was banished, he asked someone else to purchase 2,000 pieces of paper from Yuezhou (now Shaoxing City of

Figure 35 *The Coral*, Model Calligraphy in Bamboo Paper, created by Mi Fu in Northern Song Dynasty, collected in the Palace Museum in Beijing

Zhejiang Province), among which, most of them were bamboo paper. Celebrities like Wang Anshi (1021 ~ 1086) and Mi Fu (1050 ~ 1107) also preferred bamboo paper. The calligraphic work *The Coral*, created by Mi Fu, collected in the Palace Museum in Beijing, was written in bamboo paper as well (Figure 35), and Mi Fu's another work *Cold Light* was written in mix-material paper made from bamboo and paper mulberry bark. After the development of printing technology, many books of the Song Dynasty were printed on bamboo paper, such as *Bodhisattva Kusuma-mala Sutra*, from *Tibetan Tripitaka of Gushan* which was bound in the form of *Kefanjia* (a binding form) in the Year 5 of the Yuanyou Period of the Northern Song Dynasty and collected in the Beijing Library, *The History of Jiesuoying*, published in the Year 7 of the Qiandao Period of the Southern Song Dynasty (1171), and *Extensive Records of Shilin* (Figure 36), published by Jichengtang press of Zhen's family in Jianyang in the Year 6 of the Zhiyuan Period of the Yuan Dynasty (1269). *The Great Compassion Tuo Ronnie Sutra*, carved and printed by Hu Ze (minister of war) in the Year 2 of the Mingdao Period in the Northern Song Dynasty, was in excellent bamboo paper.

In the Song Dynasty, rice straw was also used to make paper for

Figure 36 *Extensive Records of Shilin*, Carved Copy on Bamboo Paper, published in Year 6 of Zhiyuan Period of Yuan Dynasty (1269), collected in Beijing National Library

different uses, such as packing paper, fire paper and toilet paper, cheaper than bamboo paper. Volume 4 of *Four Books of Stationery*, written by Su Yijian, recorded: "people in Zhejiang Province used wheat stalk and rice straw to make paper, which was brittle and thin, but wheat straw and Youteng (a plant classified as Euphorbiaceae) were better raw materials to make paper." To make the best use of used paper, it was put into fresh pulp to make new paper, called reborn paper. Through testing, the paper used for *Save Sufferings of Living Beings Sutra* was reborn paper, written in the Year 5 of the Qiande Period of the Northern Song Dynasty (967) and collected in the Museum of Chinese History. *Ancient Coin*, Volume 9 of *Examination of Historical Documents* (1309), written by Ma Duanlin (1254～1323), recorded "*Huizi*, paper money, issued in the Hubei and Hunan areas in the Song Dynasty was printed on the reborn paper, which was made by text papers of failed *Juren* (First-degree Scholars in ancient times) and wasted tea selling permits."

Compared to the paper in previous dynasties, bark paper in the Song and Yuan Dynasties was in an overall development period during which its output, breadth, and quality surpassed the past and could meet various needs, thus becoming the largest paper type. It also replaced hemp paper as the largest paper product. The dominant position of bark paper was the outstanding feature of this era. Creative works in bark paper became fashionable among calligraphers and painters, but few were handed down. *Psalm and Paintings of Three Horses*, created by Su Shi and collected in the Palace Museum, and *Rainy Landscape of Xishan Mountain* (29.5 cm × 105.5 cm), created by Huang Gongwang (1269～1345), were both in fine mulberry paper. Calligraphic works were in Chupi paper such as *Noble Mansion*, written by Li Jianzhong, *Xinsui Weihuo Tie*, written by Su Shi, *Summer Poem*, written by Zhao Ji (Emperor Huizong of the Song Dynasty), and *Ink Painting Sketch*, written by Fa Chang (1176～1239). According to incomplete statistics, the breadth of paintings handed down were 650 cm^2 in the Tang Dynasty, 2,412 cm^2 in the Song Dynasty, and 2,937 cm^2 in the Yuan Dynasty. The amplitude of paintings rose dramatically. Cursive script *Thousand Characters*, written by Emperor Huizong of the Song Dynasty and collected in the Museum of Liaoning Province, was 3 *zhang* (unit of length) in length (nearly 10 meters), which hit record in history. The development of giant landscape paintings and colorful paintings were related to the progress of production technology of bark paper.

Most books for public reading, published in the Song and Yuan Dynasties, were printed in bamboo paper, but more elaborate ones were still in bark paper. *Buddhist Akilasitva Sutra* (carved in 973), from *Open the Treasure*, published in the Northern Song Dynasty and collected in the National Library, was printed in advanced mulberry bark paper, which was dyed yellow with double-side wax. *Collection of Mr. Changli*, published in the Shicaitang press in the Southern Song Dynasty and collected in the same library, was printed in white thin mulberry bark paper. Many works were in mixed-material paper made from mulberry bark and bamboo, such as *Finest Blossoms in the Garden of Literature*, edition of Jizhou in Jiangxi Province in the first year of the Jingding Period of the Southern Song Dynasty (1260), and *Dream Stream Essays*, published in Chaling County in the Year 9 of the Dade Period of the Yuan Dynasty, used Chupi paper. *Lotus Sutra*, block-printed edition in the first year of Tianxi Period of the Northern Song Dynasty, discovered in the Ruiguang pagoda of Suzhou City in 1978. *Shilin Yanyu*, written by Ye Mengde (1107~1148), said: "among printed books, books in Hangzhou ranked first, followed by those in Sichuan area and the books in Fujian area were worst." There were many books in Hangzhou printed by Directorate of Imperial Academy in mulberry bark paper, so the quality of which ranked first. Moat books in Guangdong area were copied and carved in bamboo paper privately, so they had advantage of low price and were not for publication. The quality of books in Sichuan area, which were printed in bark paper and hemp paper, fell between that of Zhejiang area and Guangdong area.

Another new use of paper in the printing field was the issuance of paper money. In the first year of the Tiansheng Period in the Northern Song Dynasty of Emperor Renzong, paper money was firstly issued, called *jiaozi*, which was renamed as *qianyin* in the first year of the Daguan Period (1027) and renamed again as *huizi* in the Southern Song Dynasty. The Jin Dynasty, coexisting with the Northern Song Dynasty, imitated the system of the Northern Song Dynasty to print and issue paper money called *jiaochao* from the second year of the Zhenyuan Period (1154). During the Yuan Dynasty, *baochao* was issued from the first year of the Zhongtong Period (1260). Issuing paper money required large-scale printing, and its layout consisted of copper plate and bronze schrift, which were all printed in mulberry bark paper. As the revolution of money in history, issuing paper money facilitated the development of social commodity economy. More details about paper money were introduced in chapter 2 of the book. The gunpowder manufacturing industry, which developed in the Song Dynasty, was also a new industry which consumed a large amount of paper. Gunpowder flask and gunpowder lines were all made from bamboo paper and bark paper. Pyrotechnics and firearms were used in large quantities, and gunpowder lines required thin and tough paper with superior quality. The silkworms used for agriculture were generally made of coarse and heavy mulberry paper. Silk worm-egg cards for agricultural use usually adopted rough and thick mulberry bark paper.

Various paper products were used in daily life in the Tang Dynasty, such as paper clothes, paper umbrellas, paper lanterns, window paper, kites, paper cutting, playing cards, paper fans, visiting cards, paper crowns, fire paper, toilet paper, paper screens, paper armors and so on, which were developed and improved during the Song and Yuan Dynasties. The paper uses were further increased with some new creations in variety and design. Dated back to the Song Dynasty, paper quilt was recorded in the poem *Appreciating Zhu Yuanhui for the Paper Quilt*, written by Lu You (1125~1210), because the paper quilt was presented by Zhu Xi (literary name Yuan Hui, 1130~1200). In his poem, it said: "paper quilt was worn in the winter, it was whiter than fox mink and softer than cotton." *Paper Quilt*, Volume 33 of *Collection of West Mountain* pointed: in the snowy and windy winter, it was so difficult to endure cold on the frozen land in daytime, let alone at night. If there were no such paper quilt, people would curl up in the cold. Volume 4 of *Four Books of Stationery*, written by Su Yijian, also introduced a kind of health-preserving paper pillow. Most fans in previous dynasties were circular, but portable folded fans appeared in the Song and Yuan Dynasties. Most playing cards were made by printing in the Song and Yuan Dynasties. Printed in the Yuan Dynasty, playing cards were once unearthed in Xinjiang and delivered to Europe by Mongolian army.

Baixi Jiyi (《百戏技艺》, *Acrobatics Art*), in Volume 20 of *Mengliang Lu* (《梦粱录》, *A Record of Fond Dreams*, 1274), written by Wu Zimu (1231~ 1309) of the Song Dynasty, recorded the folk shadow play, which was popular in the Northern Song Dynasty. The people who first created the shadow play used thick white paper to carve figures and later changed to use sheep skin with colored decorations for protection. There were new developments of paper cutting technology in the Song Dynasty. In addition to inheriting the techniques of cutting flowers, animals, and figures of previous generations, they could also cut out famous writings vividly. Yang Wanli wrote a poem about paper cutting to the artist, and it said in the preface of the poem: "the artist used blue paper to cut out the poem of *Kuang Yishi Has Been Apart for Many Years* of Li Yishan, the character style of which vividly resembled that of Mi Yuanzhang."

That was to say, the artist used blue paper to cut out poems of Li Shangyin of the Tang Dynasty, the font of which was as ingenious as that of Mi Fu, the great calligrapher of the Song Dynasty. It was magnificent. When it came to lamp decorations, Volume 2 of *Past Things of Wulin* (1270), written by Zhou Mi, said that people in Hangzhou used five-colored wax paper to cut out figures and horses and put them in the lamp spinning like flying, which was known as hot-air zoetrope. *Lantern Festival* (on 15th day of the first lunar month) *in Wuzhong* (now Suzhou), Volume 23 of *Poems by Shihu Jushi*, written by Fan Chengda (1126~1193), had the verse that "shadows spun freely on Lantern Festival", with annotation of "Maqi Deng" (hot-air zoetrope) in private. It can be seen that people in the Song Dynasty got the new trick to play.

Hot-air zoetrope was a special paper lantern with profound scientific principle that an impeller was installed on the vertical shaft with a candle underneath, and ascending hot-air currents promoted the rotation of the impeller. Four thin wires, each pasted with five-colored wax paper cut of figure

and horse, were placed in the middle of vertical shaft horizontally. All the above were put into the paper lantern (Figure 37), and the candle was lit at night. Paper cuts spun with the spinning of impeller, and shadows cast on the paper of lantern spinning as if flying. It indicated that rotating impeller with hot air currents could turn thermal energy to kinetic energy. "It seemed extremely likely that the use of ascending hot-air currents in western countries derived from the earlier zoetropes of China," doctor Joseph deemed. In the second half of the 15th century, Europeans used the same principle to put impellers and vertical shafts into the

Figure 37　Paper Hot-air Zoetrope, selected from Liu Xianzhou (1962)

smoke pipe, and rotated the roasted fork through gears. The origin of hot-air zoetrope can also be traced back to the Tang Dynasty, but it was more widely used after the Song Dynasty.

The papermaking and paperprocessing technology of the Song and Yuan Dynasties were also promoted on the basis of the technology used in the Tang Dynasty. In places where there were a large number of hydraulic resources, instead of man power and animal power, water-powered trip hummer was commonly used to pound paper stock to improve the efficiency of beating. With the help of the force of water flow, water-powered trip hummer rotated impeller to change the rotary motion into the vertical motion through crank, connecting rod and gear system and driving stone hummer head to pound materials. As a beating method, using water-powered trip hummer to pound material could be dated back to the Han Dynasty, and it was widely used to pound rice, later for papermaking. Volume 19 of *Agricultural Book* (1313), written by Wang Zheng, introduced a battery of hydraulic trip hammers that four hammers could be driven to operate at the same time. He said:

> *People can make a kind of waterwheel, and its axle can be several feet long with a row of intersecting columns. During its operation, the waterwheel drives a row of mutually mounted columns with hammers on the axles to pound successively. Waterwheel can only be installed by the river with high water potential.*

In the book, Wang Zheng also gave the illustration of battery of hydraulic triphammers (Figure 38). For dual purposes, it could be used to pound rice and paper stocks or applied in the workshops for making bamboo paper and bark paper.

When talking about papermaking in Chengdu of Sichuan area, *Summary of Various Paper in Sichuan* (about 1360), written by Fei Zhu (about 1303~1363) of the Yuan Dynasty, recorded: "if only the raw materials were used to make paper, they must be pounded by the water-powered trip hummer and cleared by water." It referred to water-powered trip hummer used to pound paper stock. Meanwhile, in the Song and Yuan Dynasties, papermakers commonly added mucilage into the pulp during the process. In addition to Abelmoschus manihot and carambola vines, mucilage

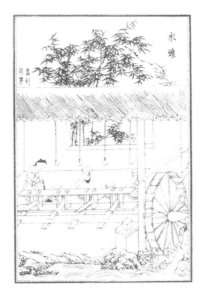

Figure 38   Illustration of Battery of Hydraulic Triphammers by Wang Zheng

could also be extracted from some other plants, such as Ampelopsis brevipedunculata, Hibiscus syriacus and so on. In the process of sizing treatment, the technique that animal glue and alum were mixed into pulp to replace starch was commonly practiced during this period. *Essay of Ten Paper* (about 1100), written by Mi Fu, said: "hemp paper, produced in Sichuan area, was treated by adding alum instead of starch to make yellow paper." Paintings like *Teaching Show of Wei Mo*, fine brush line-drawing painting of Li Gonglin (1049~1106), who was a painter of the Northern Song Dynasty, collected in the Palace Museum in Beijing, and *Field Scenery in Autumn*, fine brush color painting, were painted in bark paper, which was treated by glue and alum.

In terms of weaving curtains for papermaking, there were also innovations in the Song and Yuan Dynasties. *Summary of Various Paper in Sichuan*, written by Fei Zhu of the Yuan Dynasty said, "there were Lianer paper, Liansan paper and Liansi paper." Such kind of paper had not been seen before the Song Dynasty, and Liansi paper was nicknamed as Lianshi paper in the Ming and Qing Dynasties. It may be coincidence that the paper got the name because some predecessors deemed that it was created by brother Lian, but without documentary and material evidence. Fei Zhu once annotated Liansi paper that paper sellers called Liansi paper as "Chuanjian paper", hence it can be seen that it was named not from a family name but based on weaving and papermaking crafts. In the past, a piece of paper was manufactured by a single curtain. If a strip of cotton was sewn in the middle of the long paper curtain, its surface will be divided into two parts. As the strip of cotton prevents water filtration, wet paper layer cannot be shaped, thus it was made into two sheets of wet paper called Lianer paper. If one adds two strips of of cotton (Figure 39), Liansan paper will come into being, therefore three strips of cotton will make Liansi paper. However, the cotton strips can be increased only to four. Skilled papermaking craft was required by papermakers to make such kind of paper. Once finished, the paper kept original edge

Figure 39　Equipment for Making Liansan Paper, painted by Pan Jixing (1998)

with no need to be tailored.

In the Song and Yuan Dynasties, as comprehensive technological improvements took place in pulping, curtain weaving and manufacturing, larger scale paper could be made compared to that of previous generations, which hit the record. Volume 4 of *Four Books of Stationery* recorded:

> *A large amount of high-class paper was produced in Yi County and She County, characterized as Ningshuang paper and Chengxin paper. There was also a longer one with 50 feet in length, and people in Xi county spent several days to neaten paper mulberries and soaked them in a ship. Dozens of people held the paper curtain together to manufacture paper alongside someone to make the order for coordination. Then the wet paper was dried by surrounding ovens rather than being stoved on the wall. The evenness and thickness of entire paper was the same.*

In the Tang Dynasty and the Five Dynasties, only 1 *zhang* (equal to 3 meters) paper could be made. However, 5 *zhang* (about 15 meters) Chupi paper could be made in Yi County and She County in the southern part of present Anhui Province in the 10th century of the Northern Song Dynasty, five times as much paper as in the Tang Dynasty. Moreover, the evenness and thickness of entire paper was the same, which was a major technical achievement and reflected the comprehensive technical level of papermaking most. Volume 7 of *Treatise on Superfluous Things* (1640), written by Wen Zhenheng, also pointed: "there were a bolt of paper with different lengths from 3 *zhang* to 5 *zhang* in the Song Dynasty." *Sketch of Vegetables*, created by Fa Chang, painter of the Southern Song Dynasty, collected in the Palace Museum in Beijing, was over 3 *zhang* in length (nearly 10 meters), and *Thousand Character Classic*, written by Emperor Huizong in the Northern Song Dynasty, was over 3 *zhang* in length, and the paper of two works featured uniform thickness and smooth surface. It required boat-shape giant paper slot to make such large paper. Dozens of people held the paper curtain together during the manufacture, and someone commanded them to coordinate their actions. The wet paper was dried by ovens around and removed from the curtain after the water of which was filtered. In order to get uniform thickness of the paper, high fiber beating degree and uniform fiber suspension shall be achieved.

Various paper produced by former papermakers along with papermaking crafts continued to develop during the Song and Yuan Dynasties, such as dyed paper, yellow and white wax paper, powder wax paper, gold flower paper, embossing paper, water-marks paper, Chengxintang paper, and Xue Tao letter paper, etc. There is no need to go into details here and only some new special paper will be introduced. We can know from the woodcut seal at the end of the book — *The Spring and Autumn Annals*, officially published in the Year 3 of Chunxi Period, that the book was printed in the insect-resistant paper treated by fruit infusion of Zanthoxylum bungeanum, classified as Rutaceae. Volume 12 of *Wuzazu*, written by Xie Zhaozhe, pointed: "There was a kind of paper called Yunmu paper in Changzhou area." It referred to the coated paper, which was added with fine powder of muscovite on the surface to present sliver-white appearance. In the Song and Yuan Dynasties, colored wax paper, offered to the government, was painted with many patterns like dragon phoenix lines or others by muddy gold.

1.4.2 Papermaking Technology in the Ming and Qing Dynasties

The Ming Dynasty (1368~1644) and the Qing Dynasty (1644~1911) were the last two feudal dynasties in Chinese history. During this period, papermaking technology was developed on the basis of the Song and Yuan Dynasties and entered a summation stage of traditional technology when places of origin, production, quality, and functions of paper exceeded those of previous generations, and specialized works, which were rare in previous generations, and recorded papermaking and processing technology appeared as well. During this period, raw materials were basically inherited from previous generations without no new development. However, the production of bamboo paper ranked first, followed by bark paper. Nevertheless, the output of bark paper was still larger than that of the Song and Yuan Dynasties. Writing paper, calligraphy and painting paper and printing paper still accounted for the largest share of paper consumption, which was the same as that in previous generations. The quantity of material items now being passed down is overwhelmingly larger than any other dynasties. There is no need for specific introduction here. Although paper production places have spread all over the country, the major production areas are still concentrated in the southern provinces of Jiangxi, Fujian, Zhejiang, Anhui, Guangdong, Sichuan and mainly in the northern provinces of Shandong, Shanxi, Shaanxi, Hebei, Henan, while other provinces have low output. Only a small amount of hemp paper is produced in the northern region.

Xuande paper, made from paper mulberry in Jiangxi Province in the Ming Dynasty, and Xuan paper, made from Pteroceltis tatarinowii in Jing County in the south of Anhui Province in the Qing Dynasty, were top-grade paper used as advanced cultural paper. These two kinds of paper exceeded various famous paper in history, with various products and different processing methods, forming two series of paper and leading the technological trend of that time, thus reflecting the higher level of papermaking technology. Mulberry paper made in Zhejiang Province had high quality as well. The bamboo paper figured Lianshi paper and Maobian paper produced in Fujian and Jiangxi Provinces, and it looked white and was for printing books due to technological improvement. In the Ming and Qing Dynasties, there were great varieties of paper, all of which were named based on characteristic, appearance, and use of products. In the final analysis, the raw materials were still nothing more than bamboo, bark materials, or different types of unbleached paper and processed paper, while processing technology was almost the same.

The paper products used in daily life of the previous generations were all preserved, produced and used. In the Ming and Qing Dynasties, many new paper products were worth mentioning, such as paper inkstone, paper cup, paper flute (a vertical bamboo flute), paper-weaving picture, and wall-paper which were more popular than that of previous generation. It is worth introducing here. Volume 1 of *The Analects of Ji Yuan* (1897), written by Ji Yuan of the Qing Dynasty (1874~1941), recorded: "it is heard that paper inkstone was produced in Guizhou, and it can last for a long time, unbroken. The paper cup, appearing in Yuhang (now Yuhang District in Zhejiang Province), could be full of wine without leaking." Before it, the person with surname Cheng, in Beisi lane of Haining County of Zhejiang Province, could make paper inkstone as well. *Jianyang Congbi*, written by Zhang Qian (1733~1818), recorded: "Previously, the person with surname Cheng was good at making paper inkstone by using various stone sand and lacquer." Such paper inkstone, an enduring handicraft, the color and lustre of which were no different from that of Duanxi inkstone and Longwei inkstone, was granted as treasure. In the same way, it is expected that paper cup was made by paper and then coated with lacquer.

Zhou Lianggong (1612~1672) of the Qing Dynasty wrote in his *Note of Fujian* (about 1650): "I bought a paper flute before the Kaiyuan Temple in

Fujian, the color of which was like that of yellow jade, and the sound could be heard clearly when knocking it. People who was good at playing said that the paper flute has higher quality than super *Zhuxiao* (a vertical flute made by bamboo), with no gloss outside the surface, and the air blown in could be collected, and the sound of which was neither stagnant nor floating. Latter then, the paper flute was sent to Liu Gongyong, who thus wrote the poem *Poetry of Paper Flute*. Pan Zhong, people in Yunjian (Songjiang District of Jiangsu Province), could make *Weiqi* (a game played with black and white pieces on a board of 361 crosses) pieces by paper as well, which, with lustrous appearance, had the same shape as that of Yunnan, and the sound of knocking could be heard. The five-leaf blossom box was coated with decoration of painting, then the trace of seam of which would not be found. All the above were magnificent handcrafts." When Zhou Lianggong was in charge of administrative commissioner in Fujian Province in the Shunzhi Period, he bought the paper flute around Kaiyuan Temple of Fuzhou and said that when *Weiqi* pieces and paper flute were knocked, we could hear the sound of knocking due to the use of lacquer, and the lacquer box could be made by paper as well.

The paper-weaving painting, invented by craftsmen of Fujian during the Ming and Qing Dynasties, was much like a new flower in Chinese folk arts. After the thin and tough bark paper was dyed into various colors, it was cut into long slips of paper with the same width and then twisted into thin strings. Replicas of calligraphy and painting, woven by paper strings, were similar to the original ones with no need of ink and paint. Wang Shidian (1634~1711) wrote in his *Fen-gan Notes*: "Fujian boasts various paper-weaving paintings, which are filled with beautiful colors, and the content includes landscape, flowers, plants and feather. It is said to be created recently, but according to the record of *Liuqing Rizha*, *Kesi* (silk weaving article in ancient China), *Nasha* (a piece of clothing worn by a Buddhist monk) and several paper-waving paintings, preserved for a long time, were found in the house of Yan Song who was searched by the officials in the Jiajin Period (1565)." *Liuqing Rizha* (1579), written by Tian Yiheng (1535~1605) of the Ming Dynasty recorded that paper-weaving paintings were found in the house of Yan Song, who was a treacherous court official and searched by the officials in 1565, here is the evidence of the existence of the paper-weaving paintings in the Ming Dynasty. *Picture of Farming and Weaving* (1690), paper-weaving painting of the Qing

Dynasty, held by a collector from Stockholm in Sweden, consisted of 48 pieces all together in black and white, which was 24. 5 cm × 28. 1 cm in size for each, and illustrations, poems and inscriptions written by Emperor Kangxi were quite realistic.

During the Ming and Qing Dynasties, wall-paper was used for popular indoor decorations. The so-called wall-paper generally referred to the paper with colored images by drawings or printing, which was used to paste the wall. The bottomed paper may be white or in other colors, and sometimes coated with white powder. Such kind of paper was widely used in official residences and folk houses, even exported to Europe with high popularity. Volume 12 of *Shuyuan Notes* (1475), written by Ming Lurong (1436~1494), recorded:

> People in Quanzhou of Zhejiang Province were skilled at making paper. Paper offered to the government was wasted by the public or private without counting of expense, however, officials took it for granted. In Tianshun Period (1457 ~ 1464), an eunuch, coming back from Jiangxi Province, wept in silence when seeing that the government took official paper as wall-paper. For knowing that such paper was not made easily, he felt regrettable that it was in such reckless waste.

This historical record showed that, during the period from 1457 to 1464 in the Ming Dynasty, the high-grade processed paper including golden colored powder wax paper shipped by the local was used as wall-paper, which was a waste because such paper was quite expensive and could be replaced by ordinary wall-paper. In August of the Year 46 of Wanli Period (1618), Russian Tsar sent Cossack officer Ivan Petrin (1583~1646) to China, and he pointed in his report after return that walls of houses and pavilions in Baicheng (now around the Xuanhua District of Zhangjiakou City) were decorated with thick patterned paper in vivid colors. It was wall-paper as well. Sheng Bang (1551~1596), people in Linxiang City of Hubei and Hunan Provinces in the Ming Dynasty, appointed governor of Wanping County in Shuntianfu District (now Beijing area) in the Year 18 of Wanli Period (1590). From Volume 13 to Volume 14 of *Wanshu Notes* (1593), compiled on the basis of archives by Sheng Bang during his term (1590 ~ 1593), introduced that eight *dao* pieces (800 pieces) of window paper or wall-paper sold for 4. 8 *qian* (1 *qian* equaled to 5 grams), and mounting sold for 1. 5 *qian* in the Year 16 (1588) of Wanli Period. At the time, when Emperor Shenzong of the Ming Dynasty paid a formal visit to imperial mausoleum, numerous officials welcomed him at Fuchengmen (city gate) in Beijing and the inside of the mansion for resting was decorated by wall-paper. Dramatist Li Yu introduced a unique manufacturing method of wall-paper in his *Sketches of Liweng*:

> First of all, caramel paper is used as the base to paste the wall of study, and then the green Yunmu paper (mica) is cracked into sporadic pieces, which are flat, short, long, triangular, quadrilateral, or pentagonal, but not round, and readily pasted on caramel paper. All pieces of different sizes, in staggered arrangement, reveals the line of caramel paper. When finished, the whole room coated with crackle marks are as beautiful as Ge kiln. It is creative to write poems on the larger pieces of Yunmu paper among bits and pieces, which are seemly like inscriptions written on the ladler.

According to Li Yu, caramel paper and green Yunmu paper were used to paste walls in staggered arrangement by artistic techniques, which made the whole room look like beautiful Ge kiln. It was indeed ingenious to write poems on large pieces of paper. There are various wallpapers (Figure 40) from the 17th to the 18th centuries in old palace of the Imperial Palace in Beijing. In 1550, China's wallpapers were introduced into Europe by Spanish and Dutch merchants. After the 17th century, Europe imported a large number of Chinese wallpapers, which were then copied by Britain, France, Germany and other

Figure 40   Powder Wax Wall paper, printed with Green Flower and Bird Patterns, in the Palace of Qianlong Period of Qing Dynasty, collected by Pan Jixing

countries. Therefore, there are doubts that wallpaper originates in China, which should be amended.

Among various paper products, such as paper umbrella, playing card, paper lantern, paper kite, paper fan, window paper, visiting card, paper armor and so on, most of them had been documented and only few real collections were passed down. However, many of these collections and image materials during the Ming and Qing Dynasties can still be seen so far, which helps us to deepen our understanding of these paper products visually. Meanwhile, it was rare that people in the Ming and Qing Dynasties wrote several literature to sum up their techniques, such as *Monograph of Yezi*, monograph about playing card written by Pan Zhiheng (about 1536~1621), native of Shexian County, who lived in Nanjing in the Ming Dynasty, *Collection Of Illustrative Plates of Kite*, monograph about paper kite written by litterateur Cao Xueqin (1715~1763) in the Qing Dynasty, and *Collection of Paper Cutting*, monograph about paper cutting and character cutting written by artist Jiang Zhou (1734~1789) in Quanjiao County in Anhui Province in the Qing Dynasty and so on.

The more important was that noticeable valuable scientific works on papermaking and paper processing were left by people in the Ming and Qing Dynasties, which were first-hand records written by the authors based on field visit of papermaking production, hence reflecting the actual condition of traditional technology. There are several works on bark paper manufacturing such as *Treaties of Jiangxi Province — Paper Mulberry Book*, *Shuyuan Notes*, *Huizhou Prefecture • Property*, *Exploitation of the Works of Nature* and so on, and several works about bamboo paper manufacturing such as *Exploitation of the Works of Nature*, *The Three Provinces Frontier Defense*, *Papermaking Theory*, *Atlas of Bamboo Paper* and so on.

In the first year of the Yongle Period of Emperor Chengzu in the early Ming Dynasty (1403), a large-scale governmental paper mill was ordered to build at the site of Cuiyan Temple in Xinjian County in Nanchang Prefecture of Jiangxi Province. Emperor Chengzu dispatched officials to supervise manufacturing of high-ranking Chupi paper for the use of government. In the same year, the grand secretary Jie Xie (1369 ~ 1415) compiled *The Yongle Canon* (1408) on imperial order, and wrote it in such paper. After being tested, such paper, smooth, delicate, thick, white and receptive, seemed to take Chengxintang paper as sample. In Xuande Period (1426 ~ 1435) of

Emperor Xuanzong in the Ming Dynasty, Xishan paper evolved into famous Xuande paper. Paper mill moved to Qishan County in Guangxi Prefecture of Jiangxi Province since the Longqing and Wanli periods (in the second half of the 16th century), but paper was made by original method. *Treaties of Jiangxi Province* (1556), compiled and edited by Wang Zongmu (1523 ~ 1591), introduced porcelain, specialty of Jiangxi Province, but paper was not mentioned. Hence Lu Wan-gai (1515 ~ 1600) supplemented *Paper mulberry Book* as one of the volumes (the 8th volume) in the *Treaties of Jiangxi Province* in the Year 25 of the Wanli Period (1597). Lu Wan-gai, together with Wang Zongmu, was in the same dynasty. He was the native in Pinghu City of Zhejiang Province, ranked as imperial scholar in the Year 2 of the Longqing Period (1568), appointed as local official in Jiangxi Province, and introduced technology of governmental paper mill in Qianshan County in the province. The original text was explained as follows:

(1) Bundling the peeled paper mulberry bark and soaking it in the river for several days→(2) Removing part of the blue bark, picking it up after bundling →(3) Stewing with water→(4) Removing the outer skin and pulling the bast into silk→(5)Cutting it into small pieces with knife→(6) Soaking bark material in lime slurry, stacking over a month→(7) Stewing the material socked in lime slurry→(8)Placing the material in cloth bags, washing it with running water, and removing lime water by feet→(9) Stacking the material on the riverside, and bleaching naturally → (10) Pounding the material into fine mud → (11) Building the material up, drenching it with hot plant ash water and drying it in the shade for half a month→(12) Re-stewing→(13) Washing it in the river with the former method→(14) Stacking the material on the riverside, bleaching it again in sun and rain→(15) Removing remaining impurities and colored materials by hand→(16) Putting crispy material into a cloth bag and washing it in the river→(17) Placing slates in paper slot and making pulp by adding fresh water from the mountain→(18) Adding plant mucilage (Paper potion) and stirring→(19) According to the set paper size, the thin bamboo strips are woven into a papermaking curtain, then into a curtain bed by yellow silk thread, and the papermaking curtain is tightened on curtain bed→(20) Manufacturing paper with pulp by using papermaking curtain in paper slot. The curtain in larger size requires 6 people all together with 3 people on each side to coordinate lifting the curtain, and smaller curtain requires two people→(21) Filtering water on the paper slot after the paper is picked out, and the wet paper was layered →

(22) Extra moisture is squeezed out by wood press and the paper is left overnight→(23) Uncovering half dry paper one by one and drying it on brick wall. The brick wall is hollow and its both sides are brushed into smooth walls by fine lime. Burning wood to warm walls by smoke and fire→(24) Uncovering the dried paper and stacking it together→(25) Bundling and packaging, 1 *dao* for each hundred pieces of paper.

Volume 13 of *Shuyuan Notes* (1475) written by Lu Rong (1436～1494) recorded the production processes of Chupi paper (Figure 41～Figure 45) from Changshan County and Kaihua County in Quzhou Prefecture of Zhejiang Province as follows:

Figure 41   Picture of Grinding Bark Material, selected by Pan Jixing

Figure 42   Soaking Bark Paper in Lime Water, provided by Pan Jixing (1979)

Figure 43   Stewing, selected by Lin Yijun (1983)

A   B

Figure 44   Process Map of Manufacturing Paper, selected by Lin Yijun

A. Holding curtain for manufacturing paper B. Putting off wet paper

Figure 45   Squeezing Extra Moisture,
provided by Pan Jixing (1979)

(1) Cutting down branches of paper mulberry, peeling and bundling → (2) Stewing with water → (3) Removing the bark when it is hot →(4) Dipping the bark in lime slurry for 3 days, and removing the outer skin by stepping and rubbing → (5) Washing off lime in the river → (6) Socking the bark material in the pool for 7 days→(7) Re-stewing after it is taken out (it is dipped in the lime slurry or leached by plant ash water)→(8) Washing lime water off and colored materials in the river→(9) Stacking the material on the riverside or the hillside and exposing it to the sun for more than 10 days as daylight bleaching→ (10) Pounding it into paper material→(11) Putting the material in the cloth bag and washing it in the river→(12) Putting paper material into the paper slot, adding water to make paper pulp, and stirring→(13) Adding mucilage of Carambola vine as paper potion→(14) Manufacturing paper by swaying paper curtain→(15) Layering the wet paper and squeezing to remove moisture→ (16) Uncovering semi-dry paper successively and drying it on the brick with lime→(17) Uncovering the dried paper→(18) Finishing and packaging.

Papermaking techniques of Chupi paper in Huizhou Prefecture in Anhui province described in *Treaties of Huizhou Prefecture · Property* (1502), compiled by Wang Shunmin (1440～1507) in the Ming Dynasty, was roughly the same as that of Quzhou Prefecture in Zhejiang Province recorded in

Shuyuan Notes, while relevant techniques were introduced briefly in *Exploitation of the Works of Nature*. Many works of Qing Dynasty's were plagiarized from the works of the Ming Dynasty without new ideas. As major producing regions of bark paper, Jiangxi, Anhui, Zhejiang Provinces are made comparison in terms of their representative technologies. According to *Treaties of Jiangxi Province*, the making processes of paper made for imperial storehouse and adopted by Jiangxi provincial governmental bureaus were quite complex. There were 25 major processes, including three times of stewing: clean water for the first time, lime water for the second time, and plant ash for the third time; plus two times of natural bleaching, three times of washing and several times of removing colored skin. Due to long processing cycle, huge consumption of human and material resources and costly input, the paper produced was finest and figured natural whiteness and evenness, thus only royal products were produced in this manner.

Shuyuan Notes and *Treaties of Huizhou Prefecture* reflected that the papermaking technology of local paper mills in Zhejiang and Anhui Provinces included 18 main processes, 7 processes less than those of Jiangxi governmental bureaus. It could also produce advanced paper mulberry paper by stewing twice and bleaching once. Therefore, there was a waste in the process of papermaking of the Jiangxi governmental bureaus, and some of the repetitive processes could have been omitted. From the view of technological economics, some processes like natural bleaching "regardless of the day and the night" and "continual exposure to the sun" are unreasonable, because such a single step would greatly lengthen the whole process and the folk paper workshop would by no means do so. The most ideal and economical production plan should be a compromise between Jiangxi Province and Zhejiang Province, and Anhui Province, or the finer operation of each step of the latter two provinces. In the Ming and Qing Dynasties, Xuan paper in Jing County was made in this way by famous folk paper mill.

Xuan Paper was mainly made from bark of Pteroceltis tatarinowii, classified as Ulmaceae. It was produced in Jing County which originally belonged to Xuanzhou in Anhui Province during the Ming and Qing Dynasties, thus was also named as Xuan paper. During the Song and Yuan Dynasties, as local place was rich in Pteroceltis tatarinowii, which was similar to paper mulberry, the family Cao began to make paper. Papermaking was further developed in the Ming and Qing Dynasties and most people in papermaking industry were surnamed Cao. Scholars in the Ming and Qing Dynasties liked the paper because it could be compared with Chengxintang paper and Xuande paper. After the Qing Dynasty, the government-run paper bureau previously set up in Jiangxi Province in the Qing Dynasty declined, thus Jingxian paper became tributary paper. In the Year 38 of Qianlong Period (1773), compiling institution of *Imperial Collection of Four Divisions*, multi-volume series, was set up and it was finished in the Year of 47 (1782) in Xuan Paper of Jing County. Xuan paper naturally became the paper used by the imperial storehouse. Technically speaking, there are clear genetic transmission relations among Chengxintang paper, Xuande paper and Xuan paper of Jing County. The secret of success lies in absorbing the merits of the fine craftsmanship of Xuande paper made by the Mingguan Bureau in Jiangxi Province, simplifying its complicated processes regardless of cost, and improving the manufacturing techniques of bark paper in Zhejiang Province and Huizhou Prefecture. Its manufacturing craft is as follows:

(1) Cutting down branches, peeling and bundling→(2) Stewing with clean water→(3) Beating the bark, pulling it into silk and removing blue skin→(4) Bundling the bark and socking it in the pool→(5) Socking the bark in lime slurry and stacking for one month→(6) Putting bundled bark material into pot and stewing→(7) Washing and treading material in the water→(8) Stacking the material on the riverside or the hillside and bleaching naturally for 3 to 6 months, turning over at any time→(9) Washing and removing the impurities→(10) Pounding the material into gunk→(11) Putting the material in the cloth bag and washing and rubbing it in the river→(12) Putting the material into the paper slot, adding water from mountain to make paper pulp→(13) Adding mucilage of Carambola vine and Pubescent Holly Root or others as paper potion and mixing up→(14) Manufacturing paper by swaying paper curtain held by 2 or 4 people or several people according to the paper size→(15) Layering the wet paper after filtering water→(16) Removing moisture by squeezing and letting stand overnight→(17) Uncovering semi-dry paper and drying it on the brick wall→(18) Uncovering paper, stacking it up, cutting four sides level off, stamping, packaging, 1 *dao* for each hundred pieces of paper.

Killing Green (1637), one chapter of *Exploitation of the Works of Nature*, written by scientist Song Yingxing in the Ming Dynasty (1587~1666), made a detailed description of making techniques of bamboo paper, and provided 6 illustrations (Figure 46 ~ Figure 48), mainly introducing manufacturing processes in Fujian Province, which are as follows:

Figure 46   Illustration of Cutting，Socking and Stewing Bamboo
in *Exploitation of the Works of Nature*

Figure 47   Illustration of Manufacturing Paper by Swaying Curtain and Removing
Moisture by Squeezing in *Exploitation of the Works of Nature*

Figure 48   Illustration of Drying Paper in *Exploitation of the Works of Nature*

(1) Cutting bamboo in June and bundling→(2) Socking bamboo in the pond for 100 days→(3) Washing and beating bamboo in the river, drawing it into filamentous shape, and removing the bark→(4) Socking bamboo material in lime slurry (stacking it for 10 days)→(5) Stewing the material for 8 day→(6) Taking the material out, and washing in the river while stepping on→(7) Soaking material in plant ash waster, and stewing for 10 days→(8) Putting the material into bag or bamboo basket, and washing it in the river→(9) Pounding the material into gunk in water-powered trip-hammer→(10) Putting white material into paper slot and adding spring water from the mountain to make pulp→(11) Adding mucilage of Carambola vine into the pulp and stirring→(12) Manufacturing paper by swaying curtain→(13) After filtering moisture of wet paper, taking paper down off curtain, and piling it up to a thousand pieces→(14) Removing moisture by squeezing, letting it stand overnight→(15) Uncovering a corner of semi-dry paper by copper tweezers, then putting the paper by hands and brushes, and drying it on the brick wall→(16) Uncovering the paper, and piling up neatly→(17) Cutting four sides level off, 1 *dao* for each hundred pieces of paper, and packaging.

In the first year of the Daoguang Period (1821) in the Qing Dynasty, Yan Ruyu (1759~1826), under the Emperor's order, investigated frontier defence in the border of Shaanxi, Sichuan, and Hubei Provinces for half a year and wrote the book *The Frontier Defense of Three Provinces*, all together 14 volumes on the basis of his experience and relevant works of previous dynasties. The book was published in the following year and reprinted ten years later (1830). Volume 10 *Mountain Product* in the book recorded the papermaking situation seen by the author in Yang County, Dingyuan County and Xixiang County in southern Shaanxi Province and pointed that there were 140 to 150 paper mills in the folk. "Larger paper mill had over one hundred staffs, and smaller one had 40 to 50 people", which had the production scale of handicraft in capitalist industry. The bamboo paper produced in the northern Yellow River Valley introduced by the author can be compared with bamboo paper in the south of the Yangtze River. The manufacturing processes of bamboo paper in southern Shaanxi can be summarized as follows:

(1) Cutting bamboo in June or July, and bundling→(2) Socking bamboo for 10 days→(3) Washing bamboo in the water, beating and bundling→(4)Soaking the material in lime slurry and stacking it for more than 10 days→(5) Stewing the material for 5 to 6 days, letting it stand overnight→

(6) Washing the material in the river→(7) Stewing the material in the pot with plant ash water for 3 days→(8) Washing in the river→(9)Mixing the material with pulp made from soybeans and rice, and cooking it for the third time for 7 ~8 days→(10) Washing the material→(11) Pounding the material into gunk→ (12) Adding clean water in the material to make pulp→(13) Adding rice juice in the material, and mixing it up→(14) Holding the curtain to make paper, and filtering moisture→(15) Uncovering the wet paper from the curtain and piling it to a height of 1 foot → (16) Squeezing it to 3 inches high, overnight → (17) Uncovering the semi-dried paper one by one and putting it on the brick wall for drying→(18) Uncovering the paper and folding it up→(19) Tailoring the paper neatly and packaging. Packaging 5 to 6 *he* (unit for measurement) for each bundle, and 200 pieces for each *he*.

After comparison between discussion of Song Yingxing in the Ming Dynasty and that of Yan Ruyu in the Qing Dynasty on bamboo papermaking process of different regions, we found that they had similarities as well as differences and each had its own merits. In above mentioned two kinds of processes, cutting and socking bamboo both began in June, but the length of time of socking bamboo was different. Generally speaking, it was enough to sock bamboo for 30 days, and 100 days was too long while 10 days were a little short. They both stewed bamboo material in lime water and plant ash water twice, but at different times. It was advisable to add carambola mucilage into pulp while adding starch liquid was an outdated method. The wall for drying, woven by bamboo strips and coated with lime slurry, was not as effective as brick wall. In contrast, the technology of the Southern Shaanxi Province in the Qing Dynasty was less advanced than that of the south of Yangtze River in the Ming Dynasty. Processes of making bamboo paper in Changshan County in Zhejiang Province, recorded in *Papermaking Theory* (about 1885),written by Huang Xingsan (1850 ~1910) in the Qing Dynasty, were roughly the same as that in *Exploitation of the Works of Nature*, and were also consistent with the introduction of bamboo paper in Tingzhou in Fujian Province recorded in Yanglan's *Textual Research of Linting* (about 1885), but the process of natural bleaching was added. It seemed that he did not investigate in papermaking fields but listened to the description of the local people and natural bleaching helped improve the whiteness of bamboo paper. In the 18th century, during the reign of Emperor Qianlong of the Qing Dynasty, the artist drew 24 meticulous paintings with colors depicting the processes of making bamboo and paper, which spread across Europe and produced extensive influence.

The Ming and Qing Dynasties witnessed a peak of processed paper in history. An integrated papermaking system was formed by production area of Xuande paper in Jiangxi Province, and Xuan paper in Jing County of Anhui Province, in which the varieties and specifications of finished paper were complete, and different papermaking curtains could be made as well. In addition to the unbleached paper, a variety of dye paper and processed paper could be produced too. There were official and private processed paper workshops in the city of Beijing, Nanjing, Suzhou, Hangzhou and other places. Various kinds of processed paper and methods have already been mentioned in the aforementioned section about papermaking in the Tang and Song Dynasties. Some special papermaking methods of processed paper were found in the *Kaopan Yushi* (1600) written by Tu Long (1542~1605) in the Ming Dynasty, now with various circulation editions. Original copy of processed paper in the Ming and Qing Dynasties can be seen in major museums, libraries, and even can be purchased in the antiquities markets.

It should be pointed out that during the reign of Emperor Kangxi in the early Qing Dynasty (1662~1722), copper net was used to make water-marks paper. Xu Kang (1820~1880) wrote in his *Dust of Dreams* (1879): "during the reign of Emperor Kangxi, the old friend Chen Baijun (county magistrate) once found a few pages of ribbed paper with wide screen marks, which was decorated with dark lace, six feet long for each side. Such paper was dyed by a skilled craftsman Wang Chengzhi, who was famous for making paper in Hangzhou. To be honest, there are only narrow curtains today. Ribbed paper, though in small size, was made by bamboo curtains and wide papermaking curtains were woven by copper wires." It referred to copper net, a wide format paper curtain which was woven by copper wires and could make hidden veins. The copper wire could not be used to make patterns in bamboo curtain, but could in copper net. During the reign of Emperor Kangxi, the copper net was also made into a tube-shaped papermaking curtain, which had raised oblique screen marks made by thick copper wires, and then the so-called tube-shaped paper with oblique screen marks could be made. The dramatist Kong Shangren (1648~1718) wrote in his *Book of Xiangjin* (about 1712): "tube-shaped paper with oblique screen marks is several *zhang* in width, the pattern of which is like grinding gear, gifted by a friend." Such paper was once offered to Emperor Kangxi.

Volume 26 of *Yangjizhai Notes* (about 1863), written by Wu Zhenyu (1792 ~1871) recorded that during the Qianlong Period (1757), Emperor Gaozong got two sheets of tube-shaped paper with oblique screen marks in an inspection trip in the south, and one was for collection and the other was for writing a poem. After inspection of the imperial storehouse, another five sheets of such paper was found. In 1782, the copies of tube-shaped paper with oblique screen marks were made in Zhejiang Province and offered to the Emperor. This tube-shaped paper was sideless and seamless, thus getting the name. Emperor Qianlong wrote poems again in such paper and awarded to the officials, thus Peng Yuanrui (1731~1803) made a poem in this paper for expressing gratitude. This is to say, in 1757, Emperor Gaozong went to Zhejiang Province and was offered two sheets of tube-shaped paper with oblique screen marks made in Kangxi Period. Emperor loved the paper very much, and one was for collection and the other for writing the poem *Chant of Paper with Oblique Screen Marks*. After the inspection of the imperial storehouse, another five sheets of such paper were found. The local officials in Zhejiang Province knew that the Emperor loved the paper and ordered paper workers to make copies which were offered to the Emperor in the Year 47 of Qianlong Period (1782). Emperor Qianlong wrote other poems in such paper and awarded to the officials. Peng Yuanrui (1731 ~ 1803), ministry of works, made a poem in this paper for expressing gratitude and imitated the original style of the Emperor's inscriptions. The grand secretary Ruan Yuan (1764~1849) wrote in his *Essay of Shiqu* (1793): "In the Qianlong period, copies of tube-shaped paper with oblique screen marks were made, which was dark and heavy and over 1 *zhang* long."

In 1965, the residue of tube-shaped paper with oblique screen marks, collected by Li Zongren (1891 ~ 1969), was seen in the Museum of Sichuan University, and a complete piece of tube-shaped paper with oblique screen marks, copied in Qianlong Period in the Qing Dynasty, was seen in storage department of Chinese History Museum in 1973, which was taken out from the Forbidden City in the early republican period, previously collected by Mr. Zhang Boju (1897~1982). A careful examination of the paper showed that it was not made by bamboo curtain but machine, dark-skinned and thick, and had uneven surface and oblique veins, which took bark fiber as raw material with a high beating degree. The entire paper was like tube in shape, seamless, large in breadth and was very long when unfolded. Such paper was very appropriate to be called and translated as tube-shaped paper with oblique screen marks according to the accurate confirmation in literature.

During the Kangxi Period (1662~1722), how was the tube-shaped paper with oblique screen marks manufactured? From the perspective of technology, it was made by a tubed copper net with convex oblique veins on the surface woven by rough copper wires, and only by using such kind of papermaking curtain, can tube-shaped paper with oblique marks be made. In order to form a wet paper on such papermaking curtain, it must be in rotation, and the pulp flows from the upper slot through the duckbill-shaped discharge port into the rotating papermaking curtain, and the water is filtered while filling pulp, thus forming wet paper after rotating papermaking curtain in a circle. The filtered water is discharged in a flow channel inside the tube-shaped copper net. At the same time, another revolvable round roller, coated with soft material, stays close to the papermaking curtain and rotates in the opposite direction, so that the moisture in the wet paper can be squeezed out, and the uniform thickness of the paper can be ensured. After drying, the paper is lifted from both ends of the paper, and then peeled off with a long sheet until the entire paper is separated from the copper net.

Therefore, three key technical issues must be solved for making tube-shaped paper with oblique screen marks: (1) Design and manufacture of tube-shaped copper net; (2) Technical ideas of making paper by rotating tube-shaped papermaking machine; (3) Use of principle of sugar mill (namely sugar cart in ancient time) that squeezing moisture is by making tube-shaped papermaking machine close to another round roll, which both rotate in the opposite direction. These three points constitute the basic elements of a modern mono-cylinder paper-machine, and it was not invented until 1809 by the British John Dickinson (1782~1871), which was patented and then used throughout the world. The tube-shaped papermaking machine in China is quite similar to that of John Dickinson in terms of structural principle. It also explains why the tube-paper with oblique screen marks looks like the machined paper. This is a technological miracle that China could develop the prototype of cylinder papermaking machine during the Kangxi Period.

1.5   Spread of China's Papermaking Technique in Europe and America

1.5.1   Beginning of Papermaking in Spain, Italy and France

Though located on both ends of the old continent and being far away from each other, China and Europe have been in contact for a long time, either through the medium of central Asia and west Asia, or in a direct way by their own people, and their exchanges have been carried out via land, sea or land-sea. The spread of Chinese papermaking in Europe can be divided into two-stages: the first stage was when the Arabian people transmitted the hemp-paper technology of the Chinese Tang Dynasty to Europe in the 12th and 13th centuries; the second stage occurred in the 18th and 19th centuries when the Europeans directly introduced the technological achievements of papermaking since the Song Dynasty from China to Europe, leading to the revolution of papermaking technology. Whereas previous studies are mainly confined to the first phase, we will do a supplementary research on the first stage, make a tentative study on the second stage of the spread, and then discuss the prompting role of the Chinese traditional technology on the modernization of European papermaking. The two-stage theory on the spread of Chinese papermaking technique in Europe is a new concept presented by us.

Europe first imported papermaking technique on the eve of the Renaissance. Before that, medieval Europeans largely wrote on parchment or papyrus sheet. It was not until the Arab world started to make paper and then diffused it into Europe that the Europeans began to use paper, but they had to pay quite some gold coins every year for paper. Paper was produced locally in Europe only after the Europeans had learned the Chinese papermaking technique via the Arabs. The first European country to make paper was Spain, which was once ruled by an Arab regime. Having taken power in 750, Abu'l-Abbas (721~754), founder of the Arab Abbasid Dynasty (750~1258), ordered all royal members of the former Umayyads Dynasty (661~750) be executed. Prince Abd al-Rahman ibn-Mu'-awiyah (731~788) was the only escapee of the royal family. He led some followers to flee to North Africa, and then to Spain. In 756, he won the battle with the locals of Spain and set up an independent regime there, with Cordoba as its capital. Historically, his regime was called the Post Umayyad Dynasty (756~1036).

The 9th to 10th century Post Umayyad Dynasty (929~961) under the reign of Abd al-Rahman al-Nāsir (891~961) demonstrated great power in its strong Muslim rule of Spain, its occupation of part of Morocco, Africa across the Strait of Gibraltar, and establishing its capital Cordoba as one of the most important literature and academic centers in Europe. The earliest paper document in Spain is a 10th-century manuscript discovered in the city of Santo Domingo, written on paper made of flax fibers, smeared by starch paste, and similar to Arabian paper. The paper of the 1129 manuscript found in San Gilos Monastery, might have been imported from Morocco. Due to the surge in the amount of paper used in the Post Umayyad Dynasty, in 1150, the earliest paper mill of the regime was built in Xativa, an area in southwest Spain rich in flax. The paper mill was run by the Arabs who were then known as Moors in Europe, and the technology was introduced from Morocco. The Arabian geographer Abu Abdullah al-Idrisi (1099~1166), who lived in Spain, in his *Kitāb Nuzhat al-Mustāq fi Ikhtirāk al-Afāq* (1154), said that the town Xativa makes unparalleled paper that can not be produced in other parts of the civilized world to transport to the other countries east and west[❶].

After 1031, the Post Umayyad Dynasty's rule declined and split in Spain. The Spaniards began their fight to recover the lost territory, and gained independence in 1035, led by Ramiro I (1035~1063). After its independence, Spain also built a paper mill in Vidalon in the northwest in 1157, operated by the Spaniards themselves[❷]. As a result, there were two paper mills in 12th-century Spain, and the paper produced was exported to other countries nearby.

In the 11th ~ 12th centuries, Arabian paper was transported to Europe through the routes of Damascus-Constantinople of Byzantium, or Egypt-Morocco-Sicily Island in the Mediterranean. The papermaking technique might have been introduced into Italy by the above two sea routes. Still well kept are a few old files from the 12th century: a rescript of the Sicily King Roger I or Roger Guiscard (1031~1101) written in Latin and Arabic on the colored paper in 1109; and a 1154 paper manuscript kept in Genoa Archives. But the paper of these old documents can not prove to be made in Italy; nevertheless, the paper

---

❶ DOZY R. Description de l'Afrique et de l'Espagne par Idrisi[M]. Leiden: Brill, 1866.

❷ BLUM A. On the Origin of Paper[M]. New York: Bowker, 1934.

was imported from the Arab region. Since paper was expensive, the King of Naples and Sicily, Frederick Ⅱ (1194~1250) banned writing official documents on paper in 1221, to boycott the dumping of Arabian paper. But that did not reduce the amount of paper used, and the whole 13ᵗʰ century saw the continuous entry of Damascus paper to Europe.

In 1276, the first Italian paper mill was established in Montefano of central Italy❶, producing hemp paper. Montefano is today's Fabriano in Marche. Later, the paper mill made some technical improvements, including the production of watermark paper in 1282, with the cross and circular patterns on the metal roller to press watermark. Italy debuted the production of watermark paper in Europe❷, while this kind of paper had appeared in China as early as the 10ᵗʰ century. Italian watermark paper technique had since been copied by other European paper mills. In 1293, new paper mills were started in the cultural city of Bologna. Italy became the second European country to make paper after Spain. With the fast development of its paper industry and the gradual increase of paper producing regions, Italy had become the supplier of paper in Europe in the 14ᵗʰ century. Although its technology came from the Arabs, it had outgrown Spain and Syria in the production of paper.

France, bordering on Spain, learned the papermaking technology from Spain. Once, it was believed that the first French paper mill was built in 1189 in Lodève of the southern province Hérault, which is close to the Mediterranean and Spain, but it was proved untrue because of the misinterpretation of the literature. The paper mill, built near the city of Troyes, southeast of Paris, in 1348, may have been the earliest in France. From 1226 to 1270, France was under the reign of King Louis Ⅸ (1214~1270) of Capetien (987~1328), who carried out a series of reforms to develop the economy and strengthen the rule of the kingship. From 1348 to 1388, other paper mills were established in Essones, Saint Pierre, Saint Cloud, and Toiles. This allowed France adequate supply of paper at home and also exportation of paper to countries such as Germany.

---

❶   CARTER T. F. The Invention of Printing in China and Its Spread Westward[M]. 2ⁿᵈ ed. New York: Ronald Press, 1955.

❷   HUNTER D. Papermaking: The History and Technique of an Ancient Craft[M]. 2ⁿᵈ ed. New York: Dover, 1978.

1. 5. 2 Spread of Papermaking Technique in Other European and American Countries

Germany, located in central Europe, had been using paper since 1228, but its paper was still imported from other countries till as late as the second half of the 14th century — its north imported paper from Italy, and the Rhine region from France. At the sight of papermaking in Italy, the 14th-century Nuremberg businessman Ulman Stromer (1328~1407), decided to invest in a paper mill back in his home country. The National Museum of Germany in Nuremberg keeps two pages of Stromer's diary manuscript, written in archaic German in the 14th century, titled *Püchl von mein Gelslecht und von Abentur*, which is the earliest literature on paper technology in Europe. It relates in great detail the author's operation of the first paper mill in Germany. In 1390, Stromer encountered in the commercial port Milano, Lombardia of northern Italy some Italian papermaking artisans (Franciscus de Marchia, his brother Marcus and his apprentice Bartholomeus), and persuaded them to leave Lombardia for Nuremberg and to make paper with him.

In 1390, Stromer brought the Italian papermaking artisans to Nuremberg, employed a German named Closen Obsser as the foreman and overseer, and made them pledge to stay loyal to Stromer, and not to betray the papermaking technology secrets to outsiders. Then his paper mill (Figure 49) was built near the Pegnitz River outside the city's west gate[1][2]. Mortars and pestles were used for beating hemp raw material. Each water wheel drove 18 pestle sticks. On the hemp paper was a watermark of the letter S, representing the owner of the mill- Stromer. It was also mentioned that Stromer cared his own interest so much that the Italian workers were on a lazy strike, for which the Franciscus brothers were detained illegally in a tower as punishment on August 12, 1391. In 1392, Stromer hired a local carpenter Erhart Zymerman to repair the pestle sticks and the deckles, and also help the calendering of paper. His wife was in charge of sorting out and classification of raw materials and rags, hanging wet paper for drying, and rallying the number of paper before packaging. In the process,

---

[1] SANDERMANN W. Die Kulturgeschichte des Papiers[M]. Berlin: Springer-Verlag, 1988.

[2] HUNTER D. Papermaking: The History and Technique of an Ancient Craft[M]. 2nd ed. New York: Dover, 1978.

German workers learned a whole set of papermaking techniques from the Lombardians. From 1390 to 1394, Stromer became wealthy by papermaking, turned an MP for Nuremberg and hence stepped into politics. In 1394 he leased the paper mill to Thalmann. Stromer died in 1407.

Famous for papermaking, Nuremberg later became the printing center in Germany. Papermaking technology was soon bought by other mill owners in Germany, and paper mills mushroomed in other places. The new paper mills in Chemnitz (1398), Ravensburg (1402) and Augsbourg (1407) became strong competitors to Stromer Paper Mill. Afterwards, Strassberg (1415), Lubeck (1420), Wartenfels (1460), and Kempten (1468) all turned out to be paper producing districts. By the end of the 16th century, the number of German paper mills had reached 190, backing up the development of the printing industry effectively. In 1493, a Nuremberg native Hartmann Sohedel published *Liber Chronicarum* written in Latin, which describes the scenery in many places, has 645 illustrations, and is known as *Nuremberg Chronicle* because it was published in Nuremberg. In the book is a map of Nuremberg, in the lower right corner of which is a sketch of Stromer Paper Mill (Figure 49). This picture is the earliest woodblock printing that portrays a paper mill in the European literature.

Figure 49   Picture of Stromer Paper Mill in 1390, Nuremberg, Germany, from *Liber Chronicarum* (1493)

Because of the establishment of paper mills in many regions, papermaking was no longer a secret, so much so that it became the subject of poets and painters. In 1568, an illustrated book of Gothic Style (Gotische Schrift) in archaic German was published in Frankfurt am Main, entitled *Eygentliche Beschreibung aller Stände auf Erden, hoher und niedriger geistlicher, und weltlicher, aller Künsten, Handwercken und Händeln* or succinctly translated as *A True Description of All Trades*. The book was reprinted in 1588 in Nuremberg. The Latin version was published in 1574. In 1960, it was reprinted in Leipzig and hereinafter referred to as *Das Ständebuch* (*A True Description of All Trades*). The new edition has a total of 134 pages and 114 woodcut illustrations. Each picture is a depiction of the characters of different trades drawn by the painter Jost Amann（1539～1591）, and is accompanied with a poem by the Nuremberg cobbler and poet Hans Sachs（1494～1576）. Picture 18 in the book（Figure 50）portrays the papermaking workers. Here we translate the poem that goes with this picture by Sachs into English:

Ich brauch Hadern zu meiner Mül/
Dran treibt mirs Rad deß Wassers viel/
Daß mir die zschnittn Hadern nelt/
Das Zeug wirt in Wasser eynquellt/
Drauß mach ich Bogn/auff den Filß bring/
Durch Preßz das Wasser darauß zwing.
Denn henck ichs auff/laß drucken wern/
Schneeweiß vnd glatt/so hat mans gern.

　　　　　　З ij　　Der

Figure 50　The Earliest Picture Depicting Papermaking in Europe, followed with a Poem by Sachs, from *Das Ständebuch* (1568)

Waste rags brought into the paper mill; water wheels busy rolling.

Waste rags cut into pieces; raw material in water becoming pulp.

Paper suspension drained through a screen; the wet mat laid on top of a damp cloth.

Water removed using a press; the sheet hung to dry for packaging.

White and smooth paper thus made; everybody（not exaggerated）loving to use it[1].

In the picture, Amann drew two men. One of person is screening paper in the pulp suspension. After being screened, each piece of wet paper is put on a sheet of absorbent cloth of the same size on the board. Then, another layer of absorbent cloth is laid on the top. The process is repeated till a neat stack of wet paper is piled up. In the lower right corner of the illustration are the finished stacks of wet paper. Behind this paper artisan are the presses to remove water: the top and the bottom are thick planks, and the rotating screw presses out the excess water in the wet paper between the two planks. The other person in the

---

[1]　JOST A. A True Description of All Trades[M]. New York: Brooklyn, 1930.

picture is the apprentice, whose task is to lay a blanket or cloth on the wet paper after it is taken out from the pulp. He then presses the water out of the pile of wet paper. In the picture, he is taking the pressed paper to air drying.

In the top left hand corner are the water-powered trip-hammers ramming raw material. The big water wheel is turned by the impact of running water, and then the rotating motion is turned into an up-and-down vertical motion by the driving device. The up-and-down vertical motion then drives the hammer sticks and the hammers to pound raw material. Such device was also invented in China, as described in the Yuan Dynasty fellow Wang Zhen's *Nong Shu* (《农书》, *Book on Agriculture*, 1313), and introduced to Europe during the Renaissance. Because the picture is small, the water wheel that drives the water hammers is not drawn. Pressing wet paper with a rotating screw stick (made of iron) is a European invention, which is more advanced than the leverage device employing the gravity of stones to press paper used by the Chinese. It can be asserted that the painter painted the picture first and then the poet created a poem to match the picture. The picture is the earliest in the existing publication that portrays the papermaking process, 69 years earlier than the Chinese scientist Song Yingxing's *Tian Gong Kai Wu* (《天工开物》, *Exploitation of the Works of Nature*, 1637).

The Netherlands borders on Germany and has imported paper since the 14th century. The earliest paper document kept in the Hague Archives dates back to 1346. But it was not until 1586 that the first paper mill was built in Dordrecht, south of the famous city Rotterdam, and apparently its technology and equipment were introduced from Germany. The Dutch contribution to papermaking was the invention of the pulping machine called Hollander beater, in 1680. Unlike Germany which has rich water resources, the Netherlands is the state of windmill. The Dutch, finding it difficult for windmills to drive water wheels, tried to develop a device that needed less power and after generations' effort, invented the Hollander beater. This is an ovoid wooden groove (Figure 51), with a rotatable beater wheel of hardwood at the side in the middle of the groove, and 30 iron blades on the wheel. The bladed wheel is called the flying-knife roller.

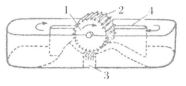

Figure 51　The structure of the Hollander beater invented by the Dutch in 1680, from *Ci Hai* (《辞海》, *Chinese Word Dictionary*, 1979)

1. Flying-knife roller　2. Flying knives
3. Bottom knives　4. Baffle

Between the bottom of the raceway and the roller is a grooved stone/metal slope, called the backfall, with immovable iron blades fixed to it (known as the bottom knives). The bottom knives face the flying knives, but do not tough them. A baffle is installed at the empty place close to the flying-knife roller and facing the center of the raceway, so that the pulp circulates in the raceway. When the flying knife rotates, the raw material will be cut into fibrillated fibers through the mechanical action of the flying knives and the bottom knives. The wet paper stock rolls along the flying-knife roller, goes over the grooved slope, flows to the end of the raceway under the work of gravity, passes the baffle, and returns to the other end of raceway — the cycle is repeated and the raw material is cut up by the knives.

The flying-knife roller can be driven by a Dutch windmill and does not require pretreatment of waste rags. The German chemist Johann Becher (1635 ~1682) in 1682 reported in his book *Närrische Weisheit und weise Narrheit* that he witnessed such pulping beater in the paper mills near Serndamm, Netherlands, and that this stuff might be worth giving further, close attention to[1]. The Hollander beater was later diffused all over the world, and the improved machines were in universal use for more than 300 years.

Switzerland, the majority of its population being German speaking, is also bordered by Germany. Paper mills were founded in Basel in 1433. Like Nuremberg of Germany, Basel became a printing center. Austria, to the south of Germany, set up paper mills in Vienna in 1498. Germany in central Europe, also worked as a medium to spread papermaking technology to Eastern Europe. In 1491, Crakow had the first paper mill in Poland; then Wilno built its paper mill in 1522, and Warsaw, in 1534. Despite its early contact with paper, Russia had its own paper mill as late as 1576 in Moscow, when the Czar Vasil'evich Ivan, Groznyi (1530 ~ 1584) was on the throne. German technicians were invited to help them run paper mills.

Although divided by the sea from continental Europe, the Great Britain has been writing documents on paper since 1309. In 1476, William Coxton (1420~ 1491), an Englishman, having learned printing in Cologne, Germany, printed books on paper made on the European continent, bringing about a strange phenomenon that in England, the paper industry developed after the printing

---

[1] SANDERMANN W. Die Kulturgeschichte des Papiers[M]. Berlin: Springer-Verlag, 1988.

industry. The first paper mill of England was set up by the London cloth dealer John Tate in Herford, north of London. The publisher Wynkyn de Worde bought Tate's paper to print books, and in 1496 he published a book which said that it was printed on Tate's paper which had watermarks with a shape of two circles and an octagonal star, resembling a wheel. There is no evidence that Tate's paper mill existed before 1494, so it was originally built in 1495❶. In 1557, paper mills also emerged in Fen Derford, and by the end of the 17th century, more than 100 paper mills had been built in Britain. For its geographical position, Nordic Europe made paper at a much later time: Sweden had its first paper mill in Klippan in 1573; Denmark, in 1635; Norway, in Oslo in 1690. Till the 17th century, the major European countries all had their own paper industries.

In the 16th-century New World, people wrote on the old writing materials of parchment and bark, and paper must be imported from Europe. Mexico became the first to establish paper mills in America after a large influx of Spaniards. At that time, the Spanish referred to Mexico as New Spain. In January 17, 1580, an article written in Spanish entitled "Relacion del Pueblo de Culhuancán desta Nueva España" mentioned the building of the earliest Mexican paper mill in Culhuancán. The paper mill was built on the land leased to Hernán Sánchez de Muñón and Juan Cornejo according to a royal contract signed on June 8, 1575, allowing them to make paper using the raw materials found in New Spain. The site of their paper mill is now at the foot of Estrella Hill, southeast of Mexico City. Papermaking in Mexico began in 1575.

Before the American Independence in 1776, the first handmade paper mill was built in 1690 by the German settler William Littenhouse (1644～1708) in Germantown near Philadelphia, Pennsylvania. Germantown was a settlement of the German immigrants formed between 1683 and 1684 in northeast Philadelphia. The name is still in use today. The literature which first talked about this paper mill is *A Short Description of Pennsylvania* written by Richard Frame published in 1692. A long poem in the book relates that Germantown, at least 1.6 km long, is home to the German and the Dutch who make linen cloth (because the place is abundant in linen) and make good paper out of waste linen rags, and that the perfect match of weaving and papermaking advantages the

---

❶   HUNTER D. Papermaking: The History and Technique of an Ancient Craft[M]. 2nd ed. New York: Dover, 1978.

construction of a paper mill❶.

In 1710 and 1727, another two paper mills were set up in Pennsylvania. In the 18th Century, paper mills were found in New Jersey, Massachusetts, Maine, Virginia, Connecticut, New York, Maryland, North Carolina, Delaware, Kentucky, etc. The scientist and statesman Benjamin Franklin (1706~1790) propelled Philadelphia to establish itself as the earliest papermaking and printing center in America. The paper which he used to print books in Philadelphia was all ordered and made in the paper mills, watermarked by the crown design and the letters BF (abbreviation of Benjamin Franklin).

Canada, north to America, first imported paper from the United States and Europe. In 1803, the first paper mill of Canada was built in Saint Andreus, Quebec, with the help of a papermaker from Massachusetts, named Newton Lawer Falls. The manual paper mill was run by Walter Ware. The paper was made to publish the *Montreal Gazette*. In 1819, R. A. Holland built the second manual paper mill in a village in the Bedford Basin near Halifax. The first paper mill in Oceania was established in 1868 near Melbourne, Australia.

By the 19th century, Chinese papermaking technique had spread throughout the five continents of the world. Reviewing the history, papermaking initially spread from China to central Asia, west Asia and the Arab world in north Africa, then travelled from the Arab world to Europe, the United States and Oceania, enabling all countries to share the achievements of this invention and promoting the development of human civilization. After all, all these should be attributed to the technology transmission of the Chinese war prisoners in the Talas Battle of 751 A. D.. These papermaking workers of Tang Dynasty in military uniforms remained anonymous, but they made Chinese papermaking technology spread to the western world. They were the unsung heroes of Chinese and Western technological exchanges in history and should always be remembered. What the Europeans first learned was the Tang Dynasty papermaking technology. Later in the Song Dynasty, Chinese papermaking technology met further advances, which were introduced to Europe and America in the 18th century, hence the second phase of the spread of Chinese papermaking technique in Europe and America.

---

❶ JONES H G. Historical Sketch of the Rittenhouse Paper Mill[J]. Pennsylvania Magazine of History and Biography, 1896(3): 315-333.

1.5.3   Introduction of Chinese Papermaking Technology into Europe and America in the 18th Century

According to the book *Baizhitu Yong*（《百职图咏》，*A True Description of All Trades*，1568）published in Frankfurt，the illustrations from the book

Theatrum   Machinarum   Novum (Figure 52) written by Georg Andreae Böckler in 1662 and from the book *Papyrus sive Ars Conficiendae Papyri* written by J. Imberdi in 1693，early European   hemp   papermaking procedure   was   as   follows：first，sorting out the rags，cleaning them and cutting them short；then water was   added   into   ferment   before limewater was admixed to steam and stew；later，the material was put into

Figure 52   Illustration of French Paper
Mill by Sandermann（1988）

a cloth bag to be rinsed in the river，with mortar and pestle used to crush the material. The crushed material then was put into a trough to be mixed into paper pulp. The trough was usually a waist-high oval barrel placed on the ground.

At first，a Chinese-style bamboo screen or a ponytail screen was used，but later，the screen was made of steel wires. The screen was fixed into a frame and it was one person who was responsible for drawing the paper out of the water and placed it on the screen. After draining，the paper was removed onto a thick coarse cotton cloth，waiting for another worker to place another cotton cloth on the paper. With one cloth upon one piece of paper for several layers and using of a press plate，water and moisture were pressed out of the paper. Then，cotton cloth was taken away and the paper was hung on a pole to dry. If the paper was for writing，it should have been placed into a size tub for surface sizing before individual pieces were hung on poles to dry and were calendered by agates and fine stones. Compared with the Chinese technology，due to different physical and humanistic circumstances，the principle and procedure were similar，only with slight differences concerning specific steps or equipment. The Europeans learned from the Arabs about the hemp papermaking technology in the northern areas from the Tang Dynasty in China，but when they started to make paper, China had already entered the Song and Yuan Dynasties when hemp paper was replaced by bast paper and bamboo paper. The Europeans had not learned the more advanced papermaking technology in China after the Song Dynasty and thus failed to be on the same level with the Chinese in making paper.

It is believed that even in the 16th to 17th century, the Europe failed to match the 11th century Song Dynasty in making paper, not to mention the Ming Dynasty of China. Technically speaking, the European paper was mainly made of hemp, so it was thick, with more visible fibers and worse color. It was not smooth to write on and could not be expanded more than an area of 78.7 cm × 134.6 cm. In contrast, the Chinese paper was of a variety of materials, including hemp, bast, bamboo, grass and a mixture of materials. It was thin and smooth to write on, with less fiber and very fine white color. A piece of Chinese paper could be unfolded to as large as 166.4 cm × 332.7 cm, two times that of the European counterparts. Generally, the Chinese method was more reasonable regarding the materials, manufacturing technique and equipment, so it was with a more promising future; yet the Europeans would be in difficulty once the material of hemp was out of supply since they only made paper out of hemp. What was worse, metal screens were quicker in filtering water, so the Europeans could only make thick and small-sized paper.

In Europe, due to the lack of mucilage or viscous additive (纸药), it was difficult for the workers to separate piles of wet paper, so they had to use cotton cloth, which was an extra and redundant procedure. Besides, it was real labor to hang the paper on poles to dry, and then calender them piece by piece. That they did not use the Chinese drying apparatus was also a blunder. The traditional papermaking industry in Europe was too complicated and overloaded in the manufacturing procedure, thus causing great waste in labor, wealth and time. In China, the workers added size in the pulp so that it was not necessary to calender the paper piece by piece. Although the expresser and beater were more advanced in Europe, they could not make up for the other deficiencies of the procedure. During the Renaissance period, Europe outran China in many aspects, yet their papermaking industry was far lagging behind China before the 18th century.

In the 18th century, along with the flourishment of economy, science, literature, education and printing industry, the Europeans were in enormous demand of paper that resulted in enormous demand of rags, causing a material crisis in every paper factory in Europe, further threatening the development of papermaking industry. On one hand, the printing and publishing industry both were affected. On the other hand, the inclination in the European art world was to paint on paper with watercolor, or to print pictures from an engraved plate onto paper, and maps were drawn onto large-size paper. As a consequence, it became a common question of how to get rid of the material crisis, how to reform the extant craft and how to improve the quality of paper. Following the so-called "Sinomania", the Europeans then focused on Chinese papermaking industry to collect relevant information about its technology. At that time, the communication between China and the Western world was rather frequent and convenient, so the Europeans could take the Chinese technology back to their hometown without any intermediary agents.

From the 17th century to the 18th century, some European countries had sent many a missionary to China, and they often mentioned Chinese papermaking industry in their report, especially in the books *Description Geographique*, *Historique*, *Chronologique*, *Politique et Physique de l'Empire de la Chine et de la Tartarie Chinoise* published in the 18th century in Paris (4 vols, 1735), *Lettres Edifiantes et Curieuses Ecrites de Missions Etrangeres par Quelques Missionaires de la Compagnie de Jesus* (34 vols, 1702～1776) and *Memoires Concernant l'Histoire*, *les Sciences*, *les Arts*, *les Moeurs*, *les Usages ets des Chinois*, *par Missionaires de Pekin* (16 vols, 1776～1814). These books are almost the literary treasure recording relevant information from China. For instance, in Volume 2 of the first book, it notes: "According to the book *Zhi Pu* (《纸谱》, *A Chronicle of Paper*, 986) written by Su Yijian (苏易简, 958～996), Shu people made paper from hemp and Emperor Tang Gaozong (唐高宗) ordered the manufacture of quality hemp paper for the recording of confidential information. Fujian people made paper from young bamboo, but people in the North made paper from mulberry bark. Zhejiang people made paper from straw, but people in the South of the Yangtze River made paper from bark. People in the Hubei area made sieve pattern paper and named it Chu paper." Reading this paragraph, it was natural for the Europeans to begin to think of enriching their papermaking materials.

Anne Robert Jacques Turgot (1727～1781), Chancellor of the Exchequer in France (1774 ～ 1776) and economist, was the first European who ever introduced Chinese papermaking technology into Europe when he was the governor for Limoges from 1761 to 1774. In 1765, when the two Beijing youths Gao Leisi (高类思, 1733～1780) and Yang Dewang (杨德望, 1734～1787) went back to China after studying in France, Turgot met with them in Paris and asked them to help solve 52 problems concerning China, some of which were relating to papermaking industry: (1) the materials and skills in making screen (Samples were required for future imitation); (2) the materials for making paper; (3) the method of making large-size paper (264 cm × 396 cm) and how to swing the screen and tear paper off it without tearing it apart; (4) 300~400 samples of bark paper suitable for copperplate etchings (132 cm × 198 cm) required for future imitation.

All the problems and requirements were urgent in Europe, so he wanted the solution badly. In 1766, after the two students went back to China, they bought all the samples, materials and screens, along with instructions, and shipped them to France. The realistic writer Honoré de Balzac (1799~1850) even wrote about the theme of learning papermaking from China in his novel *Les Illusions Perdues* (1843), with David Séchard as the protagonist, learning to use various materials other than rags to make paper and add size into pulp to simplify the procedure. The story took place in the later 18th century to the early 19th century.

In the novel, after reading a Chinese book, David Séchard managed to make paper from grass and reed, successfully adding size into the pulp; later on, unfortunately, he was undermined by an unscrupulous merchant and had to give up his patent, hence the disillusionment. After some survey, the author believes that the Chinese book that had affected David should be *Tian Gong Kai Wu*. French factories employed David's method to "make an inexpensive paper, similar to Chinese paper, which is able to diminish half the size and weight of a traditional book". Balzac wrote: "Ever since the invention of David Séchard, it is like the huge body of French papermaking industry that has been nourished with required nutrient. Thanks to the use of other materials than rags, French paper is less expensive than all the other countries in Europe." Although David was a fictional figure, it must be true that such persons existed in that period in France or Europe.

The American celebrity Franklin, a peer of Turgot, also called for the use of Chinese method of papermaking. On June 20, 1788, Franklin read a paper "Description of the process to be observed in making large sheets of paper in the Chinese manner, with one smooth surface" in the meeting held by American Philosophical Society in Philadelphia, but the paper was not published until the year of 1793. The paper begins by criticizing the European way of making paper—too complicated, too labor-consuming and wasteful of materials and time. Comparatively, the Chinese technology was much simpler and more efficient. Then it proceeds to introduce the Chinese process of making large sheets of paper with smooth surface: pulp was poured into a big sink before it was drawn onto the screen. The screen was hung onto a rope with one end of the rope fixed on the ceiling. This type of hanging screen was very efficient for two persons to swing it. To size the paper, one only needed to put the size into the pulp and the paper would come out very smooth. After the water was filtered from the wet paper, one had to brush the paper onto the wall to dry without sizing them piece by piece.

Franklin wrote: "Thus the Chinese made paper with far fewer operations." He noted that the Chinese paper was 514 cm × 171 cm. Although the details were not exactly the same, the principles and skills were basically correct, so his article actually could expand the conception of the paper makers both in America and Europe. Franklin aimed to help the Europeans "make large sheets of paper in the Chinese manner with one smooth surface" with his article, which explained his vision for that matter.

At the end of the 18th century, Michel Benoist (1715~1774), a French Jesuit in Beijing, sent Chinese fine brushwork paintings in series on bamboo papermaking procedures to Paris. Benoist was the Latin teacher of Gao Leisi and Yang Dewang, the two overseas Chinese students in France, and he participated in designing the Old Summer Palace, thus was appreciated by Emperor Qianlong (乾隆). The paintings that he had sent back to Paris consist of 24 pieces. Being of palace style, they are valuable both artistically and technically. Several copies of the paintings are now collected in the National Library in Paris, Library of France Studies and Buchmuseum Leipzig, etc. The book *Arts, Metieres et Culture de la Chine* published in 1815 in Paris made public 13 paintings, and the compiler noted that the paintings were painted on the request of some Jesuits in China and were then sent to Paris to be made in copperplate. The article written by Franklin quoted some of the paintings, and one can imagine the huge influence of them in Europe.

In 1952, Adolf Benedello (1886~1964) had published white-and-black photos of the whole set of the paintings in his book *Chinesische Papiermacherei*. The author began studying the 18th century copy of the paintings in Buchmuseum Leipzig. The setting discloses the whole procedure of making bamboo paper in China, including all the materials, tools and steps concerned, especially the form and use of the bamboo screen, the drying of the wet paper and the use of mucilage, which were all very fresh knowledge to the Europeans in that period. One can imagine how they had learned from these paintings and instructions to improve their papermaking technology and to manage to make other types of paper than merely hemp paper. In fact, David Séchard from Balzac's novel reflected the real situation in the then Europe of how papermakers were trying their best to renovate the industry under the Chinese influence.

1.5.4 Influence of China's Papermaking Technique on Europe in the 19th Century

Since the papermaking technique has been introduced from China in the 18th century, Europeans have shown growing interest after that. An obvious example was the translation of the Chinese work on papermaking technique into European languages for the first time, which enabled the local papermakers and technologists to have a direct conversation with the ancient Chinese people. The Chinese work on papermaking technique here refers to *Tian Gong Kai Wu* by Song Yingxing, a scientist and thinker of the Ming Dynasty. This illustrated encyclopedia of technology was available in France since the 18th century and was kept in the royal library of Paris (Bibliotheque Royale), now the National Library of France. In 1840, Professor Stanislas Julien (1799 ~ 1873), a sinologist from the College of France (College de France), translated some parts concerning bamboo-papermaking in the chapter "The Ending" of the book into French. The article entitled *"Description of the Chinese Procedures of Paper Fabrication"* (*Description des procedes chinois pour la fabrication du papier*) was published on pages 697 to 703 in Volume 10 of *Proceedings of the National Academy of Sciences*, the French scientific journal of top level.

After comparing the French version with the Chinese source text, it was apparent that the author conveyed the original meaning faithfully. He translated it not for historical research, but for practical purpose of making ancient things serve the present-day needs. Therefore, instead of sinology publication like *Journal of Asia* (*Journal Asiatique*), it was originally published in a scientific journal, which might arouse wide attention among the scientific and technological circle. Balzac, who has failed in running a paper mill and later devoted himself to literary creation, also read the article. It means that the article was widely publicized. In 1846, Julien changed the title of the translated version into *"Fabrication of Bamboo Paper"* (*Fabrication du papier de bamboo*), which was reprinted on *Review of the East and Algeria* (*Revue de l'Orient et de l'Algerie*) in Paris❶. This version has also been included in Volume 2 of *Contemporary China*, or *A Historical, Geographical and Academic Overview of the Great Empire Based on China's Documents* compiled by Antoine Pierre Louis Bazin (1799~1863).

Since most educated Europeans of the 19th century were able to read French, it was possible for them to get useful information directly from the translated version by Julien. It was believed that *Tian Gong Kai Wu* in China could provide the following technical information to the westerners: (1) in addition to rags, other raw materials like broussonetia papyrifera, mulberry, lotus bark, bamboo and straw could also be used in papermaking, and the scrap paper could also be recycled; (2) paper pulp could be made from mixture of different raw materials, such as 60% mulberry bark and 40% bamboo or 70% bark and 30% bamboo and straw; (3) technology and equipment for making bamboo paper and bark paper, especially the shape, preparation and use of the flectional bamboo screen device; (4) a paper drying device with a smooth surface; (5) "paper medicine" made by adding the Actinidia chinensis plant slime into the paper pulp; (6) an operating chart of making bamboo paper.

According to the research, the author found that copper wire was used to make paper in the reign of Emperor Kangxi (1662~1722) of the Qing Dynasty in China. After a significant transformation, Chinese people developed "tube-shaped paper with oblique screen marks", a cylinder more than one *zhang* (330 cm) long. The production of this must be based on the following 3 techniques: (1) the design of a cylindrical papermaking device made of copper wire; (2) the production of paper by rotating the cylindrical device; (3) squeezing out the water by pressing two rotating cylinders in the opposite direction. In 1782, it was imitated again in Zhejiang Province. In the 18th century, the Chinese people took the lead in proposing the new technique, later called "revolving endless wire cloth" by the westerners, and put it into practice. The structure of two paper machines in the 18th to 19th century, a symbol of modern papermaking revolution in the world, shared the similar principles with the above ideas, but were a century later than China❷. This has attracted the attention of the western Jesuit missionaries and businessmen who stayed in China. Meanwhile,

---

❶ JULIEN S. Fabrication du papier de bambou[M]. Paris: Editions Belin, 1846.
❷ 潘吉星. 从圆筒侧理纸的制造到圆网造纸机的发明[J]. 文物,1994(7):91-93.

China's new techniques such as dividing the large screen into several sections with non-filtration material and making several paper sheets at a time as well as brushing the paper with a mixture of the white mineral powder and glue as a surface coating on the paper were introduced to Europe.

To sum up, the technologies and technical ideas about papermaking introduced by Europeans from China till the first half of the 19th century included at least the following 10 aspects: (1) diversification of raw materials for papermaking; (2) technologies of making bark paper, bamboo paper and straw paper; (3) production of paper pulp by mixed raw materials; (4) production of large sheets of paper with large, flectional bamboo screen; (5) usage of plant slime; (6) technique of adding the glue into the paper pulp; (7) technique of drying paper with a smooth paper drying device; (8) technique of making paper with rotating cylindrical copper wire and squeezing out water by rotating circular rollers; (9) technique of dividing the large screen into several sections with non-filtration material and making several paper sheets at a time; (10) technique of brushing the paper with a surface coating. Europe was short of the above 10 Chinese techniques before the 17th century, so the introduction of these techniques certainly would bring innovations in European traditional papermaking industry in terms of raw materials, manufacturing process and basic equipment, and adjust the route of technology development. Countries such as France, Germany and Britain were ahead of other European countries in this aspect.

Westerners have done a lot of experiments on the diversification of raw materials. France and the United States have introduced broussonetia papyrifera from China in the hope of making mulberry paper but failed because of different climate and soil conditions. French scientist Jean Etienne Guettard (1715~1786) and German botanist Jakob Christian Schsffer (1718~1790) and other experts have carried out similar experiments with David, the protagonist of the novel written by Balzac. Guettard published some articles such as *Observations About Different Materials Used for Papermaking*, *Researches About Materials Which May be Used for Papermaking* and so on. Schaffer, who has visited Asia, published six volumes of *Experiment and Samples About Making Paper with Rags and a Few Additive Rather than Only Rags*[1] from 1765 to 1772

---

[1] GUETTARD J E. Observations sur différentes matières dont on fabrique le papier[M]//Recherches sur les matières qui peuvent servir faire du papier. Paris: Berger-Levrault, 1768:45-52.

(Figure 53). Volume 2 (1765) introduced paper pulp made from rags and mixtures like cannabis sativa, broussonetia papyrifera bark and straw used in China and a sample was attached to the book❶. In 1800, Matthias Koops managed to make paper with wood and straw in London and printed his work on straw paper❷, and reprinted it with recycled paper in the following year. In 1856, the British man Charles Thomas Davis listed 950 raw materials of papermaking in *The Manufacture of Paper*. During 1857 to 1860, another British man Thomas Routledge succeeded in making paper with wide Stip tenacissms in the Gramineae group from Spain and north Africa and printing *Illustrated London News* on the paper. France obtained this grass from French Algeria for papermaking.

Figure 53 Title Page of Volume One, Monograph on Papermaking Raw Material by German Schaffer

In 1875, Routledge made bamboo paper from bamboo for printing *Bamboo as a Papermaking Material*❸. Britain imported bamboo from India for papermaking. In 1876, *Bamboo and Bagasse as Foundations for Paper Preparation*（*Bamboe en Ampas als Grondstaffen voor Papierbereiding*）published in Arnhem City of Holland was also printed on bamboo paper. The above examples showed that the Europeans have drawn on Chinese experience and succeeded in making paper with raw materials in addition to rags and mixed materials, which eased the raw material crisis since the 18th century. Therefore, China has made a great contribution to modern papermaking industry in Europe. Since the middle of the 18th century, Europe has introduced flectional bamboo screen as a replacement of traditional fixed mould. Named as "budge

---

❶ SCHAFFER J C. Versuche und Muster ohne alle Lumpen oder doch mit einem geringen Zasatze derselben Papier zu machen[M]. Regensburg, 1765.

❷ KOOPS M. Historical Account of the Substances which Have Been Used to Describe Events and to Convey Ideas from the Earliest Date to the Invention of Paper[M]. London: T. Burton, 1800.

❸ ROUTLEDGE T. Bamboo as a Papermaking Material[M]. London, 1875.

type screen (type de velin)" by French people, this type of Chinese papermaking device was composed of bamboo screen and screen frame, which was movable and could be torn open. In 1826, James Whatman, the owner of a paper mill in Kent shire of England, for the first time produced eight sheets of paper for writing letters with Chinese techniques.

The flectional Chinese bamboo screen was a symbol of the advanced papermaking ideas, which served as a necessary step to the modern paper machine. American paper historian Dard Hunter (1883~1966) once said: "The great paper industry of today is built upon the original Oriental (China) bamboo mould which came into being almost two thousand years ago."❶ This statement was logical, but it was necessary to explain the implication. As mentioned above, European traditional papermaking technology has exposed such shortcomings as trivial procedures, old equipment and being unable to produce large smooth sheets till the 18th century. After the problem of raw material was solved, the next step was to simplify the operating procedures and reform the bamboo mould. The way out was to realize mechanization in papermaking process.

Nicolas-Louis Robert (1765~1828) from France first managed to realize mechanization of papermaking. In 1797, he succeeded in producing two large paper sheets (12 m to 15 m long) with the help of machine and related in the application for a patent for invention in 1798 that, "it has been my dream to simplify the operation of making paper by forming it with infinite less expense, and above all, in making sheets of an extraordinary length ... using only mechanical means." Robert's description in part read: "At the end of the cloth wire extending on the vat there is a fly-wheel, or cylinder, fitted with little buckets which plunge into the paper stock, or liquid pulp. This cylinder, by its rapid movement, raises the material and throws it into a shallow reservoir in the interior of the head, which recovers it, and thus pours, without interruption, like a sheet of water upon the endless wire cloth. As the material settles on the cloth, it receives a side-to-side movement, the wire retaining the fibers and the water draining into the vat beneath. A crank turns the machine and causes the wire cloth to advance, the sheet of newly formed paper finally running under a felt-covered roller. When the paper leaves the first felt roller it is no longer

---

❶ HUNTER D. Papermaking: The History and Technique of an Ancient Craft[M]. 2nd ed. New York: Dover, 1978.

saturated with water, but can be removed from the machine, just as a sheet of handmade paper is taken from the felting after pressing in a press." This was the first long wire (long screen) paper machine in modern Europe, which was quite simple in construction.

Due to the social unrest resulting from the Great French Revolution, paper machine has made little progress. This patent was later sold to a British factory owner and finally to a London businessman Henry Fourdrinier (1766~1854) who invested in production. A mechanic named Bryan Donkin (1768~1855) managed to produce the machine and apply for the patent in 1803. It was improved in 1804 and called Fourdrinier. That was how the French invention turned into the British patent. Later, the bamboo screen was replaced by the copper wire, and other components like couch roller, compression roller, steam drying roller and paper roller were added. The large and complicated machine could realize continuous mechanical operation from paper pulp to paper products. Up to the middle of the 19th century, Europe has achieved the mechanical production of paper and realized the modernization of papermaking technology. After the long curtain machine was put into production, another type of machine, the cylinder, was originated and perfected in England and became operative in 1809 in the mill of John Dickinson (1782~1869). This machine differed somewhat from the Robert machine in the method of forming the web of paper. "A cylinder covered with a woven metal wire screen, and half immersed, revolves in a vat of pulp, and by means of a vacuum within the cylinder, the pulp is made to adhere to the screen on the periphery of the cylinder, thereby forming the paper, which is then detached and passed on to a cylinder with felting." This mono-cylinder paper machine was improved later. These two types of paper machines can make paper of infinite length.

The production of long wire machine and cylinder machine completely solved the technical transformation problem of traditional European technology. The invention of the chemical pulp making method, which used wood as the raw material, also made a vital role in the development of paper machine. The reasons why European papermaking could realize mechanization and modernization were the diversification of raw materials and the invention of papermaking machine, which ultimately depended on the following 6 principles: (1) diversity of raw materials; (2) making large paper with flectional screen; (3) making paper by rotating cylindrical screen in the paper pulp; (4) squeezing water out of the wet paper with rotating circular rollers;

(5) drying the wet paper with heat producer; (6) replacing handmade paper with machine production. Steps (1) to (6) had a direct relation with the invention of paper machine, and steps (2) to (5) laid a solid foundation for designing the crucial components of paper machine. For example, the flectional bamboo screen of long wire machine was made into a ring, and the key of cylinder machine was cylindrical mould. The idea of drying the wet paper with heat producer was fundamental to the design of drying roller. The application of rotating circular roller was the precondition of installing the couch roller, compression roller and paper roller.

However, flectional curtain, cylindrical mould, rotating circular roller and drying device have already been used by Chinese papermakers long ahead of Europeans, and relevant information or materials has been introduced into Europe in 1797 when the first paper machine appeared in Europe. After a detailed examination into the techniques of traditional European papermaking in the 18th century, we noticed that all the techniques have not developed into modern technologies except the pulping machine in the 17th century. In contrast, more than half of the 10 technical elements relating to modern technologies originated from China and introduced to Europe, which directly transformed into the components of modern technology. This is another great contribution China made to the development of modern papermaking in Europe. The replacement of handmade paper by machine production was the only progress happening in Europe. In the 18th and 19th centuries, the introduction of new Chinese papermaking techniques and technical ideas played an important role in the transition from manual production to machine production in Europe. European paper machines would not appear without the techniques and technical ideas from China[1].

---

❶ 潘吉星. 从造纸史看传统文化与近代化的接轨[J]. 传统文化与现代化,1996 (1):74-83.

Chapter 2　Invention, Development and Spread
of China's Printing Technique

2.1　Invention of Woodblock Printing
and Copper Plate Printing

2.1.1　Woodblock Printing Originated in the Sui Dynasty

2.1.1.1　Origin Time of Printing: Upper and Lower Limits

In the previous investigation of printing origin, people usually made a point according to one or two sentences in ancient books rather than taking into consideration several factors like archaeological excavations, social economy, cultural background and development law of printing technique. Due to different understandings of thoughts expressed in ancient books, once there was no agreement among many views. Over the past 50 years, with the deepening of research and the accumulation of information, it was time to end this situation. First of all, woodblock printing did not appear overnight. It should be considered as the product of people's exploration into a new type of replication technique as a replacement of handwritten work, or the process of evolution from classical to mechanical replication technique. After reaching a certain stage, the printing technique appeared naturally. Therefore, it was improper to set the origin time of printing in a particular year. Instead, it would be better to delimit an appropriate period of time, find out the upper limit and the lower limit, and then identify the origin time according to documentary records, unearthed objects and technical reasoning. It might be more reliable to study the origin of printing than the previous practice.

From a technical point of view, the upper limit of origin time of printing could be the Northern and Southern Dynasties (the fifth to sixth centuries), because at that moment the papermaking and ink-making techniques were mature enough to provide paper and ink which could meet the needs of printing. At the same time, the technique transiting from seal and stone inscription rubbing to printing was ready to mature. As a stimulation to printing, Buddhism has developed greatly in the Northern and Southern Dynasties. History has suggested that printing technique was originally invented by the general public and had close relationship with religious beliefs of the Buddhist believers, so most of the early printed matters were Buddhist sutras, charms and figures of Buddha. Almost all Buddhist scriptures conveyed the instructions of the Buddha: reciting, transcribing, and supporting sutras and charms could set good roots, eliminate the evil fortune and summon good luck, for fear that they might suffer in the hell after death. People were much tired of transcribing sutras and charms. In order to express their faithfulness, many illiterate people had to pay money by inviting classics scholars to write on their behalf, with tens of thousands of Buddhist scriptures found in the Dunhuang Caves as strong evidence. Reduplicating the scriptures and figures with the help of block printing technique can provide a lot of cheap printed copies, on which the believers just had to fill in their names or their vows and in this way they could be blessed by the Bodhisattva, so they were willing to do that. This was how printing became popular among the general public. Many predecessors in China and abroad prefer the view that the printing technique originated during this period of time. Although the historical materials can be further verified, the readers had better not easily deny the conclusion since the possibility of origin time before the Northern and Southern Dynasties was very slim. However, more direct evidence of printing activities in the Northern and Southern Dynasties need to be collected, so it is not yet possible to make a conclusion at present, which remains to be proven by archaeological discoveries in the future. The practical approach to study the origin time was to start at the lower limit of time, since there has already been a set of evidences. These evidences show that the 50 years between the Sui and the early Tang Dynasties or the turn of the centuries (590~640) were the critical period leading to the emergence of early printing.

2.1.1.2　Records Related to Printing in the Sui Dynasty

Let's start with the documentary records. As early as the Ming Dynasty, it was suggested that woodblock printing began in the Sui Dynasty (581~618). Lu Shen (1477~1544) wrote in *Hefen Yanxian Lu* (《河汾燕闲录》, *Yen Hsien Lu Fol*):

> *Under the Emperor Wen of the Sui Dynasty, in the thirteenth year of Kaihuang, the eighth day of the 12th month (January 5, 594), on orders from the Emperor, all neglected hsiang (the word means*

either images or pictures) and scattered ching (classic texts or sutras) were carved and composed. This is the beginning of the printing of books. It was thus earlier than Feng Yingwang (i. e., Feng Dao, 882～954).

Hu Yinglin (1551～1602), textual bibliographer of the Ming Dynasty, in *Shaoshi Shanfang Bicong* (《少室山房笔丛》, *Shaoshi Shanfang Pits'ung*, about 1598), included in *Jiabu · Jingji Huitong Si* (《甲部·经籍会通四》, the Forth Volume of *Confucian Classics Research in Jiabu*), also believed that:

> *As quoted from Lu Shen's statement in his book Hefen Yanxian Lu, under the Emperor Wen of the Sui Dynasty, in the 13th year of Kaihuang, the eighth day of the 12th month, on orders from the Emperor, all neglected hsiang and scattered ching were carved and composed. This is the beginning of the printing of books. Judging from this, the printing originated in the Sui, earlier than Liu Pien (848～898) as well as Feng Dao and Wu Chao-i (about 902～967). In my opinion, what was carved in the Sui Dynasty were mainly sculptures and statues of Stupa instead of other books, because many people believed in Buddhism in the Six Dynasties. After the middle of the Tang Dynasty, all kinds of books were gradually printed by the similar method. The printing developed in the Five Dynasties, gained wide popularity in the Song Dynasty and reached the peak at present (in the Ming Dynasty) ... In a word, block printing had its birth at the beginning of the Sui Dynasty. It expanded greatly in the Tang Dynasty, took a lead forward in the Five Dynasties and finally came to its fullest development in the Song Dynasty. This is my summary based on previous scholars' research, which should be convincing.*

In fact, Lu and Hu put forward their views on the basis of the Sui scholar Fei Changfang (557～610)'s *Lidai Sanbao Ji* (《历代三宝记》, *Three Treasures of the Past Dynasties*), Volume 12:

> *On the eighth of December, the Emperor of Sui, a disciple of Buddhism, named Yang Jian (541～604) respectfully indicated that, the Northern Zhou Dynasty (557～581) brought disasters and chaos by insulting the holy relics. Pagodas and temples were destroyed and abandoned. Confucian classics and Buddha statues disappeared. As*

the official, I am anxious to save the general public by rebuilding the Buddha statue to show respect for the Buddha. The destroyed temples should be restored. The abandoned Buddhism scriptures and statues should be recarved and rebuilt. Special rooms are set for believers to worship the scriptures and statues. About 100,000 people would bathe the Buddha statues with special water containing fragrant spice everyday. ❶

"Disasters and chaos of the Northern Zhou Dynasty" in original records by Fei Changfang, referred to the accident which happened in the year of 575 A.D., Emperor Wu ordered to ban the Buddhism and Taoism, destroy temples, scriptures and statues, force Sramana and priests to resume secular life. Many religious relics and publications were wiped out. Later, Yang Jian, a devout Buddhism believer, overthrew the regime of the Northern Zhou, founded the Sui Dynasty and united China. On January 5, 594, after a recovery of national economy and further development of social culture and religion, he respectfully made a wish in front of Buddha to revive the Buddhism by recovering all the destroyed pagodas and temples, damaged Buddhist scriptures and statues. A public commitment was made that the abandoned Buddhism scriptures and statues should be recarved and rebuilt. The Emperor and empress also did charity work like offering and donation. When the casting of copper statue was finished, he called in the general public to bathe the Buddha statue with special water containing fragrant spice. Therefore, *hsing* here mainly referred to copper statue of Buddha, and *ching* mainly referred to Buddhism scriptures. "Recarved" meant "engraved" and "rebuilt" meant "cast"❷. Undoubtedly, "recarved and rebuilt" here should be explained as "recarved and recast". It was improper to interpret "rebuilt" in this context as "written". In fact, "the abandoned Buddhism scriptures and statues should be recarved and rebuilt" was "the abandoned Buddhism scriptures and statues should be recarved and recast". "Casting" referred to casting the Buddha statue as well as recarving the block to print the Buddhist scriptures.

In the writing of Fei Changfang, "recarved" and "rebuilt" were two transitive verbs whose objects were Buddha statue and Buddhist scripture. In

---

❶ 费长房.历代三宝记[M]//高楠顺次郎.大正新修大藏:第 49 册.东京:大正一切刊行会,1924.

❷ 张玉书.康熙字典[M].北京:中华书局,1958.

terms of Buddha statue, "rebuilt" meant to show its original appearance by recasting. As for Buddhist scripture, "recarved" meant to engrave the block and print for holy purpose. Therefore, statements of the Ming Dynasty scholars Lu Shen and Hu Yinglin were believable that sentences of *Lidai Sanbao Ji* could be regarded as literary records of block printing in the Sui Dynasty and it was inappropriate to judge it by its brief words. This record should be analyzed in social background at that time, and the connection between the printing document and the printed matter which appeared later in the early Tang Dynasty should also be taken into account. As mentioned above, both technical and material preparations for printing have been made in the Northern and Southern Dynasties, so it was better prepared in the Sui Dynasty. During the reign of Emperor Wen when the country has reached the stage of the national reunification, wealth accumulation and social stability, the ruling class showed vigorous support for Buddhism. It was possible to adopt the most convenient method by carving the block and printing in order to spread the previously damaged Buddhist scriptures as soon as possible.

However, the Qing Dynasty scholar Wang Shizhen (1634 ～ 1711) mentioned Lu Shen's view in Volume 25 of his works *Juyi Lu* (《居易录》, *Record of Juyi*): "I am familiar with his writing, because what was recarved was Buddha statue and what was rebuilt was the Buddhist scriptures. Lu Shen made mistakes by reading the original words together." Some modern scholars also pointed that statue and scriptures were different things and suspected that the block printing of Buddhist scriptures was done in the Sui Dynasty[1]. It was unconvincing to understand the original words of *Lidai Sanbao Ji* in this way and it might bring new misapprehension. If, as the sceptic put it, "recarved" specifically meant to recarve the Buddha statue, the problem arose. Because Zan Ning (919～1001), in *Seng Shi Lüe* (《僧史略》, *The Monk's Brief History*), explained the bath for Buddha, most for copper statues, which could only be done by casting instead of carving. In "to bathe the Buddha statues with special water containing fragrant spice" in *Lidai Sanbao Ji*, the copper Buddha statues could not have been made by carving. "What was recarved was Buddha statue" was incorrect because it did not conform to the original meaning. The readers had better have an overall understanding of the original text of Fei Changfang instead of taking the statement out of the context.

---

[1]   张秀民.中国印刷术的发明及其影响[M].北京:人民出版社,1958.

Another historical material of the Sui Dynasty was also noteworthy. As recorded in Volume 78 of *Sui Shu* (《隋书》, *the History of the Sui Dynasty*), Lu Taiyi (548~618), with a style name Xiezhao, a native of Hejian (now Hejian in Hebei Province), has read many books and known well about Buddhism and Taoism, thus winning recognition of Emperor Wen of the Sui Dynasty, "afterwards, he was blind but he could recognize the word by touching the book". He arrived at the east Liao area with Emperor Yang in the year of 613 A. D. and died at Luoyang several years later[1]. Wang Renjun (1866~1914) in his *Gezhi Jinghua Lu* (《格致精华录》, *Record of Essence of Gezhi*), explained it like this: "He could recognize the word by touching the book, so what he touched was the block printing ... It is clear that there was plate of book at that time, so we can see that he touched the plate"[2]. On the plate, the word was in reverse direction. By touching the words in reverse direction, he could guess the standard form, which proved that Lu was smart. Some people said that he touched inscriptions on a stone tablet, and it was improper to deny this historical material. The historical material said clearly that Lu Taiyi touched the book, but the stone tablet could not be called a book, and most inscriptions on it were standard words, which could not reflect the special skill of Lu. There was a reason to interpret the book touched by Lu as a plate. That was how printing appeared in the Sui Dynasty.

2.1.2 Literary Records and Material Evidence Concerning Printing in the Early Tang Dynasty

Since the early Tang Dynasty, there were a large number of printing records. According to Volume 10 of *Daci 'ensi Sanzangfashi Zhuan* (《大慈恩寺三藏法师传》, *Biography of Master Xuanzang of the Great Temple of Pity*, 688) by Xuanzang's disciple Yanzong (625~690), due to Emperor Taizong's deep respect for Xuanzang, Emperor Gaozong, after inheriting the throne, also thought highly of Xuanzang and sent imperial envoys to greet him with lots of silk and cassocks. Then Xuanzang distributed them to the poor and foreign Buddhists. During the years between 658 and 663 in his old age, Xuanzang

---

[1] 魏徵.隋书:卢太翼传[M].上海:上海古籍出版社,1986.

[2] 王仁俊.格致精华录[M].上海石印本,1896.

"vowed to make ten koti figures and finally realized his goal"❶. "Koti" in Sanskrit means a hundred thousand or ten million. Judging from the technical point of view, "koti" here means one hundred thousand and "ten koti" refers to one million. The word "make" in the language of the Tang Dynasty means "print". For example, it says in the preface of *Jingang Jing* (《金刚经》, *the Diamond Sutra*) in 868: "Wang Chieh had ordered the printing to honor his parents". Therefore, according to Yanzong's records, during the year 658 to 663, Xuanzang vowed to make millions of single sheet Buddha figures and printed them.

As recorded in Volume 5 of *Yunxian San Lu* (《云仙散录》, *Scattered Remains of Clouded Immortals*, 926) by Feng Zhi: "Every year, Master Xuanzang printed five packs of single sheets containing the portrait of Bodhisattva Samantabhadra with the huifeng paper and distributed them to Buddhism believers and there was no sheet left. "❷ Feng related the same thing with Yan by further pointing out that the Buddha figure printed by Xuanzang was the portrait of Bodhisattva Samantabhadra, the right guarder of the Buddha in charge of morality. Together with the left guarder in charge of wisdom — Manjusri, Samantabhadra was one of the four great Bodhisattvas of Buddhism in China. "Buddhism believers" or the Fourfold Assembly included millions of monks and nuns as well as faithful men and women, who had received the portrait of Bodhisattva Samantabhadra donated by Xuanzang. The records of Yanzong and Feng Zhi could complement and verify each other. In the past, it was suspected that *Scattered Remains of Clouded Immortals* was a counterfeit by Wang Zhi (1090~1161) in the Northern Song Dynasty. But no strong counter-evidence was provided. Besides, the carving copy of this book by Guo Yingxiang in the year of 1205 A. D. was found and the author's preface written in the year of 926 A. D. was carved in the preface. Earlier than that, this book was cited by Kong Chuan, a descendant of Confucius, in his works *Kongshi Liu Tie* (《孔氏六帖》, *Six Posts of Kong Family*, 1161), which also recorded the event of printing the Buddhism figures by Xuanzang, thus being credible.

Wu Zetian (624~705), the ruler of the early Tang Dynasty, sincerely

❶ 慧立,彦悰. 大慈恩寺三藏法师传[M]//高楠顺次郎. 大正新修大藏: 第 50 册. 东京: 大正一切刊行会,1927.

❷ 冯贽. 云散录: 卷五[M]. 台北: 商务印书馆,1983.

believed in the Buddhism. During her reign between 689 and 704, she published some Buddhist sutras and also printed documents with printing techniques. According to *Datang Xin Yu* (《大唐新语》, *Miscellanies and Anecdotes of the Great Tang Empire*, 807) by Liu Su (770~830), at the early days of her reign, in September of the year 691 A. D., Zhang Jiafu, the official responsible for drafting imperial edicts, instigated Wang Qingzhi, a native of Luoyang and others to submit a petition in the request of replacing the Empress's nephew Wu Chengsi for her son Li Dan as the crown prince, thus turning the regime of Li into that of Wu. But it was rejected by Empress Wu. Then Wang Qingzhi "fell to the ground and pled with death. So the Empress had to send him back by giving him 'nei yin yin zhi' (paper sheet printed by the imperial palace) and said: show this to the doorkeeper then you can come in. Qingzhi took the printed paper to meet Empress Wu at the court for several times, thus irritating the Empress. "[1] Volume 204 of *Zizhi Tongjian* (《资治通鉴》, *History as a Mirror*, 1084) titled *Twenty Records in the Tang Dynasty* also recorded the same story on the basis of the memoir in the palace[2]. The so-called "paper sheet printed by the imperial palace" here refers to a special pass made of paper for entering the court. In the Tang Dynasty, "printed paper" refers to printed matter with a specific purpose, like tax receipts used in the year of 783 A.D. as mentioned in *Jiu Tang Shu · Shi Huo Zhi* (《旧唐书·食货志》, *The Old Book of the Tang Dynasty · Farming, Cloth and Currency*), which showed that the system of printing paper in the Tang Dynasty was still implemented in the early Song Dynasty.

In the Wuzhou Dynasty, Empress Wu ordered Fa Zang (Dharmakara), the founder of theoretical system of Hua-Yen Buddhism, to preach inside and outside the palace in Luoyang the 60 volumes of *Dafangguangfo Huayan Jing* (《大方广佛华严经》, *Buddha-vatam-saka-maha-vaipulya-sūtra*) translated in the year of 421 A. D.. At that time, there were different opinions among several sects on the formation mechanism of the content of the "eight meetings", namely, after the Buddha achieved God-realization, he expounded Buddhist doctrines to the disciples. Tiantai Sect believed that the first seven of "eight meetings" were doctrines made on the first three "seven days" after the Buddha achieved God-realization. What were expounded on the eighth meeting were

---

[1] 刘肃. 大唐新语：笔记小说大观本[M]. 扬州：广陵古籍刻印社，1983.

[2] 司马光. 资治通鉴：唐纪二十[M]. 上海：上海古籍出版社，1987.

the doctrines afterwards. Fa Zang disagreed and held that the Buddha didn't expound any doctrine in the first seven days, but finished it in the second seven days. To prove his understanding, he offered a variety of evidence and also made a simile with printing to illustrate his point of view. He wrote in his *Huayan Wujiao Zhang* (《华严五教章》, *General Introduction to Hua-Yen Teachings*) around the year of 677 A.D.:

> *Therefore, all the Buddhist doctrines were expounded at the same time on the second seven days. It is just like printing in the world. We read the text sentence by sentence, but when copied, the sentences are printed at the same time. "At the same time" is not contradicting with "before and after", so it is the same thing with doctrine.* ❶

In his *Huayan Jing Tanxuan Ji* (《华严经探玄记》, *Secrets About Hua-Yen Sutra*) during the years of 687 A.D. to 692 A.D., Fa Zang mentioned this again and restated that ... the doctrines were expounded at the same time. If so, why would it be before and after? The answer is like this: in reading the text, we read one sentence after another, but in the printing, the sentences all appear in the meantime. ❷

These two important historical materials were first discovered by the Japanese printing historian Dr. Kanda Kiichiro. He also examined the exact time of the two books❸. Obviously, in the view of Fa Zang, various scriptures of *Huayan Jing* (《华严经》, *Hua-Yen Sutra*) was arranged in sequence, but all the doctrines were understood by the Buddha at the same time after he achieved the God-realization, and expounded on the second seven days. As in the printed book, we read the sentences in sequence, but during such procedures as the inking and printing plate, the sentences appeared on the same sheet of paper. Therefore, there was no contradiction but a dialectical relationship between "before and after" and "at the same time". The Buddhist monk Fa Zang made a comparison of printing to illustrate the doctrines of *Huayan Jing* (《华严经》,

---

❶ 法藏.华严经探玄记:卷二[M]//高楠顺次郎.大正新修大藏:第 35 册.东京:大正一切刊行会,1926.

❷ 法藏.华严五教章:卷一[M]//高楠顺次郎.大正新修大藏:第 42 册.东京:大正一切刊行会,1926.

❸ 神田喜一郎.中国における印刷术の起源について[J].日本士院纪要,1981,34(2):89-102.

Hua-Yen Sutra or *Avatamsaka Sutra*) in the early Tang Dynasty，which indicated that the woodblock printing technique was quite popular in China in the seventh century.

The above-mentioned documentary records were also verified by archaeological findings. In the year of 1906 A. D. ，several remaining volumes of *Miaofa Lianhua Jing* (《妙法莲华经》, *Lotus Sutra* or *Saddharma Pundarik Sūtra*) printed in the Tang were unearthed at Turpan，Xinjiang，including Volume 5 in scroll binding. Extant words were arranged in 194 lines，19 words in each line. The plate was 13 cm high. The words were about 5 mm to 7 mm in diameter，thus called the small letter edition. They were printed on yellow hemp paper and written in standard style of handwriting. There was no publication record and exact year，but the scroll was found in Gaochang area，which was isolated from Central China after the tenth century of the Five Dynasties. Therefore，relics unearthed in Gaochang were remains in or before the Tang Dynasty. This copy of *Lotus Sutra* was sure to be the edition printed in the Tang Dynasty.

At first，this unearthed copy was privately collected by Wang Shunan (1851 ～1936)，Governor of Xinjiang in the Qing Dynasty for a short period of time. Later，it was kept by Eto Toyu who came to Xinjiang in search for antiques，and finally bought at a high price by Tokyo painter and art collector Nakamura Fuori (1868～1943)，then preserved in the Museum of Calligraphy founded by him in Eastern District of Tokyo. It was reported that Nakamura had photocopied the sutra in 1936，but it had been judged as a Sui Dynasty copy by mistake[1]，because he did not notice the words made in the Wuzhou Dynasty on the sutra. These special words included 18 characters created and issued by the Empress Wu Zetian during 689 to 698 in the reign from 689 to 704，such as "heaven" "earth" "day" "month" "year"，etc. After her son Li Xian ascended to the throne，he banned these words and other regulations made by Empress Wu in the year of 705 A. D.. Therefore，in writings and printings of the Tang Dynasty，these words made in the Wuzhou were one of the distinctive symbols to judge whether the product belonged to this Dynasty. In 1952，textual bibliographer Dr. Nakasawa Kikuya (1902～1980) made a careful study into this copy of *Miaofa Lianhua Jing* (《妙法莲华经》, *Lotus Sutra* or *Saddharma Pundarik Sūtra*) and wrote：

---

❶　秃氏祐祥. 东洋印刷研究[M]. 东京：青裳堂书店,1981.

Therefore, *one volume of* Miaofa Lianhua Jing (《妙法莲华经》, Lotus Sutra or Saddharma Pundarik Sūtra) *collected by Nakamura was claimed to be the archaeological discovery of Turpan. It was printed on antique yellow hemp paper*, *19 letters in each line*, *4 inches* (13 cm) *in height. The words on the text were quite small*, *mixed with variant Chinese characters created by Wu Zetian. Although there was no publication record*, *it is the most recent edition from the Wuzhou Dynasty. If so*, *this should be the earliest existing edition.* ❶

Dr. Nakasawa's judgment was correct. As the main classic of Tiantai Sect of Buddhism in China, this sutra was translated very early and the doctrines were elucidated, which made this sutra become one of the most popular Buddhist classics in the early Tang Dynasty since Wu Zetian assisted Emperor Gaozong in participating in state affairs. In the reign of Empress Wu, Buddhism flourished and printing developed. Keen on the Buddhist scriptures, she ordered to print them. This copy might be printed and published in Luoyang during the early Wuzhou Dynasty (689~696) rather than the later period (697 ~704), which was the earliest existing printed matter in scroll. Several decades later, the Esoteric Buddhist classic — *Wugou Jingguang Datuoluoni Jing* (《无垢 净光大陀罗尼经》, *Aryarasmi-vimalvi suddha-prabha nama-dharani sutra*) printed in the year of 702 A. D. in Luoyang was found at a stupa of the Buddhist temple Pulguksa in Kyongju, Korea in 1966, which was a material evidence of printing activities which happened in the Wuzhou Dynasty.

In recent years, an early Tang edition even ahead of Wuzhou edition has been unearthed in China. In 1974, a single sheet of a dharani charm of Vidya-dharani-yana in Sanskrit was found in a Tang tomb in a diesel machinery factory in the western suburbs of Xi'an City. Put in the copper arm bracelet worn by the dead, it was square with a large size of 27 cm by 26 cm, which was printed on hemp paper and was broken after unfolding. In the middle of the paper, there was a blank square frame, 7 cm by 6 cm, and in the top right corner of the paper, the vertical lines by ink said "Wude (afterlife) happiness", indicating that it was the tomb of Wude (Figure 54). Outside of the square frame, dharani or charm of Vidya-dharani-yana classics in Sanskrit was sealed in 13 lines of each side. The sealing words were surrounded by four borders, a

---

❶   长泽规矩也.和汉书の印刷とその历[M].东京:吉川弘文,1952.

space of 3 cm with the inner square frame. There were such figures as lotus, buds, musical instruments and hand knotted printings (mudra) within the spaces, with a rope tied at the outer frame on the block. These patterns perfectly coincided with the Buddha statue of Vidya-dharani-yana in the Sui and Tang Dynasties of China, because during these periods, most Bodhisattva statues held in hand musical instruments, flowers, ropes, arm bracelets, and mudras❶.

Figure 54　A Single Sheet of a Dharani Charm of Vidya-dharani-yana in Sanskrit found in a diesel machinery factory of Xi'an, 1974, from Han Baoquan (1987)

Figure 55　Illustration of Bronze Mirror found in a diesel machinery factory of Xi'an, 1974, from Han Baoquan (1987)

Historical relics like arm bracelet made of copper (chuan) and tetrarchy bronze mirror (Figure 55) were also unearthed. The former was worn on the arm by the dead people, which was seen frequently in the Tang tombs at suburbs of Xi'an City. The image of arm bracelet was also seen on the stone statues of the early Tang Dynasty in Lotus Cave of Guangyuan Buddhas in Sichuan Province and statues of thousand-hand Bodhisattva on the northern cliff of Wanfo valley in Dongshan, Longmen in Henan province. Tetrarchy bronze mirror was 19.5 cm in diameter, 0.3 cm thick, with the height of 0.8 cm and inscriptions. It shared similar shape and characteristics of the burial mirror in the Sui and Tang Dynasties. Therefore, according to the research by

---

❶　吕建福.中国密教史[M].北京:中国社会科学出版社,1995.

archaeologist Han Baoquan, they were relics in the early Tang Dynasty (at the beginning of the seventh century), and this printed copy of dharani charm of Vidya-dharani-yana in Sanskrit was "the oldest known printed matter in the world"❶. On November 20ᵗʰ in 1996, Shaanxi Cultural Relics Appraisal Committee has invited experts for a collective appraisal and confirmed this conclusion. Later on, the Sanskrit expert Jiang Zhongxin has made a comparative study of the Sanskrit on this printed copy and the Sanskrit unearthed at the known ages, confirming that the typeface used on it had been widely spread at latest since the 6ᵗʰ century. After examining the printed paper, we thought it was the hemp paper of the early Tang, and clarified the origin time or the exact year of patterns on the printed copy in China, which proved the periodization of the printed copy to be correct. This dharani charm was printed at the beginning of the seventh century in Chang'an, capital of the early Tang, and it was a product of the long-term development of Esoteric Buddhism in China.

After a further study, it was found that the sutra mantras were derived from the original texts in Sanskrit of *Da Shen Zhou* (《大身咒》, *Great Charms*) in *Qianshou Guanyin Tuoluoni Jing* (《千手观音陀罗尼经》, *Sabacrabhuja-sahasraneta-aralokitesavara-dharani-sūtra*) from India between the years 618 A. D. and 626 A. D.. Moreover, they were arranged in a square shape and placed in the arm bracelets, which was a tradition at the beginning of the Tang Dynasty, from Emperor Gaozu, to Taizong and to Wuzhou. The sutra mantras in Sanskrit unearthed in Xi'an in 1974 were cut blocks for printing into single sheets on the basis of the Sanskrit script from India in the early Tang Dynasty, mainly as talisman for the living and the dead. Charms are arranged into square matrix with a different number of bars around it as a selection of an altar method. As stated in the translated version of *Tuoluoni Ji Jing* (《陀罗尼集经》, a sutra) in the year of 653 A. D., the printed copy unearthed in Xi'an adopted the altar methods of four cubits and triple court. The blank at the middle of the frame was called the "essentials of charm" and could be used to draw different things for fulfilling various wishes of believers. For example, a nine-head dragon would be drawn in order to pray for rain and a woman praying for a child would draw a little boy. If nothing was drawn, it would protect the body,

---

❶ 韩保全. 世界最早的印刷品：西安唐墓出土印本陀罗尼经咒[M]//石兴邦. 中国考古学研究论集. 西安：三秦出版社，1987.

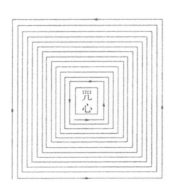

Figure 56   Arrangement of charms on a printed copy of *Tuoluoni Ji Jing* (《陀罗尼集经》, a sutra) in Sanskrit unearthed in a Tang Tomb in Xi'an, recovered by Pan Jixing (2000)

and the dead wearing the arm bracelet with charm would be brought into heaven. We have also clarified the arrangement of charms in Sanskrit and made a recovery (Figure 56). Due to certain limitations, the Sanskrit letters could not be displayed one by one. After a comprehensive study, it was believed that this printing copy was printed and published in Chang'an sometime between 640 A. D. and 660 A. D.❶, almost at the same time or a bit earlier than previously mentioned single-sheet Buddha figure by Xuanzang. Naturally, it was evaluated as a first-class cultural relic by Shaanxi Cultural Relics Appraisal Committee.

The above-discussed documentary records and archaeological discoveries have proven that printing activities appeared in China as early as the beginning of the seventh century. The order of "the abandoned Buddhism scriptures and statues should be recarved and rebuilt" was made in the year of 594 A.D., less than half a century ago. According to the development law of technique, there might be a period of preparation for printing activities in the early Tang Dynasty. Therefore, the view of "block printing had its birth at the beginning of the Sui Dynasty and expanded greatly in the Tang Dynasty" by the Ming scholar Hu Yinglin could be confirmed by new information today. However, the idea that printing technique was originated during the reign of Emperor Tang Xuanzong (712~755), between the seventh and eighth centuries, or the year 824 A.D., needed to be corrected.

Shortly after the development of woodblock printing in the early Tang Dynasty, a similar technique, copper plate printing, appeared in China. Both of them belonged to mono-block printing: printing with ink on a whole printing plate. Although the copper plate was more expensive than the woodblock, it was strong and durable. Made by casting, it could be recast in a new version when not in use, which is better than the woodblock. Ye Changchi (1847~

❶   潘吉星. 1974 年西安发现的唐初梵文陀罗尼印本研究[J]. 广东印刷 (广州), 2000(6): 56-58; 2001(1): 63-64.

1917) spent 20 years collecting more than 8000 stone inscription rubbings of past Dynasties and making intensive study and then wrote 10 volumes of *Yu Shi* (《语石》, *Language Stone*, 1909). He mentioned stone inscription in reverse direction in Volume 9: "In addition, copper plates *Xin Jing* (《心经》, *Heart Sutra*) in the year of 1075 A.D. and between 713 to 741 as well as book printing of Han Yu's writing in Sichuan are all in reverse words". *Xin Jing* was an abbreviation of the translated version of *Prajna-paramitahrdaya-sūtra* by Xuanzang. As a summary, *Xin Jing* (《心经》, *Heart Sutra*) was in a single roll with a small length. The study of the remains recorded by Ye Changchi found that these reverse characters cut in relief were on a copper plate directly used to print *Xin Jing* (《心经》, *Heart Sutra*) rather than a book version❶. It was the earliest material of Chinese copper plate printing in the first half of the eighth century.

The copper plate used for printing Buddha statues in the Tang Dynasty was discovered more than 20 years ago in Baoji, Shaanxi Province. It was a copper plate for printing a thousand-Buddha statue in the year of 834 A.D. (Figure 57) in square shape, 14.8 cm in height, 11.5 cm in length, 0.7 cm in thickness and 455.8 g in weight❷. There were three main Buddha statues in the middle of the plate and 105 small Buddha statues all around in 9 layers. The total number of Buddha statues was 108. On the back of the plate, there were Jin Gang's real words and vows carved on the handle in bow shape: *Jingang Diming Zhenyan* (《金

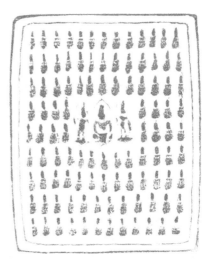

Figure 57   Rubbings of Copper Plate for a Thousand-Buddha Statue in the year of 834 A.D., from Gao Ciruo (1986)

刚抵命真言》, *The Real Words of Diamond*), "to seek help from Bodhisattva, to pray for getting rid of calamities and bringing happiness. On April 18th, 834, believers in the Ren family vowed and made the statues carved in an eternal support for Buddha." The plate was as large as a size of today's thirty-six mo,

❶ 庄葳.唐开元心经铜范系铜版辨[J].社会科学,1979(4):151-153.
❷ 高次若.宝鸡市博物馆收藏铜造像介绍[J].考古与文物,1986(4):71-73.

and it was 9. 4 cm higher than the plate copy carved in Hangzhou of *Dhatū-karanda-dharani-sūtra* at Wuyue in the Five Dynasties and Ten Kingdoms. This was the earliest extant copper plate material in the world, indicating that not only the woodblock printing, but also the copper plate printing was invented in China.

Although there were a large number of images on the above copper plate, the lines were clear and each Buddha statue was clearly cut, which require fairly skilled casting techniques. It was still in good condition and could be used to print figures. According to the author's research, this was cast on the basis of the stories about the translated version of Esoteric Buddhism classic — *Jingang Ding Jing* (《金刚顶经》, *Vajrasekhara-sūtra*) related by the Tang monk Amoghavajra (705～774). It was said that Maha-vairocana, the leader of Esoteric Buddhism, had 108 Dharmakaya, so there were 108 statues of Buddha, Maha-vairocana in the center, with Puxian and Guanyin at each side, small statues Dharmakaya around. *Jingang Diming Zhenyan* (《金刚抵命真言》, *The Real Words of Diamond*) was composed of nine characters, which were transliterated into the Chinese characters from Sanskrit. The meanings were incomprehensible, but the main idea was to pray for good luck. We have explained some of these words but will not go into details since it was a special issue.

2.2 Development of Block Printing from the Tang Dynasty to the Northern Song Dynasty

2.2.1 Printing from the Middle to Later Tang Dynasty

The period between the middle of the Tang Dynasty and the Five Dynasties (from the 8th century to the 10th century) should be considered as the early stage of printing, and there are many related documentary records and unearthed objects. In 1975, in a metallurgical machinery factory in the west suburb of Xi'an, a single sheet of the god mantra from *Foshuo Suiqiu Jide Da Zizai Shenzhou Jing* (《佛说随求即得大自在神咒经》, *Sutra by Buddha*) was found in a Tang tomb. When being excavated, it was placed in a small box and had been glued together. It was in square shape after being spread, 35 cm by 35 cm, and printed on hemp paper, with a box of 5.3 cm by 4.6 cm in the center, two coloured drawing people inside, one standing and the other kneeling down. The outer rim of the frame was engraved with the inscriptions of the gods, 18 lines on each side, 72 lines in total, and there was a boundary between the lines. There were border lines outside the mantra sheet, surrounded by mudra, which were used to attract different bodhisattvas for blessing (Figure 58).

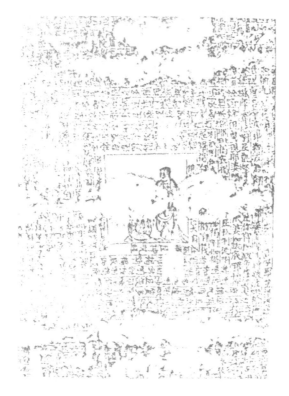

Figure 58   A single sheet of God Mantra from *Foshuo Suiqiu Jide Da Zizai Shenzhou Jing* (《佛说随求即得大自在神咒经》, *Sutra by Buddha*) in the Middle Tang Dynasty unearthed in Xi'an, from Han Baoquan (1987)

The paper was yellowish and the charm text was incomplete. There were only 8 Chinese characters left in the title, and the 4 missing characters should be "Shuo Sui Qiu Ji". The sutra was translated by Ratnacina (625~721) from Kashmir in the year of 693 A.D. in Tiangong Temple, Luoyang, whose name and text were different from the version translated by Amoghavajra (705~774). Archaeologists have confirmed this version as a relic of the glorious age of the Tang Dynasty (713~779)[1].

---

[1]   韩保全. 世界最早的印刷品：西安唐墓出土印本陀罗尼经咒[M]//石兴邦. 中国考古学研究论集. 西安, 1987.

After a further research, it was revealed that the standing person in the middle was Vajradhara, the appearing form of the Buddha when expounding the doctrines, who put one hand on the head of a Buddhist monk on his knee and sent him up to heaven after death. Therefore, the owner of this Tang tomb might be a monk. The monk's head was unseen since parts of the picture had come off. The outside of the frame was printed with various mudras, implying that different Bodhisattvas would be invited. This cultural relic could help us to understand the popular secret law of the Tang Dynasty. There were various functions of this charm, and by drawing different images of the charm, respective wishes could be satisfied. The upper limit of printing time was the year of 693 A. D. , and the lower limit would not be later than the reign of Emperor Xuanzong. Because in the year of 758 A. D. during the reign of Emperor Suzong, Amoghavajra also translated the same sutra from the original text in Sanskrit, but changed the title with 27 characters. Since then, this new translated version was widely spread. Therefore, the unearthed copy should be published before the emergence of Amoghavajra's version, and the time of publication should be during the reign of Emperor Xuanzong. The printing of the sutra should adopt the new version instead of Ratnacina's old version translated more than 60 years ago. Some people said that copies of Esoteric Buddhism appeared only after "three great monks came to China during the years 713 to 741 of Kaiyuan", but the unearthed copy proved that it was not correct.

There should have been more printed matters during the flourishing Kaiyuan period (713～741) in the mid-Tang Dynasty. Owing to Emperor Wuzong's anti-Buddhism in the year of 845 A. D. , a large number of Buddhist scriptures printed before were destroyed, and most early printed matters were Buddhist publications. However, printing was also used for non-religious purposes, for example, the "printed paper" issued by the Ministry of Revenue in the year of 783 A. D. was used for economic activities. As recorded in *Shi Huo Zhi* in *Jiu Tang Shu*, in order to relieve economic difficulties, two kinds of taxing — housing tax and income tax were put forward, and printed pieces of paper were used as tax receipts. According to *Suanchumo Fa* (《算除陌法》, *Taxing Law*), for any public or private income as well as trade income, in every *min* (1000 pence), 50 pence should be retained by the government as the tax. City officials in charge of the financial affairs would hand out the printed paper, columns such as taxing items, revenues, taxes and taxpayer printed on the paper in advance. When the taxes were paid, the above information would be filled in and the taxing was testified by the official seal covered on the paper❶. Those who evaded tax would be fined a large sum of money and punished by severe flogging. After this law was launched, complaints were heard everywhere, so it was abolished two years later. Therefore, the "flying money", something similar to bill of exchange issued from the early years of 806 to 820❷ was also made by printing. Because the money was used in every province, and "the payee could get the cash as long as he provides the correct bill of exchange", those bills with the same format were in large demand. The printing would be the best choice.

In the late Tang Dynasty, there were a variety of printed matters. Besides the Buddhist scriptures, some Chinese reference books like Chinese dictionaries and Chinese phonology as well as books for the mass market including geomancy, fortune-telling and the calendars were published in succession. The calendars were usually issued by the Ministry of Rites in feudal China, but they were often printed in private for the sake of profits. Therefore, on December 29, 835, Emperor Wenzong "forbids the private printing of calendars by woodblock at all levels of governments"❸. Chengdu, Huainan and Yangzhou were the nongovernmental printing centers of the calendars, from where merchants bought the calenders and sold them out all over the country. Each year, before the Bureau of Astronomy has memorialized the Throne suggesting the promulgation of the new calendar, these privately printed calendars were already everywhere. The fragments of the calendar (Figure 59) in the year of 877 A. D., unearthed in the Dunhuang Caves, are now preserved in the British Library. Designed in a complicated layout, it was quite fine and elegant with figures and tables, and each item was bounded by the crossbar. The remaining pages included only information from April to August. Apart from calendar days, solar terms, there were fortune-tellings like "laws of removing ill fortune relating to twelve Chinese Zodiac signs" as well as the zodiac chart and the fengshui chart. This calendar with rich content was almost similar to that of the Qing Dynasty.

------

❶　刘昫.旧唐书:食货志下[M].上海:上海古籍出版社,1986.
❷　欧阳修.新唐书:食货志[M].上海:上海古籍出版社,1986.
❸　刘昫.旧唐书:文宗纪[M].上海:上海古籍出版社,1986.

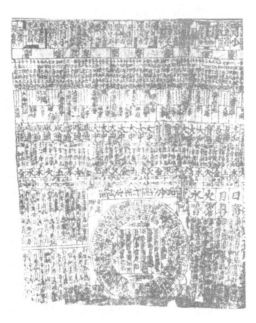

Figure 59   Fragments of Calendar in 877 unearthed in dunhuang Caves,
now preserved in British Library

As recorded in *Tangyu Lin* (《唐语林》, *Anecdotes of the Tang Dynasty*) by Wang Dang (1075~1145) of the Song Dynasty, in the first month of the lunar year 881, to avoid attack of Huang Chao, Emperor Xizong left Chang'an and fled to Chengdu, when the merchants seized the chance to print calendars in private, including natives of Sichuan and the east of Jiangnan district. ❶ Those versions were earlier than those of court historian. Jiangnan district almost covered areas of today's south Jiangsu, Zhejiang, Fujian and Taiwan. Its main printing centers included Yangzhou, Suzhou, Hangzhou and Yuezhou (now Shaoxing). Due to the differences in the calculation methods of private calendars, there was a dispute among merchants in regard to the first and last day of the month as well as season and climate. In the previous regime of Emperor Xizong, during 860 to 873, the woodblock printing technique has been highly developed, and the pictures printed from engraved plates were the best to display the advanced printing skills. In 1907, Dr. Stein discovered the whole volumes of *Jingang Jing* printed in 868 in the Dunhuang Caves. This sutra with the full name *Vajracchedik-prajna-paramita-sūtra* in a scroll binding (Figure 60), is now preserved in the British Museum, London.

---

❶   王谠.唐语林:卷七[M].上海:上海古籍出版社,1978.

Figure 60   *Jingang Jing* printed in 868 in Dunhuang Caves,
unearthed in Dunhuang Caves in 1907, now preserved in British Museum

This sutra was 525 cm long with 7 sheets of paper. In the preface, a beautiful illustration reflected that the Buddha Sakyamuni sitting on a lotus was expounding Buddhist doctrine to his disciple Subhuti in the lonely park. The carving was exquisite and the way of cutting was proficient. The following 6 sheets, 26.67 cm by 75 cm in size, were printed with scriptures in regular script[1]. In the end, printed into the text was the statement that the book was "reverently made for universal free distribution by Wang Jie on behalf of his two parents on the 15th of the fourth moon of the ninth year of Xiantong (May 11, 868) on May 11 of 868". When staying in the UK in October 1982, the author tested this sutra with paper and confirmed that it was hemp paper. It was white and smooth with tight fiber. This was an excellent work in terms of paper, craving and ink brushing, and also the earliest illustrated printed matter with a complete inscription showing the date of printing.

At the end of the Tang Dynasty, the scholar Sikong Tu (a style name called Biaoshen, 837~908) wrote in Volume 9 of *Sikong Biaosheng Wenji* (《司空表圣文集》, *Collected Works of Sikong Biaosheng*):

> ... *In the reign of Emperor Wuzong, temples were destroyed and monks expelled in Luoyang. Buddhist sutras were burnt and the printed copies like old Doctrines of Sifenlv in Riguang Temple were lost. Now it's time for recarving and reprinting.*

---

[1]   CARTER T F. The Invention of Printing in China and Its Spread Westward[M]. New York: Columbia University Press, 1925.

Here，"old Doctrines in Riguang Temple" referred to *Sifenlü Shu* (《四分律疏》, *Doctrines of Sifenlv*) by the monk Fa Li (569～635) of the Sunlight Temple at Xiangzhou (now Anyang，Henan) in the early Tang Dynasty[1]. Emperor Wuzong launched movement against Buddhism in 845. As stated by Sikong Tu，there was a printed copy of *Sifenlü Shu* (《四分律疏》, *Doctrines of Sifenlv*) before that，which was sermoned by the monk Hui Que in Jing-ai Temple at Luoyang. In 845，this sutra was burned and the temple destroyed. During the reign of Emperor Xuanzong (847～859)，the ban of Buddhism came to an end. Then Hui Que requested Sikong Tu to write a letter of raising donations. The letter was written around 874，in which there were words of "making 800 printed copies"，which meant that 800 leaflets were widely distributed. The mantra of *Yiqie Rulai Zunsheng Foding Tuoluoni* (《一切如来尊胜佛顶陀罗尼》, a sutra) printed at the end of the Tang Dynasty (the ninth century) was found in the Dunhuang Caves and is now preserved in the national library of Paris (Figure 61).

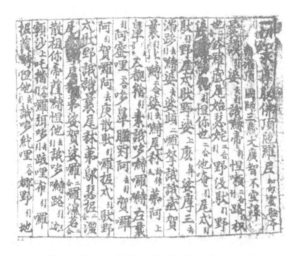

Figure 61　A Tang Copy of *Yiqie Rulai Zunsheng Foding Tuoluoni*
(《一切如来尊胜佛顶陀罗尼》, a sutra) in Ninth Century found
in Dunhuang Caves，now preserved in National Library of Paris

It was recorded in the second volume of *Yunxi You Yi* (《云溪友议》, *Short Stories of the Tang Dynasty*，about 870) by Fan Shu (840～912)，at the same time，He Ganzhong (817～884)，inspector of Jiangxi，who has studied alchemy

---

[1]　周一良.纸与印刷术[M]//李光璧,钱君晔.中国科技发明和科技人物论集.北京:三联书店,1955.

for many years, published thousands of *Liuhong Zhuan* (《刘弘传》, *Biography of Liu Hong*) and distributed to his fellows between 847 to 859.

During the reign of Emperor Yizong, Chinese reference books such as *Tang Yun* (《唐韵》, a rhyming dictionary published originally in 751) and *Yu Pian* (《玉篇》, an old dictionary compiled in 543) were also printed in Chengdu, Sichuan and were spread to Japan. A Japanese monk and scholar Zong Rui came to China for study in 862, and three years later, returned to Japan with the ship of Tang merchant Li Yanxiao. He brought many books from China, which were listed on *Shuxie Qinglai Famen Deng Mulu* (《书写请来法门等目录》, a catalogue):

> ... a 5-roll Xichuan printing of Tang Yun, a 30-roll Xichuan printing of Yu Pian. Miscellaneous books, although not Buddhism classics, can satisfy the needs of the general public. ❶

"Xichuan printings" were the copies printed in Sichuan, which were collected by the Japanese monk Yuan Zai in his study in Chang'an and delivered to Zong Rui and later brought to Nara Todaiji temple in the year of 865 A.D.. As recorded in *Liushi Jiaxun* (《柳氏家训》, *Family Instructions of Liu*) by Liu Pin (848~898), in the book market of Chengdu — a second capital, he has seen many books concerning yin and yang, divination of dreams, geomancy, the nine heavenly palaces and the five planets which were engraved on blocks and printed paper. Those books were in large quantities, only second to dictionaries and other books of lexicography. The remaining volumes of Tang carving copy of *Datang Kanmiu Buque Qieyun* (《大唐刊谬补缺切韵》, *Acta absurdum implevit Qieyun*, No. P-5531) unearthed in the Dunhuang Caves, now preserved in the National Library of Paris, belonged to dictionaries and lexicons.

2.2.2　Printing in the Five Dynasties and the Northern Song Dynasty

The Ming scholar Hu Yinglin (1551~1602) summarized the history of Chinese woodblock printing during the first 600 years in very concise and accurate words: "Block printing had its birth at the beginning of the Sui Dynasty, expanded greatly in the Tang Dynasty, took a lead forward under the Five Dynasties and finally came to its fullest development under the Dynasty of

---

❶　木宫泰彦. 日中文化交流史[M]. 胡锡年,译. 北京:商务印书馆,1980.

Song". Indeed, as a link between the Tang and the Song Dynasties, the Five Dynasties were the extending period of printing. Let's first talk about the printing activities in the north. One important turning point was *Jiu Jing*, which was printed under the organization of the government. All the scholars were required to read these classics, thus making the printed books presentable and acceptable. The printing was initiated by the Later Tang (923~936) Prime Minister Feng Dao (882 ~ 954). Having received the imperial order, the Northern Song scholar Wang Qinruo (962~1025) wrote Volume 608 of *Cefu Yuangui* (《册府元龟》, *Reference for Governing the Country*, 1013), and it was recorded that Feng Dao proposed to Emperor Mingzong in the year of 932 A. D. that:

> *We have seen, however, men from Wu and Shu who sold books that were printed from blocks of wood. There were many different texts, but there were among them no orthodox Classics. If the Classics could be revised and thus cut in wood and published, it would be a very great boon to the study of literature.*

Since many other books were printed in Wu and Shu in the south, and the Confucian classics were not covered, Feng Dao decided to revitalize culture and education. Taking *Kaicheng Shi Jing* (《开成石经》, *Kaicheng Stone Classics*) as the original text, he suggested all the scholars called together by the Imperial Academy to edit the text for the purpose of producing the plates of *Jiu Jing*. The petition was granted and the civil official Ma Gao (854~about 938) was appointed as the leader. It was worth noting that although the north had witnessed four dynastic changes, Feng Dao had always held his position as the Prime Minister, and the printing activities he advocated never stopped. Some officials had retired or died, but Tian Min (about 881~972) remained in the post. It took 21 years (932~953) to complete the task. *Jiu Jing* including *Yi Jing* (《易经》, *The Book of Changes*), *Shang Shu* (《尚书》, *The Book of History*), *Shi Jing* (《诗经》, *The Book of Songs*), *Yi Li* (《仪礼》, *Etiquette*), *Li Ji* (《礼记》, *The Book of Rites*), *Zhou Li* (《周礼》, *The Rites of the Zhou Dynasty*) and *Chun Qiu* (《春秋》, *The Spring and Autumn Annals*), a total of 130 volumes were printed. To make the copies in the same layout and typeface, the Imperial Academy published *Wujing Wenzi* (《五经文字》, *Characters of the Five Classics*) and *Jiujing Ziyang* (《九经字样》, *Typeface of the Nine Classics of Confucianism*) in the year of 946 A. D., specifying the regulations of different procedures in the printing and publishing as well as standardized typeface.

Those two books were of vital importance to the printing history, but unfortunately lost after the Song Dynasty. The printing copies of the nine classics by the government in the Five Dynasties had a profound influence on the future printing.

At that moment, more books were printed by individuals. A large package of Buddhism figures discovered in the Dunhuang Caves, was now preserved in the national library of Paris (No. P-4514): (1) 5 single-sheet copies of portraits of Guanyin Bodhisattva, carved by craftsman Lei Yanmei and published by the military officer Cao Yuanzhong (about 905~980) of Gui Yijun in the year of 947; (2) 11 copies of figure of Dashen pishawen heavenly king by Cao Yuanzhong in 947; (3) 11 single-sheet copies of portrait of Dashen Wenshu Shili Bodhisattva, 31 cm by 20 cm for each; (4) 5 copies of figure of Amida Bodhisattva; (5) 1 copy of figure of the Tibetan Bodhisattva. In the upper part of the copies is a figure and characters in the lower part (Figure 62). They were

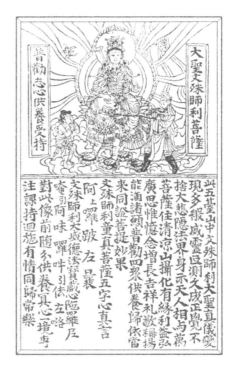

printed in Dunhuang. In the year of 947 A.D., Cao Yuanzhong also issued *Vajracchedika-prajna-pa-ramita-sutra*, carved by Lei Yanmei, in leaf binding, now also preserved in the National Library of Paris (No. P-4515). No. P-4515 and No. P-4516 were two parts of the same Buddhist scripture. As a local official in Longyou (now Gansu), Cao Yuanzhong made a significant contribution to the development of papermaking, printing and protection of Dunhuang grottoes.

As recorded in Volume 127 *Hening Zhuan* (《和凝传》, *Biography of He Ning*) of *Jiu Wudai Shi* (《旧五代史》, *Old History of the Five Dynasties*), the musician He Ning (898 ~ 955) was appointed as the Prime Minister and his works spread widely in the two capitals: Kaifeng

Figure 62   A Single-Sheet Copy of Portrait of Dashen Wenshu Shili Bodhisattva in 950 found in Dunhuang Caves, preserved in National Library of China in Beijing

and Luoyang. "Up to 100 rolls of works were carved on plates and printed into hundreds of copies by himself and were then distributed to others." Volume 127 *Gaozu Ji* (《高祖纪》, *Record of Gaozu*) has related that, Emperor Gaozu of the Later Jin Dynasty (936～947), a faithful believer of Taoism, ordered the priest Zhang Jianming to carve the Taoism classic — *Daode Jing* (《道德经》, *The Tao Te Ching*) "onto the plate, then required He Ning to write a new preface and published all over the country".

The Five Dynasties refer to the 5 successive Dynasties in the south and north along the Yellow River in North China. At the same time, 10 regimes were established in South China, known as the Ten Countries. Comparatively speaking, there were fewer wars among those co-existing ten countries in different areas, so the booming economy and culture have created favorable conditions for printing activities. The Former Shu (907～925) and the Later Shu (934～965) at Sichuang, the printing center of the Tang Dynasty, have all kinds of books on Confucianism, Buddhism and Taoism. The Former Shu has printed 30 volumes of *Daode Jing Guangshengyi* (《道德经广圣义》, *Doctrines of The Tao Te Ching*) by the priest Du Guangting on more than 460 plates during 909 to 913. In 923, *Chanyue Ji* (《禅月集》, *Collected Works of Chanyue*) by the Buddhist monk Guan Xiu was published in Chengdu by his disciple Tan Yu, including almost 1,000 poems. Almanacs were also published in the Former Shu.

According to Volume 2 of *Huizhu Yu Hua* (《挥麈馀话》, *Casual Literary Notes*) by a Song scholar Wang Mingqing (1127～1216), When Wu Zhao'i was poor, he made the practice of borrowing *Wen Xuan* (《文选》, *The Selection of Literary Works*) from his friends. When they showed their annoyance he made up his mind that, "if at some future date he came into high office, he would then print the work from woodblocks which might be within the reach of scholars. In due course he became minister in Shu under the family Wang and was at last able to carry out his proposal and printed it." After taking the post of Prime Minister in the Later Shu, he "ordered his disciples Gou Zhongzheng and Sun Fengji to carve texts of *Wen Xuan* and others on the plates"[1] and published them in the whole country. The Wuyue State (907～978), located in today's South Jiangsu, Zhejiang and East Fujian, developed printing mainly in the capital Hangzhou. Emperor of Wu Yue, Qian Chu (929～988) has printed

---

[1] 脱脱.宋史:毋守素传[M].上海:上海古籍出版社,1986.

84,000 copies of *Baoqieyin Tuoluoni Jing* (《宝箧印陀罗尼经》, *Dhatū-Karanda-Dhrani-sūtra*) with bark paper and bamboo paper. This sutra with a single volume has 3 extant printed copies❶.

The first edition was unearthed in 1917 in Tianning temple in Huzhou, Zhejiang Province. Each sheet was 5 cm by 60 cm in size, with 341 lines, 8 or 9 characters in each line. There were illustrations of the Buddha and guardians on each side, and figures of people paying respect for the Buddha. Most lines and appearances were simple and clear. In front of the illustrations, there was an inscription: "84,000 copies of *Dhatū-Karanda-Dhrani-sūtra*, printed by Qian Chu, the Grand marshal and Emperor of Wuyue, were worshiped in the pagoda. Recorded in 956 of Bingchen Year". Xiande was the reign title of the Later Zhou Dynasty. In 1971, the second edition was found in a stupa sealed with gold and mud in Shaoxing, Zhejiang Province. It was placed in a bamboo tube of 10 cm long. There were 11 or 12 characters in each line and similar illustrations and inscription with the first edition, printed on the white bark paper. The carving was finer. There was no reign title in the inscription, only Yichou Year, which referred to the year 965 A. D.. The third edition printed on the bamboo paper was found in Leifeng pagoda in Hangzhou in 1925. It was 3. 6 cm high, 190. 5 cm long, in 271 lines, 10 or 11 characters in each line. In the preface, there was an illustration of the Queen and her maids paying respect for the Buddha. The inscription was: "84,000 copies of *Dhatū-Karanda-Dhrani-sūtra*, printed by Qian Chu, the Grand marshal and Emperor of Wuyue, are worshiped in the brick pagoda of Xiguan (Figure 63)." Recorded in August, 975.

From 956 to 975, Qian Chu has spent 19 years printing *Baoqieyin Tuoluoni Jing* (《宝箧印陀罗尼经》, *Dhatū-Karanda-Dhrani-sūtra*) and put them in different Buddhist pagodas all over the state. Moreover, the monk Yan Shou (904~975), from Lingyin temple in Hangzhou, has printed more than 10 kinds, all together 400,000 copies of scriptures, charms and Buddha figures, of which 160,000 were printed on silk❷. Nan Tang (937~975), to the west of Wu Yue, located in today's Anhui, Jiangxi, West Fujian and most parts of Jiangsu, with Jiangning (now Nanjing) as the capital, was a vast and populated state. It

---

❶　钱存训. 纸和印刷[M]//李约瑟. 中国科学技术史: 卷 5(第 1 册). 北京: 科学出版社, 1990.

❷　张秀民. 五代吴越国的印刷[J]. 文物, 1978(12): 74.

was famous for its paper in Chengxin Hall. The book collector of the Ming Dynasty Feng Fang (1510~1567) recorded in his *Zhenshangzhai Fu* (《真赏斋赋》, *Ode to Zhenshang Hall*) that *Shi Tong* (《史通》, *Historical Theories and Criticism*) and *Yutai Xin Yong* (《玉台新咏》, *New Songs from the Jade Terrace*) were among the first published books of Nan Tang. It was remarked in the notes that there were words like "Jianye seal for collection of Nan Tang" on these books. Nan Tang was established after the end of Wu (919~936). As recorded in the proposal by Feng Dao, "we have seen men from Wu and Shu who sold books that were printed from blocks of wood. There were many different texts", it indicates that Wu has also published many books and sold them out to Luoyang and other places in the Later Tang Dynasty.

Figure 63　A Copy of *Baoqieyin Tuoluoni Jing* (《宝箧印陀罗尼经》,
Dhatū-Karanda-Dhrani-sūtra) Printed in 975, found in Hangzhou in 1925,
preserved in National Library of China in Beijing

After the regimes of the Five Dynasties and Ten States came to an end, the Northern Song Dynasty (960~1126), a new Dynasty uniting the country, was established. The south and north had more economic, scientific and cultural communication, thus achieving integrative development. The printing technique, based on the development of previous Dynasties, now entered a golden period. The printed books have already played a dominant role and woodblock printing has obtained unprecedented development. Almost all the areas including Confucianism, Buddhism, Taoism and various schools of thought were covered. The printing was even applied to the economic field, such as issuing the paper money. In the Song Dynasty, there were many publishing centers all over the country, forming a printing network of governments, workshops and individuals. On the basis of woodblock printing, movable type appeared; In addition to single-color printing, there was multicoloured printing. The minority areas have also developed printing, and China has become the world's largest producer of paper and printing. Due to the popularization and progress of printing technology, the printed matters in this period tended to be more perfect, and could be regarded as the model for the future generations. There was also a new breakthrough in binding from the scroll to new binding forms suitable for typography. Printing of the Yuan, Ming and Qing Dynasties were mainly developed on the basis of the pattern of the Song. The Song edition (Figure 64 and Figure 65) is famous for the preciseness on carving, printing and proofreading, thus considered as a reliable text by the later generations, and many copies are handed down. More details of the woodblock printing after the Song can be found in relevant works.

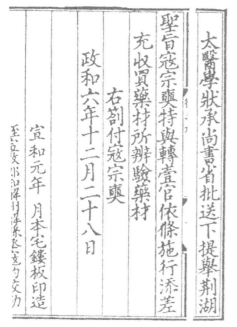

Figure 64   *Bencao Yanyi* (*Expanded Materia Medica*), emendated by Kou Yue in 1119, from Nakayama Kushiro (1930)

Figure 65   *Chunqiu Guliang Zhuan*, published by Yu Renzhong in 1191, from Nakayama Kushiro (1930)

2.2.3   Traditional Block Printing Technique

2.2.3.1   Raw Materials in Printing: Woodblock, Paper and Ink

The main raw materials in the woodblock printing were plate, ink and paper. These materials should be processed through a series of procedures by various tools before the final printed matters are made. The previous chapter of this book has discussed the paper. However, it should be pointed out that the printing paper had to meet certain technical requirement such as smoothness, good ink acceptability with less fiber bundle. It should also have enough whiteness, compactness and moderate thickness. Chinese people were used to printing on single side and the early printed copies were bound in scrolls, so the paper should be a little thicker. After the Song Dynasty, it was printed in the fold, then bound together in book form. Therefore, the paper could be thinner, neither too hard nor too thick. In general, paper with 0. 1 mm~0. 15 mm in thickness was suitable for printing. Hemp paper was widely used for printing in the Sui, Tang and Five Dynasties, and the second choice was bark paper. More bark paper and bamboo paper were used after the Song Dynasty. Because the printing paper was in large demand, it was unnecessary to use good paper, which was different from the paper for painting and calligraphy. The Imperial College copies in the Five Dynasties and the Song were printed by publishers with enough financial resources, so better paper and ink were used. For individual printing, suitable paper could be chosen based on different economic situations. The printing factories usually had to store a large amount of paper to ensure that the books were printed with the same paper and it was also convenient for pricing.

The wood for making plate were mainly strong and tall trees in order to ensure the enough large space. The hardness of wood should be moderate so it was easy to carve and also had sufficient strength. The wood should be fine-grained with regular texture and without or with only a few scars, and it was easy to accept ink. Trees should be widely distributed, in abundant supply and not too expensive. It was a waste to use rare trees for woodblock. In view of these technical and economic conditions, catalpa wood, pear wood and common jujube were used. Other trees could also be used according to the local situation, but the pine and cypress were not suitable for printing in spite of a wide distribution and cheapness, because the wood were soft and pine trees contained too much resin.

The catalpa wood was widely used in woodblock, which was recorded in the preface of the carving copies in and prior to the Song and Yuan Dynasties. For example, the eastern hall edition of *Shiji Jijie Suoyin* (《史记集解索隐》, *Index of Shihchi Variorum*) which was engraved by Cai Mengbi in Jianyang, Fujian in the year of 1171, claimed in the preface of the second volume: "Cai Mengbi, a style name Fuqing, did the proofreading and carving at the eastern hall by himself". In the year of 1250 at Jun county in Shangrao, Jiangxi, the carving copy was revised by Zhuxi. The epilogue of the second volume of his disciple Cai Jiufeng's works *Shu Ji Zhuan* (《书集传》, *Researches on Shu Ching*) was written by Lü Yulong: "so I consulted and queried his works, and printed them by woodblock and published them at school." These two records showed that catalpa wood was used for carving plate❶. In the early Jin Dynasty, sometime between 1140 and 1178, the preface of the thirty-first volume of the book *Zhaocheng Zang* (《赵城藏》, *Buddhist Scriptures of Zhaocheng*) mentioned that Yang Chang and others in Jing village at Wanquan County had donated 50 pear trees for carving plate. Two printing plates of woodblock in the Northern Song Dynasty, which were kept in the Museum of Chinese History, were made of date wood. Catalpa (zi), pear (li) and date (zao) woods were so popular in the printing industry that such phrases as zixing (梓行), fuzi (付梓) and fu zhi lizao (付之梨枣) became synonyms words for "publish".

Catalpa, or Catalpa ovata, a kind of deciduous tree, belonged to Bignoniaceae, 6 meters high and distributed in a large area from the southern part of the northeast in China to the Yangtze River. It grew fast and it was hard wood with straight texture. It was not easy to decay so it could be used for the coffin. That was why the Emperor's coffin was called Zi Gong (梓宫, Catalpa palace). Pear, or Pyrus sinensis, a kind of deciduous tree, also belonged to Bignoniaceae. It has become one of the fruit trees in China over two thousand years ago and was widely distributed. Date, or Ziziphus vugaris, a kind of deciduous tree, belonged to Rhamnaceae. It originated in China and widely distributed especially in such provinces as Hebei, Shandong, Henan, Shaanxi, Gansu and Shaanxi. Apricot or Prunus armeniaca, which belonged to the same plant family with pear and originates in China, could also be used for platemaking. Occasionally, banyan or Ficus microcarpa in the Moraceae plant family was used. Red birch or Betula albo-sinensis, a kind of deciduous tree belonging to Betulaceae, was used for platemaking in Dege Tibetan area in Sichuan.

Carbon black, the main component of printing ink, was the light-black powder produced by incomplete combustion of materials containing carbon. Known as amorphous carbon in Chemistry, it was composed of many small graphite crystals with a complex microstructure. China had a long history of

---

❶   北京图书馆.中国版刻图录:第 3 册[M].北京:文物出版社,1961.

making ink from carbon black. At first, Chinese ink was made from a mixture of carbon black and glue. Then, ink stick was made on the basis of Chinese ink. As reported in *Yinxu Buci Zongshu* (《殷墟卜辞综述》, *Review of Archaeology and Cultural Relics*, 1956) by Chen Mengjia, there were black words on oracle bone pieces unearthed in Yinxu, Anyang City in Henan Province, and this kind of oracle bone scripts were popular between year 1250 B.C. and 1192 B.C.. The micro-chemical analysis of these black words has proven that the component was carbon black[1], which indicated the entry into the prehistoric period of inking. Since the Western Zhou Dynasty (1046 B.C. ~771 B.C.) and the Spring and Autumn (770 B.C. ~477 B.C.), there were more and more ink printings and writings[2].

It was stated in *Zhuangzi* by the philosopher of the Warring States Period Zhuang Zhou (369 B.C. ~286 B.C.), that the ruler of the Song, Emperor Yuangong (530 B.C. ~516 B.C.) wanted to draw a picture, "the officials went to court and worshiped on bent knees, making a bow with hands folded in front, and then stood, moistening the pen and mixing up the ink". It referred to mixing the ink rather than grinding the ink stick. It still needed further discussion on whether the unearthed silk manuscripts and bamboo scripts in the Spring and Autumn and Warring States Period were written with the ink or ink stick. The ink stick was obviously easy to keep and carry, but required more complicated manufacturing process. From 1975 to 1976, ink sticks made during the late Warring States Period to the Qin (the fourth to the third centuries B.C.) were found in Yunmeng, Hubei[3]. In the Western Han Dynasty, there were more ink sticks and ink stones[4]. In 1953, ink stones and ink sticks used by the master of Rolls were found on the wall paintings of a Western Han tomb in Wangdu, Hebei[5], and there were ink marks on the unearthed paper. Recorded in *Hou Han Shu · Baiguan Zhi* (《后汉书·百官志》, *History of the Eastern Han Dynasty · Records of Officials*) were such words as "... paper, pen, ink ... and seal." Ying Xun (140~206), in the Han Dynasty, related in his *Hanguan Yi*

[1] BENEDETTI PICHLER A. Microchemical analysis of pigments used in the fossae of the incisions of Chinese oracle bones[J]. Industrial and Engineering Chemistry: Analytical Edition, 1937(9):149-152.

[2] 蔡运章. 洛阳北窑西周墓墨书文字略论[J]. 文物,1994(7):64-69.

[3] 孝感地区考古短训班. 湖北云梦睡虎地十一座秦墓发掘简报[J]. 文物,1976(9):53.

[4] 麦英豪,黄展岳. 西汉南越王墓:上册[M]. 北京:文物出版社,1991.

[5] 河北省博物馆. 望都汉墓壁画[M]. 北京:中国古典艺术出版社,1955.

(《汉官仪》, *Rules for Officials in the Han Dynasty*, 197): "every month, ministers and subordinates are given one large and one small piece of Yumi ink". "Piece" indicated that the ink referred to ink sticks. Yumi ink was produced in today's Qianyang, Shaanxi, and was famous for its excellent quality. So Yumi was considered synonymous with "ink" by the later generations.

Chinese carbon black was made of pine wood and tung oil. Since the Han and Tang, pine wood was burnt into pine soot carbon black and made into ink, known as pine-soot ink. The raw material of Yumi ink was pine wood at Zhongnan Mountains. In *Mo Jing* (《墨经》, *Mohist Canon*, about 1100) by Chao Guanzhi (1050~1120) in the Song Dynasty, a list of pine woods were mentioned, which were rich in resin and can be made into carbon black. The cost of pine-soot ink was low, thus making it easy for mass production. Tung oil was burnt into lampblack and made into lampblack ink. It was generally believed that it was developed from the Song Dynasty, and the methods and equipment used were different from those of pine-soot ink. Since after the Tang and Song Dynasties, more and more pine-soot ink were used in the printing, the preparation of pine-soot ink was discussed in detail. As introduced by Chao Guanzhi, two kinds of kilns were used to burn the pine soot: the vertical type and the horizontal type. More vertical kilns were used before the Song Dynasty. They were more than 1 *zhang* (310 cm) high with a wide chamber and a small opening. A large jar instead of a chimney was covered on the chamber, which was connected by 5 smaller jars in different sizes. The jar at the lowest level was the largest and the one at the top was the smallest. Each jar had a small hole at the bottom to connect the next jar, and the joint part was sealed with mud. The pine wood was burnt in the chamber, so the airstream and pine soot flew upward. The airstream could be controlled to some extent. The cooling pin-soot particles were kept in each jar, which could be collected in turn.

The anoxic zones in the whole vertical were in a state of incomplete combustion. The airstream went up through six jars, each like a baffle and cooling equipment. After a thick covering of pine soot was accumulated in each urn, the burning fire was put out. After some time of cooling, the carbon black would be cleaned with a feather duster. In the top jar, the particles were the finest with the best quality. The particles in the lowest jar were the largest, which could be used for second-class ink or black pigment. The finest particles could be made into the best ink. In this method, the equipment was easy to operate, but the funnel was too short. The particles of carbon black were easy to dissipate, so with a low productivity, it was difficult to get a large amount of carbon black.

After the Song Dynasty, the horizontal kiln appeared, to which "horizontal kiln in current use" in Chao's *Mo Jing* referred. Built along the terrain on the hill, the sloping horizontal kiln was 100 *chi* (31 m) long, 3 *chi* (93 cm) high, 5 *chi* (155 cm) wide. Several chambers were put together with a baffle at each joint part. At the bottom was the stove with an opening of 1 *chi* (31 cm) in diameter, where pine wood was burnt. Between the stove and the smoke chambers, there was a hole of 2 *chi* square. The airstream flew upward along the funnels of each chamber till the end. Every time, 3 to 5 pieces of pine wood were burnt in the stove. Then more pieces were added and the burning would last 7 days. After cooling, the pine soot could be collected in the kiln. The scanning electron microscopic analysis of the pine-soot ink marks on the painting in the Yuan Dynasty has proven the carbon black particle to be less than 0. 1 μm, reaching the modern level of carbon black particle size[1].

The scientist of the Ming Dynasty, Song Yingxing also introduced horizontal kiln in his work *Tian Gong Kai Wu · Zhu Mo* (《天工开物·朱墨》, *Cinnabar Ink* in *Exploitation of the Works of Nature*, 1637) which was similar to the description by scholars of the Song Dynasty. In the Ming Dynasty, 90% of ink were the pine-soot ink and oil-soot ink accounted for 10%. The kiln was made into huts in doom shape with bamboo battens, like a rainshed on the ship. Several huts were put together, more than 10 zhang long. All the joint parts were fixed by pasting mat and paper. The joint part with the ground was sealed with mud. The baffle inside the hut was built by laying bricks, and a hole for the funnel was set aside. It could reduce the airflow resistance and dissipate heat equally in all directions. Leaving a hole at intervals could control the airstream and flowing velocity, which was beneficial to the sedimentation of carbon black at different levels. The height and width of the kiln as well as the size of the stove opening were not mentioned in this book, which should be the same with that of the Song Dynasty. Judging from the illustration of the original book, the stove opening was too large, and it should be a small opening in 1 *chi* square. We have changed it a little bit here (Figure 66). After burning 7 days, the fire

---

[1]   WINTER J. Preliminary investigation on Chinese ink in Far Eastern paintings[J]. Advances in Chemistry, 1975(138):213-214.

would be put out. After cooling, the pine soot could be collected in the kiln. Pine soot from the two chambers near the fire were called "soot particles", which referred to carbon black in large particles and were used to printing books after ground into smaller particles. "Puffs of soot" referred to pine soot in the two chambers far way from the fire and these finest particles were made into first-class ink. Pine soot in between were called "mixed soot" and were made into common ink or used for printing books.

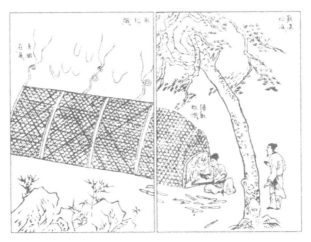

Figure 66   Picture of Collecting Pine Soot by Burning, from *Tian Gong Kai Wu* (《天工开物》, *Cinnabar Ink* in *Exploitation of the Works of Nature*, 1637)

In order to make solid ink, carbon black and additives like glue were mixed and mashed, then took shape. The method was recorded in *Qimin Yaoshu* (《齐民要术》, *Ancient China's Agriculture Encyclopedia*, about 538) as follows: to grind the pine-soot particles and sieve the fine particles in case of dissipating. Mix 1 kg of pine soot and 100 g of animal glue. The weight ratio was 100 : 31, with 67% to 77% carbon black and 23% to 33% animal glue. This ratio has been used for a long time in the past. Later on, the quantity of glue has been increased or decreased, and the general weight ratio of carbon black and glue was between 100 : 30 to 100 : 50. This book also mentioned that ash bark sap, albumen, cinnabar, musk and other things were added in order to improve ink performance and adjust ink color.

2.2.3.2   Block Cutting and Brushing Techniques

The wood was carefully chosen and then the bark was peeled off. Several blocks were divided and finely planed along the grain of wood into rectangular boards. The size of printing board was determined by the layout of a printed book, almost two pages of a book. Boards were usually 2.5 cm thick, 20 cm wide and 30 cm long. All the surfaces should be polished and the board was soaked in the water for a period of time so as to remove the gum. The soaking time depended on the season, usually about a month, and the board would be stewed if necessary. After the board was dried in the shade, the surface should be planed smoothly and rubbed with vegetable oils such as soybean oil and rapeseed oil, and then be polished with the stem of Alternanthera sessilis in the Amaranthaceae group. If there was a hard section on the board, it must be cut out and inserted with small pieces of wood and then planed.

The manuscript was neatly transcribed within the red lines on the bark paper by a calligraphy master, with stated typeface and size of characters. The lines in every direction as well as the center line were drawn. Then the sample should be proofread by marking the mistakes and omissions. If there were too many changes, another transcription and proofreading were required. The surface of the board was rubbed over with a paste made from boiled rice, then the sample was united to this surface so that they adhere, but in a reversed position.

The fine brush was first used to make the surface smooth. Then the rough brush was used on the backside of the paper and the coarse hair was rubbed off and dried. So a clear impression in ink of the reversed writing was left on the board.

The next step carving was done by skilled block cutter, who used different kinds of gravers (Figure 67) to cut and remove all that portion of the wooden surface which was not covered by the ink, leaving characters with the ink marks in fairly high relief. He first used the bevel knife and the flat knife to cut straight and transverse lines in square shape around each character. With a ruler

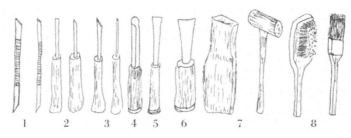

Figure 67　Block Cutting Tools

1. Double-edged graver, 2. Single-edged graver, 3. Single-edged chisel, 4. Semicircle chisel, 5. Flat chisel, 6. Scraping knife, 7. Mallet, 8. Hairbrush, drawn by Pan Jixing (2001)

in the left hand and a graver in the right hand, he cut from the distant marks to the near ones, from the left to the right of the board. For each character, different strokes like transverse line (—), leftfalling (╯), rightfalling (╲), straight (|) and point (╮) should be carefully cut (Figure 68). The gravers should be carried neither upright nor flat, but with a certain bevel angle.

Figure 68   Procedures of Block Cutting
1. Transcribing the manuscript, 2. Unite the sample in a reversed position to the surface of block,
3. Carving characters, from Pan Jixing (2001)

When cutting lines in various shapes, all kinds of gravers with different sizes and shapes were used, some with wide and flat or narrow and flat edge, some with bevels and others with two edges at the opposite extremities of the same handle. The gravers were made of steel with wooden handles. After the characters were carved, the rest on the wooden surface should be removed with different knives, so that characters were in high relief. The large blank could be shoveled with round nose chisel. The cutter hits the handle of the chisel with a mallet in order to push the chisel forward and removes the useless wood dust. Then the character was 1~2 mm higher than the surface of the board. At last, the flat nose chisel was used to remove along the lines of each character the portion of the wooden surface which was not covered. After the procedure of carving out, the frames of the board were cut with a saw, and polished with a scraper so that the lines were in symmetrical pattern. The lines of characters were marked with a graver along the edge of a ruler. After the carving was finished, the woodblock was washed with hot water in order to remove the residue like wood dust.

The next procedure was printing the sample. A professional person would ink the face of the characters and the paper then being laid on the block. A number of copies were printed out for proofreading. If there were too many mistakes, the cutter had to do it again, which was a rare case. Any slight error may be corrected, as in the woodcuts, by taking out the wrong characters and inserting small pieces of wood with new characters in high relief. After proofreading, it moved to the step of printing. The ratio of carbon black and glue was between 100 : 20 and 100 : 25. Carbon black and glue were mixed and stirred into something like thick cream and were added by a small amount of wine. The liquid might turn into black paste after half a month and then be kept in a vat for fermentation❶.

When printing the book, the plate was fixed on the wooden table and the paper in neat order, ink and brush were put next to the plate. A pile of paper were fixed with a clip along the left side of the plate (Figure 69). The printer dipped a cylindrical flat brush in ink and brushed on the plate, laid a sheet of paper on the plate, then ran the dry brush over it so as to take the impression. He turned the printed paper over and put another sheet on the plate. The ink was added when necessary. This was repeated until the required number of copies was printed, and then the printed sheets were bound into a book in correct order (Figure 70).

Figure 69   Procedures of Printing, from Pan Jixing (2001)

---

❶ 卢前.书林别话[M]//张静庐.中国现代出版史料:丁编(上卷).北京:中华书局,1959.

Figure 70   Procedures of Binding, from Pan Jixing (2001)

It was required that the ink color should be the same and any spot near the character was not allowed. The procedure of printing was crucial to the printed matter. Generally speaking, one person could print 1500 to 2000 sheets every day and each plate could be used for ten thousand times[1]. After printing, the plate should be stored properly in case of reprinting.

In summary, the woodblock printing included the following procedures: (1) preparing the plate; (2) transcribing the manuscript and printing a sample; (3) proofreading; (4) adhering the sample to the plate in a reversed position; (5) carving the characters on the plate; (6) clearing the plate; (7) printing a sample page; (8) proofreading; (9) correcting the mistakes; (10) proofreading again; (11) inking the plate; (12) putting the printed page in correct order; (13) binding; (14) packing for shipment. Some procedures might include several steps. The process of making a plate covers more than half of the whole procedures, so it played a vital role in printing.

2.2.3.3   Binding Technique of Printed Books

Printed paper sheets should be bound into a finished book. The binding forms had undergone several changes before the final fixed form appeared[2]. In the Sui, Tang Dynasties and the early period of Five Dynasties, a single sheet without binding could be used directly or mounted for reinforcement. But a multi-sheet book should be bound into a scroll, which was the earliest form of

---

[1]   NEEDHAM J. Science and Civilization in China: vol. 5[M]. Cambridge: Cambridge University Press, 1985.

[2]   蒋元卿. 中国书籍装订技术的发展[J]. 图书馆学通讯, 1957 (6): 20-25.

binding. Such copies as *Miaofa Lianhua Jing* (《妙法莲华经》, *Lotus Sutra* or *Saddharma Pundarik Sūtra*) printed in the Wuzhou of the Tang Dynasty unearthed at Turpan, Xinjiang in 1903, *Wugou Jingguang Da Tuoluoni Jing* (《无垢净光大陀罗尼经》, *Aryarasmi-vimalvi suddha-prabha nama-dharani sutra*) printed in the year of 702 A.D. in Luoyang found in Kyongju, Korea in 1966, *Jingang Jing* printed in 868 found in the Dunhuang Caves in 1907, *Tang Yun* and *Yu Pian* printed during the reign of Emperor Yizong in Chengdu, Sichuan and *Jiu Jing* printed in the Five Dynasties were all in scrolls. The scroll form originated from the binding form of hand-written books before the Tang Dynasty.

The printed paper sheets in the same size were put into a long roll with a special paste. The roll could be as long as a few meters to a dozen meters. A scroll or a long stick as a central axis was put on the blank sheet at the end of the roll, which was made mostly from wood, sometimes painted with red lacquer, or from jade, ivory and other materials. The long roll was wound up from left to right along the axis, and then a scroll book was made. On the right was the beginning of the roll, wrapped with paper or silk for protection and pasted with a string of pin, which was used to tie up the scroll. A bookmark was placed on the cover page, and the scroll could be laid on the bookshelf. Another bookmark could also be hung on the end of the axis (Figure 71). There

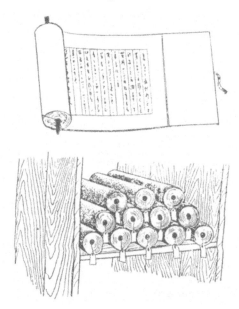

Figure 71   Scroll Binding, from Liu Guojun (1962)

were edges of paper around the printed page, more than 10 words to 20 words in each line and 20 to 30 lines on each page. A line as a boundary among characters could be found or sometimes not found on the page. Such information as the title of the book, the volume number, the name of the chapter and the author were printed at the preface, and then the text. The notes were smaller characters in double lines. The end of the book was inscribed with the inscriptions, indicating the name of the cutter and exact year, which were sometimes printed at the beginning of the book.

At first, the printed book took the entire form of a handwritten book. However, in order to read text at the end of the scroll, the entire scroll must be unwound, which was very inconvenient. Then another binding form emerged at the right moment. The long roll was folded into a stack of paper sheets in the same size (about 10 cm wide). For protection, the covered pages or the preface and end were attached to stiff boards, which was pasted with silk cover. Then a bookmark recording the book title and volume number was stuck on the cover. Therefore, the book in the original cylindrical shape became flat like a long and narrow cube. This binding form appeared approximately at the end of the Tang Dynasty (mid-ninth century), which were still in existence in the Song and Yuan Dynasties and lasted until the Ming and Qing Dynasties. The extant materials were Buddhist scriptures of the Five Dynasties discovered in the Dunhuang Caves. This form was related to Buddhism because Buddhist scriptures were often bound by this method, called a leaf, and also known as accordion binding or Sanskrit-clip binding (Figure 72). It might be an imitation of palm leaf manuscript in India and was designed by Chinese monks engaged in printing. But it was only similar in appearance, and the actual binding method was different. The folded books by accordion binding were composed of several volumes, each of which was folded up. The books were then covered with a

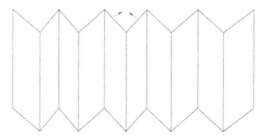

Figure 72　Accordion Binding, copied by Pan Jixing on Basis
of Collection in National Library of China (1998)

folding case made of thick paperboard and wrapped with cloth or silk. The books with a bookmark could be laid or kept upright.

The folded books, like an accordion, could be read page by page, which was much more convenient than the scroll. Once reading off some parts, the reader could put a bookmark here and continue reading next time. Many Buddhist scriptures in the Song, Yuan and Ming Dynasties were bound in this way. Some rubbing copies from a stone inscription were also bound in this form, but accordion binding was rarely used in non-religious writings. However, a major disadvantage was that the folding parts were exposed to the outside, thus easy to break, which seldom appeared in the scrolling form.

In the Tang Dynasty (mid-ninth century), a whirlwind binding form (Figure 73) appeared as an improved form of the scroll. As recorded in Volume 3 of *Mozhuang Man Lu* (《墨庄漫录》, *Notes of The Song*) by the Song Dynasty scholar Zhang Bangji (1090~1166), "a Tang Dynasty scholar Wu Cailuan, who was good at dictionaries and lexicons, once transcribed *Tang Yun* (《唐韵》, a rhyming dictionary published originally in 751) and sold out the copies ... Some extant copies of *Tang Yun* were also bound in whirlwind form." The recorded copy by Wu was now preserved in the Palace Museum in Beijing. The paper of this copy was thicker, waxed and written on both sides. Each volume consisted of 4 to 5 paper sheets, pasted on a long thick paper from the left to the right, just like fish scales. A wooden axis was put at the beginning of the long roll and the roll was tied up around the axis. When the roll opened, the reader could turn page by page and read texts on both sides of the paper sheet, which was like a whirlwind. But it was not sure whether the printed book was also bound by this method, since there was no real printed matter. Another version was the long roll, which was folded in accordion binding and wrapped by thick paper as the cover. This has also not proven right because no material object has been found yet. In fact, whirlwind binding was in name only and there has been a huge controversy over the specific forms of "whirlwind binding" of the ancient

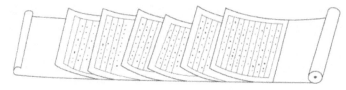

Figure 73   Whirlwind Binding, copied by Pan Jixing on Basis
of Collection in National Library of China (1998)

books.

The binding of printed book should follow the development trend from cylindrical to flat shape, and the appearance of butterfly binding adapted to this trend. Each printed sheet was folded along the middle of the page by leaving the two half pages with characters inside and the blank pages to the outside. All the folds in neat order were pasted together and wrapped by thick paper. This was the original form of the album, with thick paper covered on the top and the bottom of the book, sometimes wrapped with silk for protection. The title tag was attached to the upper left corner of the cover. The paper sheets were folded up, thus overcoming the disadvantage of accordion binding. When you opened the book, you might come across a page with characters or no character, and turning the page was like the wing of a butterfly, so it was called a butterfly binding (Figure 74). The written copies in butterfly binding appeared between the late Tang Dynasty and the Five Dynasties. The printed copies in this form came out in the Song Dynasty. The copy of *Wenyuan Yinghua* (《文苑英华》, *Collection of Poems in Ancient China*), printed in the year of 1260 and now preserved in the National Library of China, was in butterfly binding form. It indicated that printed copies in butterfly binding appeared in the Northern Song Dynasty at the latest.

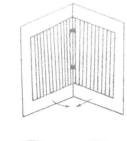

Figure 74   Butterfly Binding, from Liu Guojun (1962)

The butterfly binding was more suitable for printing because binding was more convenient for reading than the roll, and it was easy to store and carry. Besides, it overcame the shortcoming of easy breakage of folded books. However, it needed more time to consult certain texts since there were many blank pages. Therefore, it was replaced by the wrapped-ridge binding after the Yuan Dynasty. The wrapped-ridge binding originated from the Song Dynasty. The printed sheets were folded into folio, but opposite to the butterfly binding, the two half pages with characters were put outside and the blank pages inside. The folded sheets were piled in neat order and pasted or bound with paper string, and the unnecessary edges were cut. Then the book, including preface, book ridge and back, was wrapped by thick paper. That was why it was called

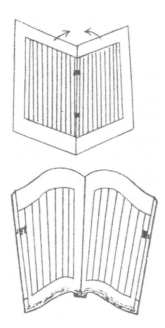

Figure 75 Wrapped-ridge Binding, from Liu Guojun (1962)

wrapped-ridge binding (Figure 75). There was a silk cover with title tag on it for some exquisite books. It was very convenient for reading and needed to be laid on the shelf. Sometimes, several volumes of a book were placed in a folding case made of thick paperboard for protection and could be kept upright. This type of book had been prevalent till the middle of the Ming Dynasty (mid-16th century) and was then replaced by a stitched book in the Qing Dynasty.

A sample of this binding was *Zhuwengong Jiao Changli Wenji* (《朱文公校昌黎文集》, *Discussion with Changli by Zhuxi*) reprinted by Rixin Hall in 1281. The paper string of the wrapped-ridge binding was easy to break and the sheets would be scattered. So the binder used the silk thread or hemp rope instead by punching small holes and stringing the sheets. Therefore, the wrapped-ridge binding evolved into stitch binding (Figure 76). In this way, the ridge of the book was not pasted with thick paper. Only a soft cover of yellow or blue paper was covered in the front and back of the book. The title tag was pasted on the top left corner of the cover. Of course, a number of volumes could also be enclosed in a folding case for protection. It was believed that the stitched book originated from the middle of the Ming Dynasty. Actually, it had been very popular at that time. The author had seen the original stitched book among

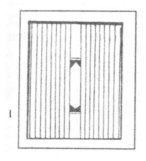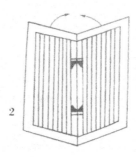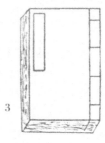

Figure 76 Stitch Binding
1. Printing, 2. Folding along the middle of the page, 3. Binding, from Qian Cunxun (1990)

manuscripts in the Five Dynasties unearthed in the Dunhuang Caves. It was bound into several volumes with hemp strings and there were characters on both sides of the paper. This showed that the idea of stitch binding occurred a long time ago. It had become the dominant form of binding since the Ming and Qing Dynasties.

According to the paper by Jiang Yuanqing, the binding techniques were summarized in the following steps: (1) folding along the middle of the page by leaving the two half pages with characters outside; (2) keeping the folds neat in sequence; (3) adjusting the pages along the middle part and pressing from both sides; (4) adding 2 or 3 blank pages before and after each volume; (5) binding the pages with paper string; (6) covering the binding form with yellow or blue paper, with a thin sheet inside; (7) cutting away the edges; (8) sanding paper; (9) packing the book corners; (10) punching 4 to 6 small holes near the book ridge; (11) putting silk thread or hemp rope across holes; (12) pasting a bookmark. In wrapped-ridge binding, the first five steps were followed by pasting the book ridge and covering. In order to be kept upright, the book could be clipped with two pieces of wood and then strapped. Besides, the book could be covered with a folding case made of thick paperboard and wrapped with cloth or silk.

2.3   Invention of Non-Metal Type Printing

2.3.1   Invention of Wooden Movable Type Printing

In view of the very important role of typography in the history of world civilization, it is necessary to study it in two chapters: the previous chapter mainly discusses the invention of wooden type as the earliest form of printing, and also covers the origin of copper-plate printing; in turn, this chapter will discuss the invention of movable type printing, including both non-metal type and movable metal type printing. It took China 400 to 500 years to develop from woodblock printing to movable type printing, with the latter appearing in the Northern Song Dynasty (11th century) as an inevitable product of the former. If we regard the technological development as a ladder, there are three rungs in Chinese typography: block printing →movable non-metal type printing →movable metal type printing. The three rungs were not completed until the 11th century. We can also regard the ladder as a typical example for the rest of the world since in some countries in East Asia and Western Europe, although they had similar ladders, the time needed to cross from one rung to another was greatly shortened because they might have learned from Chinese experience. It not only reflects the technological laws within the development and evolution of the early printing history, but also fully embodies the achievements of the technological exchanges between China and foreign countries over the past few centuries.

The major difference between block printing and movable type printing lies in different methods of plate making, albeit similar technologies relating to inking, brush marking and bookbinding. Plate making is a key process in printing technology, which requires totally different tools and operations, therefore, movable type printing can be regarded as an invention, but not simply as a technical innovation. In other words, movable type printing is another milestone in the history of printing technology after block printing. Movable metal type printing, the advanced form of movable type printing, is the starting point of modern world printing, so it is of epoch-making significance. Concerning woodblock printing and copper-plate printing, China was ahead of the rest of the world for hundreds of years, so it is quite natural that both movable non-metal type and metal type printing were invented first in this country.

Woodblock printing ruled over the printing industry before the 11th century. It might be far superior to hand-copied books, although it still wasted a generous number of woodblocks and time considering the fact that it was a one-time setup. To the dealers in the printing business, block printing took a much longer production cycle and it was a problem concerning the storage and management of piles of blocks. In the Northern Song Dynasty, woodblock printing came to its heyday, yet its inadequacies also became prominent. There was an urgent need for new printing technology. Fretsaw was used to cut each character out of the woodblocks, thus it was possible to set different characters onto one woodblock more freely, and one set of characters would accomplish the work of more than one block. From then on, a "dead" printing technology turned into a "live" one—the movable type printing.

As a matter of fact, the idea of movable type printing had long since been put into practice—it could be traced back to the casting of bronze in the Shang and Zhou Period. For instance, in 1920, Qintong Gui (青铜簋, a bronze eating utensil) was unearthed in Gansu Province. There are 50 characters inscribed on it. The rubbings showed that for each character, there is a *fan*（范，mould）; for each article, several *fans* are required. In the inscription, the boundaries between the characters are crystal-clear, so it is quite obvious that individual types（单个字的模）was pressed into the pottery *fan*, and then were casted. At the end of the 19th century, the pottery unearthed in Linzi, Shandong Province, was inscribed with the imperial edict of the First Emperor of Qin in 221 B.C.. The inscription consists of ten *fans*, with two characters per line, and two lines per *fan*. These practices in the pre-Qin period may provide ideas for the use of movable type in printing in the Song Dynasty, but they are not movable type printing, for the social and technical requirements for printing were immature.

Movable type printing can only be invented after woodblock printing, for it can improve the platemaking efficiency, save plates and costs. After printing, the movable types can be removed and be reused in the future. Thanks to its small size, movable types were easy to store and move. Considering the laws of technical development, wooden type is one of the earliest movable types—it might be directly born from wood engraving. Since wooden type can be regarded as a single wood engraving, we can cut one wooden type just as we can cut a wood engraving with many characters on it, except that we use wooden types to print. It is still a question about when China began to print books from wooden types, but it is quite certain that it would not be very late. It is no later than the end of the Five Dynasties to the beginning of the Northern Song Dynasty（11th century）that the earliest movable wooden types were put into experiment while engraving printing was undergoing major development. The skill had entered the practical stage in the first half of 11th century.

Since the beginning of the Northern Song Dynasty, all the prefectures and counties were issuing a considerable number of land deeds. The government set a uniform format for all the deeds and printed them with wooden plates. Ma Duanlin（1254~1323）, a historian in the Yuan Dynasty, wrote in Scroll 19, *Wenxian Tong Kao*（《文献通考》, *Examination of Historical Documents*, 1307）, that in Shaoxing Five Years（1135）:

> *At the beginning, the local magistrates of all the prefectures were ordered to print and sell land deeds ... But the county magistrates often printed extra deeds and put them on illicit sale, so now the land deeds must be printed on a monthly basis with Qian Zi Wen（千字文, One Thousand Characters）as the serial number. When the people bought the deeds, the officials handed them over.*

In other words, if the people intended to buy and sell land and real estate, they had to buy land deeds from local governments. After the parties concerned filled in the deeds, the government had to check and stamp the deeds to make them valid. The truth was that, however, almost every local government printed extra deeds to make profits. In order to put an end to the malpractice, it was decided that the vice-prefects of each prefecture be responsible for printing land deeds with one thousand characters as the serial number on the paper, and the deeds would be issued to each prefecture and county each month, thereby to control the amount of printing. The classic article of *Qian Zi Wen* originates from Zhou Xingsi's (周兴嗣, 470~521 A. D., Liang Dynasty) book *Thousand Character Classic* (515 A. D.). The book contains exactly one thousand characters, each one used only once. The article was used as a primer for teaching Chinese characters to children from the sixth century onward, so with different combinations of the characters, we can have different serial number for different deeds.

Xie Shenfu (谢深甫, 1145~1210, Song Dynasty) wrote in Scroll 30 of *The Total System* (《经总制》), from the book *Qingyuan Tiaofashi Lei* (《庆元条法事类》, *A Collection of Rules and Regulations in Qingyuan Period*, 1202):

> For each household that applied to buy deeds, the old traditions should be observed, that the prefects in every prefecture should set up Liaoli (料例, *standards of material*) with the thousand characters printed on them as serial number. The deeds should be collected by the subordinate county magistrates and handed out to applicants.

The so-called *Liaoli* (料例), *Liaohao* (料号), *Ziliao* (字料) *or Zihao* (字号) refer to the serial number of the thousand characters on the deeds. In the book *Song Hui Yao* (《宋会要》, *The Collection of the Song Dynasty Official Documents*) about *Shi Huo* (《食货》, *Farming, Cloth and Currency*), in 1171 (乾道七年, the seventh year during the reign of Qian Dao) notes:

> The rule is that the deeds must be printed by every prefect and the serial number should be the thousand characters before they are sold by the counties. The personnel officers should set up Liaoli (料例, *standards of material*) with the thousand characters printed on them as serial number. The deeds should be issued to subordinate prefectures and sold to applicants.

This shows that all the deeds with serial number were all registered in the government book before they could be issued to the subordinate prefectures or counties, so that they were all traceable afterwards. The above mentioned policies can be traced back to the early Northern Song Dynasty （北宋） with Emperors Taizong （太宗，976~997）, Yongxi （雍熙，984~987）, and Duangong （端拱，988~989） when the license for sale of salt and tea were issued thus. Later on, in 1048 （庆历八年, the eighth year during the reign of Qingli） and 1074 （熙宁七年, the seventh year during the reign of Xining） the laws were also observed in issuing government license for sale of salt and tea. After the merchants paid the governments, they were given licenses （引, yin） with thousands of characters serial number on them and the licenses were registered officially. Since the 11th century, all the deeds were printed with wooden blocks, so it was most convenient to leave indentations on the plates to fill in corresponding serial number types, therefore, there was no need to reprint each deed character for character. In conclusion, real estate deeds or licenses for sale of salt and tea were in fact the products of combining wooden block printing and movable wooden type printing.

Besides official deeds, movable wooden type printing could also be used to print Buddhist sutras and other texts. The skill was soon spread to Xixia （西夏，1032~1227）. Xixia was founded by Dangxiang ethnic group （党项族） in northwest China, where Han （汉）, Tufan （吐番）, Huihu （回鹘） and Qidan （契丹） ethnic groups living together. When Li Yuanhao （李元昊，1003~1048） took throne, he claimed to be King following the Han customs, and he named his state Da Xia （大夏, the Great Xia）. People called it Xixia because it was located northwest to the Song Dynasty. Ye Li Renrong （野利任荣）, the head of secretariat and *Moning Ling* （谟宁令, an honorary title）, coined Fan Wen or Xixia Wen （蕃文,西夏文, Xixia characters） and ordered the use of the new characters along with Chinese characters within borders in 1036. The papermaking and printing industry had developed well in the country and Buddhism also thrived there. Thanks to the frequent communication with the Northern Song Dynasty, its level of productivity was close to that of the Central Plains of the Northern Song Dynasty. There were papermaking institutes and engraving departments within the country, responsible for producing paper for officials to use and to print. They put emphasis not only on wooden block printing, but also on wooden movable type printing, which has been verified by unearthed remains. The remains are very helpful to understand the early moveable type printing in China from the 11th century to the 13th century.

In September, 1991, on the old haunts of Xixia, in the Baisigou Square Pagoda （拜寺沟西夏方塔） in Helan County, Ningxia Province, the printed book *Mahā-laksmi-dhārani-sūtra* （《吉祥遍至口和本续》）was unearthed. This is a Tibetan version of the Buddhist tantra, the counterpart of Chinese *Daharani-sūtra* （《大吉祥陀罗尼经》）. It was printed on white hemp paper （白麻纸）（Figure 77）, with 100,000 characters and 220 pages altogether. There are 9 scrolls on accordion binding （经折装）. Every half page is 23.6 cm in height and 15.5 cm in width, with two lines all around （四周双边）, with white central seam （版心白口）, without tail （无鱼尾）, but with pages. The size of the big characters is 20 mm, and the size of the small ones is 6 mm～7 mm. The strokes vary in size; the color of the ink is unevenly, and individual characters were printed upside down. For

Figure 77   Printed Book *Mahā-laksmi-dhārani-sūtra* （《吉祥遍至口和本续》） from Niu Dasheng, 1994

the box, the lines do not join each other on each corner, and for the version of the heart, the lines differ in length. For the pages, there are quite a few misprints and haplographies. There are traces of bamboo pieces serving as the spacing of boundary. All these signs show that this sutra is a copy printed with wooden movable types. Within the text, there printed such Xixia characters as "the persons in charge for the print are Śramana Gao （沙门释子高）and Fa Hui （法慧）", yet on the pages and in the central seam, there are Chinese numerals "four" "twenty-seven" and "twenty-two", etc. Some characters were printed upside down. These signs show that the persons in charge of the printing are Monk Zi Gao （子高）and Fa Hui （法慧）, the printers Hans （汉人）, and Dangxiang clansmen were in charge of setting types for printing. The printing took place in the latter half of the 12th century （1150～1180）, about the Xixia Renzong period （西夏仁宗时期, 1140～1193）. This conclusion was confirmed by the experts from Ministry of Culture and Technology on 16, November, 1996. It is believed that this copy is the extant earliest wooden movable type printing.

In 1907, the Russian Peter Kuzmich Kozlov (1863~1935) discovered more than 2,000 manuscripts and printed books in both Xixia and Chinese characters in the site of Xixia Kharahoto along the bank of Erginar River in Inner Mongolia. The texts are now in the Institute Vostokovedeniia. Among the texts, the book *Sandai Xiangzhaoyan Wen Ji* (《三代相照言文集》, No. 4166) printed in the 12th and the 13th centuries was identified as wooden movable type printing. This book is a non-religious book with 41 leaves (82 pages) in butterfly binding style, and each half page is 24 cm in height and 15.5 cm in width. The printing chase (版框) is 17 cm in height and 11.5 cm in width, two lines all around (四周双边), 16 characters per line and 17 lines per half page. Within the white central seam, there are Xixia and Chinese characters; at the end of the scroll, there is a vowing article, reading "the vower is a upasaka and *prince* (节亲主, the peculiar way of naming a prince in Xixia) …//the printer who used wooden movable type printing method is *Chen Jijin* (陈集金)". It is clear that the printer Chen was a Han Chinese people. Shi Jinbo (史金波), a Xixia character expert, claimed that since *Jieqinzhu* (节亲主, prince) is a peculiar Xixia title, the person who vowed to print this book should be Hui Zhao (慧照) from the royal imperial family, and his name was Wei Ming (嵬名氏).

In St. Petersburg Branch of the Institute of Oriental Studies, one can find the book *Dexing Ji* (《德行集》, *The Book of Virtues*) (No. 799, 3947) from the Xixia period. The book is also a wooden movable type printed book in butterfly binding with 26 leaves (52 pages), 7 lines per half page, and 14 characters per line. As for the printing chase, there is one line all around, with white central seam and both Chinese and Xixia characters in the seam. At the end of the scroll, there is a publisher's note that "the sponsor for the printing and proofreading is the director of Fan-xue College (番学大学院, a college for foreign students in the ancient time)", academician and prince Wen Gao (文高). In 1917, the Xixia version of the printed book *Buddhā-vatam-saka-mahā-vaipulya-sūtra* (《大方广佛华严经》) was unearthed in Lingwu County (灵武县), Ningxia Province, and later it was spread to all over the world. Experts from both home and abroad believe that this is a wooden movable type printed book, though they disagree with each other about the year of the publication. Shi Jinbo (史金波) believes this book was published in the late Xixia period (1162~1227), and according to the collection in National Library in Beijing, there are the names of Sheng Lü (盛律), Meng Neng (美能) and Hui Gong (慧共),

monks responsible for the printing of the book❶. Lingwu County was close to Zhongxing Fu（中兴府，now Yinchuan City，Ningxia province），the capital of Xixia，so this old tome should be printed in Zhongxing Fu.

In 1970，Ningxia Museum had received two packages of printed Buddhist sutras in Xixia characters. Wang Jingru（王静如），the expert of Xixia characters，claimed that these sutras were translated from the remaining chapters of Chinese Buddhā-vatam-saka-mahā-vaipulya-sūtra（《大方广佛华严经》），which contains the regnal name of Xixia Renzong（《西夏仁宗》，1140～1193）. Mr. Wang also read the following vows from the end of Scroll Five collected by the Institute for Research Humanities，Kyoto University："all vow to use wooden movable type；and Duluo Huixing（都罗慧性，a Dang Xiang man），the one responsible for the printing，also vows that all the people rejoice with us shall all attain Buddhahood." This sutra was claimed to be printed and published in Da-de Year（大德年间，1297～1307），the beginning of the Yuan Dynasty❷. In short，it is quite certain that the Xixia Buddhā-vatam-saka-mahā-vaipulya-sūtra（《大方广佛华严经》）was published between the end of the 12th century and the beginning of the 14th century，and it is certainly an early wooden movable type printed book.

The excavations of the above mentioned Xixia wooden movable type texts have provided important physical materials for the study of early wooden type technology in China. This skill was learned from the Northern Song Dynasty，so the Northern Song Dynasty must also have applied the skill to printing books besides land deeds，yet we still have to wait for the excavations of such texts and copies. In the Southern Song Dynasty and the Yuan Dynasty，the skill would be further developed. The scientist Wang Zhen（王祯，1260～1330，the beginning of the Yuan Dynasty），in his article Zaohuozi Yinshufa（《造活字印书法》，The Practice of Typography，1298）had drawn a systematic conclusion and given full explanation of the skill since the Song Dynasty. Wang Zhen included this article at the end of his well-known book Nong Shu（《农书》，Book on Agriculture，1313）as an appendix. From 1295 to 1298，when he was the governor for Jing De County，he had 30,000 wooden types manufactured to print the book Jingdexian Zhi（《旌德县志》，Jing De Annal）that counts 60,000 Chinese characters and at the same time he had invented a turntable storage device to

---

❶ 史金波，黄润华. 北京图书馆藏西夏文佛经整理记[J]. 文献，1985(4)：238-251.

❷ 王静如. 西夏文木活字版佛经与铜牌[M]. 文物，1972(11)：8-18.

store all the types.

In China, except for Xixia, the Uygur area was also leading in developing woodblock and wooden movable type printing. In 1908, a French named Paul Pelliot (1878~1945) discovered a barrel of wooden movable type printed Uygur types (altogether 960 pieces)[1] in a vault of the Thousand-Buddha Cave (千佛洞, Dunhuang Mogao Grottoes) in Gansu Province. According to Carter: "Pelliot claimed these types to be from the year of 1300 since they were stored in No. 181 cave (now No. 464), among other factors. There are hundreds of them, mostly in good condition. The wooden movable types were carved from hard wood and sawed to the same height and width, which was exactly what Wang Zhen had described in his article."[2] Now the wooden types are collected in the Musée Guimet in Paris, with a few going to the Metropolitan Museum of Art in New York City or into private hands. It is said that the Uygurs had used these types (Figure 78).

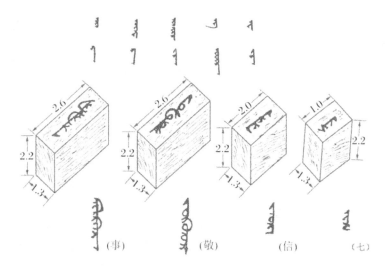

Figure 78   Wooden Movable type Printed Uygur Types from the 12th to the 13th
Centuries unearthed in Dun Huang in 1908

The above-mentioned Uygur types are 2.2 cm in height, 1.3 cm in width and 1.0 cm~2.6 cm in length, so they are a bit different from the Chinese

[1]   PELLIOT P. Une bibliothèque médiéval retrouveée au Kansou[J]. Bulletin de l'Ecole Française d'Extrême-Orient (Hanoi), 1908(8): 525-527.

[2]   CARTER T F. The Invention of Printing in China and Its Spread Westward[M]. New York: Columbia University Press, 1925.

ones. The Chinese characters are ideographs, so every type is of the same size; but the Uygur language is alphabetic, so each type contains various letters with different lengths. In order to save wood, every Uygur type is carved with characters on two sides. After studying the types in the Musée Guimet, the Chinese expert of the Uygur language, Ya-sen Wu-shou-er (雅森・吾守尔) found out that there are small clamping strips with punctuations, lines and blanks carved in them. Considering the changes in grammar, there are even prefixes and suffixes for some words. These types should have been completed between the end of the 12th century and the early part of the 13th century. The Uygur types have provided references for how the movable type printing had transferred from the Chinese types to alphabetic types, and for how the skill was introduced into the Western World through Xinjiang Uygur Autonomous Region.

In the central seams of Xixia wooden movable type printed books, it is more than often to find a few Chinese characters there, which suggests that there must have been Chinese printers at work along with local workers. Unfortunately, fewer Chinese wooden movable type printed books survive the time. In March, 1999, on the exhibition of the books from the Song Dynasty in the National Palace Museum in Taiwan in China, I noticed that in the book (It is of Hui Zhou edition printed in 1252) *Yili Yaoyi* (《仪礼要义》, *The Essence of Etiquette*), the characters are not neatly aligned and the color of the ink is not even, so the book might be a wooden movable type print. This book is one of the series of *Jiujing Yaoyi* (《九经要义》, *The Essence of Nine Classics of Confucianism*) of the same edition. Miao Quansun (缪荃孙, 1844~1919), the expert of Textual Bibliography from the Qing Dynasty, had written an epigraph in his own collection of *Di Xue* (《帝学》, *The Emperor's Education*) written by Fan Zuyu (范祖禹, 1041~1098) and printed in 1221. In the epigraph, he claimed that the book was a wooden movable type print and was printed by Fan Ze (范择), the fifth grandson of Fan Zuyu. Other experts from both home and abroad agreed with the conclusion[1]. Although certain opinions are that "there are overlaps between some characters", so it can't be a wooden movable type print[2], Miao's conclusion can't be altered, for overlaps are quite common in

---

[1] HUMMEL A W. Movable type printing in China[J]. The Library of Congress Quarterly Journal of Current Acquisition, 1944, 1(2): 13.

[2] 张秀民. 中国印刷史[M]. 上海:上海人民出版社,1989.

wooden movable type printed books. For instance, in the clay type printed book *Foshuo Wuliang Shou Jing* (《佛说无量寿经》, *Buddha Speaks Infinite Life Sutra*) from the Song Dynasty (published in 1103 and unearthed in the White Elephant Tower in Wenzhou City, Zhejiang province, in 1965), and in the copper type printed books in the Korean area, similar overlaps are also common.

2.3.2   Wooden Movable Type Printing After the Yuan Dynasty

In his book *Jingdexian Zhi* (《旌德县志》, *Jingde Annal*), Wang Zhen (王祯), who lived in the early years of the Yuan Dynasty, wrote:

> *When I was the governor in Jingde County (1295), I had the printers carve many (wooden) movable types. The job was done within two years (in 1297); two years later (1299), when I was maneuvered to Yongfeng County, Xinzhou, I took these types with me. At that time, I had just finished my book Nong Shu (《农书》, Book on Agriculture, 1313), and I wanted to print it with wooden movable types. Now in Jiangxi Province, I have the knowledge that there are already printed editions of the book, so I have the newly carved types stored for future use* ❶.

Since there was already the skill of movable type printing in the Song Dynasty, so Wang shouldn't be regarded as the inventor of movable type printing, but his contributions are that he created the turntable for storing and selecting types—before him, people used cabinets (活字字柜) to store types.

In China, it is the wooden movable types that have been recorded to the fullest degree. There are two recordings about wooden movable type printing, one is the article *Zao Huo Zi Yin Shu Fa* appendixed to the book *The Agricultural Book* (1313) written by Wang Zhen, and the other is *Wuyingdian Juzhenban Chengshi* (《武英殿聚珍版程式》, *The Illustration of Movable type Printing Written at the Wu Yin Palace*) written by Jin Jian (金简, 1724~1794), a scientist in the Qing Dynasty. Both texts were detailed narrations of wooden movable type printing, based on the authors' personal experience in printing and thus both books are credible, with the latter an illustrated edition. Now, I will start the following passage introducing the technology recorded by Wang

---

❶   王祯. 农书:造活字印书法[M]. 上海:上海古籍出版社,1994.

Zhen, and then I will discuss what Jin Jian had written down, and lastly, I will compare the former with the latter. Wang Zhen's article is explained in vernacular Chinese as follows.

In the first place, one marks the paper with squares of the same size, and then a person who was good at calligraphy chose the characters according to the official rhyming system issued by the Imperial College, and he wrote down the regular scripts in the sequence of five-tone head vowel. The printer pasted the paper on a polished board and carved the characters into raised types in reverse direction. There are gaps around the four sides of the characters so that it is easy to be saw-cut. Commonly used characters were classified into different groups and should be printed in abundance. These characters were cut by fretsaws along the gridlines into wooden movable types, and there were altogether 30,000 characters. All the types were smoothed by small knives and were put into rectangular sheaths to see whether they fit the criterion. If they didn't fit, they should be remade.

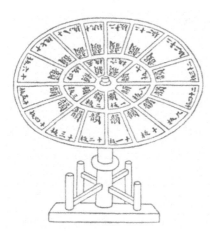

Figure 79   A Turntable to Store Types Invented by Wang Zhen in 1298, from *The Agricultural Book Four*

In the Song Dynasty, standing wooden cabinets were used to store types. The types were put into small cases within little drawers in the cabinets. The drawers were labeled from outside with the head vowels of the types so that it was easy to fetch them. Relative illustrations can be referred to in Jin Jian's book. Wang Zhen, nevertheless, had considered the inconvenience of such cabinets, so he invented a turntable (Figure 79) to store types. The turntable was round and made of balsa wood (轻木). It was 215 cm in diameter and its shaft was 92.2 cm in the height. The rotating table was supported by the shaft and the shaft was supported by horizontal support frames. There were altogether 24 word-boxes arranged in two lines on the turntable. Each box had its own number and vowel to which types were stored. During printing, there were two sets of turntables for printers to fetch types without having to move around cabinets. This is exactly why Wang had invented it in the first place, but it was only efficient to print books with fewer words; for tomes or books with more words, cabinets were still the only wise choice.

In order to quickly fetch the types, the printers had to copy the rhyming book that they used to choose the characters, and each line and page of the book should be numbered to correspond with the numbers of the types in the word-boxes. The indexers couldn't memorize all the numbers of the types, so another printer was needed to read out loud the numbers of the types in the word-boxes, and then the indexers needn't waste time in referring to the book. The chosen types were put onto an indexing plate (检字盘) before they were put onto the printing plate into composition. With two people working together, it was more efficient (Figure 80).

Figure 80   Wooden Movable Type Printing Process depicted
by *The Agricultural Book*, from Liu Guojun (1962)

The printing plate was made of wood. A dry wooden board of the size of the page of a printed book was smoothed before wooden lines were installed on all four sides. The lines were of the same size of the types. The four sides were covered with copper sheets. The types were placed from left to right, with bamboo slips being placed vertically alongside the types, a wooden wedge stuck from the right to fixate the bamboo slips. Since the alignment of ancient Chinese writing is upside down, and from left to right, the bamboo slips were always placed to the right of the types, from left to right. After the whole plate was full, another plate was needed to flatten the printing one. Then, a silk cotton cloth was dipped into the ink and would be used to color the printing plate. Finally, a piece of paper was put onto the colored printing plate and a brown brush was used to brush the inked characters onto the paper. Thus, pages were bound into books. Of course, proofreading was always the most important procedure to rule out all possible mistakes no matter whether it was during the carving of the types, planting of the types, or during the process of printing.

Generally, two sets of printing plates were prerequisite to printing a book so that when one plate was ready to be inked, types were aligned into the other one. After printing, the types would be taken down from the plates and cleaned for the next round. All the types were used for more than one time. Since the types were made of wood, the strokes of the characters were clear and easy to read; but it was also possible that the lines printed were sometimes not straight, or the color of the ink might somehow not evenly put, which, however, has become a feature for wooden movable type printed books. In China, from the Song Dynasty to the Yuan, Ming and Qing Dynasties, wooden movable type printing had prospered along with woodblock printing.

In the National Library of Beijing, there is the tome *Tangshi Lei Yuan* (《唐诗类苑》, *The Collection of Tang Poetry*), compiled by Zhuo Mingqing (卓明卿, 1552~1620), printed in the year of 1586. At the bottom part of the central seam, there are such characters of "*Song Zhai Diao Mu* (崧斋雕木)" which means Zhuo Mingqing printed the tome with wooden movable type printing technology. In other words, more than a compiler, Zhuo was also the publisher. In the Nanjing Library, there is the book *Bishui Qunying Daiwen Huiyuan* (《璧水群英待问会元》, a textbook for the students of the imperial college in the Song Dynasty), compiled by Liu Dake (刘达可), and was printed around 1515~1530. At the end of the book, there are four lines: "Printed and published by Li Ze House (丽泽堂) // Written by Hu Shengshan (胡昇缮) from Gu Su (Su Zhou area) // Carved by Zhang Feng (章凤) // Printed by Zhao Ang (赵昂)". This is also a wooden movable type printed book. These two books are both from the Ming Dynasty.

During the Wan Li Years (万历年, 1573~1620), there was another well-known wooden movable type printed book, *Shimiaoshi Yu Lu* (《世庙识馀录》, a book of a variety of anecdotes between 1522 and 1566). It was published by Xu Zhaoji (徐兆稷), the son of the author Xu Xuemo (徐学谟, 1522~1593). The book consists of 26 scrolls and about one hundred copies were printed. Within the book, there are such lines:

> The book has been completed for more than ten years, yet since I cannot afford it, merely one hundred copies have been printed. All the copies are for family collection only. Even so, the printing has costed me much, so I cannot promise to print more. May the reader forgive me. Written by Xu Zhaoji (徐兆稷).

To conclude, what Xu wanted to convey is that he couldn't afford the woodblock printing, so he chose the relatively cheaper wooden movable type printing. He took advantage of the present types of the bookstores and copied one hundred books. The National Library of Beijing has collected another wooden movable type printed book *Jiaofeng Xiansheng Wenji* (《蛟峰先生文集》, *The Collection of the Articles Written by Mr. Jiao Feng*). The author is Fang Fengchen (方逢辰，1221～1291) from the Song Dynasty, but the book was printed by his 11th grandson Fang Shide (方世德) during the Wanli Years and was published in Chun'an, Zhejiang Province. Besides, in 1574, Zhu Yuyin (朱翊鈏, of the Ming royal family) had published the book *Bianhuo Bian* (《辨惑编》, *A Book of Confucianism*) written by Xie Yingfang (谢应芳) of the Yuan Dynasty. On the last page of the book, a line reads "wooden movable type printed version, published by Yi Fan (益藩, the fiefdom of Yi Prince)"❶. Judging from the extant heritage from the Ming Dynasty, the Wanli Years had witnessed the appearance of the most wooden movable type printed books, followed by the Zhengde Years (1506～1521) and the Jia Jin Years.

In Scroll Eight of the book *Jie'an Man Bi* (《戒庵漫笔》, *The Notebook of Jie'an*) written by Li Yi (李翊，1505～1593) in the Ming Dynasty, the author noted that Qian Mengyu (钱梦玉) had printed the paper of the imperial examinations compiled by his teacher Xue Yingqi (薛应旂，1509～1569) by using the wooden types from Donghu Academy (东湖书院). Mr. Xue became a *Juren* (举人, provincial graduate) in the year of 1535, and obtained the *Jinshi* degree (进士, the highest degree in the national civil service examination in imperial China) the next year, from which, it can be judged that Qian Mengyu printed the paper during the Jiajin and Zhengde Years (嘉靖，正德). Qian Fan (钱璠，1500～1557), also from Qian Mengyu's hometown, compiled *Xu Guwen Huibian* (《续古文汇编》, *A Sequel to Compilation of Ancient Articles*) and printed the book through Donghu Academy. In the article *Yu Gongsusheng Shu* (《与公肃甥书》, *A Letter to the Nephew of Gong Su*) from the book *Tinglin Wenji* (《亭林文集》, *The Collected Works of Ting Lin*), Gu Yanwu (顾炎武) noted that it was until the year of 1638 that the court aviso was starting to be printed with wooden movable type printing skill, but before, people had to hand-copy the aviso. A court aviso was the news briefing issued by the government. At the end of the Ming Dynasty, with the technology of movable type printing, news could be quickly spread around the country.

---

❶　张秀民. 中国印刷史[M]. 上海：上海人民出版社，1989.

In the Qing Dynasty, wooden movable type printing was in its flourishment with the governmental support. There were official printed books, printed books from workshops and private printed books. When the series *Wuyingdian Juzhenban Congshu*（《武英殿聚珍版丛书》, *The Series of Movable type Printing Written at the Wu Yin Palace*) was issued during the Qianlong Years, a large scale of wooden movable type printing was in operation. This series were published by the printing office and bookbindery at *the Wuyingdian*（武英殿, the Wuying Palace）, and the office was subordinate to the Imperial Household Department（内务府）. In the year of 1690, Wenshu Guan（文书馆, the Paperwork Office) was transformed into the printing office and bookbindery**❶**, and that was the beginning of the institute. In 1773, Emperor Gaozong in the Qing Dynasty Hongli（清高宗弘历, 1711～1799) commanded the establishment of the library of the *Si Ku Quan Shu*（《四库全书》, *Complete Library in the Four Branches of Literature*) to have Confucian officials proofread the lost ancient articles or books from *Yongle Encyclopedia*（《永乐大典》, 1408) compiled by secretaries（辑明）, seeking for the scattered books from private hands, and collecting books published at that moment, and thus the book *Si Ku Quan Shu* was finished in 1781 with 36,000 scrolls, hand-copied. The book was too huge to be printed, but at the very beginning of its compilation, Emperor Gaozong had ordered the publication of some of the works to present to the people who were eager to study.

Jin Jian（金简）, manager of the printing office and bookbindery at the Hall of Martial Valor and vice-president of the library of *Si Ku Quan Shu*, submitted memorials to the throne on 22nd, January, 1774 that the copper movable types which once were used to print *Gujin Tushu Jicheng*（《古今图书集成》, *The Collection of Ancient and Modern Illustrated Books*) during the Yongzheng Years（雍正年间）had been melted to mint coins, and since wood carving plates were too costly, he suggested using wooden movable types to print the *Si Ku Quan Shu*. Gaozong agreed with the suggestion and changed the name of *Huo Zi Ban*（活字版, wooden movable types）into *Ju Zhen Ban*（聚珍版, collecting treasures) because the Emperor believed the latter was a nice name. In May of 1774, Jin Jian had finished making 253,000 wooden movable types made from date wood, among other tools and materials, costing only 2,339 *liang*（两,

---

❶ 赵尔巽. 清史稿：职官志[M]. 上海：上海古籍出版社,1986.

50 g) of silver❶. According to Tao Xiang's (陶湘，1870～1940) catalog，there were altogether 134 categories and 2,389 scrolls❷. For all the printing，a common format was used：21 words per line and 9 lines per half-page；at the top of each page，there are characters "Ti Wu Ying Dian Ju Zhen Ban Shi Yun (题武英殿聚珍版十韵)"，and underneath each first line of the first page，there are characters "Wu Ying Dian Ju Zhen Ban (武英殿聚珍版)".

Five to twenty copies of this series of books were printed using luxurious Lian Shi paper (连史纸，fine paper made from bamboo) for the Imperial Palace，and another 300 copies were printed by using regular paper made from bamboo to be put for sale on the market. Now what have left of the printed copies are all of regular paper with a pale-yellow color. This series has collected the lost works on humanities and science since the Song Dynasty，so academically，it is very precious. In 1775 (Qianlong 41 Years)，the wooden movable type printed version of the series was conferred to the five provinces of Su，Zhe，Min，Gan and Yue (苏、浙、闽、赣、粤) and they were allowed to reprint it. This set of wooden movable types were also used to print other separate editions of works，such as *Baxun Wanshou Shengdian* (《八旬万寿盛典》，*The Octogenarian Ceremony*)，*Qiansouyan Shi* (《千叟宴诗》，*Poetry for the Banquet of Thousand Aged People*) and *Xixun Shengdian* (《西巡盛典》，*The Grand Tour to West China*)，all written by Gaozong himself，though with various page-layout for each book. In order to conclude this experience，Jin Jian had printed and published the book *Wuyingdian Juzhenban Chengshi* in 1776. The book elucidates all the following items：wooden movable types，wood carving，cabinets for types，holding strip，blanks (顶木)，center column (中心木)，indexing plate (类盘/捡字盘)，preliminary printing plate (套格)，setting types for printing (摆书/植字)，backing board (垫板)，proofreading，brush printing，clearing (拆版/归类)，contiguous arrangement of working round (逐日轮转) etc.，including all the operations and processes of producing types，composing，brush-printing，clearing，among others. There are 16 illustrations printed in it.

Jin Jian's method was more sophisticated than the previous ones. First of all，he sawed the date wood into centre columns with appropriate thickness，then the centre columns were cut vertically into narrow squares. All the squares

---

❶　国朝宫史续编：卷九十四[Z]. 北平：故宫博物院，1932.

❷　陶湘. 武英殿聚珍版丛书目录[J]. 图书馆学季刊，1929，1(2)：205-217.

were planed after drying in the shade, and finally were cut into wooden types. Then, several of the types were placed tightly in grooves or slots made from hard wood and were planed to the exactly same size: 0.9 cm in thickness, 1.28 cm in width and 2.24 cm in length. Small types were 0.64 cm in thickness, 1.28 cm in width and 2.24 cm in length. A copper square funnel was used to check whether all the types were of the same size. Secondly, characters were written on a thin paper and the paper would be turn over to stick onto the types, thus reversed characters were carved onto them. The types were sorted into twelve sections according to the *Kangxi Dictionary* (《康熙字典》, 1716) — *Zi*, *Chou*, *Yin*, *Mao* (子丑寅卯), etc. There were twelve cabinets corresponding to the twelve sections, and each cabinet obtained 200 drawers, each drawer 8 cases, each case 4 characters. Outside the drawers, section, character or stroke were marked (Figure 81).

Figure 81　Wooden Movable Type Printing Process depicted
in Book *Wuyingdian Juzhenban Chengshi*

In order to pick out the right types quickly, a list was compiled and the arrangement of the characters on the list corresponded to that of the *Kangxi Dictionary*. The indexers (捡字工, workers who pick out the characters from the cabinets) picked out the right types from the cabinets and put them on the plates; typesetters (排字工) placed the types into a wooden plate with slots in it, fixing centre columns needed for the page composition. For the big types, two pages could be composed for each day; for the small ones, only one page could be done.

Unlike Wang Zhen's (王祯) method, Jin Jian (金简) produced individual types first, so the size of each type was guaranteed. Secondly, Wang used a turntable to store and fetch types while Jin used cabinets and could store much more types. Thirdly, Wang composed the whole page on one printing plate and printed the page at one time; Jin used overprinting method to avoid a blank-out spot along the lines of a page. Jin's method enabled a quicker and clearer printing. Lastly, Jin knew how to make workers work by turns and could print different books at the same time. His method enabled a timely and orderly setting and clearing of the types. In a word, Jin's method is a sophisticated version of the previous ones. Jin's books were of the Imperial Edition (殿版), so they were authoritative and influential. They were the yardstick for most of the wooden movable type printed books at that time.

2.3.3　Invention of Clay Type or Porcelain Type Printing

Compared with woodblock printing technology, wooden movable type printing required less wooden materials. However, with a smaller size, the wooden materials had to be harder; and if one book was to be printed, at least 100,000~200,000 types were prerequisite, so it was still costly. Bi Sheng (毕昇, 990~1051), a printer in the Northern Song Dynasty, thought that carving types were too costly both in money and time, and they could become out of shape when damp. Therefore, he invented the clay movable type printing — the wooden types were used as molds to make clay ones, and then the clay types were baked in the fire to harden. This new method saved an enormous amount of money and time. Shen Kuo (沈括, 1031~1095), a scientist in the Northern Song Dynasty, had recorded this method in Scroll 18 of his book *Mengxi Bitan* (《梦溪笔谈》, *Dream Pool Essays*) (Figure 82) as follows:

> *During the reign of Chingli, (庆历, 1041~1048) Bi Sheng, a man of unofficial position, made movable type. His method was as follows: he took sticky clay and cut in it characters as thin as the edge of a coin. Each character formed, as it were, a single type. He baked them in the fire to make them hard. He had previously*

prepared an iron plate and he had covered his plate with a mixture of pine resin, wax, and paper ashes. When he wished to print, he took an iron frame and set it on the iron plate. In this he placed the types, set close together. When the frame was full, the whole made one solid block of type. He then placed it near the fire to warm it. When the paste [at the back] was slightly melted, he took a smooth board and pressed it over the surface, so that the block of type became as even as a whetstone. The more books to print, the more efficient the method was. There were two plates adopted to place the types in turn so that one needn't wait at all for the types to be set. For each character there were several types, and for certain common characters there were twenty or more types each, in order to be prepared for the repetition of characters on the same page. When the characters were not in use he had them arranged with paper labels, one label for each rhyme-group, and kept them in wooden cases. For the rarely used characters, one could carve and bake the types quickly enough. Unlike the wooden types, the clay ones could be cleaned by warming it over fire—ink was easily wiped after being warmed. After Bi Sheng's death, all his printings have been preserved well by his nephews until today. ❶

Stanislas Julien（1799 ~ 1873）, from College de France, translated the above-mentioned paragraphs into French in 1847. In 1925, the Sinologist Carter in Columbia University translated the French words into English, thereby helped spread the knowledge all around the world.

To explain the details, what does "as thin as the edge of a coin" refer to? In the ancient times, the edge of a coin was as thin as 1.5 mm ~ 2 mm. The Koreans used to understand this figure as the height of the type, which is obviously incorrect, for a 2 mm type was too difficult to bake and was almost impossible to be inked or set. It was too fragile to print. In fact, as Carter had pointed out❷, 2 mm should be the depth for the carving, and the height of each type should be at least 1 cm to be reasonable.

Also, the "iron frame" referred to four iron bars welded onto an iron

---

❶　沈括. 梦溪笔谈：技艺[M]. 北京：文物出版社，1975.

❷　CARTER T F. The Invention of Printing in China and Its Spread Westward[M]. 2nd ed. New York: Ronald Press, 1955.

plate, which was the frame for printing. After melting sticky stuff onto it, types were set. If the spacing of boundary was required, then battens were placed and set to straighten out the printing. Bi Sheng had successfully printed many books by using this method, and all the types were preserved well (Figure 82).

From Shen Kuo's accounts, we get to know that Bi Sheng's invention included manufacturing, storing and picking out the types, setting the plates, inking, brush-printing and clearing, etc. The whole process was more complicated than wooden movable type printing. What is more, since the clay was turned into porcelain after the high temperature baking (600 ℃ ~ 800 ℃), the so-called clay type should better be called porcelain type, and since its color was black, a black porcelain type should be its appropriate denotation. However, as the Chinese are used to calling it the clay type, we needn't bother to change its name. These clay types were solid within, so they have a certain mechanical strength and good water-absorbing quality. Considering the physical features of such types, Bi Sheng had used some sticky stuff to stick the types onto the plates. Besides, without boundary-battens, one could even set the types into circles. There are printed convolution sutras from the Song Dynasty unearthed recently.

Figure 82　A Copy of Bi Sheng's Clay Movable Type Plates, collected by Museum of Chinese History

Thanks to the practicality of the technology, the Chinese had used the clay movable type printing till the Qing Dynasty in the 19th century. Since clay is easy to obtain and pottery kilns in which clay types can be baked were not uncommon, it was convenient and less costly to mass-produce the clay movable types. The historical contribution of Bi Sheng should be fully confirmed. Thanks to Shen Kuo's recording, this technology became well-known all around the world. Later, another scholar Jiang Shaoyu (江少虞，1101 ~ 1061), from Zhejiang Province, had recorded Bi Sheng's technology in Scroll 52 of his book *Huangchao Shishi Lei Yuan* (《皇朝事实类苑》, *The Categorical List of Historical Facts in Song Dynasty*), too.

Archaeological discoveries have provided more information about the development of Bi Sheng's technology after the Northern Song Dynasty. In

February, 1965, while dismantling the White Elephant Tower in Wenzhou City, fragments of the printed sutra—*Amitayurdhyana Sutra* (《佛说观无量寿佛经》) were discovered on the second floor of the tower. According to *Xiejing Yuanqi* (《写经缘起》, *The Causality of Writing the Sutra*), an ink manuscript written in 1103 (in the Northern Song Dynasty), the sutra was published around the year of 1103. *Aparimitāyur-sūtra* is one of the three classics of the Pure Land Buddhism, translated by Kang Senkai from Sanscrit in the Three Kingdom period. There were altogether two scrolls of it. Later in the Tang Dynasty, Master Shan-tao (善导) annotated the sutra and expanded it into four scrolls. The handwritten version from the Tang Dynasty was unearthed in the Dun Huang grottoes in 1900. The fragments found in Wenzhou City was 13 cm in width and 8.5 cm~10.5 cm in height. They were written in Song typeface and set in convolution style with the symbol "o" to signal the end of each line. There are 166 recognizable remaining characters, with irregular intervals between each other. The characters are of different sizes and strokes, the ink-color uneven, and the typeface childlike. The handwritings look slightly sunken, a few of which are missing. The character "色 (color)" is horizontally printed. Regarding all the facts, archaeologists have decided that this fragment is a clay movable type printed version from the Northern Song Dynasty in 1103.

The Wen Zhou fragments were published 50 years after Bi Sheng's time, and they were the earliest extant clay movable type print from the Northern Song Dynasty. They are also the historical evidence for Bi Sheng's invention. As I have pointed out earlier, by sticking types onto the plates, one could set the types into straight lines, oblique lines or even circles; and these fragments were printed with convolution setting. The fonts are different in size, with big characters 0.5 cm wide and 0.45 cm long, mid-sized characters 0.4 cm wide and 0.3 cm long, and small ones 0.3 cm wide and 0.15 cm long. All the characters are somehow accumbent and rectangular, whose height is 1.5 times the width. Compared with other prints from the Song Dynasty, these are the smaller types. They might be specially baked for the sutra itself, for the other prints are usually set with larger types and the fonts are often square shaped.

Considering the overlaps of some strokes of the neighboring characters, some opinions argue that the fragments can't be movable type prints; and the inverted characters are to indicate the ending of each sentence, so they have to be woodblock prints. Dr. Qian Cunxun (钱存训) disagrees with the opinions. He thought that it is more likely for movable type printing to overlap the strokes because the types were of different sizes and it is more possible for characters to be left out during the setting of types. The inverted characters are not indication for the ending of each sentence, for there are the symbols "o" to signal it. The inversion of the character "色" is an important proof that the fragments were movable type prints. I agree with him. Since the types were specially designed to print the peculiar pattern of the characters, the printers had to tightly set all the types together to form the curves, so it was impossible to avoid the overlaps of the strokes. It is typical for movable type printed books to have some overlaps between certain strokes, which can be proved by the copper movable type prints discovered in North Korea. As for the Wenzhou prints, there are obvious differences between woodblock prints and wooden movable type prints, so they should be clay movable type prints. Yet, is it possible that the printers were printing the fonts by pressing each type on the paper? It is the least likely happening because it must be a serious technological setback.

Zhou Bida （周必大，1126 ~ 1204）, a scholar in the Song Dynasty, published books with Bi Sheng's technology. In 1193, in a letter to his friend Cheng Yuancheng（程元成）who had passed the civil service examination in the same year with him, Zhou wrote："Recently, I have used the Shen Kuo's recording of the clay movable type printing method to print some books. Now, I've finished printing the book *Yutang Zaji* （《玉堂杂记》, *Notes of Yu Tang*）, and I present one copy for you to read. There are still a dozen of the anecdotes that I'd like to note down in the future. I think you might also want to have a look into the passing times and share with me the nostalgic emotions." Yu Tang is the name for Hanlin Academy（翰林院，the Imperial Academy）. Zhou passed the civil service examination in 1150 and became an academician in the Hanlin Academy in the Xiao Zong period（1163~1189）. The book *Yutang Zaji* （《玉堂杂记》, *Notes of Yu Tang*）is a memory of the years spent in the academy. He resigned in 1191 when he was 67 years old, and from then on, he started to print his books with clay movable types so that he could share the copies with his friends.

The clay movable type printing was exported into the Xixia Kingdom（西夏）. In May 1989, the remaining 54 pages of the printed copy of *Vimalakīrti-nirdeśa-sūtra* （《维摩诘所说经》）was unearthed in Wuwei（武威），Gansu Province. The copy was in accordion binding, 17 characters per line and 7 lines per page. Each page is 28 cm in height and 12 cm in width; each character is of the size of 1. 4 cm × 1. 6 cm. The lines are not straight and the ink are not evenly put. Some of the strokes are stiff and deformed with broken edges or color-peeling. With the sutra, there unearthed other books from the same period from the year of 1224 to 1226, and this confirms the publishing time of the sutra. Archaeologists believe the sutra is a clay movable type print in the first half of the 13th century. In 1907, the Russian Kozlov (科兹洛夫) also discovered a sutra with the same name from the ruins of the Black Water City in the Xixia Period. It is in accordion binding, too, with single boundary lines on each of the four edges of a page. For each page, the height is 27. 5 cm~28. 7 cm and the width is 11. 5 cm~11. 8 cm, 17 characters per line and 7 lines per page. It is now collected by the St. Petersburg Branch of the Institute of Oriental Studies (No. 223,737). After careful study, experts believe that it is also a clay movable type print published from the middle part of the 12th century to early 13th century.

Since the Yuan Dynasty, the clay movable type printing technology had developed further, thanks to the advocacy of Yao Shu (姚枢, 1201~1278), the counsellor for Kublai Khan. Yao was one of the Prime Ministers and a scholar in the Imperial Academy. When he died, his nephew Yao Sui (姚燧, 1239~ 1314) wrote *Zhongshu Zuocheng Yaowenxiangong Shendaobei* (《中书左丞姚文献 公神道碑》, *The Tablet for the Left Prime Minister Yao Wenxian*, 1278); in 1235, the imperial edict of the Emperor Taizong had announced the second Prince to attack the Southern Song Dynasty and commanded Yao and Yang Weizhong (杨唯中) to follow the army to search for an assortment of talents; in 1241, Yao was rewarded farmland for his task and then carried his family to the Hui County in Henan Province to enjoy the countryside life. In the countryside, in order to cultivate the common people, Yao had published some books on Chinese exegetics, *Yu Meng Huo Wen* (《语孟或问》, The Questions and Answers about The Analects of Confucius and Mencius) and *Jia Li* (《家礼》, Family Etiquette). He also encouraged Yang Zhongshu to print and publish *The Four Books* (《四书》) and Tian Heqing's (田和卿) version of The Book of History ... All of the books were published in Beijing. Considering that books on Chinese exegetics were still not very popular, he asked his pupil Yang Gu (杨 古, 1216~1281) to use Shen Kuo's printing method to print the books *Jin Si Lu* (《近思录》, Reflections on Things of Hand) and *Donglai Jingshi Lunshuo* (《东 莱经史论说》, Mr. Dong Lai's Comments on History) so that the copies could be diffused among people.

The books were printed and published between 1241 and 1250 in Henan Province or Beijing City. Yang Gu's contemporary Wang Zhen said:

> *Someone takes another course: they use iron as the printing plates, pouring thin pitch into the boundaries. After the pitch cools down, they flatten the surface and carefully put it on soft fire to soften the pitch. Then, clay movable types are set on the plates to serve as movable type printing blocks.*

In Bi Sheng's time, the plates were filled with rosin and wax, but now they were replaced by pitch. Pitch is organic gelatinous material, so it is of high binding property. Using pitch is an improvement. It also proves that Bi Sheng's technology was very influential in China and had been successfully applied to practice (Figure 83). But there are always such claims that Bi Sheng's invention was "impractical, for it is too fragile to live long"[1]. All the clay movable type

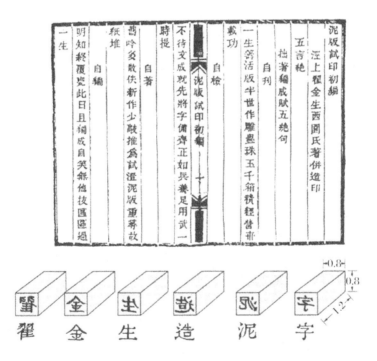

Figure 83   Clay Movable Types and a Clay Movable Type Printed Book produced by Zhai Jinsheng from Jing County, Anhui Province in 1844 (unit: cm)

---

[1]   Sohn Pow-key. Invention of movable metaltype printing in Korea: Speech at the International Symposium on Printing History in the East and West[R]. Seoul, 1997-9-29.

prints from the 12th to 13th centuries unearthed have proved that such claims are incorrect. Simulations have indicated that Bi Sheng's clay movable types are hard and solid; they were intact even when hurled down from 2 m high onto the marble ground. I witnessed the experiments. That was why Bao Shichen (包世臣, 1775~1855) observed that the clay movable types made by his friend Zhai Jinsheng (翟金生, 1775~1860) in the year of 1844 were "as unbreakable as bones and horns". Thus, people should stop doubting the applicability of the clay movable types, instead, the types can be described as a unique invention in China.

In the early Qing Dynasty, there appeared white pottery types made from porcelain clay or kaolin, and those are the second unique invention in China. Wang Shizhen (王士禎, 1634~1711) wrote in Scroll 23 from the book *Chi Bei Ou Tan* (《池北偶谈》, *Casual Remarks from Chibei*): "Zhai Jinshi (翟进士, Scholar Zhai, the *Jinshi* degree holder) from Yi Du is the assistant to the ranking official in Raozhou State (饶州府) and is rather tyrannical. He once ordered the potters to print *The Book of Changes* (《易经》). The porcelain types were of the finest regular scripts. Just like the stone inscription of *The Thirteen Confucian Classics* (《十三经》), the porcelain print were also modified many times before it was finally completed." According to *The Record of Raozhou Government* (《饶州府志》, 1872), Zhai Shiqi (翟世琪, 1625~1670) were from Yizhou County, Shandong Province. He passed the imperial examination and became a scholar in 1659; then in 1667, he went to Jiangxi Province to be the assistant to the ranking official in Raozhou Government. He was said to be "learned and kind, beloved by his people", and we believe he was Zhai Jinshi (翟进士, Scholar Zhai, the *Jinshi* degree holder) recorded by Wang Shizhen, for he was the one who ordered the print of *The Book of Changes* in Jingde Town in Raozhou State.

In the book *Jinxiang Shuo* (《巾厢说》, *A Small Book*) written by Jin Zhi (金埴, 1730~1795) from Zhejiang Province, the author noted that "Between the year 1717 and 1718, there was a scholar in Taian who could bake the clay into types, but I forget his name." This scholar was actually the provincial graduate Xu Zhiding (徐志定, 1690~1773). National Library in Beijing has collected the book *Zhouyi Shuo Lue* (《周易说略》, *On the Book of Changes*) printed and published by him in 1719. In the preface, the author noted that "In the winter of 1717, I happened to have made pottery movable types which are more solid than wood, so I had to correct the wrong types for several times to manage to print the book in the spring of 1718. " The question is what exactly are the types made of? Some thought they were pottery movable types, for the lines in the book are almost arc-shaped, characters skew and with different sizes and ink-color; there might be glaze covering the types during baking. Still others believe that Xu had baked a whole pottery block to print since there are breaks in the strokes and page-setting, which is typical of block printing.

Obviously it is wrong to claim that Xu had used whole-block printing skill because it is quite often to observe breaks in the strokes or overlaps of the characters in movable type prints; but it is worthwhile to discuss the possibilities of the above two opinions if we are trying to study the print from Jin Zhi's record and the technology adopted during printing. The problem here should be: what materials did he use to bake the types? What did it look like when using the types? How to name such prints? We think Xu Zhiding was using kaolin as his baking material just like Zhai Shiding did, and he must have baked the types as Bi Sheng did before he printed the book. The reason is that kaolin contains less than 2% of iron, so it can be baked into white hard porcelain at 900℃, and the moisture absorption ratio is as high as 10% to enable it to be inked. Now, such pottery is named white pottery, to distinguish it from Bi Sheng's black pottery.

Although Xu Zhiding named his types as "Tai Shan Pottery Block" and Wang Shizhen named Zhai Shiqi's print as "Celadon Book of Changes", it is because they couldn't strictly tell the difference between porcelain and pottery as we do today. Rather, we shouldn't ignore the difference now that the technology is so sophisticated. Technically speaking, the pottery movable types are definitely forbidden to be glazed, they should be baked through biscuit firing, so they are not celadon movable types, but only pottery movable types.

2.4   Invention of Metal Movable Type Printing

2.4.1   Copper Movable Type Printing Originated from the Northern Song Dynasty

The woodblock printing is the forefather of the printing technology, and the metal movable type printing is a sophisticated development of traditional printing technology. Between the two periods, there existed a transition time of copper block printing and non-metal movable type printing. When metal was used to replace wood to produce printing blocks, the blocks were solid and durable and the printing lasted longer. Non-metal movable types, on the other hand, unlike the wooden and copper ones, turned the blocks into detachable and "recombinable" movable settings, hence promoting efficiency of plate-making and material utilization. By combining the advantages of both the copper movable type printing and non-metal movable type printing, the metal movable type printing was born; the earliest metal movable types were made of copper, which proves that they were the product of the grafting of the two above-mentioned technologies.

The copperblock printing skill can be dated back to the glorious age of the Tang Dynasty, yet the people then couldn't bake copper movable types, for the technology was not possible until the Northern Song Dynasty in the 11ᵗʰ century. In the Song Dynasty (including the Northern Song and the Southern Song Period from 960 to 1279), the technologies of incusing and coin-casting, papermaking, copperblock printing and non-metal movable type printing all entered a new age, so it was likely for them to influence each other. Besides the technological consideration, there were external factors such as politics, economics and culture that had matured the printing industry into adopting metal movable type printing. At the same time, however, the other countries did not have the technologies and could not beat China in inventing movable type printing skill. Therefore, it is true that China had witnessed the invention of metal movable type printing and had it spread out to the East and the West.

Considering China had always been a big country with a large population, printed books always found themselves in huge demand and a variety of types, the printers had to go to great lengths to meet the massive market, so for a short period, it was really difficult to tell the difference between block printing and movable type printing; as long as the printed books were inexpensive, consumers wouldn't fuss about the fact whether the books were printed with blocks or types. That was why movable type printing could never replace block printing completely. In China, it might be true that politics, education and religion had played very important roles in developing the printing technologies, but more precisely, it was the newly presented requirements for politics and economics that had stimulated the development of metal movable type printing.

In 960, after the establishment of the Song Dynasty, the 50-years' chaos of the Five Dynasties and Ten Kingdoms were ended, and China was united once more. The rulers of the Northern Song Dynasty had executed a complete set of laws and rules all over the country. Thanks to the policies to restore and develop production, all the industries around the country were developed quickly and tremendously when the Four Great Invention entered a new era of flourish. All the Emperors of the early Song Dynasty, especially Emperor Taizong (太宗, 976~997), were fully aware of the significance of printing, so they paid special attention to the printing and publishing industry. The metal movable type printing was invented to cater for a new economic maneuver — the issuing of paper money.

In the Northern Song Dynasty, thanks to the rapid development of commodity economy, the copperblock printing had been widely used in the economic sphere, which was a new direction for the printing kill. In the Museum of Chinese History, there is a collection of a square copper block for Liu's Needle Store from Jinan State (济南府, now Jinan City in Shandong Province) in the Northern Song Dynasty. The block was used to print advertisement paper for product packaging. It is 12.4 cm in height and 13.2 cm in width, with a rabbit graph in the middle; above the graph there is a line of characters cut in intaglio— "Ji Nan Liu House Kongfu Needle Store". On the left and right side of the graph, there are characters "认门前白,兔儿为记" (The White Rabbit on the Front Door is Our Trademark). Below the graph there are characters cut in relief meaning "We purchase quality steel bar to produce fine needles; it doesn't matter whether the steel bars have been used or not. If you come to resell steel bars, we will raise the price." (Figure 84) Considering the rabbit graph and all the characters both cut in intaglio and relief, it was of great difficulty to set such a printing block. This is the earliest copper block we've

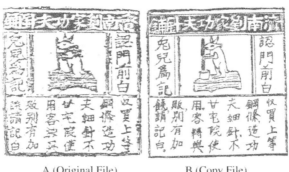

A (Original File)   B (Copy File)

Figure 84   Square Copper Block for Liu's Needle Store from Jinan State
(10th C. ~11th C.), collected by Museum of Chinese History

ever discovered.

Technically, it is the trend for metal movable type printing to be invented in the 11th and 12th centuries since both copper block printing and non-metal movable type printing had already at their maturity. ❶ From the Northern Song Dynasty, China had issued the earliest version of paper money around the world, which is a revolutionary work in the monetary history. Meanwhile, the issuing of paper money also provided a new motive for the promotion of the printing skill. The great progress of the urban industry, commerce and transportation made it possible for merchants to travel around the country, but it was very convenient for them to carry metal money with them, so the people then decided to draw on the experience of "Flying Money (飞钱, remittance)" from the Tang Dynasty. During the Zhenzong Period (1008~1017), 16 rich and powerful families in the Chengdu City, Sichuan Province jointly printed paper exchange certificate "*jiaozi* (交子)". Zhang Yong (张咏), the governor of Shu (蜀, now Sichuan Province), supported this system and stamped the exchange certificates with his official seal. *Jiaozi* thus became the predecessor of modern paper money.

In the book *Songchao Shishi* (《宋朝事实》, *The Categorical List of Historical Facts in the Song Dynasty*), Li You (李攸) wrote in Scroll 15 *Cai-yong* (财用, *Material and Tools*): "The rich families agreed to print the exchange certificates with the same paper. On the certificates there are graphs of houses, people and stores, with peculiar inking marks of red and black color as secret codes. The receiver has to fill in the paper the exact number of money that he has received as the initial deposit." Thus we know that some precautions were taken in the design of the exchange certificates and a direct influence had been formed on the later system of paper money issuing. During the Emperor Zhenzong Tianxi Years (真宗天禧年间, 1017~1021), due to the fact that the stores of the exchange certificates were running short of cash, a bank run took place and lawsuits followed, so the magistrate of Yizhou (益州) Koujian (寇瑊) called a halt, and submitted memorials to the Emperor, suggesting that the exchange certificates should be placed under the government's management and an office should be especially designed for it.

In the year of 1023 when Emperor Renzong (宋仁宗, 1023 ~ 1063)

---

❶ Pan Jixing. On the origin of movable metal type technique[J]. Chinese Science Bulletin, 1998, 43(20): 1681-1692.

succeeded to the throne, Kou Jian left Sichuan Province, but the Commissioner of Yizhou Road (益州路转运使) Zhang Ruogu (张若谷) and the new magistrate Xuan Tian (薛田) were ordered to prepare for the setting of the office of exchange certificates. They suggested that:

It is better for the government to manage the exchange certificates system. Now we still ask for the business to be printed with copperplates and Yizhou should be in charge of it. There should remain the graphs and codes on the certificates, but the number of money should be limited from 1 guan (贯, a unit of Chinese monetary system, about 1,000 copper coins) to 10 guan. Also, there should be supervisors to supervise the contracts and cash. All the business should be done for the convenience of the customers.

Emperor Renzong ratified this suggestion and thus the official *jiaozi* now was in print. The iron coins were the reserve fund and three years was one bound for the replacement of old coins by new ones. These batches of *jiaozi* were the earliest type of paper money. The book *Song Shi · Shi Huo Zhi* (《宋史·食货志》, *The History of the Song Dynasty · Farming, Cloth and Currency*) also noted that the earliest official paper money was authorized from the year of 1023, Renzong Tian-Sheng-Yuan-nian (the first year of the Saint Years). *Jiaozi* had been in circulation from 1023 to 1105 in the Northern Song Dynasty, only within Sichuan and Shaanxi areas. From 1023 to 1038, it was discovered that the different denominations and notes made it very complex to fill in, so later, from 1039 to 1068 the denominations were reduced to two types, issued respectively in 5 *guan* and 10 *guan*, with 20% printed in 5 *guan* and 80% in 10 *guan*, and one needn't fill in the number, instead, printing became the norm. Generally, for each bound of issuing, 1,250,000 *guan* of *jiaozi* was printed. From 1069 to 1105, the denomination was once again changed into 5 *guan* and 500 *wen* (文, a unit of Chinese monetary system), with 60% of the *jiaozi* printed in 5 *guan* and 40% in 500 *wen*. In the back of the *jiaozi*, one could find a rectangle stamp of "One Guan Back Contract (壹贯背合同)", which was to rectify the denomination printed on the front page of the *jiaozi* so that one could not tamper with the number. There were also serial numbers printed on the *jiaozi* for security check so that it was difficult to fake. The issuing system in the Song Dynasty was stricter and more complete than that of the former periods.

After the *jiaozi* practice, a new type of paper notes was issued with the name of *qianyin* (钱引). According to the book *The History of the Song Dynasty*

• *Monograph on Food and Currency*, "in the year of 1105, the system of *qianyin* was introduced into some areas in China. It was printed in a new style, but the old style was still permitted in Sichuan Province ... Except for Min, Zhe and Huguang (闽，浙，湖广) areas, all the rest provinces were using *qianyin* ... In 1107, Sichuan Province was ordered to replace *jiaozi* by *qianyin*." Within the *qianyin* system, the reserve fund was still iron coins and one bound was three years, the denomination 1 *guan* and 500 *wen*, popular in Jingdong, Jingxi, Huainan and the capital city (now Kaifeng, Henan). Sichuan joined in the new system in 1107. From 1109 to 1234, there were altogether 56 bounds of *qianyin*, which made it the paper note that had been used for the longest time. From 1160, the Southern Song Dynasty issued *huizi* that had been in circulation from 1160 to 1279 in the south eastern area of China, including Hangzhou, Huai, Zhe, Hubei and Jingxi (杭州，淮，浙，湖北，京西). For this system, copper coins were used as the reserve fund and the denominations were 1 *guan*, 2 *guan* and 3 *guan*, with 300 *wen* and 500 *wen* added later on. One bound was still three years. The first batch of *huizi* was issued in a total of 10,000,000 *guan*.

Since the paper notes needed to be printed in large quantities and of vital importance to national security, it was inappropriate to print them with woodblock printing skill. woodblocks tend to be deformed when the temperature is changing and they are not wear-resistant. What is worse, they were easy to fake. Therefore, since the Song Dynasty, it had been common in China to use copperblock printing skill and special made paper to print paper notes. This has also been proved by documents and unearthed relics. For instance, in *Wenxian Tong Kao • Qianbi Kao* (《文献通考 • 钱币考》, *Examination of Historical Documents • On Coins*), Ma Duanlin (马端临，1254 ~1323) noted that when Song Xiaozong took throne in 1163, he ordered the Logistic Office of Finances and Taxes to collect the copperplates for printing *huizi* in the Huguang area, and the blocks were kept by the Department of State Affairs. According to Scroll Nine of *Book of General*, in 1176, the third and fourth bounds of *huizi* would be out of circulation three years after the issuing; then the Huizi Office belonging to the Tea Trade and Tax Office (都茶场会子库) would continue to print 2,000,000 copies of the fourth bound of *huizi* to be kept in the Southern Treasury Granary (南库). Therefore, we can draw the conclusion that all the *jiaozi*, *qianyin* and *huizi* were all printed with copperplates and there are unearthed relics to support the presupposition.

In 1936, Chen Rentao (陈仁涛) who surveyed ancient coins had received a copper block for printing *huizi* of the Southern Song Dynasty. The block has been stored in the Museum of Chinese History since 1958. It is 17. 8 cm in height, 12 cm in width and 1. 7 cm in thickness, 2,700 g in weight, with rust on it (Figure 85). In the middle of the block, there are characters "行在会子库"(Xin Zai Huizi Ku, the Huizi Bank of Hangzhou), above which there are ten characters "大壹贯文省"(Da Yi Guan Wen Sheng, One Guan's Worth) and "第壹佰拾料"(Di Yi Bai Shi Liao, The One Hundredth Issue). Between the ten characters, there are seven lines of characters meaning "One who

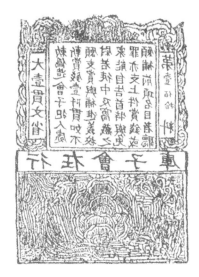

Figure 85   Copper Movable Type Printed Version of *Huizi* of One Guan (壹贯会子) Printed and Circulated around 1161 to 1168 in Southern Song Dynasty, collected by Museum of Chinese History, drawn by the Author

fake the *huizi* will be executed". One who turns the culprit in will be awarded 1,000 Guan. If one is not interested in money award, he shall be promoted to be a military officer. If one who has harbored the culprit now turns him in, the guilt can be absolved and money award is still valid. Or he might also be promoted to be an officer at his own will." Below on the block, there is a graph of mountain spring and decorative patterns are surrounding the design. Actually, there is another similar copper block for printing *huizi*, with a different graph of city walls, but this one is a fake and should be recognized so.

The above-mentioned characters and design conform to what was depicted in the books *The History of the Song Dynasty · Monograph on Food and Currency* and *Book of General · On Coins*. The former especially noted that, in 1160, the assistant minister of the Ministry of Revenue Qian Duanli (钱端礼,1109~1177) was ordered to print *huizi*. All the paper money was to be circulated around Hangzhou City, which was supervised by the Tea Trade Office. In 1161, a law of counterfeiting *huizi* was formulated and implemented. At that time, the material paper for printing *huizi* was taken from Huizhou, Chizhou and Jizhou (徽州、池州和绩州) and the printing took place in Chengdu and

Lin'an（成都和临安）. At first，*huizi* was confined within the boundaries of Qiantang River Zhejiang area（两浙），but later it was circulated to Huai area（淮）as well ... in 1168 ... the Minister of Revenue Zeng Huai（曾怀）was ordered to print *huizi* in a special office named Huizi Bank（会子库），and a stamp was made with the characters of "提领措置会子库"（Tiling Cuozhi Huizi Ku）. Xing Zai Huizi Ku was another name for Du Cha Chang Huizi Bank（都茶场会子库）. Du Cha Chang reported to Xin Zai Trading and Taxing Office，responsible for Tea License，but in 1160，Gaozong had ordered the setting of the Huizi Bank to print *huizi*. The copper block in question was printed between 1161 and 1168.

What is worth noticing is that of all the five characters "第壹佰拾料"（Di Yi Bai Shi Liao），the two characters "第"（Di）and "料"（Liao）are obviously of different font from that of the rest，and they are slightly askew，so we think that it should be printed with copper movable types. This is very similar to copper movable type printed paper money in the Jin Dynasty（1115～1234），therefore，this block printed between 1161 and 1168 is the earliest copper printed *huizi* and offers us a valuable opportunity to study the history of paper money in the Song Dynasty and the history of metal movable type printing skill.

So far，there have been no findings concerning *jiaozi* in the Northern Song Dynasty and the copper block of *qianyin*（钱引，paper currency），but there are documentaries related to the shape and structure of the money. In the book *Zhubi Pu*（《楮币谱》，*Catalog of Paper Money*，1360），Fei Zhu（费著，1303～1363）in the Yuan Dynasty noted：

> *In May of 1107*，jiaozi *was renamed as* qianyin，*and there were six types of stamps：Chi Zi*（敕字，*imperial edict*），*Da Liao Li*（大料例，*serial number*），*the printing year and the circulation year*，*the denomination on the back*，*the patterned edges and decorative designs*. *The first four stamps were all printed in ink*，*but the patterned edges were printed in blue while the decorative designs were printed in red*.

Here，it is generally misunderstood that the six types of stamps were made from six separate printing blocks，but technically，it was very difficult to combine all the blocks to print one piece of banknote；and since there were three different colors printed on the *qianyin*，it was impossible to use six printing blocks at the same time. It is only possible that it was one printing block with three colors to print *qianyin*，just like the printed *Notes to the Diamond Sutra*（《金刚

经注》) in 1341 in Zhongxinglu (today's Jianglin, Hubei Province).

The book *Catalog of Paper Money* did not offer any illustration of *qianyin*, instead, all the details were depicted with words, especially the *qianyin* issued from 1161 to 1179, so that it is barely likely for us to reproduce the paper money. *Qianyin* was issued after *jiaozi*, so it should imitate the design of *jiaozi* though the latter might not be printed in colors.

Since the paper money could be printed by using copper movable types, it was also possible to use them to print books. Sun Congtian (孙从添, 1769 ~ 1840), the well-known scholar of edition in the Qing Dynasty, wrote in his book *Cangshu Jiyao* (《藏书纪要》, *Summary of A Collection of Books*), "The printed books in the Song Dynasty are very rare now ... There are several types of printed books then: Shu edition, Taiping edition, Lin'an Shu-peng edition (a book store in Lin'an), the Academy edition, fine gentry edition, private printed edition, imperial printed edition ... copper printed edition and (wooden) movable type printed edition, of which only the Shu edition, Lin'an edition and imperial edition are the best. ❶" He had witnessed all the editions. We believe that the copper printed edition referred to copper movable type printing. Unfortunately, it is really rare for copper printed books from the Song Dynasty to survive till now.

2.4.2　Copper Movable Type Printing after the Song Dynasty

2.4.2.1　Copper Movable Type Printing in the Jin Dynasty

The Jin Dynasty employed coins cast in the Northern Song Dynasty and the Liao Dynasty (916~1125) at the early time (1115~1153). In 1150, Emperor Hailing Wanyan Liang (完颜亮) overthrew Emperor Xizong of Jin Dynasty (金熙宗) to claim the throne, and he moved the capital city to Yanjing (now Beijing City) that he renamed Zhongdu. Most of his officials were of the Han nationality, and the monetary policy was similar to that of the Northern Song Dynasty. According to *Jin Shi · Shi Huo Zhi* (《金史·食货志》, *The History of Jin · Farming, Cloth and Currency*):

> *In 1154, after the capital city was moved to Yanjing, the Minister of Revenue Cai Songnian (蔡松年, 1127 ~ 1159) had reintroduced the monetary issuing system to cast jiaochao (交钞, jiaozi) to be used along with coins ... At the beginning, from 1153 to*

---

❶　孙从添. 藏书纪要[Z]. 1805.

1155, there established printing houses and store rooms with relative officialdom being set up. Large-denomination notes were printed ... Seven years were regarded as one bound for the old notes to be replaced with the new ones, so the Sichuan method of issuing and storing jiaozi *was observed.*

This *jiaochao* was issued with denominations of 1 *guan*, 2 *guan*, 3 *guan*, 5 *guan* and 10 *guan*. The notes were printed by special printing houses and the whole process was under the supervision of the Ministry of Revenue. In 1160, with the Southward Migration to the Nanjing City (now Kaifeng City in Henan Province), all the relative institutes were moved to the South. It was also noted that the *jiaochao* was circulated within certain geographical limits, but from 1189 to 1215, the limits were gradually broken and the seven-year bound was no longer observed, and even local governments were permitted to print their own notes. From 1197, small-denomination notes had been printed. Again, according to *The History of Jin • Farming, Cloth and Currency*, there were descriptions concerning the look of the *jiaochao*:

The look of jiaochao: *patterns decorating the four margins of the notes, with denomination printed in the above space, serial number in the left and the name of the shop in the right. Besides, there were seal characters of "whoever forge the notes shall be beheaded" and "whoever turned the forgers in shall be rewarded 300* guan *of the money" Also, such characters of "on imperial orders we print* jiaochao. *The notes can be exchanged at certain shops and are in currency for both official and private use. All the notes can be used without time limit ..." All relative officials would stamp the note.*

The above documents were listed under the item of December in the year of 1189, but except for the single new policy of "*the notes can be used without time limit*", all the other contents were mere observance of old rules and policies concerning *jiaochao*. Therefore, Mr. Liu Sen, an expert on the history of paper notes of the Song and Jin Dynasties, holds that "considering that in the Song Dynasty, salt-selling licenses, *jiaozi* and *huizi* all have either the names of the shops or Qian Zi Wen as the serial number, and also regarding the layout of the *jiaochao* after the year of 1189 along with other historical notes, on the paper notes of the *jiaochao* from 1154 to 1189, there should have been such contents as 'the name of the shop', denomination, prohibition to forge and reward for reporting forge, circulation bound and exchange spots, printing institutes and release registration, the officials responsible for management and their stamps, etc. There were either patterns or images of dragons and cranes surrounding the characters." This statement is absolutely correct. In addition to it, not only the early *jiaochao* from the Jin Dynasty, but all the notes of *jiaozi* and *huizi* from the Northern Song Dynasty were similarly in design.

To sum up, on the page of the copper block of the paper notes from the Song and Jin Dynasties, there were characters of the name of the notes, denomination, circulation area, printing institutes, printing time, reward as well as decorative patterns, which needed to be carved beforehand. Besides, in order to prevent forgery and strengthen the supervision of the printing, other measures on top of official stamps were taken, such as the additional contents of "*Liaohao* (料号, part number)" "*Zihao* (字号, the name of the shop)" and the signatures of the officials responsible for the printing and issuing of the paper notes. All the additional contents would not be carved out at the beginning; instead, they would be carved and printed until the last moment to be confidential. There were altogether two ways of numbering the part number or serial number: one was the combinations of One Thousand Characters or other characters with numbers, such as "Heaven No. 50" "Earth No. 1", etc.; the other was a two-character combination selected from the One Thousand Characters, such as "Heaven and Earth" "Sun and Moon", etc. Both ways were proved true by unearthed relics.

As for the officials' signatures, the officials themselves were to take turns in printing their signatures on the notes, thus the signatures turned out to be a variable. Technically, slots should have been saved on the printing plates first to fill in movable types so that the variable contents could be set and printed. And the movable types ought to have been copper to match the copper printing plates. Consequently, around the $11^{th} \sim 13^{th}$ centuries, out of the requirement of politics and economics, both copper block printing and copper movable type printing were widely employed in printing paper notes, which was a great step ahead of the other countries in the world. Not only documents, but also excavations can provide evidence for this.

Since the 20th century, the printing plates for *jiaochao* from the Jin Dynasty have been excavated in succession. Luo Zhenyu (186 ~ 1940), a heritage archeologist, had included in the book *Sichao Chaobi Tulu* (《四朝钞币图录》, *A Compilation of Pictures of the Ancient Paper Money from Four*

Dynasties, 1914) the rubbings of the printing plates for 5 *guan* of the *Zhenyou Baoquan* (paper currency circulated during the reign of Zhenyou) from the Jin Dynasty, which was the collection of Xu Ziyin (1792~1855, from Taicang, Jiangsu Province). A bronze type with a character "輶" (*you*) (Figure 86-A) was placed on the plate, and we assume that there should be another type which is now missing above the character "輶" because according to *Qian Zi Wen*, the original complete sentence should be "易輶攸畏" (*Yi You You Wei*), so the missing character is mostly likely to be the character "易" (*yi*). Of all the nine marks on the pate, at least eight of them needed to be printed with copper types, plus the above-mentioned two characters, there should be ten copper types employed to complete the printing. The *baoquan* under discussion was printed and circulated during the years of 1214~1215, covering Jingzhao State to Pingliang State (from today's Xi'an, Shanxi Province to Pingliang, Gansu Province). On the outside margins of the plates, there printed such characters of "京兆府合同" (Banknote from Jingzhao State) and "平凉府合同" (Banknote from Pingliang State) to prevent possible obliteration and reassure the circulation areas. The copper types on the plates can be dated back to 1215~1216. The book *Zhongguo Guchao Tulu* (《中国古钞图录》, *A Compilation of Pictures of Chinese Ancient Paper Money*) transported the pictures of the plates

A   B

A. The Type of "*you*" on the Plate,   B. A Plate without Any Copper Movable Type Set.
   from Luo Zhenyu (1914)   collected by Shanghai Museum

Figure 86   Copper Movable Type Printed Version of Zhenyou Baoquan Currency
of Five Guan around 1215 to 1216

and claimed that "there is a type of '*you*' on the plate"❶.

In Shanghai Museum, there is another printing plate for 5 *guan* of *Zhenyou Baoquan* (Figure 86-B). Comparing this with that of Xu Ziyin's collection, one will find out that the former betrays the original state of the plate with two hollow blanks for filling in bronze types, for the sizes of the blanks matched those of the bronze-types at that time.

In 1956, the Inner Mongolian Museum acquired a copperplate for 10 *guan jiaochao* circulated in the Jin Dynasty, and it was dated back to the year of 1215. From the picture (Figure 87), it is obvious that there are six hollow blanks for filling in bronze types. And it is the same with the Wang Yuxi's collection of the 11 *guan jiaochao* copperplate circulated from 1213 to 1214. The rubbing collected by Yan Jingyan (颜敬颜) shows there are 14 types employed on the copperplate for *jiaochao* in the reign of Taihe (1201~1209), and this rubbing is the extant earliest relic for the employment of copper type in printing paper money (Figure 88).

Figure 87   Copperplate for 10
Guan Jiaochao in 1215,
from Wei Yuewang (1992)

Figure 88   Remaining Part of Copperplate
for 300 *Guan Jiaochao* around
1201 to 1209, restored by author (1998)

---

❶ 卫月望,乔晓金. 中国古钞图辑[M]. 2 版. 北京:中国金融出版社,1992. Wei Yuewang, Qiao Xiaojing, *et al*., ed. A Compilation of Pictures of Chinese Ancient Paper Money. Cai Mingxin, tr. Beijing: China Finance Publishing House, 1992.114.

The Jin Dynasty had started to use copper type in printing paper money since 1154, but the history can be traced back to the Northern Song Dynasty with the *qianyin*（钱引）system. All the extant relics have shown that the copper type was used in both the Southern Song Dynasty and the Jin Dynasty while the custom of numbering paper money started from the Northern Song Dynasty. Since the latter half of the 11th century, all the paper money was printed with copper type, without handwriting fillings on the layout of the printed page. This technique had then spread all over the south and the north in the early half of the 12th century.

Therefore, it is without doubt that since the 12th century, China has led the world in casting large amounts of copper type in printing paper money. Some foreign scholars might argue that paper money printing cannot be an actual adoption of metal movable type printing on the ground that most of the characters on the plates were previously casted, but not set. As a matter of fact, this argument is quite weak theoretically in that it is type-setting when one is putting dozens of copper type into the plates. It is unlikely for one to print paper money without the step of type-setting first. Before printing paper money, the Chinese had already invented the technique of using non-metal movable type to print books, and it is the same, both technically and operationally, to print books and paper money with movable type. Besides, one shouldn't equal the history of printing to the history of printing books because the former concept ought to include such behaviors as printing paper money, receipts, contracts and advertisements, among others, and they should all be considered as a part and parcel of the printing history. To print books with copper type was merely an extension of the technique of printing paper money in the 11th century in China. Therefore, it is reasonable to regard paper money printing as the origin of metal movable type printing; it is the earliest form of type-casting printing. The chemical components for Chinese copper type is similar to that of the copper coins with 64% of copper, 9% of tin, 23% of lead and very little iron and zinc.

2.4.2.2 Metal Movable Type Printing in the Yuan and Ming Dynasties

For all the 800 years from the Song Dynasty to the Qing Dynasty (11th century ～ 19th century), it had become a special printing business for the government in China to print paper money with bronze plates and bronze type, supervised by the Emperors themselves, therefore, the metal movable type printing had been supported by both political and economic powers from the very beginning. In the Song and Jin Dynasties, each different batch of paper money was with a different denomination, and the money was numbered by characters from the book *One Thousand Character Primer*. Since the issuing of paper money was always in enormous amount, a great number of printing plates were in demand, so the casting of the types must have been carried out on a large scale, which means only the government was powerful enough to do so.

In 1271, Kublai Khan (1215~1294) founded the Yuan Dynasty, but in 1260 when he took the throne, he had started to print *Zhongtong Yuanbao Jiaochao* (《中统元宝交钞》) in Kaiping State (today's Inner Mongolia), imitating the Song Dynasty's *jiaochao* system, with silver as the unit of measurement — 2 *guan* for 1 *liang* of silver. There was no time limit to the circulation of the money. Later, with more and more money being printed, it was greatly devalued. In 1287, the law of money was altered by issuing *Zhiyuan Tongxing Baochao* (《至元通行宝钞》), but the old paper money was still in circulation; in 1309, the old *Zhongtong Yuanbao* currency was out of circulation; in 1311, Emperor Yuan Renzong (元仁宗) resumed the printing and circulation of *Zhongtong Yuanbao* currency; in 1350, the new *Zhongtong* currency was printed, with 1 *guan* for a thousand *wen* of the coins. For more information, one may refer to the book *The History of the Yuan Dynasty • Farming, Cloth and Currency*.

In 1287, the domination for *Zhiyuan Tongxing Baochao* were cut into 11 categories, from 2 *guan* to 500 *wen*. 1 *guan* of the new currency equaled to 5 *guan* of *Zhongtong Yuanbao*; 2 *guan* equaled to 1 *liang* of silver and 20 *guan* 1 *liang* of pure gold. Throughout the Yuan Dynasty, paper money was the major form of currency, observing the form of paper money from the Jin Dynasty. At the beginning (1260~1274), the money was printed by wooden plates, but later in 1275, bronze plates were employed to replace the wooden ones. The unearthed relics that we see today were all printed with bronze plates, with movable type (33.2 cm×25.5 cm) set in. The paper was thick, grey mulberry paper. On the front page, there are the title of the money, denomination, the name of the issuing store, serial number, the printing and issuing office, rules for rewarding and punishing whoever concerned and the age. You can assume the filling blanks left on the printing plates for the names of the stores and the serial numbers to be filled in with bronze movable type. It is estimated that tens of thousands of bronze movable type were employed.

From 1907 to 1908, the Russian Peter Kuzmich Kozlov discovered a piece of paper money in Black Water City (today's Inner Mongolia). It is *Zhongtong Yuanbao Jiaochao* with a face value of 1 Guan. There are 3 lines of characters on it: "中统元宝交钞" (*Zhongtong Yuanbao Jiaochao*) on the top, "壹贯文省" (*Yiguanwen Sheng*) in the middle, and the bottom line is the name of the printing office, time and relative regulations. It is now exhibited by the Ermitazh Gosudarstrennyi, Khudozhestvennyi and the Istoriko-kul'turni Muzei. In the back of the money, there is a rectangular ink remark of "至正印造元宝交钞" (*Zhizheng Yinzao Yuanbao Jiaochao*, *yuanbao* paper money printed during the reign of Emperor Zhizheng from 1341 to 1368). From this, we can assume that it was the new issue of Zhongtong currency with bronze types employed to print the serial number.

In 1909, the 2 *guan* paper money unearthed in Xinjiang Province was printed with two bronze types; in 1965, the 1 *guan* Zhongtong currency unearthed in Xianyang City was printed with two bronze types[1]; the 2 *guan* *Zhiyuan Tongxing Baochao* collected in the Cultural Relics Management Institute in Inner Mongolia was printed with two bronze types. In numerable examples can be cited about the use of bronze types in printing paper money from the 13th century to the 14th century. *Baochao* in the Yuan Dynasty was not only in circulation across China but also used in its affiliated countries as Gaoli and Vietnam, even in Russia; it was the most widely used paper money in the world.

Besides all the relics, there are loads of literary documentation of the bronze movable type printing in the Yuan Dynasty. For instance, Huang Jin (黄潛, 1277~1357), a member of the Imperial Academy in the Yuan Dynasty, once mentioned the printing of *Dazang Jing* (《大藏经》, *Siddham Sutra*) in the bibliography that he wrote for the monk Zhiyan of the Chanzong Temple in Beijing.

> *Each time when Yuan Renzong came to participate in the Buddhist services in Qing Shou Temple, he made it a routine to meet the Chan master Zhiyan and talked with him. After Yingzong took throne, he inherited the custom by paying visits to Chan masters and studied sutras with them. The Emperor ordered Zhiyan, along with others, to proofread the Siddham Sutra and print it with copper plates. Later, when they were copying the sutra with golden paint,*

---

❶　卫月望. 中国古钞图辑[M]. 2 版. 北京：中国金融出版社，1992.

they discovered several mistakes and blanks from the early editions，so they decided to print an improved version of it. It must be pointed out that it is impossible for them to print with copper plates considering the immense volumes of the sutra，so it must have been the copper movable type to be employed to print. The ancient Chinese were not very cautious in adopting technical terms；this careless habit can still be located in the Ming and Qing Dynasties.

Emperor Renzong was very eager in Buddhism and Confucianism; he often took part in the Buddhist services in Qingshou Temple. He even conferred 170 *qin* of land to the temple. In 1314，he ordered the establishment of an office responsible for printing sutras; in 1316，the office was promoted to a higher rank to be Guangfu Jian❶. The successor Emperor Yingzong also observed this respect to Buddhism by financing the translation of more sutras with the printing technology. Literally，the two Emperors of the Yuan Dynasty were practically involved in printing sutras with copper type printing.

This statement can also be supported by the historical documents from Korea. According to *The History of Korea*，in the May of 1321，during the reign of Yuan Yingzong and Gaoli King Zhong Su（高丽忠肃王），"the former Lord Yi Cheng Hong Yue（洪瀹）has come to ask for paper to print the Siddham Sutra with an imperial order". The record proves that Yuan Yingzong had decided to publish the Siddham Sutra and ordered the Gaoli paper to be the printing material. In the February，1323，the proofreading job was completed and the Left Prime Minister Bai Zhu（拜住，1298～1323）was ordered to preside over the manuscript of the sutra in golden paint as the master copy. During the manuscript，it came to the notice that "it might be the fact that all the previous sutras are filled with false conceptions and abuse of certain creeds，so it is unavoidable to delete certain contents"，therefore，copper movable type should be prepared beforehand in order to make corrections to the sutra. After all the careful preparations，in August of 1323，unfortunately，a coup d'état took place，and both Emperor Yingzong and Baizhu were killed by Tieshi，the Censor; thus the printing was halted. Had it not been for the event，the printed sutra might have been published in 1323 and brought about a climax in the massive copper movable type printing in the Yingzong reign.

In 1374，the beginning of the Ming Dynasty，Zhu Yuanzhang（1328～

---

❶　宋濂.元史:仁宗纪二[M].上海:上海古籍出版社,1986.

1398) set up *Baochao Tiju Si* (宝钞提举司，an office for issuing paper money) in Nanjing. In 1376, under the supervision of Executive Secretariat, *Daming Tongxing Baochao* (大明通行宝钞，*Daming Tongxing Currency*) was printed and issued❶. There were six different denominations, from 1 *guan* to 100 *wen*, "1 *guan* was about 1,000 *wen*, and 4 *guan* was about 1 *liang* of gold". The look mainly follows that of the Yuan Dynasty but the characters were much reduced. The copper plates and movable type from 1375 to 1398 were cast in Nanjing and the paper materials were still grey, thick mulberry paper. Unlike the Yuan version, the serial numbers were moved from the back side to the front in the Ming currency (Figure 89).

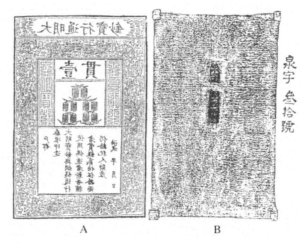

Figure 89   Da Ming Tong Xing Currency Cast in 1376
A is Front Page, B is Back Page, collected by Museum of Guizhou

From the plate, it can be observed that there are two rectangular filling blanks reserved in casting it, so the five copper movable type of "*Quanzi* (泉字)" and "*Sanshi Hao* (叁拾号)" could be put into them to print down the serial numbers. The size is 1. 3 cm × 1. 3 cm × 0. 5 cm in height (Figure 90). According to the rubbings offered by Jia Jingyan (贾敬彦), a historian, there are two other filling blanks in the back of the plate for another 5 copper movable type, namely "*Yongzi* (永字)" and "*Wushi Hao* (伍拾号)". Thus, from the beginning of the Ming Dynasty, there were different groups of copper movable type employed for printing all the six different denominations of paper money. With gold and silver as the standard and controlling the issue of the

❶   张廷玉. 明史:食货志五(钱钞)[M].上海:上海古籍出版社,1986.

money, the paper money was basically stable during the reign of Hongwu (洪武). Each issue was confined to millions of pieces. With the circulation of the currency, the nation now witnessed the heyday of printing paper money with copper movable type. There are not rare unearthed relics to support the conception, and the copper movable type were of the font of the printed book *Zhenguan Zhengyao* (《贞观政要》, *The Political Program of Zhenguan*) in the early Ming Dynasty of the Jingchang version (经厂本).

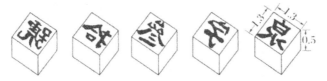

Figure 90　Copper Types Used to Print *Daming Baochao Currency* (cast in 1376), drawn by the Author (unit: cm)

Since the Southern Song Dynasty, only the government had both the wealth and people to print paper money with copper plates and movable type, and it ought to be a special monopoly for the government to do so. As for other types of metal movable type printing, it depended on the considerations of realistic and financial demands. It had been the national conditions that decided the nation to focus on wooden plate printing, with non-metal and metal movable type printing as supplementary means, which had been totally different from other countries in the world. The tradition had survived into the Qing Dynasty. Only when we understand this feature can we avoid misunderstanding of the history of Chinese metal movable type printing.

Among the non-governmental fields, though, as far as the finance permitted, copper movable type printing was possible. The folk printers in the Ming Dynasty were especially active in this type or printing: there was a well-known metal movable type printing house in the Wuxi area, represented by Hua Sui (1439~1513) from the Hua family. Thanks to the introduction in Volume 8 from the book *Shulin Qinghua* (《书林清话》,1911), scholars began to pay heed to the printing house. The earliest bibliographical author was Shao Bao (邵宝, 1460~1527), a scholar from Wu Xi, who wrote in the book *Huitong Huajun Zhuan* (《会通华君传》):

> *Hua Sui, the Huitong Gentleman, with the courtesy name Wen Hui* (文辉)*, is from Wuxi. He was proficient in both history and Confucian classics in his youth and is a keen reader ever since. In his*

mature time，*he was good at proofreading books and like discussing the contents with other scholars*．*Later on*，*he began to print books with metal movable type*，*and claimed to be an all-round man*，*hence the nickname Huitong Gentleman*（会通君，*an all-round man*）．

Qiao Yu（乔宇，1457～1524），Hua's peer，pointed out that："In order to stop the books and manuscripts from missing，he used metal movable type printing to print all the rare books and manuscripts." Hua Sui might have printed 15 copies of books with names to be referred to，11 of which are collected in the National Library in Beijing. The earlier print is *Song Zhuchen Zouyi*（《宋诸臣奏议》） in 1490，with the five characters of " *Huitong Guan Yinzheng*（会通馆印正，printed and proofread by Huitong House)"；the

Figure 91　Copper Movable Type Printed Book，
Notes Taken in Rong Study（《容斋随笔》）
in 1495 printed and proofread
by Huitong House in Wuxi，
collected by National Library of China

next print is *Jinxiu Wanhuagu*（《锦绣万花谷》） in 1494 with 160 volumes and *Rongzhai Suibi*（《容斋随笔》，*Notes Taken in Rong Study*） （Figure 91） in the year after.

Hua's print was mainly tomes. He noted in the book *Song Zhuchen Zouyi*（《宋诸臣奏议》）："At first，I decided to print books to save the trouble of hand copying them，but now I decided to publish them … I was born into a civilized period，and the copper movable type printing is a gift from high above." It seems that he started casting movable type in 1465 to 1487 and succeeded in 1489. Some of his early prints were not of high quality，but the later ones were obviously improved. According to the book *Album of Chinese Wood Engravings* （1961），Hua was using copper movable type，but some people believe that he was using tin movable type on a copper plate. Mr. Qian Cunxun（钱存训） thought that the movable type should have been a copper alloy，like "copper blended with tin or lead". We agree with Mr. Qian — the types might have been a copper-tin-lead alloy.

2.4.2.3　Copper Movable Type Printing in the Qing Dynasty

After the establishment of the Qing Dynasty，the rulers were not interested in issuing paper money；instead，they still cast copper coins for circulation. At first，the government printed and published the tome *Gujin Tushu Jicheng* by using copper movable type. During the reign of Kangxi（康熙，1662～1772）the scholar Chen Menglei（1651～1741）managed to compile the tome *Gujin Tushu Huibian*（《古今图书汇编》，*The Compilations of Books in Ancient and Modern Times*）with 5 years' effort. The tome was completed in 1706；but it was not until 1716 that the tome was ready for the Emperor to read and officially name it，and then it was to be printed and published. Bao Shichen（包世臣，1775～1855）claimed："During the reign of Kangxi，the imperial storehouse had cast millions of fine copper movable type to print books." Unfortunately，the Emperor passed away before the publication of the tome. After Yinzhen （胤禛）took throne，he changed the reign to Yongzheng（雍正，1723～1735）and ordered Jiang Tingxi（蒋廷锡，1669～1732）to recompile the tome. The task was completed in 1726 and 66 copies of it was published in 1728 by the Wuying Palace（武英殿修书处），employing copper movable type printing （ Figure 92 ）. Consisting of more than 10,000 rolls with 160 million characters and 5,020 volumes，the tome was the largest tome in the world，four times the content of the 11ᵗʰ edition of Encyclopaedia Britannica in 1911. It was printed with two sizes of types of the Song typeface（the large type was 1 cm×1 cm，and the small type was 0.5 cm×0.5 cm）.

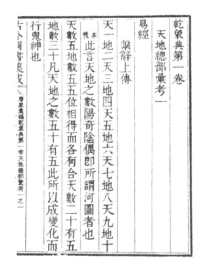

Figure 92   Copper Movable Type Printed Book，*The Collection of Ancient and Modern Books*（《古今图书集成》）in 1726，by Imperial Storehouse of Qing Dynasty

In the book *Chenyuan Shilüe*（《宸垣识略》，1788），the author Wu Changyuan（吴长元，1743～1800）noted："Wuying Palace has cast copper movable type since the printing of the tome." However，there was an opinion that "The copper movable type was carved one by one❶ to print the tome *Gugin Tushu Jicheng* during the reign of Kangxi" according to the book *Ti Wuyingdian Juzhenban Shiyun* appointed by Qing Gaozong（清高宗）. Besides，

---

❶   GILES L. An Alphabetical Index to the Chinese Encyclopedia[M]. London：British Museum，1911.

someone found out that within the same page, for the same character, the organization of its parts can be different from each other, so the types should have been hand carved instead of being cast in bunches. Here, we agree with Wu Changyuan's conception that the copper movable type cannot have been hand carved one by one; they must have been cast. In the history, only wooden movable type was hand carved because it was easier and more convenient. Nevertheless, it would be very difficult and time-consuming to hand carve tens of thousands or even millions of copper types. It is impossible both technically and economically. It is quite likely, however, for one mold to be used several times to cast the types yet with slightly different organization of the parts of one character. One mold was to be carved by one person, but once the casting mold was completed, it is much easier to form types out of the molds by casting.

Li Guijing (李圭景, 1788~1862), a North Korean scholar once observed:

> The best movable types in China are made by Wuying Palace. For each type, the back is flat but not dented. When printing, an iron thread is to fixate all the types by going through holes in the middle of the types so that all the types won't move or be set out of line. In North Korea, the types are of various sizes, the backs dented, no thread used to fixate the types, so it is rather nonstandard to the eye.

Li heard about the Chinese movable type printing from members of the Korean diplomatic corps visiting China. What he had described tallied with what Wang Zhen (王祯) in the Yuan Dynasty had described about the types in the Song Dynasty in employing tin movable type in printing. Therefore, Chinese printing technology had come down in one continuous line through generations. Li thought that it was the shape and structure of the types that decided the elegance of a printed book.

This technology was also promoted in Taiwan area. When Wu Longa (武隆阿, 1765~1831) was the commander-in-chief in Taiwan area in 1806, he once printed and published the book *Shengyu Guangxun Zhu* (《圣谕广训注》, *Notes on Sacred Edict*) by adopting Wuying Palace's pattern. Yao Ying (姚莹, 1785~1853), a *Jinshi* (进士, *Jinshi* degree holder), had witnessed the book and the types in Taiwan area when he was an official on the island. He wrote a letter to his friend and noted:

> Here (Taiwan) the Wu family (commander-in-chief Wu Longa,

武隆阿) *were also involved in casting copper movable types with Song typeface, yet there are only 8 lines per plate, which is to my dislike. A Mr. Lin from the Min area (闽) carves wooden movable types with 10 lines per plate and 11 characters per line, which is better than Wu's way of printing.*

Just before the imperial storehouse cast its copper movable types during the reign of Kangxi, there published a book *Wenyuan Yinghua Lüfu Xuan* (《文苑英华律赋选》, *Finest Ones in the Garden of Literature*) in the regions south of the Yangtze River. This book is comprised of 4 rolls, black mouth, single lines around the four margins of a page, double fish tails, with 10 lines per half page and 18 characters per line, handwritten and elegant. On the title page, there are characters of "Mr. Qian Xiangling (钱湘灵), from Yu Shan (虞山), selected the articles" and "Chuili Ge (吹藜阁) copper plates". The preface was written by the compiler Qian Lulan (钱陆灿) in the 25th year of Kang Xi reign when he was 75 years old, claiming that "the movable types are used to print the simplified version for publication". Qian Lulan (1612~1710), his courtesy name Xiangling, alternative name Yuan Sha (圆沙), from Yushan (today's Changshu, Jiangsu Province), was a provincial graduate during the reign of Shunzhi, interested in collecting books and taught in Changzhou and Jinlin (today's Nanjing). His masterpiece was *Tiaoyunzhai Ji* (《调运斋集》). The book *Wenyuan Yinghua Lüfu Xuan* was completed when he was teaching in Changzhou and was proofread by his disciple Liu Shihong (刘士弘). Now the book is collected by the National Library in Beijing.

In the first half of the 19th century, the rich in Fujian, Zhejiang and Guangdong Provinces were also into casting and printing books. According to Roll 9 of the book *Yi Shi Jishi* (《壹是纪始》) written by Wei Song (魏崧) in 1834, it noted: "Movable type printing began in the Song Dynasty ... and copper and lead have been used recently." It was during the reign of Daoguang (道光, 1821~1850), and the copper should be copper alloy and it was tin and its alloy rather than lead that was used at that time since the people in the Qing Dynasty couldn't tell the difference between tin and lead. Lin Chunqi (林春祺, courtesy name Yizhai, 怡斋, 1808~1873), a Fujian person, went to study in Su and Hang area, and started casting copper movable types to print books at the age of 18. After twenty years' effort, he consumed 200,000 *liang* silver to cast 400,000 pieces of copper movable types in large and small sizes, observing the fonts of the book *Hongwu Zhengyun* (《洪武正韵》). Among his prints, 12 volumes of *Yinxue Wu Shu* (《音学五书》, *Five Books on Chinese Phonology*) written by Gu Yanwu (顾炎武) and 2 volumes of *Jun Zhong Yifang Beiyao* (《军中医方备要》, *Essentials on Surgical Methods for the Army*) are handed down to us today. Lin was from Longtian Town, Fuqing County, Fujian Province, so his print was named *Futian Shuhai* (《福田书海》, books from Futian area), and there are the four characters printed in the middle of the type area of each book.

2.4.3　Origin and Development of Tin Movable Type Printing

The copper movable types were not actually pure copper because it was very expensive to use the metal, so to be more precise, it was the alloy of copper, tin

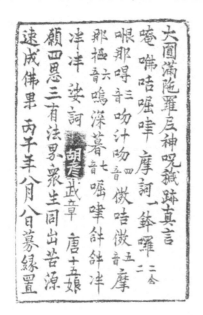

and lead that were used to cast movable types. It should be more exact to name it bronze movable types. Ever since the Southern Song Dynasty, copper had been in short supply because the Jin government had occupied almost all the copper mines found in the Northern Song Dynasty, and therefore, people in the Southern Song Dynasty had to use tin instead, which was reflected both in casting coins and in printing. For instance, in 1957, a tin printing plate was unearthed in West Lake in Hangzhou. It is rectangular (27.5 cm × 6.5 cm × 0.65 cm) (Figure 93). On both sides of the plate, there are characters cut in relief with Song typeface. The square types are of two

Figure 93　Tin Movable-Type Printed Book (《大圆满陀罗尼神咒秒迹真言》, *Dzogchen Dharani*) in 1186, unearthed in 1957 in Xihu, Hangzhou, from Jin Baidong (1996)

sizes: 1.4 cm and 0.7 cm. On one side are the 47 characters from the *Dayuanman Tuoluoni Shenzhou Huiji Zhenyan* (《大圆满陀罗尼神咒秒迹真言》, *Dzogchen Dharani*), with "Hu Yan (胡彦)" among the sutra, which is regarded to be the name of the plate maker. At the end of the plate is the vow: "Wu Zhang (伍章) and Tang Shiwu Niang (唐十五娘) wish that all the people and beings, though suffering the same, may all attain their Buddhahood soon. Made on 8, August, the Year of Bingwu (丙午年)".

Hu Yan was a printer in the reign of Emperor Xiaozong in the Southern Song Dynasty (1163~1189), so he must have made the plate in 1186. The assumption is that he made the plate on 22, September, 1186. On the other side of the plate, there are 36 characters from a mantra. This sutra came from *Guangming Tuoluoni* (《光明陀罗尼》, a dharani) taught by Amoghapāśa and Mahāvairocana and was translated by Bod-hiruci. After the mantra, there is a dharani (《文殊五字陀罗尼》) recorded by the Buddhists Suo chengyun (琐承悸), Wang Yancheng (王彦珵) and Shen Zhirong (沈志荣). This mantra is the earliest tin-alloy movable type printing with exact date in the history.

In 1983, there unearthed a printing plate of "*Guanzi Currency* (金银见钱关子)" from the dictatorship of the Prime Minister Jia Sidao (贾似道, 1213~1275) in the reign of Emperor Lizong (理宗) in the Southern Song Dynasty in 1264. The plate is claimed to be of lead-iron alloy, but it is never officially analyzed. We assume the material to be the alloy of tin, lead and slight iron. It is rectangular, 22.5 cm in height, 13.5 cm in width and 0.4 cm in thickness, 1 kg in weight (Figure 94). On the above and bottom margins, there are patterns. The line of "*Hangzaique Huowuduichong Jinyinxianqianguanzi* (行在権货务对椿金银见钱关子)" appears on the top, beneath which

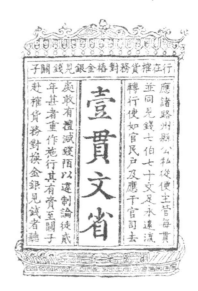

Figure 94　Tinplate of *Guanzi Currency* Issued in 1264 in Southern Song Dynasty, discovered in 1983 in Dongzhi County, Anhui Province, from Wang Benchu (1987)

there are 3 more lines of characters. Altogether, there are 94 characters on the plate, all cut in relief. With the plate, there unearthed 8 types of plates, including an imperial edition, enacted edition and an Aquarius edition, with patterns and characters carved in them. The 474 Roll *Jia Sidao Zhuan* (《贾似道传》) of *The History of the Song Dynasty* (《宋史》) and Roll 7 of *Xu Wenxian Tongkao* (《续文献通考》, *A Sequel to Examination of Historical Documents*, 1586) written by Wang Qi (王圻, 1540~1615) both record the issuing of Guan Zi in the Southern Song Dynasty.

Some argue that all the 8 plates are parts of a full set printing plate and they can be put into a Jia-shape plate in accordance to the recording in the book *The History of Song*. Nevertheless, it is rather doubtful. In the Song Dynasty, it was common to use copper plates to print paper currency, so it was rare for them to use a tin plate; besides, if we put the 8 plates together, it would be more than 50 cm in height and it would be too tall to print any paper currency. Thus, the plates might not be regarded as authoritative printing plates, instead, they might be samples. And we shouldn't assume the 8 plates to be a full set since there are no serial numbers on them. They are dated back to the Southern Song Dynasty for certain, but they might not be what they had been designed to be at the very beginning but were re-cast after some tests and checks.

Since in the Southern Song Dynasty, tin-alloy printing blocks were made to print books, it may be possible for people to use tin-alloy types to print books. At the beginning of the Yuan Dynasty, Wang Zhen（王祯）, the scientist, had looked back into the Chinese printing history to study wooden movable type printing in 1298. He pointed out that in the 10th century, when *Nine Confucianist Classics* was printed with wooden blocks, it was quite a time-consuming and painstaking task. "Though the book is printed, the effort put into it was rather shocking and thus fail to be popularized among the followers. Somebody had struck upon an idea ... to set clay movable type into the tray to print." Here, "somebody" was actually Bi Sheng（毕昇）or someone who replaced wooden printing blocks by clay movable type in the 11th and 12th centuries. Then, Wang Zhen continued to note:

> So far, tin movable types have been cast and bound by an iron bar, set in a plate to print books. However, the types were hard to be inked, so it is not a solid means to print.
>
> Now there is another tangible way to print books: to use wood as a form tray, within which bamboo pieces are set to demarcate the characters. Wooden movable type is carved individually ... and set in the tray line after line, with bamboo demarcation lines in between.

When Wang Zhen claimed that "now there is another tangible way to print books", he actually meant that it was he who had improved the wooden movable type printing technology inherited from the 13th century (the Yuan Dynasty), and the specific improvement was to introduce the revolving table for type storing to replace the huge type cases. It is wrong for us to take it for granted that Wang Zhen had invented the wooden movable type printing technology. Then how to make sense of the phrase "so far"? Someone holds that it referred to the 14th century when Wang published his book *Nong Shu* (《农书》, *Book on Agriculture*), but again they are making a mistake by postponing Chinese tin movable type printing technology for a whole century. ❶

To understand Wang Zhen's statement, one has to take into consideration the meanings of ancient Chinese. Here, what Wang referred to as "generations to come" should be read as the Song Dynasty, but not the Yuan Dynasty, for he had already referred to the Yuan Dynasty as "this world". Besides, he wrote *Zao Huozi Yinshu Fa* in the 13th century, so one cannot interpret movable type printing as an invention in the 14th century. From Wang's context, it is obvious that he indicated that there was tin-alloy movable type printing technology in employment as late as the Southern Song Dynasty❷, and it was an alteration to the copper movable type printing in the Northern Song Dynasty in order to lower the cost by reducing the consumption of copper, which was also a trend in the Southern Song Dynasty.

Wang also mentioned that in the Southern Song Dynasty, the tin movable types were bound together by iron through a hole in the middle of each type, and thin bamboo demarcation lines were set in between to fixate the characters. This method was still in use in the Qing Dynasty, for instance, the book *Gujin Tushu Jicheng* published in the reign of Emperor Yongzheng (雍正) was printed by copper movable types, precisely following the method of printing with tin movable types in the Southern Song Dynasty. Nevertheless, what Wang failed to observe is that as early as the Song and the Jin Dynasties that the metal movable types were improved in the inking technology and were successfully employed in printing paper currency. In 1998, Guangling Ancient Books Printing Press (广陵古籍刻印社) in Jiangsu Province had succeeded in using Wang Zhen's method to print books with tin-alloy movable types, so it has been

---

❶ Chon Hye-bong. Development process of movable metal type printing in Korea: Speech at the International Symposium on Printing History in the East and West[R]. Seoul, 1997-9-30.

❷ 潘吉星. 中国、韩国和欧洲早期金属活字印刷的比较研究[J]. 传统文化与现代化, 1998, (1):71-80.
Pan Jixing. A comparative research of early printing technique in China, Korea and Europe: Speech at the International Symposium on Printing History in the East and West[R]. Seoul, 1997-09-29.

proved true that regular ink can be used to print books and the author of this book had witnessed the experiment. Someone claimed that before the 14th century, metal movable type printing ended in failure in China on the ground that Wang Zhen mentioned that tin movable types "will not last long enough", "and that it was not until the 15th century that metal movable type printing began to be in use in the Ming Dynasty", but obviously it is against the truth and is denied by the documents and relics exemplified in this chapter in the book.

When Wang Zhen claimed that the tin movable type printing would not last long enough, he meant that since China was re-united in the Yuan Dynasty, copper was no longer in short supply, so copper could be used in casting movable types and tin would not be used as a replacement. In the Southern Song Dynasty, however, tin movable type printing was not a flash in a pan, it was still in employment during the Ming and the Qing Dynasties. For instance, during the reign of Daoguang (道光, 1850), the Tang family in Foshan Town in Guangdong Province had invested 10,000 *liang* silver to cast three sets of tin movable types (altogether 30,000 pieces), including types of large flat size, large long size and small long size. The large types were 1 cm × 1 cm, the small ones were 0.6 cm × 0.8 cm, both of which were 1.32 cm in height. They are shorter than regular copper movable types in order to save the material. The large types are of handwriting typeface and were used in printing colored notes. Of the tin movable types, the main component was tin of course, including lead and slight copper and zin, among others. In 1852, these tin movable types were used to print the book *Wenxian Tongkao* written by Ma Duanlin (马端临) in the Yuan Dynasty, which consists of 348 rolls, nearly 20,000 pages and 120 volumes. Other books were also printed by using these types. The scripts are all of good clarity and fine paper.

Figure 95　Tin Movable-Types Cast by Tang Family in Foshan, Guangdong Province in 1850 from Williams (1850)

An American clergyman S. W. Williams (1818~1884) was in Guangdong province then, and he had witnessed the types cast by Tang, so he introduced this printing by publishing an article in *The Chinese Repository* with illustration of some of the types (Figure 95). Tang's types were cast according to Chinese traditional method: firstly, wooden movable types were carved as master patterns; secondly, the wooden types were impressed upon clay to form molds; thirdly, melted tin alloy was poured into the molds, with four characters forming one tin-mold. After the types were cast, workers used knives to trim the surface and the sides. When setting the types, Ormosia henryi was used as a tray and brass strips were placed as demarcation lines. After setting the types, workers proofread the plate and then ink them to print the books. There is an opinion that the Tang family was actually using a Western printing method❶, but that is not the truth.

2.5　Spread of China's Printing Technique in Europe

2.5.1　Beginning of Woodblock Printing in Europe

In the 12th to 13th centuries, with the introduction of Chinese papermaking technique through the Arab region, European countries such as Spain, Italy and France had built paper mills, but all kinds of reading materials were still copied by hand. After the 14th and 15th centuries, namely, the Renaissance period of Western Europe, owing to the development of social economy, urban industry and commerce, science, culture, the demand for reading materials was rapidly increasing. The supply of manuscripts could not meet the needs of the society, thus there was the breeding ground to stimulate the appearance of printing technique. The rise of printing in turn promoted the development of society. At that time, the Yuan and Ming Dynasties in China were at a new stage of all-round development of woodblock, copperplate and movable type printing. This period was also unprecedentedly active in the aspects of Sino-European direct contact, so Europeans were likely to get technical information of printing from China and develop printing techniques.

As mentioned above, the Mongol Ⅱ-Khanate close to Europe printed and issued paper money with the Chinese printing technique in 1294, which was described in its Persian scholar Rashidal-Din's *The Jami' al-tawarikh* in 1311. The printing technique was rapidly developed in the reign of Ghazan Khan Mahmud (1295~1304), and also spread to Egypt from 1300 to 1350. Printing

---

❶　HIRTH F. Western appliances in the Chinese printing industry[J]. Journal of the North China Branch of the Boyal Asiatic Society, 1886(20): 166-167.

was then available in Western Europe and North Africa. It was not difficult for Europeans to get printing materials and relevant technical information. On the other hand, the Western expedition by the Mongols in the 13th century reopened the previously blocked Silk Road between Asian and European continents, thus providing conditions for the technical, economic and personnel exchanges between the East and the West. There was no obstruction to traffic from the Great Capital of the Yuan Dynasty (now Beijing) to European cities like Rome, Paris and so on. China and Europe had direct contacts, so Europeans could introduce first-hand printing technique from China. The proverb "All roads lead to Rome" was a proper illustration of various means of spreading Chinese printing to Europe.

In the Yuan Dynasty, Chinese and European messengers, businessmen, religion believers, tourists, craftsmen and scholars exchanged visits through the Silk Road, so Europeans in China were likely to get close contact with printing materials and relevant technical knowledge. For example, in 1245, the Pope Innocent Ⅳ (1243～1254) sent Italian Jean Plano de Carpini (1182～1252), archbishop of Germanic in 1228 and archbishop of Spain in 1230, to Mongolia, whose entourages included the Polish priest Benedict and Austrian businessmen. He arrived at Kara and Lin and was received by Yuandingzong Alacon Kaan from Mongolia in 1246. After returning to Lyon, France in 1247, in his Latin writings *Bulletin of the Oriental Experiences* (*Libellus Historicus*), he introduced China, saying that Chinese people were skilled in their matchless technologies, and there were written words and records of detailed ancestral history❶. He also found sacred books like *Bible* in China, which might refer to the printed books of Buddhist scriptures.

At the same time, King Louis Ⅸ of France (1214～1270) sent French Franciscan Guillaume de Rubrouck (1215～1270) to the east, accompanied by Italian Bartolomeo da Cremona and others. They arrived at Kara and Lin and were received by Mongon Kaan from Mongolia in 1253, and returned to Paris, France in 1255. In *Journey to the East* (*Itinerarium ad Orientals*, the travel notes written by Rubrouck), he made a detailed introduction of printed money in the Yuan Dynasty as follows: "The ordinary money of Cathay is a piece of paper made out of cotton, a handbreadth in width and length, and on this they stamp lines like Mangu's seal. They write with a brush like those with which

---

❶ DAWSON C. The Mongol Mission[M]. London: Sheed & Ward, 1955.

painters paint and in a single character they make several letters which form one word."❶"Cotton paper" here might refer to mulberry bark paper. Rubrouck was the first European to point out China's distributing paper money by the printing technique. Many Germans, Russians, French, Hungarians and British with professional skills can be seen in Kara and Lin.

During the Yuan Dynasty, the Italian merchants of Venice and Genoa were keen on trade with China. In the reign of Emperor Yuanshizhu (1260～1294), the merchants from Venice, Nicolo Polo and Maffeo Polo came to China and were summoned by the Emperor. According to the command, they led royal court envoys to Europe and handed over letters to the Pope. After completing his mission in 1271, he came to China again, accompanied by the young Marco Polo (1254～1324), who was kept in the palace by Emperor Yuanshizhu and was later entrusted with important post. Spending 17 years in China, he left from Quanzhou in 1292 and returned to Venice in 1296. *Travels of Marco Polo*, written in 1299, opened the horizon of Europeans by introducing more about China's richness and material civilization. The book mentioned that there were banknote-printing factories in Beijing, where banknotes were printed on mulberry bark papers and distributed nationwide. Worn-out banknotes can be replaced with new ones❷.

In 1952, silver coins from Venice were unearthed in Guangzhou City. Latin tombstones of two businessmen from Genoa in the year of 1342 and 1399 were unearthed in Yangzhou City. The Venice municipal archive also recorded a lawsuit filed in 1341 by a local defendant who did business in China with money. So many businessmen came to China that Francesco Balducci Pegolotti (1305～1365), in a special section of his work *Trade Guide* (*Practica Della Mercantura*), introduced how to do business in China❸. Between China and Europe, to the west of Ural Mountains and the Caspian Sea, there were two traffic routes: southern route and northern route. Along the northern route, they could start from Xinjiang of China, went westward by the way of Kiptchac Khanate, through Russia, Poland and Bohemia, then reached Germany. This was almost similar to the traditional Silk Road on land. Along the southern

---

❶ DAWSON C. The Mongol Mission[M]. London: Sheed & Ward, 1955.

❷ POLO M. 马可·波罗游记[M].李季,译.上海:亚东图书馆,1936.

❸ YULE H. Cathay and the Way Thither: Being a Collection of Medieval Notices of China[M]. London: Hakhuyt Society Publications, 1914.
张星烺.中西交通史料汇编:第 2 册[M].北平:京城印书局,1930.

route, a combination of land and sea, they left from China's Quanzhou and Guangzhou, then sailed westward by the way of Ⅱ-Khanate, through Armenia, Persia and Turkey, finally arrived at Italy. Most travelers of the East and the West made a long journey on land and on the sea along these two routes.

While the Europeans came to the east in the 13th century, there were Chinese who traveled in the opposite direction to Europe. The Mongol rulers in Beijing often sent envoys, officials and scholars to the Russian state of Ilkhanate. For example, it was recorded in the third volume of *Yuan Shi* (《元史》, *The History of the Yuan Dynasty*), namely, *Xianzong Ji* (《宪宗纪》, *Records of Emperor Xianzong*), that the officials were sent to Russia to check household registration in 1253. Russia's history also says that in 1257, Mongolian officers went to the country of Riazan, Suzdal and Murom to create positions for collecting taxes. In 1259, the Mongolian officers' family and troops to Valkhov for household registration. The registered permanent residence of the Yuan Dynasty were printed in a form and prepared for recording and binding in volumes on file, which enabled Russians to often see prints from Beijing. It was also recorded in the 27th volume of *Yuan Shi* (《元史》, *The History of the Yuan Dynasty*), namely, *Yingzong Ji* (《英宗纪》, *Records of Emperor Yingzong*) that, after Russia had followed the Yuan court in 1320, a large sum money were granted and the officials were sent back. It indicates that banknotes of Great Khanate were circulated in Russia. Russians might know how the paper money was printed and pass it to the Western Europeans.

From the 13th to the 14th centuries, Chinese also paid visits to some Western European countries. It was once recorded that from 1275 to 1276, Uighur Nestorian Rabban Bar Sauma (1225 ~ 1293), born in Beijing, and his Mongolian disciple Marcos (1244~1317) left Beijing and went west through Xinjiang for the pilgrimage to Jerusalem❶. After arriving at Baghdad in 1280, Bar Sauma was appointed as chief inspector of Nestorianism, and Marcos the bishop of Baghdad. Aru muddy khan of Ⅱ-Khanate sent Marcos in 1285 and Bar Sauma in 1287 to Rome. Bar Sauma also visited Genoa and Paris. Received by the French King Philip Ⅳ le Bel (1268~1314), he visited such places as University of Paris. Then he went to Bordeaux, meeting British King Edward

---

❶ CHABOT J B. Relations du Roi Argoun avec l Occident. Revue l Orient Latin (Paris), 1894.

Ⅰ (1239～1307). In 1288, he presented credentials to the Pope Nicholas Ⅳ (1288～1292) in Rome. After returning to Baghdad, he wrote his travel notes in Persian, later translated into Syrian and French in the 19th century and into English in the 20th century. Bar Sauma was the first Chinese to arrive in Western Europe. Having been studying in Beijing in his childhood, he knew much about the printing technique. Before leaving for Europe, he passed through the printing center at his ancestral home of Xinjiang. He would be ready to introduce to Europeans how Chinese books were printed.

It is worth noting that in 1288, the next year of his interviewing, Bar Sauma, the Pope sent the Italian Franciscan Giovanni da Monte Corvino (1247 ～1328) to China, accompanied by Christian Nicholas de Pistoia and the Italian businessman Pietro da Lucalonga. Nicholas died on the way from Rome through Ⅱ-Khanate to India. Giovanni and Pietro arrived at Quanzhou in 1293. They arrived at Beijing in 1294, the time of Yuanshizu Kublai's demise, thus he was received by Emperor Yuan Chengzong (1295～1307). After sending in the sealed book of the Pope, they got approval of missionary work. Two Latin letters from Giovanni to Vatican were kept in the National Library of Paris. The first letter, which was written on May 18, 1305, mentioned that he was excluded by Nestorians at first, and "do[es] not publish any doctrines different from the faith of Nestorianism". Thanks to the protection of Yuan Chengzong, he was able to preach alone. Two churches were established in Beijing during 1298 and 1305. In 1303, Arnold, a friar from Cologne in Germany, came to Beijing to assist preaching. 6000 people were baptized. 40 children were recruited to learn Latin and religious etiquette, and formed a chorus, singing a hymn in the Mass. By that time, proficient in Mongolian, Arnold had translated *The New Testament* and *The Psalter* into Mongolian❶.

The second letter, written on February 13, 1306, said: "I draw 6 pictures of the holy image in *The New Old Testament* for those less educated people. The pictures are followed by Latin, Tursic and Persian, so that anyone who knows one of them can read"❷. Dr. Carter regarded religious paintings with words, which Giovanni provided for less educated Christians, as printings. He said: "In China at that time, the printing of any important work has become very

---

❶   MOULD A C. Christians in China Before the Year 1550[M]. London: Society for Promotion of Christian Knowledge, 1930.

❷   MOULE A C.1550 年以前的中国基督教史[M].郝镇华,译.北京:中华书局,1984.

natural." He noted that 50 years after Giovanni published religious painting in Beijing, a similar religious painting appeared in Europe, which "may not be altogether a coincidence."[1] That is to say, Europeans had used the same method of Giovanni in China.

Dr. Carter was right. Since Giovanni had arrived in Beijing, he found copies of Nestorianism, but he was not allowed to print Christian copies. After getting support from Emperor Yuan Chengzong and Prince of Mongolia Korguz (1234~1298), Giovanni began to print religious paintings by himself. With three kinds of characters, these paintings were intended to compete with Nestorianism and were distributed to the masses of believers within the Chinese territory. It was impossible to copy thousands of paintings by manual work. They can only be printed and published by cutting blocks. During Dade era (1297~1307), the Chinese people did the job, almost at the same time as the churches were established. It is worth noting that Latin, Persian and Tursic were printed on the religious paintings, even without the Chinese language. But the Han people accounted for more than 90 percent of China's population, which seemed inexplicable. The Tursic character referred to "Tursicis" in the Vatican manuscript. What was it exactly? It should be Mongolian, especially the old Mongolian, which was spelled with ancient Uighur language. Credentials sent to Vatican in the Yuan Dynasty were written in this language. Most Christian believers in the early Yuan Dynasty were Mongolians instead of the Han people[2], and target readers of printing materials were mainly Christians, so Mongolian, the national language at that time, was chosen. Giovanni and Arnold had completed the pioneering work in China that westerners had never done. Those printing materials were easy to reach Europe and were copied by Europeans. The intermediary agents of communication were European merchants and priests who came to China from 1300 to 1368.

During the 13th to 14th centuries, besides paper money, religious paintings and printed books, Europeans came into contact with playing cards for popular entertainment. The cards were invented by the Chinese. In the early 20th century, playing cards (Figure 96) dated back to the 14th century were unearthed in Turpan, Xinjiang. During the Western expedition by the

---

[1] CARTER T F. 中国印刷术的发明和它的西传[M]. 吴泽炎,译. 北京:商务印书馆, 1957.

[2] 陈垣. 元也里可温考[M]//陈垣. 陈垣学术论文集. 北京:中华书局,1980.

Mongols, the Mongol army introduced playing cards to Europe, which soon became popular in some countries. The above Chinese printing materials were a guide of printing technique to Europe. At the first stage of European printing, from 1350 to 1400, early products were cards and religious paintings for the public. Germany and Italy were considered as the earliest European countries to produce such prints. According to the early municipal records of Augsburg and Nuremberg in southern Germany, "cards maker (Kartenmacher)" was mentioned several times in 1418, 1420, 1433, 1435 and 1438. Cards were printed with different methods, such as hand-drawing, stamp sealing and wood printing. It was obvious that the cost of wood printing was the lowest, which should be a more common method.

Figure 96   Playing Card in 14th Century China, unearthed in Xinjiang on the Silk Road

Some earliest extant Italian cards were printed, but it was hard to pinpoint the exact year. A clear piece of historical data was a decree issued by the Municipal Council of Venice, Italy in 1441, saying: "Whereas, the art and mystery of making cards and printed figures, which is in use at Venice, has fallen to decay, and this in consequence of the great quantity of printed playing cards and colored figures which are made out of Venice, to which evil it is necessary to apply some remedy … from this time in future, no work of the said art that is printed or painted on cloth or paper — that is to say, altar-pieces, or images, or playing cards, or any other thing … shall be allowed to be brought or imported into this city."❶ This showed that Venice was one of the centers of printing cards and religious paintings, and the business began as early as in 1441. The municipal authority took protective measures after it was threatened by foreign competition for local product markets. The move could be a counter measure to the dumping of German products. As recorded in the German city of Ulm, in the same period, the city put the printed cards into casks, which were then shipped to Sicily and Italy.

---

❶   Carter T F. 中国印刷术的发明和它的西传[M]. 吴泽炎,译. 北京:商务印书馆,1957.

According to a 17ᵗʰ-century Italian author, Valere Zani (1621～1696), Venetian cards were introduced directly from China. He said: "The Abbe Tressan (a French missionary to Palestine, 1618～1684) showed me when I was in Paris a pack of Chinese cards and told me that a Venetian was the first who brought cards from China to Venice, and that city was the first place in Europe where they were unknown." Judging from the frequent exchanges between Chinese and Venetians in the Yuan Dynasty, Zani's statement was well-founded and Venetian businessmen might have been exposed to this kind of entertainment earlier. But the cards as well as printing technique could be introduced to Europe in a variety of ways. Religious painting could also be printed at the places where cards were made. Shape and structure of religious paintings should be similar to those of Giovanni and Arnold in Beijing.

It was no coincidence that these two early prints appeared in Italy and Germany. The Pope was in Italy, the religious center and the source of the Renaissance, where foreign trades were flourishing, and personnel exchanges with Chinese in the Yuan Dynasty were frequent. It was quite natural that Italy was the first to introduce the printing technique. Located in central Europe, accessible from all directions, very close to Italy and Mongolia, Germany collected information from everywhere. A strong evidence was that the first Europeans borrowing Chinese technology to print religious paintings in Beijing were from Italy and Germany. The earliest known woodblock religious painting in Europe was the portrayal of St. Christopher and Jesus in 1423 (Figure 97). It was found pasted on the cover of manuscript copy in a monastery library in Augsburg, Germany, and now it is preserved in the Rylands Library, Manchester, England❶.

Figure 97　*Wading of St. Christopher and Jesus*, a German Woodblock Religious Painting, 1423, from de Vinne (1875)

---

❶　Oswald J C. 西洋印刷文化史[M]. 玉城肇，译. 东京：鲇书房，1943.

In the picture, the young Jesus waded on St. Christopher's back, with a cross in hand. Two rhythmical lines of inscription means: "whenever you see the sacred image, you can avoid the death," which was similar to the mantra of Buddhist prints. It is worth noting that waterwheel imported from China are also found at the bottom left of the picture. From 1400 to 1450, woodblock printing was popular in Germany, Italy, the Netherlands, and Flanders, now in Belgium. During this period, on the inventory list of (study) German priest Jean de Hinsberg (1419~1455) and his sister in Bethany monastery in the city of Liege, there was "a tool of printing calligraphy and paintings", and "9 woodblocks for printing images and 14 flagstones for printing others (Novem printe lignee ad imprimendas ymagines cum quatuordecim aliis lapideis printis)", which specified the use of woodblock to print sacred images.

Most early prints were roughly carved religious paintings, sometimes filled with color, with brief handwritten characters in the picture. If there were a number of related pictures, they would be bound into a volume. The British Library in London has a number of collections, many of which have not marked the exact years, places and names of cutter. The sequence could only be inferred from the shape and the structure, ways of cutting, thickness of paintings and so on like. The most famous printing is *Ways to Future Life* (*Ars Moriendi*), published in Germany in 1450, a collection of 24 sheets about how to leave the world in peace and happiness. Besides, *Apocalypse* (Figure 98) was printed as early as about 1425, place unknown.

Figure 98   *Apocalypse*, *an* European Woodblock Painting, about 1425, from de Vinne (1875)

Early prints also include a Dutch publication — *Bible of the Poor* (*Biblia Pauperum*). At the end of 15th century, more and more woodblock prints with pictures and characters were found, and there were prints full of characters, such as *The Latin Grammar* (*Ars Grammatica*) by Roman Aelius Donatus (320 ~370) in the 4th century.

Early European woodblock prints are similar to Chinese prints in the Yuan Dynasty in the shape, structure and manufacturing technique. According to the American printing historian, Trevor Vinny (Theodore Law de Vinne, 1828~ 1914), Europeans first drew the manuscripts on the paper and got the ink mark with boiled rice in a reversed position on the woodblock. Then the cutter carved the ink mark along the texture of wood. A middle line was cut to form the two pages of a plate. After that, the worker laid a sheet of paper on the plate with ink and ran the brush over it so as to take the impression on one side of the paper. At last, the sheets of paper were folded along the middle of the page by leaving the two half pages with characters outside, and were bound with thread into small holes along one side[1]. Thus it can be seen that, in terms of such processes as shape and structure, blocks cutting, inking, brushing printing and binding, early European woodblock prints completely adopted the technical methods in China, similar to thread-bound books in the Yuan Dynasty. The only difference was horizontal words instead of vertical ones.

In European woodblock printing practice, two pages were engraved on one plate, the impressions were taken by means of friction on one side of paper with a brush or something like this, and the double pages were put together two by two with the blank sides folded inside. These procedures were typical Chinese methods, but were contrary to the European tradition. Before Trevor Vinny, other European scholars had noted the similarity between the prints of Chinese and European prints, and believed that the technology was introduced from China. For example, British Orientalist Robert Curzon (1810~1873) pointed out in 1860 that, Chinese and European woodblock prints had a lot in common, saying: "We must suppose that the process of printing them must have been copied from ancient Chinese specimens, brought from the country by some early travelers, whose names have not been handed down to our times."[2] These

---

[1] DE VINNE T L. The Invention of Printing[M]. New York: F. Hart, 1876.
[2] CURZON R. The history of printing in China and Europe[J]. Philobiblon Society Miscellanies, 1860,6 (1):23.

travelers were Europeans who came to China during the Yuan Dynasty, no matter who they were. As early as Roman times, there were seals and textile printings, but they failed to transform them into printing technique over a long period of time until the printing technique in China from 1350 to 1400. The real printing industry appeared. Therefore, Carter said, "Chinese influence was the final determining factor in the ushering in of European block printing." This was an objective view that conforms to historical facts.

2.5.2   Beginning of Wooden Type Printing in Europe

It has been largely recognized in academic circles that European woodblock printing technique was introduced from China. However, many people were not yet clear about whether European movable type printing was influenced by China. Few treatises had covered this topic, so it needed to be studied carefully. According to the general development law of the history of printing technique, the woodblock printing would be followed by the movable type printing sooner or later, which was an inevitable trend in both China and abroad. China has spent four or five hundred years developing from woodblock printing to movable-type printing, while Europe has taken a shortcut to this transition in a few decades, as a result of drawing on China's prior experience. There were many circles and turns in European written forms, such as a, e, o, d, g, p, q, s, etc. It was not easy to carve woodblock with a knife. European words were more suitable for movable-type printing than Chinese characters, because Europeans could spell all the words and sentences with no more than twenty or thirty letters. Since they introduced woodblock printing technique from China in the Yuan Dynasty when convenient transportation was available, they must have heard about the Chinese movable-type printing technique that appeared hundreds of years ago and shown great interest in it.

In the Yuan Dynasty, earthenware type, wooden type and metal movable type continued to develop on the basis of printing techniques in the Song Dynasty, and coexisted with the woodblock printing in China. Wang Zhen, a scientist in the early Yuan Dynasty, published a monograph on wooden type printing technique in 1313 with his masterpiece *Nong Shu* (《农书》, *Book on Agriculture*). From their first step on China's territory in Xinjiang, Europeans would have seen movable wooden Uygur letters and the related workshops in Turpan. They would also have seen in other places of mainland China material objects about movable type or heard relevant information, and then introduced them into Europe. Wooden type was the earliest movable type in Europe, but it was definitely the product of Chinese printing culture. The influence of Chinese movable type technique on Europe was primarily manifested in the use of wooden movable type in Europe. Theodor Buchmann (1500 ~ 1564), a professor of theology and orientalist at the University of Zurich in Switzerland, wrote in 1548 that the earliest European movable type was wooden type, saying in Europe "people first engraved scripts on woodblock of a whole page space, but it took a lot of work and was expensive in this way. Therefore, they made movable wooden type to set them together in a plate for printing."

This was an important account of the use of wooden movable type printing in Europe. At the early stage, European wooden type would undoubtedly be made and typeset with Chinese technical methods, and there was no other approach. Buchmann's academic activities appeared less than a few decades from the first time for Europeans to use movable type, therefore his record should be credible reflection of the early European type printers' trial manufacture of movable type by copying China's movable type. The wooden type is a bridge from the woodblock to the metal movable type. The use of wooden type made the Europeans for the first time understand the idea of movable type. Such countries with well-developed woodblock printing as Italy, Netherlands and Germany, first engaged in the printing of wooden movable type. The 19th century British orientalist Curzon reported that in 1426, Pamfilo Castaldi (1398~1490), an Italian doctor and printer, printed some large folio editions with large-sized wooden types in Venice, which was said to have been once kept in the archives in his hometown Feltre Belluno (called Feltria)❶.

Castaldi was born in 1398 in Feltria, to the northwest of Venice, and was later engaged in printing in Venice. He was once considered to be the founder of the European movable type technique, and in 1868, the city of Lombardia in Italy built a bronze statue in memory of him. In 1908, the French sinologist Pelliot discovered in Dunhuang along the Silk Road 960 pieces of Uighur wooden types used by Xinjiang uygurs in the Yuan Dynasty, thus revealing the route of movable type technique spreading from inland to Xinjiang of China,

---

❶ CURZON R. A short account of libraries in Italy[J]. Philobiblon Society Miscellanies, 1854(1):6.
    YULE H. The Book of Ser Marco Polo the Venetian, Concerning the Kingdom and Marvels of the East. 3rd ed. London: Murry, 1903.

and then to the west. Consequently, French historian Pierre Gusman holds that, China's movable type technique in the Yuan Dynasty (the 13th century to the 14th century) was introduced into Europe by two routes: firstly, Armenians in contact with uygurs of Mongol Chagatai Khanate and later lived in Holland, introduced movable type technique into Europe in the era of Castaldi's printing activity; secondly, the German Johannes Gutenberg (1400~1468), in Prague, the capital of Bohemia, learned the movable type technique spreading from central Asia and Russia to Europe. From today's perspective, both possibilities are in line with the records of early European authors.

But the idea that Europe experienced wooden movable type printing and was influenced by Chinese technique was once criticized in the West. If this idea is accepted, the firm view of regarding movable type as an "independent invention" in Europe did not hold water. It should also be mentioned that the argument of China's influence on Europe has some questions unsettled and needs to be improved. For example, Castaldi saw in Venice China's copies brought by Marco Polo from China, which stimulated his printing, or Gutenberg's wife, Ennel Gutenberg from Contarini family in Venice, has seen China's copy and then inspired her husband. Actually, Marco Polo might have brought back to hometown in 1296 some information of printing rather than China's copy. Other businessmen of Venice could have brought the copies back. Besides, Gutenberg had broken his marriage with Ennel before the experiment of movable type.

But the idea that European movable type technique was influenced by China was correct and could be established in general, because, before 1420, there was no movable type printing in Europe, and the ideas of relevant movable type printing and movable type setting technique are products of China. As previously mentioned, in the Yuan Dynasty when exchanges between Chinese and Europeans frequently occurred, the European wooden type technique must have been from China. The early Europeans were strictly confidential on their movable type experiment, so there was no need to investigate any details which were difficult to find. Some people were reluctant to admit that Europe had experienced the similar development from woodblock and movable non-metal type to movable metal type like China. They believed that Europe jumped directly from the woodblock to movable metal type, and they did not admit that Europe in the era of Castaldi and Gutenberg had carried out experiments with wooden movable type, saying that simulation experiments of manufacturing small-sized western wooden movable type by modern precision equipment were unsuccessful❶. But the authors of these simulation experiments could not deny the fact that large-sized western wooden movable type could be produced even without precision equipment. And that is what their predecessors did. They were denying their history if they didn't admit that they could print books with large-sized wooden movable type, and historical facts should always be respected by future generations.

In addition to the Italians, the Dutch people have carried out similar activities. The Dutch doctor Hadrian Junius (1511~1575) from Haarlem city, a contemporary of the Swiss Buchmann, Laurens Janszoon (1395 ~ 1439), a native of the city, printed books like *Latin Grammar* and *Horn Book* with large-sized movable wooden type in 1440❷. Janszoon held the post of property management committee member in Catholic district of Haarlem, which was called koster in Dutch and marguillier in French. He was called as koster or Coster by some Dutchmen, similar to Sima Qian's name "*Taishigong*" given by the Chinese. Thus the Netherlands also claimed to be one of the first countries in Europe to develop movable type printing. Anyhow, the history of European imitation of China's printing books with a wooden movable type cannot be ignored, and it has laid a technical foundation for the future emergence of the European metal movable type.

2.5.3 Beginning of Metal Type Printing in Europe

As mentioned above, European have successfully type-set and printed books with large-sized wooden movable type in the middle of the 15th century, which was a quick and decisive step from woodblock to movable type. The innovation spirit of the people who conducted the initial experiment really deserved our affirmation. But it was worth noting that the layout of large-sized wooden type would waste more printing pages and sheets of paper, and the books were also too heavy. What is more, paper in European counties at that time was very expensive. In order to print more words on one sheet of paper, the movable type should be reduced in size, so the Europeans met technical difficulties in making the small-sized wooden type. Although the wooden type was cheap, but it was difficult to carve small-sized type with a knife, and the carved type did

❶ REED T B. A History of the Old English Letter Foundries[M]. London, 1887.
❷ 庄司浅水. 世界印刷文化史年表[M]. 东京:ブックドム社,1936.

not have sufficient mechanical strength. This was different from the Chinese wooden type. So the development of wooden type was restricted. In this case, the Chinese metal type of the Yuan Dynasty was soon favored by Europeans. After they had the experience of wooden type printing, they could develop metal type printing as long as they solved the problem of casting metal type.

After Europeans were aware that the Chinese had carved wooden type and cast metal type for printing hundreds of years ago, they began to do exploratory experiments in this aspect. It was said that Janszoon, the native of Haarlem city of the Netherlands, had tried to make a movable type from lead and tin while making wooden type. Besides, a German silversmith Prokop Waldfoghel (1367 ~1444) has also done the same experiments. He first settled in Prague, the capital of Bohemia (now Czech Republic) ruled by the Emperor Charles IV von Luxemberg (1347 ~ 1378) of Luxembourg of the German Empire (the Holy Roman Empire). From 1367 to 1418, he was famous in Prague for making tableware. From 1433 to 1441, he moved to Nuremberg and worked in a metallurgical plant after the outburst of Hussite War (1419~1434), which was led by the Bohemian against the German monarchy. In 1439, he became a citizen of Lucerne, not far from Basel, Switzerland. In 1441, he moved to Avignon, now in the southeast of France, the place where the Pope stationed and the center of book selling from 1309 to 1417 and from 1439 to 1449❶.

According to the recent study by technical historian Dr. Wolfgang von Stromer of Nuremberg, during living in Prague, Waldfoghel may have obtained from the East some technical information about typecasting and printing books, because Prague was one of the principal terminal points of shipping silk from China, and also the transfer point of technical information about metal type to Nuremberg, Strasbourg, Mainz and other industrial and commercial cities. After he moved to Avignon, he had an idea of making a living in a new industry by combining his technical expertise in metal technique and the technical information he had acquired in Prague. From 1441 to 1444, he developed an "art for writing artificially (ars scribendi artificialiter)" for producing books, and passed it down to his cooperators David Caderousse, a Jew, Manuel Vitalis from the diocese of Dax, his friend Arnd de Coselhac and Avignon businessman George de la Jardin.

---

❶   MARTIN H J. The History and Power of Writing[M]. Chicago: University of Chicago Press, 1994.

Some documents provided by French Abbe Pierre Henri Requin in the 19th century recorded that in 1446 Caderousse custom-made 27 Hebrew letters (scissac in ferro) in cast iron and some tools made by wood, tin and iron. For two years he possessed 408 Latin letters as security for a loan. According to the data on July 4, 1444, Caderousse and Waldfoghel kept 2 steel words (duc abecedaria calibia), 2 iron types (formes ferreas) and 48 tin types and "other things related to art for writing artificially", then sold them later[1]. This showed that such things were valuable at the time, so the so-called "art for writing artificially" meant to print written language like handwriting by combination of letters made of metal instead of handwriting. In other words, Waldfoghel and his cooperators had already attempted to make metal types in Avignon from 1441 to 1446, apparently for printing use. But they did not continue, and soon they broke up and sold out the words and equipment.

Ten years later, another German Johannes Gensfleisch zum Gutenberg (1400~1468), did similar work but obtained great success. In 1400, he was born in Mainz, the industrial and commercial city at the confluence of the Rhine river and the Main river, and studied at the University of Erfurt from 1418 to 1420. He dropped out of school because of his father's death and went back to his hometown to learn metal working. From 1434 to 1444, he moved to Strasbourg for a living, and cooperated with natives Andreas Drizehn, Hans Riffe, Andreas Heilmann and others to process precious stone together and make mirror with a new method. Gutenberg provided techniques and others invested, and profits were shared by them. After Drizehn died in 1436, his younger brother, as his inheritor, required Gutenberg to hand over the technical secrets but was turned down, so Gutenberg was sued. The archives claimed that in 1436, for things to do with printing (das zu dem Drucken gehort), Gutenberg paid Hans Dunne, a goldsmith from Frankfurt, 100 Gulden gold, and there were such words as type in the testimony.

This suggested that Gutenberg suddenly switched from making mirror and processing stone to printing experiment in secret, but did not succeed. From 1444 to 1448 he went on a journey probably in the Netherlands, Basel in Switzerland or Venice in Italy to do the technical investigation for solving the problems he had faced. Someone said he had also been to Prague. This journey

---

[1]   REQUIN P H. Documents inédits sur les origines de la typographie[J]. Bulletin de Philologie et d Histoire du Ministère de l Instruction Publique, 1890(1):328-350.

widened the field of his vision, and he seemed to get the suitable way to solve the printing problem. In 1448, he returned to Mainz and got a loan from a rich merchant Johann Fust (1400~1466) for supporting his experiment. The loan shall be secured by his technique and equipment, and the profits shall be equally divided within the five-year term of the contract. On expiration of the term, the borrower shall pay the creditor the sum total of the principal and interest. The experiment made a breakthrough. The large-sized metal type cast by Gutenberg was in the practical stage, used for publishing the *Thirty-six Line Bible* (36 *Linne Bible*) in Latin, with manuscripts gothic (Gotisch Schrift) in bold type. Later in 1454, the indulgences issued by Pope Nicholas Ⅴ (1447~1455) were published.

In 1455, the publication of *Forty-two Line Bible* (42 *Line Bible*) in Latin with metal type of 20 point marked the greatest achievement in the technical career of Gutenberg. The plate size was 30.5 cm by 40.6 cm. It consisted of 1286 pages with both sides printed and was bound into two volumes (Figure 99).

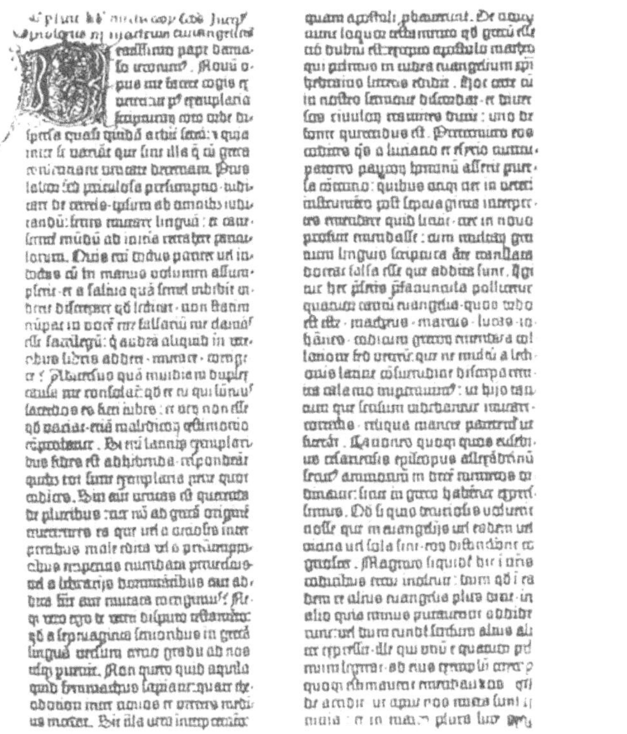

Figure 99　*Forty-two Line Bible* in Latin Printed with Lead type
by Gutenberg in Mainz, 1455, from Ostwald (1928)

Flowers and grass were engraved on the borders of each plate. It was a precious edition because there were both woodblock and movable type in it[1]. However, after the contract expired, Gutenberg was unable to pay his debts. According to the verdict by the government, Fust owned the printing plant and continued to employ the original technicians, workers, including the German Peter Schoffer (1425~1502), a graduate from the University of Paris. Schoffer was good at calligraphy and created many word forms for Gutenberg, and later became the son-in-law and inheritor of Fust. They co-published a number of books and made improvement in the aspects of type face, layout design and typecasting.

After the breakup with Fust, Gutenberg borrowed money from other people and built a new printing plant in the suburbs of Mainz in 1456. Albert Pfister (1400~1465), an original employer of Fust, returned to help Gutenberg and worked as a chief assistant. In 1462, due to the riot which broke out in Mainz, Fust's plant was destroyed by the flames of war, and the printers fled to Strasbourg, Cologne, Bamberg, Nuremberg and other places. Since then, his metal type technique was spread to other German cities and then the whole Europe, becoming the principal way of printing in Europe. During the half century from 1450 to 1500, altogether 250 printing plants adopting the technique of Gutenberg had been established throughout Europe[2], with 25,000 kinds of books published. If 300 volumes were printed for each kind of book, 6 million volumes in total were printed across Europe at that time. Gutenberg died in 1468 after witnessing the application of his technique. Printed volumes tripled in the 16th century and the technique was introduced to the United States in the 17th century. Before the advent of modern machine printing in the 18th century, printing in Europe and the United States was actually a continuation and improvement of Gutenberg's technique with the same mode, thus still belonging to the early stage of printing.

Gutenberg's metal type printing technique was undoubtedly at the top level in the world, but from the entire printing history, it was not the earliest, because China in the Northern Song and Jin Dynasties (from the 11th century to the 12th century) had cast out metal type for printing in a large scale, then the printing technique went on developing in the 19th century. Comparing the European technique represented by Gutenberg with traditional Chinese

---

[1] OSWALD J C. 西洋印刷文化史[M]. 玉城肇, 译. 东京: 鮎书房, 1943.
[2] 潘吉星. 中国金属活字印刷技术史[M]. 沈阳: 辽宁科学技术出版社, 2001.

technique, we will find the following similarities and differences:

First, both selected three-part alloy as movable type material and the type forms were exactly the same. In China, three-part alloy of copper-tin-lead was used to cast type, but three-part alloy of lead-tin-antimony was used in Europe. Chinese type was cast in the form of long cube, and there was a small hole in its body in order to string it together with an iron wire to fasten it on the plate, and so was the early European type (Figure 100).

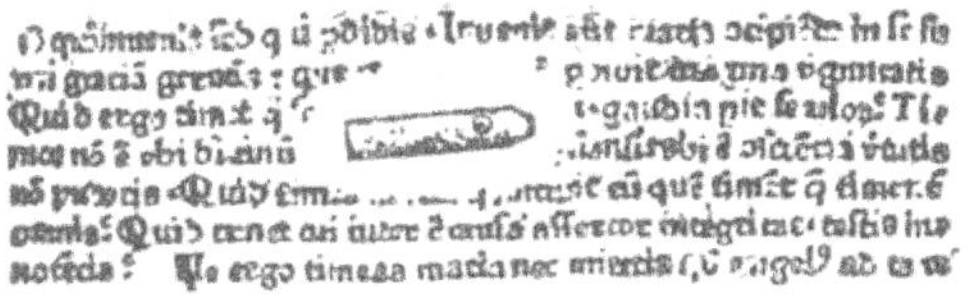

Figure 100   Images of Movable Type on *Happy Girl* Printed
in Cologne, 1468, from Ostwald (1928)

Second, both cast the model first and made it into mould. Then liquid alloy was poured so as to cast the movable type. The principles and procedures of casting movable type were the same. In China, wooden type was used as casting-model for making casting mould of earth, and metal type was cast with sand-casting method. Some American scholars held that the same method was used in Early European printing❶. But German scholars maintained that Gutenberg cut steel type as casting model for making casting mould of copper and molten metal was poured into brass box. In fact, metal-mould had also been used during the Han Dynasty in China for coining casting. Since the Song Dynasty, earthen mould had been used for coin and type casting in China, because it could save more labor, time and money.

Third, Chinese ink was made of carbon black from pine-soot and oil-soot and animal glue in the ratio of $100 : 30$ to $100 : 50$, sometimes vegetable oil was also mixed with them. After fermentation for a time, it could be used for printing. Europeans boiled linseed oil and mixed it with terebene, carbon black and other things, then through fermentation oil-based ink was prepared, which was an innovation.

Fourth, the principles of brushing printing were similar but the tools used were different. The Chinese laid a sheet of paper over the plate painted with ink, and then brushed the reverse side of paper with coir brush. Only one side of

---

❶   CARTER T F. 中国印刷术的发明和它的西传[M]. 吴泽炎, 译. 北京: 商务印书馆, 1957.

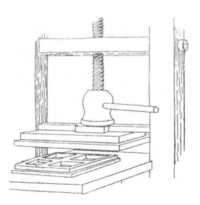

Figure 101   A Screw Press Device Used in Movable Type in Europe, from Ostwald (1928)

the paper was printed and the two half pages were put together with the blank sides folded inside and then bound into a volume. Europeans brushed ink on the plate, and laid a sheet of paper over the plate, but the impression process was completed with a special press of screw device (Figure 101). Both sides of paper were printed and then bound. This could save paper since European paper was more expensive than Chinese paper.

From the comparisons mentioned above, the principle of movable type technique and the fundamental procedures of printing in China continued to be used by Gutenberg❶. But the materials of type, casting model and mould, and the composition of coloring agent were varied by him in his own way according to the local condition. A new device, the delicate screw imprinter was used, which should be regarded as one of his inventions. Therefore, he innovated the East Asian traditional technique represented by China, making it more suitable for the Latin cultural area and the Christian world of alphabetic writing where paper was insufficient. Although other Europeans before Gutenberg did early exploratory experiments, but his technique was the most systematic and advanced, and had been successfully used in mass production. A large group of technical personnel were trained and made his technique available in other European countries. As the founder of the European metal type printing technique, he has made an outstanding contribution to the development of printing technique. We should fully recognize his contribution.

However, it has long been a popular view in the West that the metal type technique was independently invented by Gutenberg without any external influence. Some westerners even regarded him as the inventor of movable type. This view of Eurocentrism is still reflected in the works of some non-professionals. They knew little about the history of printing in the East and the West, especially the Chinese printing history and the history of Sino-European

❶ 潘吉星. 从元大都到美因茨：谷腾堡技术活动的中国背景[J]. 中国科技史料，1998，19 (3)：21-30.

relationship. Dr. Joseph Needham, in 1954, called on Europeans to learn more about Asia and talk to Asia, so that they could learn more about Europe and the whole world❶. How correct this advice was! With the passage of time and the in-depth study of scholars from both the East and the West, the idea of "European independent invention of metal type" has aroused suspicion of western experts, and they have paid close attention to East Asia and China. This is an encouraging sign, which shows that the understanding between the East and the West is deepening and the age of the Eurasian academic dialogue has arrived.

The latest example was the international symposium on printing history in the East and the West held in South Korea on September 29 to October 2, 1997. The curator of Gutenberg Museum in Germany, Dr. Eva Hanebutt Benz related the eastern background of Gutenberg's technical activities in a speech: "Was Gutenberg aware of existent achievements of East Asian movable type from the 12th century? It might be a difficult question ... but when one realized the extent of contacts between East Asia and Europe during the 12th and 13th centuries, it is my personal belief that travelers who had made their ways along the Silk Road knew about the movable type, and that this knowledge was spreading by mouth, if not by any written notice. So personally I cannot imagine that Gutenberg had never heard about this way of printing texts. I think this idea made his mind eager to find an adequate way of reaching a solution for this problem so as to adapt to the cases he met in his country."❷

Professor Henri-Jean Martin, the printing historian of Sorbonne University, said in a speech: "Printing historians in France and other European countries have always shown great interest in oriental printing technique and Asian leading invention in this field. Hence, forty years ago, I invited lady Roberte Guignard, an expert on Chinese issues, to write a chapter of oriental printing history for *The Appearance of Book* (*L'apparition du Livre*) co-authored by Lucien Febvre and me ... In an attempt to counter the Eurocentrism vision which has far so long dominated the thinking of non-specialists, I suggested that as inventors of communication technique Gutenberg and

---

❶　Needham J. Dialogue entre l Europe et l Asie. Comprendre (Paris), 1954(12); Syntheses (Helsinki), 1958, vol.143；潘吉星.李约瑟集[M].天津：天津人民出版社,1998.

❷　HANEBUTT B E. Features of Gutenberg printing process: Speech at the International Forum on the Printing Culture[R]. Seoul, 1997-10-2.

Waldfoghel were little more than local demiurge, when compared with the great wise gods of the Far East." ❶

Martin also referred to a letter written by Professor Von Turfan zum Karlstein, the German technical historian, and handed out his paper. The letter said, "he believed that the West has definitely mastered the techniques of the East. In terms of this, Emperor Charles Ⅳ has made significant contribution in the palace of Prague. Prague was one of the principal terminal points of shipping silk from the East to the West, and relevant technical information could have spread from Prague to Nuremberg, Strasbourg, Mainz and other industrial and commercial cities in Germany."

As mentioned above, from 1434 to 1444 in Strasbourg, Gutenberg suddenly changed his profession and turned to printing experiment in secret. Instead of a sudden impulse, he must have been kindled by some kind of external factor, at least he heard of the experiment of metal type by Waldfoghel in Avignon. Hans Riffe, his cooperator of mirror making, had a relative, Walter Riffe by name, who was a goldsmith in Strasbourg and resided close to Gutenberg. During this time, Walter Riffe usually visited Avignon ❷. He might have told Gutenberg about the casting and printing, which had attracted the attention of Gutenberg. Waldfogel's "Art for writing artificially" has been passed on to some people and his products have been sold in a semi-overt situation. Such information was available to Gutenberg. His success in 1450 gained after the failures from 1444 to 1445 was due to the work done and problems faced by other European forerunners during his travels, which gave him useful instructions.

While recognizing the historical contribution of Gutenberg, we cannot ignore the early work of other European forerunners. Where did they get the idea of type printing as well as the manufacturing and type setting techniques? To answer this question, like the above-mentioned German and French scholars, we should trace the source of the metal type technical information brought back from China by travelers along the Silk Road. Gutenberg and other European forerunners gained such information directly or indirectly from those travelers, otherwise it would be hard to imagine how they could suddenly do the

---

❶ MARTIN H J. The development, spread and impact of printing from movable type in 15th and 16th century Europe: Speech at the International Symposium on Printing History in East and West[R]. Seoul, 1997-9-29.

❷ MARTIN H J. The History and Power of Printing[M]. Chicago: University of Chicago Press, 1994.

casting and printing experiment. The metal type technique in Gutenberg was similar in the principles, procedures of technical operation and movable type forms as Chinese. But it is not a mere coincidence. This convergence phenomenon can only be explained by technical communication at the age of frequent contacts between the East and the West.

The technical history suggested that the news that technical process had successfully been accomplished in some far-away part of the world might encourage certain people to solve the problem in their own ways, and set off a train of development. Needham called the phenomenon "Stimulus Diffusion" and explained it with the spread of printing in Europe as an example: "As for transmission, I am satisfied that Gutenberg knew of Chinese movable type printing, at least by hearsay."❶ As long as he ever heard that "Chinese make movable type from metal alloys, with a hole in its body in order to string it together with an iron wire, which is used for typography", Gutenberg could probably make a series of development in his own way. Contemporary British, American, German and French printing historians' views on the influence of Chinese metal type technique on Europe were consistent with those of early European scholars. In 1546, the Italian historian Polo Giovio (1483~1552) wrote in his *The Modern History* (*Historia sui temperis*) published in Latin like this:

> *Quod maxime mirandum videtur, ibi (Canton) esse typographos artifices, qui libros historias et sacrorum ceremonias continentes, more nostro imprimant: quorum longissima folia introrsus quadrata serie complicen tur ut hinc facile credamus cius artis exempla antequam Lusitani in Indiam penetrarint per Scythas et Moscos ad incomparabile litterarum praesitium ad nos pervenisse.* ❷

> *There are typographers (typographos artifices) in Canton, who print with our method (more nostro, i.e., metal type), books containing histories and rites on a very long folio which is folded inwards into square pages ... so that from this we can easily believe that examples of this kind, before the Portuguese had reached India (the 14ᵗʰ century), came to us through the Scythians (Scythas) and*

---

❶　潘吉星. 李约瑟文集[M]. 沈阳：辽宁科学技术出版社，1986.

❷　CARTER T F. The Invention of Printing in China and Its Spread Westward[M]. 2ⁿᵈ ed. New York：Ronald，1955.

Muscovites (Moscos) as an incomparable aid to letters.

In this text, Scythas referred to Sythia, the name of an ancient state at the Eurasian border between the Caspian Sea and the Black Sea, which was known as Armenia of the Mongol Ⅱ-Khanate. Moscos referred to Moscovites, then known as Russ or ancient Russia in the control of the Mongol Kipchak Khanate. These places nearing West Europe were parts of the northern and southern routes of the Chinese movable type printing technique spreading westward. The expression "typographos artifices" by Giovio referred to metal type printers, and "more nostro" (our method) referred to the European metal type method for printing books. For the sake of better understanding, the Latin text was published alongside our translated version. Giovio's record was confirmed by his contemporary, the Spanish scholar Juan Gonzeles de Mendoza (1540~1620), who wrote in 1585 in his Spanish work *History of the Great Empire of China (Historia del Gran Regno de China)*:

> It doth plainly appear by the vulgar opinion, that the invention of metal type printing did beginne in Europe in the year of 1458, the which was attributed vnto Toscan (German), called John Gutenbergo ... But the Chinos do affirme, that the first beginning was in their countries, and the inventour was a man whom they reverence for a saint: whereby it is evident that manie years after that they had the use thereof, it was brought into Almaine (Germany) by the way of Ruscia and Moscovia, from whence, as it is certain, they may come by land, and that some merchants that came from thence into this kingdom, by the Red Sea, and from Arabia Felix (Arabian Peninsula), might bring some books, from whence this John Gutenbergo, whom the histories dooth make author, had his first foundation. ❶

In the age of Mendoza, metal type printing was already dominant in Europe, so the "printing" he talked about was mainly metal type printing. In this first comprehensive monograph on China by the European, he consulted many early historical records and also took relevant original data from manuscripts of *A trip to Fujian (Narrativo de Mision a Fukien)* written by

❶ DE MENDOZA J G. The History of the Great Empire and Mighty Kingdom of China[M]. London: Hakluyt Society, 1853.

Spanish Martin de Rada (1533~1578), who traveled to Fujian, China in 1576. Martin Rada mentioned his dialogue with local officials in Fujian: "The Chinese official was greatly surprised to learn that we likewise had a script and that we used the art of printing for our books, as they do, because they used it many centuries before we did."❶ In the Ming Dynasty, there was large copper type printing in Fujian, and the printed copies were marked in "copper type". Rada bought some books and brought them back to Spain in the presence of Mendoza. The above mentioned "Chinese movable types" by Giovio and Mendoza all refer to the handwritten bronze type. The studies of contemporary Chinese and foreign scholars have confirmed that the records of Giovio and Mendoza over 400 years ago were correct, that is to say, the metal type technique of the Yuan Dynasty was introduced to Europe by travelers visiting China. The Chinese background of technical activities of Gutenberg has become clearer.

---

❶ BOXER C R. South China in the 16th Century[M]. London: Hakluyt Society, 1953.

Chapter 3 Invention, Development and Spread of China's Gunpowder Technique

 ## 3.1 Invention of Gunpowder and Early Firearms in China

3.1.1 Reasons Why Gunpowder was Invented in China

3.1.1.1 Historical Facts of Early Use and Purification of Saltpetre in China

A large body of reliable documents and physical evidences prove that China is the birthplace of gunpowder and early gunpowder weapons. Why was gunpowder first invented in China rather than in other countries or regions? That is what needs to be clarified. To clarify this question will also reveal the historical background of the invention of gunpowder in China. According to the gunpowder combustion theory in ancient China, of all the ingredients in the gunpowder, saltpetre and sulfur are the main ones: saltpetre works as the king in the kingdom of gunpowder with exceptional importance, while sulfur functions as a minister to assist it. J. R. Partington once said, "The first step in the invention of gunpowder must have been the discovery of an efficient process for making purified saltpetre."❶China is the first country in the world to use and purify saltpetre. It is a long historical process in China from the discovery, utilization, and purification of saltpetre and sulfur, to the understanding of their performance, to the finding of the combustion and explosiveness of the saltpetre-sulfur-charcoal mixture, till they finally use and improve these properties to make military firearms. This process, which fully demonstrates the characteristics of Chinese traditional culture, serves as the main reason for the invention of gunpowder.

Saltpetre, scientifically called *potassium nitrate*, KNO_3, with a molecular weight of 101.11, belongs to the orthorhombic system. In nature, it is found as

---

❶ PARTINGTON J. R. A History of Greek Fire and Gunpowder[M]. Cambridge: Heffer & Sons, Ltd. , 1960.

acicular or hair-like aggregates; artificially, it is a pseudo-hexagonal columnar crystalloid made of fine white crystal grains. Saltpetre, transparent, crisp, having a cold, salty and spicy taste, with a specific gravity of 2.1 to 2.2, a hardness of 2, a melting point of 333 ℃, a decomposition temperature of 400 ℃, releases oxygen after decomposition, and thus is an oxidant. It is soluble in water and has a pH of 7[1]. Since saltpetre dissolves in water, and melts with certain metal or substance, the ancient Chinese initially referred to it as *xiaoshi* (消石, solvestone). Modern scientific research has identified the mechanism for natural saltpetre formation. It is known that 17% of the protein of all living organisms is composed of nitrogen and nitrates, the raw materials for the production of proteins, are converted from nitrogen in the air by plants. Plants are eaten by herbivores; herbivores are preyed on by carnivores. The excretion of herbivores and carnivores, their dead bodies, and withered plants, when rotten, undergo complex biochemical changes and the absorbed nitrogen is returned to the soil in the form of ammonia (NH_3) and ammonium (NH_4) salts, which, in the presence of nitrosomonas and nitrobacter, are transformed back into nitrates[2] in the soil. This is how saltpetre is formed in nature.

Natural saltpetre occurs in the places where there are human and livestock droppings, at the lower part of the house, at the foot of walls, around the filthy wet corners, on the ground near the residential areas and in the livestock circles, etc. Every spring, autumn and winter, in the above areas appears white frost of natural nitrite, which, if taken, is the crude saltpetre. Because it contains calcium carbonatc ($CaCO_3$), salt (NaCl) and dirt, it must be processed and purified to be used. Li Shizhen (李时珍, 1518~1593), in Volume 11, *Bencao Gangmu* (《本草纲目》, *Compendium of Materia Medica*, 1596), wrote, "Raw *xiaoshi* is found in all the halogenous areas, particularly in large amount in Hebei (河北), Qingyang (庆阳) and Shuzhong (蜀中). In these areas, autumn and winter see white frost of crude saltpetre in every corner. Taken and refined, it becomes purified saltpetre."[3] Fang Yizhi (方以智, 1611~1671), in Volume 7, *Wuli Xiaoshi* (《物理小识》, *Little Notes on the Principles of the Phenomena*, 1643), said, "Saltpetre is mostly taken from the ground and soil ... Men and animals' urine erodes the soil and other substances. With time going,

---

[1]　刘友樑. 矿物药与丹药[M]. 上海：上海科学技术出版社，1962.

[2]　NEKRASOV B V. 普通化学教程：中册[M]. 张青莲，等译. 上海：商务印书馆，1954.

[3]　李时珍. 本草纲目：石部（上册）[M]. 北京：人民卫生出版社，1982.

the inter-reaction forms a new substance which can be refined into saltpetre. Taking the above process into consideration, it can be inferred that saltpetre is derived from salty matters. People can collect saltpetre when sweeping the floor early in the morning. Walls of toilets are the most common places to find saltpetre."[1] Although saltpetre can be found all over China, the Yellow River Reaches and Huabei (华北, North China) produce better-quality saltpetre and have the longest history of saltpetre production.

Besides China, natural saltpetre is found in India, Persia, and Europe as well. But China is the first country to identify saltpetre from multitudes of natural mineral materials, to know its physical and chemical properties, and finally to find its practical use. The discovery of saltpetre must have been the result of scientific research and rational thinking after experiment. Unlike other substances such as gold, silver or sulfur, saltpetre could not have been discovered by the ancient people spontaneously, and thus no record of saltpetre could be found in the Chinese Pre-Qin (先秦) books such as *Shan Hai Jing* (《山海经》, *The Classic of Mountains and Seas*), nor in the ancient books of other countries. After the Qin (秦) and Han (汉) Dynasties, due to the development of medicine, pharmacology and alchemy, medical researchers, having found the medicinal value of saltpetre while looking for new drugs, took it boldly. Alchemists also found that saltpetre helped to make *danyao* (丹药, elixir of life). As a result, saltpetre has been the research subject of Chinese medicine and alchemy in two thousand years.

It is not yet clear about the exact time when China discovered saltpetre, but there is evidence that it was used as a drug no later than the Qin and Han Dynasties at the end of the third century B. C.. In 1973, in Mawangdui Han Tomb, No. 3 (马王堆三号汉墓, built in 168 A. D.[2]), Changsha (长沙), Hu'nan Province (湖南) of China was unearthed a collection of medical prescriptions written on a piece of silk cloth (24 cm long), with a record of 270 odd recipes for 52 diseases. Nameless, it was entitled *Wushier Bingfang* (《五十二病方》, *Treatments for Fifty-two Ailments*) by the researchers (Figure 102). The experts interpreted the words "稍(消)石直(置)温汤中,以洒痈"[3] taken

---

[1] 方以智. 物理小识:卷七(下册)[M].上海:商务印书馆,1937.

[2] 湖南省博物馆,中国社会科学院考古所.长沙马王堆二、三号汉墓发掘简报[J].文物,1974(7):39-48.

[3] 马王堆汉墓帛书整理小组.马王堆汉墓出土医书释文(二)[J].文物,1975(9):36.

from *Zhi Zhushang Fang*（《治诸伤方》，Ways of Treating Various Wounds）in *Wushier Bingfang*（《五十二病方》，*Treatments for Fifty-two Ailments*）as "putting saltpetre in lukewarm water and using the solution to clean abscesses". This ancient prescription is also found in the Tang Dynasty medical expert Wang Tao's（王焘）treatise *Waitai Miyao*（《外台秘要》，*Arcane Essentials from the Imperial Library*，752）. *Wushier Bingfang*，written in *zhuanzi*（篆字，a kind of Chinese calligraphy similar to the characters used in the Chu Kingdom（楚国）of the Warring States Period）during the Qin and Han Dynasties（the end of the third century B.C.）❶, is the earliest medical prescription book yet discovered in China and also the earliest reliable record documenting saltpetre.

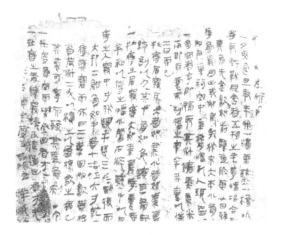

Figure 102   *Wushier Bingfang*（《五十二病方》，*Treatments for Fifty-two Ailments*），
unearthed in 1973 from Mawangdui Han Tomb，No. 3（马王堆三号汉墓），
Changsha（长沙），taken from *Wenwu*（《文物》，*Cultural Relics*）

In *Bianque Canggong Liezhuan*（《扁鹊仓公列传》，*Biographies of Two Famous Doctors Bianque and Canggong*）of *Shi Ji*（《史记》，*Records of the Historian*，90 B.C.），Volume 105，the Han Dynasty historian Sima Qian（司马迁，145 B.C.～86 B.C.）kept a record of Canggong［仓公，also called Chunyu Yi（淳于意），about 216 B.C.～150 B.C.，a renowned doctor at the end of the Warring States Period（战国）］applying saltpetre to cure the disease of Wang Meiren（王美人，a woman surnamed Wang）in Linzi［临淄，the capital of the Qi Kingdom（齐国），now Zibo（淄博）in Shangdong（山东）Province of China］：since Wang Meiren produced no breast-milk after the birth of her baby，Chunyu Yi first used henbane seed and then applied saltpetre to cure her

❶   马继兴，李学勤.五十二病方［M］.北京：文物出版社，1979.

illness❶. This is another example of applying saltpetre in the treatment of diseases by Chinese doctors in the third and second centuries B. C.. The tradition of using saltpetre to treat diseases is seen again in some prescriptions kept in *Jingui Yaolüe*（《金匮要略》，*Synopsis of Preions of the Golden Chamber*，219）by the Eastern Han（东汉）medical expert Zhang Ji（张机），also named Zhang Zhongjing（张仲景，150～219）：e. g.，"*dahuang xiaoshi Tang*"（大黄消石汤，Rhubarb-Saltpetre Soup，treating abdominal distension and dysuria caused by jaundice）and "*xiaoshi fanshi San*"（硝石矾石散，Saltpetre-Aluminite Powder，treating woman's jaundice and blood stagnation）❷. The former prescription has rhubarb，phellodendron，saltpetre and gardenia boiled in water；the latter contains saltpetre and aluminite $[KAl(SO_4)_2]$ taken together with barley porridge or juice.

Moreover，the Western Han alchemists were engaged in a large number of chemical experiments designed to synthesize *danyao*，which were calculated to be able to prolong people's life. They found that，as a result of the oxidation of saltpetre，the solution of saltpetre dissolved in concentrated vinegar（acetic acid，CH_3COOH），in the presence of some microorganisms，could improve the solubility of certain substances. The above-mentioned aqueous solution reaction is kept in *Sanshiliu Shui Fa*（《三十六水法》，*Thirty-six Aqueous Solutions*，an early mediaeval Chinese alchemical text on aqueous solutions），passed down by eight alchemists [collectively referred to as Bagong（八公，Eight Seniors)] such as Zuo Wu（左吴）and Li Shang（李商）recruited by Liu An [刘安，179～122 B. C.，King of Huainan（淮南王)] to provide *danyao* for the King. J. Needham et al. discovered that over half of the prescriptions in *Sanshiliu Shui Fa* include saltpetre and vinegar. According to Chen Guofu's textual research，the book dates back to the Han Dynasty. Some of the methods kept in it were used by the Chinese alchemists in the second century B. C..

The version of *Sanshiliu Shui Fa* we see today was collected in *Dao Zang*（道藏，*Taoist Canon*，a collection of Taoist classics，1445）and photocopied by Shanghai Hanfenlou Library（上海涵芬楼）in 1926. The book enlists the ways to make *fanshi shui*（矾石水，aluminite solution），*xionghuang shui*（雄黄水，realgar solution），*cihuang shui*（雌黄水，orpiment solution），*dansha shui*（丹砂水，cinnabar solution），*zengqing shui* and *baiqing shui*（曾青水 & 白青水，

---

❶　司马迁. 史记：扁鹊仓公列传[M]. 上海：上海古籍出版社，1986.
❷　张机. 金匮心释[M]. 王渭川，注. 成都：四川人民出版社，1982.

azurite solution), *danfan shui* (胆矾水, chalcanthite solution), *cishi shui* (慈石水, magnetite solution), *liuhuang shui* (硫黄水, sulfur solution), *xiaoshi shui* (消石水, saltpetre solution), etc (Figure 103). *Xionghuang Shui Fa* (《雄黄水法》, *The Method to Make Realgar Solution*) goes: "Put a *jin* (斤, a measurement of weight in ancient China) of realgar into a bottle made of fresh bamboo, then add four *liang* (两,1/16 of a *jin*) of saltpetre (KNO_3), seal the bottle with lacquer, immerse it into vinegar, wait for 30 days, and realgar solution is done." ("雄黄水:取雄黄一斤,纳生竹筒中,硝石四两,漆固筒口,如上纳华池醋中,三十日成水。") *Cihuang Shui Fa* (《雌黄水法》, *The Method to Make Orpiment Solution*) goes: "Put a *jin* of orpiment (As_2S_3) into a bottle made of fresh bamboo, then add four *liang* of saltpetre (KNO_3), seal the bottle with lacquer, immerse it into vinegar, wait for 30 days, and orpiment solution is done." ("雌黄水:取雌黄一斤,纳生竹筒中,硝石四两,漆固筒口,如上纳华池醋中,三十日成水。") In both recipes, saltpetre is mixed with realgar or orpiment, and put into vinegar, which will infiltrate the bamboo bottle to work on the mixture. But whether realgar or orpiment dissolves is another matter. Here the prescriptions in *Sanshiliu Shui Fa* are quoted for the purpose of proving that saltpetre was often used by the Han alchemists as a reagent for chemical experiments.

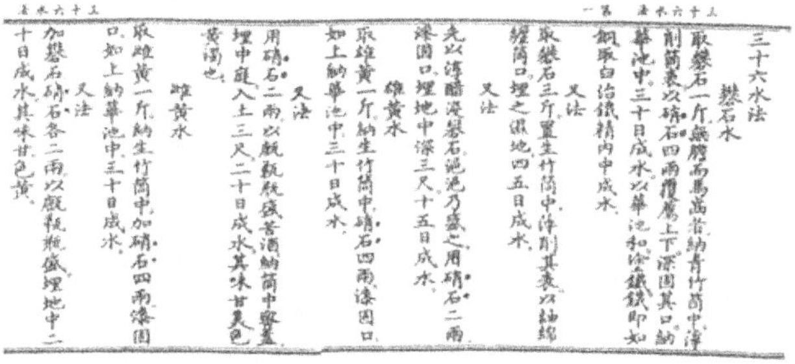
三十六水法
礬石水
取礬石一斤，無胕而馬齒者，納青竹筒中，薄削筒表，以硝石四兩覆薦上下，深固其口，納華池中，三十日成水。以華池和，塗鐵，鐵即如銅。取白治鐵精內中成水。
又法
取礬石三斤，置生竹筒中，薄削其表，以紬絳縵筒口，埋之濕地，四五日成水。
又法
先以淳醋潰礬石，泡泡乃盛之，用硝石二兩，漆固口，埋地中，深三尺，十五日成水。
雄黃水
取雄黃一斤，納生竹筒中，硝石四兩，漆固口，如上納華池中，三十日成水。
又法
用硝石二兩，以甑瓿瓴盛，苦酒納筒中，密蓋，埋中庭，入土三尺，二十日成水，其味甘美，色黃濁也。
雌黃水
取雌黃一斤，納生竹筒中，加硝石四兩，漆固口，如上納華池中，三十日成水。
又法
加礬石、硝石各二兩，以甑瓴瓶盛，埋地中二十日成水，其味甘，色黃。

Figure 103 Prescriptions Including Saltpetre in Han Alchemist Classic *Sanshiliu Shui Fa*, collected in *Dao Zang* (《道藏》, *Taoist Canon*, a Collection of Taoist Classics, 1445)

Prescriptions with saltpetre as an oral drug are still found in unearthed Han Dynasty *mujian* (木简, inscribed wooden slips). In an early Eastern Han (25 A.D. ~88 A.D.) tomb found at Hantanpo (旱滩坡), Baishu Town (柏树乡), Wuwei County (武威县), Gansu Province (甘肃) in November, 1972, were discovered 91 inscribed wooden slips or wooden cards (木椟) with medical information (including 77 wooden slips of 23 cm ~23.4 cm long and 1 cm or 0.5 cm wide, and 14 wooden cards of 22.7 cm~23.9 cm long and 1.1 cm~

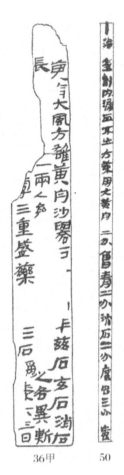

36甲   50

Figure 104   Medical Prescriptions Including Saltpetre Inscribed on Han Dynasty Wooden Slips unearthed in 1972 in Wuwei County, Gansu Province, taken from *Wuwei Handai Yijian* (《武威汉代医简》, *Wuwei Han Dynasty Wooden Slips of Medical Prescriptions*, 1975)

4 cm wide), *Zhi Baibing Fang* (《治百病方》, *Prescriptions for Diseases*) within. In the book, many prescriptions contain saltpetre as ingredients[1]. One of the prescriptions inscribed on *Wooden Slips No.* 46 & 47 goes, "《治伏梁裹脓在胃肠之外方》:大黄、黄芩、勺（芍）药各一两,消石二两 [i. e., two *liang* (两, a measurement of weight in ancient China) of saltpetre],桂一尺//。桑卑肖（桑螵蛸）十四枚,（蚕）虫三枚。凡七物,皆父且（将要）渍以醇酒五升,卒时（一昼夜煮之。三……)"（Figure 104）. Besides, on the front of *Wooden Card No.* 86 is a recipe for leprosy.

As discussed above, saltpetre used by the medics in ancient China since the Qin and Han Dynasties (the third century B.C.) has been recorded as oral medicine and surgical medicine in history. *Danyao* refined by the alchemists was taken orally in the wish of improving people's health and prolonging their lives. The fact that saltpetre was used medically and taken orally means the necessity of purified saltpetre free of any extraneous constituents. Therefore, since the third century B.C., after the discovery of the special properties of saltpetre and the application of saltpetre as drugs, the Chinese must have mastered the purification technology of saltpetre,

---

❶   甘肃省博物馆,甘肃武威县文化馆.武威旱滩坡汉墓发掘简报[J].文物,1973(12):18-21.

i. e., the technology of boiling the crude saltpetre solution and then making pure saltpetre by the recrystallization method. Since then, the purification technology of saltpetre has been passed down for two thousand years, thus forming a unique view of Chinese medicine culture, which can be traced in the herbalist works of various dynasties. The Eastern Han Dynasty witnessed such a large scale of saltpetre-purification workshops that the government banned the production of saltpetre from the *Xiazhi*（夏至, the summer solstice, on June 21 or 22) until the *Liqiu*（立秋, the beginning of autumn, on August 7 or 8). *Liyi Zhi*（《礼仪志》, *A Record of Etiquette*), Volume 15 of *Hou Han Shu* describes the regulation as such, "When the *Xiazhi*（夏至, the summer solstice, on June 21 or 22) arrives, no fire to boil the crude saltpetre solution is allowed, the production of pure saltpetre prohibited until the *Liqiu*（立秋, the beginning of autumn, on August 7 or 8)." ("日夏至,禁大火。止炭鼓铸,消石冶皆绝止,至立秋如故事。") The ban, whether it could be carried out in the private sector then, at least manifests the widespread production of purified saltpetre in China around 1 A.D..

The earliest work of pharmacology in China is *Shennong Bencao Jing*（《神农本草经》, *Shennong's Classic of Materia Medica*) edited in the first century B. C.. *Lou Hu Zhuan*（《楼护传》, *Biography of Lou Hu*), Volume 92 of *Han Shu* (《汉书》, *History of the Western Han Dynasty*) says, Lou Hu（楼护), also named Junqing（君卿), a native of Qi [齐, now Shandong（山东)], born in a doctor's family, "read *Yi Jing*（《医经》, *A Medical Classic*), *Shen Nong Ben Cao Jing* and *Fang Shu*（《方术》, *Medical Prescriptions*) in his youth" ("少诵医经、本草、方术数十万言"), and was ennobled Marquis of Xixiang（息乡侯) in the Xinmang Dynasty（新莽, 8 A.D.～23 A.D.). It can be inferred from the above document that, *Shennong Bencao Jing* came into being at the end of the Western Han Dynasty[1]. *Shennong Bencao Jing*, having collected 365 kinds of medicine, classifies saltpetre into the category of Jade & Stone（玉石部). But in the book, saltpetre（KNO_3) is called *puxiao*（朴硝), a different name from the common title of *xiaoshi* used in the Western Han Dynasty, while *xiaoshi* in the book refers to sodium sulphate（$Na_2SO_4 \cdot 10H_2O$), thus causing confusion in the naming of saltpetre. *Shennong Bencao Jing* writes about saltpetre and sodium sulphate（$Na_2SO_4 \cdot 10H_2O$) like this:

> Puxiao（朴硝): *bitter, cold in nature, non-toxic; cures*

[1]　俞慎初. 中国医学简史[M]. 福州:福建科学技术出版社,1983.

*hundreds of diseases; helps to dissolve 72 stony substances; lightens
the body, if taken as refined danyao.*

*Xiaoshi (消石): bitter, cold in nature; quickens metabolism;
cures indigestion; becomes paste like if smelted; lightens the body if
taken in the paste form.*

As can be seen from the above accounts, the book has summarized the
research results of some drugs in herbalism and alchemy of the Qin and Han
Dynasties. To identify the chemical compositions of the two drugs quoted here,
we may just take into consideration the description of their characteristics and
ignore the names used. "Helping to dissolve 72 stony substances (能化七十二种
石)" might have two denotations: (1) through aqueous reactions, assisting the
solution of some (not necessarily 72) substances difficult to be dissolved;
(2) melting with other metals or stones as a fluxing agent through pyrogenic
reactions. Only potassium nitrate has the functions listed in both cases, while
sodium sulphate does not play such roles. Therefore, *puxiao* in the book must
be potassium nitrate. "Becoming paste like if smelted (炼之如膏)" can mean the
melting process of potassium nitrate at above 330 ℃, or the lost of the
crystallized water in sodium sulphate at above 306 ℃. But through the
comparison of the medicinal properties and chemical properties of both
substances, "becoming paste-like if smelted (炼之如膏)" must be referring to
the latter process. To conclude, in *Shennong Bencao Jing*, *xiaoshi* must be
sodium sulphate; *puxiao* potassium nitrate.

Naming saltpetre *puxiao* as in *Shen Nong Ben Cao Jing* was not common in
the Han Dynasty when saltpetre was still referred to as *xiaoshi* by some doctors
such as Chunyu Yi (淳于意) and Zhang Ji (张机), and in the prescriptions
written on unearthed silk books or inscribed on unearthed wooden slips.
Potassium nitrate was found before sodium sulphate which bears some
semblance with the former. China has a vast territory and it is unavoidable that
people in various regions have different names for potassium nitrate. Anyway,
it is certain that the doctors used potassium nitrate in ancient China.
Nevertheless, the confusion on the naming of the drug needs to be clarified and
further study is needed to interpret the account of it in *Shen Nong Ben Cao Jing*.
In this regard, the sentences "*xiaoshi* is also called *mangxiao* (芒消, Natrii
Sulfas)" and "*puxiao* is also called *xiaoshipu* (消石朴)" taken from *Mingyi
Bielu* (《名医别录》, *Records of Renowned Doctors' Prescriptions*, the third
century A.D.) edited by some doctors of the Wei and Jin Dynasties and from

Bencao (《本草》, *Materia Medica*, about 235) written by Wu Pu (吴普, 150~ 230) haven't made clear the naming for potassium nitrate and sodium sulphate.

It was not until the publishing of *Bencao Jingji Zhu* (《本草经集注》, *Collective Notes to the Canon of Materia Medica*, 500) by Tao Hongjing (陶弘景, 456~536, a renowned medical expert of the Liang Dynasty) that the differentiation of the naming for potassium nitrate and sodium sulphate put on a new look. Tao Hongjing questioned the reference of the word *xiaoshi* in *Shennong Bencao Jing* and the explanation of it provided by the Wei and Jin people shown in the previous paragraph. Through scientific experiments and field survey, Tao Hongjing was aware of the fact that the Han Dynasty people used the word *xiaoshi* to refer to potassium nitrate, which, though with similar appearance with *puxiao*, is different from the latter in nature, and thus the names for the two chemicals cannot be confused. Tao Hongjing wrote:

> Xiaoshi (KNO_3) *is used to treat diseases in a similar process as* puxiao ($Na_2SO_4 \cdot 10H_2O$) *during which* xiaoshi *assists the dissolving of various stones. People of today can't identify* xiaoshi. *They still hold that* xiaoshi *has the same origin as* puxiao *and that* puxiao *can be named* xiaoshipu. *In fact,* xiaoshi *and* puxiao *do not denote the same substance. A substance acquired recently (with the similar appearance as* puxiao, *shining, snow-like, but not freezing), is believed to be the real* xiaoshi, *for it produced violet flame, turned into ash, didn't stop boiling like* puxiao *when burnt at high temperature. Someone else says* xiaoshi *can still be called* mangxiao. *The naming is unreasonable, since* mangxiao *is the remains after* puxiao *being burnt (the belief is held by the scholar and medical expert Huangfu Mi (皇甫谧, 215~282), too, but hasn't been confirmed in my experiments). The procedure of treating* xiaoshi *is recorded in* Sanshiliu Shui Fang. Xiaoshi *can be found in Long (陇), Xishu (西蜀), Qinzhou (秦州), Chang'an (长安), Xiqiang (西羌), and in the salty-soil areas in the mountains north of Dangchang (宕昌).*

The above quotation of *Bencao Jingji Zhu* was taken from the eighth century version of *Xinxiu Bencao* (《新修本草》, *Newly Revised Materia Medica*, 659) and the 1205 version of *Zhenglei Bencao* (《证类本草》, *Classified Materia Medica*, 1108) and then proofread and revised by the author. Tao Hongjing, an alchemist of rich chemical knowledge as well, roasted potassium nitrate and sodium sulphate respectively to help people distinguish the two. He found that the former (the true *xiaoshi*, potassium nitrate) smelted at 333 ℃, kept bubbling and released gas at 400 ℃, glowed with violet flame if burnt at higher temperatures, and turned into ashes at last. Nevertheless, the latter (sodium sulphate) liquefied immediately after being heated. With the crystal water gone, if burnt continuously, it did not bubble or emit gases, and naturally, no flame appeared. Because sodium sulphate melts at 885 ℃ and boils at 1430 ℃, these temperatures are not easy to arrive at. That it "didn't stop boiling like *puxiao*" (不停沸如朴消) means that after intense burning, potassium nitrate did not stop bubbling like sodium sulphate, but continued to emit gas until it became ash (K_2O). When the crystal water of sodium sulphate was removed and anhydrous sodium sulphate came into being, no bubbles would appear. It is known by all that the potassium salt flame is violet and the sodium salt flame is yellow. In the roasting experiments by Tao Hongjing, temperatures higher than 1430 ℃ could not have been reached. Then, what he saw could only be the purple flames of the potassium salt.

The experiment of Tao Hongjing, an outstanding breakthrough in the history of qualitative chemical analysis, is the debut of the flame analysis and testing method. Having differentiated in a strict manner the different physical and chemical properties of potassium nitrate and sodium sulphate, he went on to correct the naming of the two salts in *Shennong Bencao Jing*: *puxiao* should be revised as *xiaoshi*; *xiaoshi* as *puxiao*. Both salts should be pure. Besides, the names *xiaoshipu* for *puxiao* and *mangxiao* for *xiaoshi*, somewhat redundant, must be regarded as aliases of the two salts and can't be listed as new drugs. *Xinxiu Bencao*, composed by Su Jing (苏敬, 620~680) of the Tang Dynasty by order of Emperor Gaozong of the Tang Dynasty (唐高宗), is the first pharmacopoeia issued by the government, which pushed herbalism to a new stage of development with great achievement. Nevertheless, in the respect of the correct naming of potassium nitrate and sodium sulphate, *Xinxiu Bencao* failed to employ the research results of *Bencao Jingji Zhu*, for it followed the naming in *Shennong Bencao Jing*. Fortunately, the alchemists in the Northern and Southern Dynasties and in the Sui and Tang Dynasties were not affected by these herbalism works, and continued to use *xiaoshi* to refer to potassium nitrate with which they made *danyao*. The alchemists were not confused by the various names of potassium nitrate and sodium sulphate, for they knew well the difference between the two chemicals.

Later on，Ma Zhi （马志，935～1004），a great alchemist and medical pharmacist of the Northern Song Dynasty with the same intellectual background as Tao Hongjing，made contribution to the name correction of potassium nitrate and sodium sulphate. Just as Tao Hongjing clarified in his book *Bencao Jingji Zhu* the confusion of the naming of the two drugs in *Shennong Bencao Jing*，Ma Zhi eliminated that of *Xinxiu Bencao* in his book *Kaibao Bencao* （《开宝本草》，*Kaibao Materia Medica*，974）(Figure 105). Ma Zhi wrote in 974：

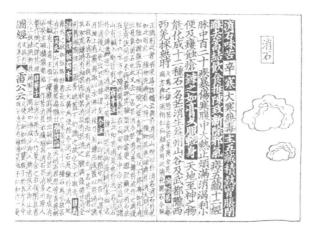

Figure 105   On Differences of *xiaoshi*，*mangxiao* and *puxiao* quoted from Ma Zhi's *Kaibao Bencao* （《开宝本草》，*Kaibao Materia Medica*，974），kept in *Zhenglei Bencao* （《证类本草》，*classified Materia Medica*，1108）

Xiaoshi，*i. e.*，dishuang （地霜，*ground frost*），*is formed through the decoction and refining of the aqueous solution of the white frost of natural nitrite taken from the ground of mountainous and watery areas in winter. Probably because potassium nitrate can help the solution of many stony substances，it is called* xiaoshi （*in Chinese means* "disappear，dissolve，or make something disappear or dissolve"）. *It's not in the same family as* puxiao （*sodium sulphate，*Na_2SO_4）*and* mangxiao，*a variation of sodium sulphate*）. *The reason why Mingyi Bielu* （《名医别录》，Records of Renowned Doctors' Prescriptions，*the third century*）*named potassium nitrate* mangxiao *may be that the aqueous solution of the white frost of natural nitrite，when first decocted，produces needle-like substance* （mang （芒）*in Chinese means needle-like thing*）*bearing a resemblance with* puxiao （*sodium sulphate*）*and* mangxiao （*Natrii*

Sulfas). *Therefore*, mangxiao *in Mingyi Bielu*, *denoting potassium nitrate*, *is absolutely different from* mangxiao *in the following entry.*

Mangxiao (*Natrii Sulfas*) *is produced like this*: *pour warm water on* puxiao, *decoct the solution till half is left*, *leave the condensed solution in the basin overnight and then thin-needle-like substance will be found*. *Because of the needle-like appearance*, *it is called* mangxiao. *There is still* yingxiao (英消, *also called* mayaxiao (马牙消)), *a crystobalite-like*, *four-or-five-sided substance*, *white in color*, *sparkling and crystal-clear*, *having the same healing function as* mangxiao. *It is also derived from* puxiao, *but produced in a different way from* mangxiao. Xinxiu Bencao *of the Tang Dynasty is wrong in classifying* mangxiao (*Natrii Sulfas*) *into the same category as* xiaoshi.

Puxiao (*sodium sulphate*) *is produced in Yizhou (today's Sichuang and Chongqing) where the natives mine the raw material*, *pour water over it*, *and then decoct and refine the solution to make* puxiao. Puxiao *is named* xiaoshipu, *too*: *in the name*, xiao (消, *soluble salt*) *literally refers to the raw material*; shi (石, *stone*), *solid whitish thing*; pu (朴, *crudeness*), *the state of being original and unprocessed*. *As* mangxiao *and* yingxiao *are derived from it*, puxiao, *in a comparatively original unprocessed state*, *is then called* xiaoshipu.

Ma Zhi's (马志) wonderful statement concludes clearly that *xiaoshi* is potassium nitrate, *puxiao* is sodium sulphate, and *yingxiao*, *mangxiao* and *mayaxiao* are all aliases of *puxiao*. He called *xiaoshi* as *dishuang*, which distinguishes it from sulfates, and suggests the source of saltpetre of his time. That saltpetre is "formed through the decoction and refining of the aqueous solution of the white frost of natural nitrite (以水淋汁后,乃煎炼而成)" describes the recrystallization method to produce purified saltpetre. He ended the long-term confusion of names for potassium nitrate and sodium sulphate by exploring the reason why *Mingyi Bielu* "named potassium nitrate *mangxiao*" ("一名芒消") and said that this *mangxiao* is not sodium sulphate whose variant name is also *mangxiao*. Li Shizhen (李时珍) had a high praise of Ma Zhi's (马志) work:

The soluble salts (potassium nitrate, sodium sulphate, etc.) had varied names causing confusion from the Jin to Tang Dynasties.

People merely guessed the referents of the names but reached no consensus. It is Ma Zhi's (马志) Kai Bao Ben Cao that ended the confusion. He stated that xiaoshi *is refined from* dishuang *and that* mangxiao *and* mayaxiao *are derived from* puxiao. *The other doctors were stumped by the varied names for they didn't know that the two soluble salts are similar in appearance but completely different in properties: one (sodium sulphate) is aqueous; the other (potassium nitrate) is igneous.*

Soon after its debut, *Kaibao Bencao* by Ma Zhi (马志) replaced the Tang-Dynasty *Xinxiu Bencao* as the mainstream materia medica work since the Song Dynasty, based on which other similar works have been compiled. Later, *xiaoshi* has been called *huoxiao* (火消, fire solvestone), *yanxiao* (焰消, flame solvestone), *dishuang* and after the Ming Dynasty *xiaoshi*, but no confusion of naming has been found with *puxiao* any longer. The purification of saltpetre, as mentioned earlier, which had already begun in the Han Dynasty, was recorded in many books of the Tang and Song Dynasties. Till the Ming Dynasty, more detailed information about the purification of saltpetre was seen, and the procedures were perfected on the basis of previous research results. Mao Yuanyi (茅元仪, about 1570~1637) wrote in *Wu Bei Zhi* (《武备志》, *Records About Armament and Military Provisions*, 1621):

The purification of saltpetre needs spring water, river water or lake water, fresh well water acceptable if the former three are unavailable. Fill 7/10 of a big pot with water, pour in a hundred jin *(1 jin = 16 liang = 596.82 grams) of crude saltpetre, heat the solution till it boils, wait for it to cool, repeat the boiling and cooling process twice and then pour in a* jin *of plant ash solution. (If the pot is not as big, pour in 50* jin *of crude saltpetre and then add half a* jin *of plant ash solution.) The calcium salts and magnesium salts in the crude saltpetre, in the presence of the plant ash solution, turn into reddish water that do not precipitate. Heat the rest of the solution till it boils, pour it into a porcelain container; and then the muddy matters will be at the bottom, while purified saltpetre will be in the middle. Leave the whole bulk there for one or two days, pour out the reddish water, get rid of the muddy bottom, and dry the whitish paste in the sun. Thus, purified saltpetre is made. The purification of saltpetre should be carried out in the second, third, eighth and ninth months of the Chinese Lunar calendar (the rest of the year is not suitable for the operation, for these months are either too cold or too hot).*

Potassium carbonate (K_2CO_3) in the plant ash solution mixed with the aqueous solution of saltpetre, reacts with the small amount of soluble calcium and magnesium salts in crude saltpetre to form carbonates that precipitate and separate from the mother liquor. The crude saltpetre solution, repeatedly boiled and crystallized, finally becomes pure saltpetre, which is held in a porcelain or black ceramic container. The water used to boil the crude saltpetre must be clean. The impurities that float on the surface of the crude saltpetre solution must be removed with a cloth bag. In *Tiangong Kai Wu*, Song Yingxing (宋应星, 1587~about 1666) said, "Immerse the crude saltpetre in water for a night. Skim the dirty impurities on the water surface. Put the crude saltpetre into the heater, add water, decoct the mixture till the saltpetre dissolves and the water dries, pour the rest into another container, and the saltpetre will be crystallized overnight ... To obtain purified saltpetre, the crystallized saltpetre must be boiled in water together with some roots of raphanus sativus and then poured into a basin. After a night, white snow-like crystallized saltpetre is formed. Since the purified saltpetre is contained in a basin (盆, pronounced *pen* in Chinese), people call it *penxiao*." The roots of raphanus sativus contain unsaturated organic acid and raphanusin which help to remove coloured impurities.

A Yangzhou (扬州) fellow Li Pan (李盘, 1590~1645) of the Ming Dynasty provided a more specified account of purifying saltpetre in *Jintang Jie Zhu Shier Chou* (《金汤借箸十二筹》, *Twelve Schemes for the Defense of a City*, 1630):

Half a pot of crude saltpetre and half a pot of fresh water are mixed and boiled together till saltpetre dissolves. Plunge four or five pieces of red raphanus sativus root into the boiling solution. Let the red raphanus sativus root boil with the solution for a while and then take it out. Mix three egg whites with two or three bowls of water, pour the mixture into the saltpetre solution, stir it with an iron spoon, and then skim the impurities that float to the surface. Take two liang (1 liang = 1/16 jin = 37.3 grams) of refined animal gelatin, dissolve it in water, pour the animal gelatin solution into the saltpetre solution, wait until the mixture boils for three to five times, pour all the solution into a porcelain basin, lid the basin and let it be in a cool place for a night. If in the next morning, the saltpetre is fine-needle-like and bright, then it is applicable in the manufacture of gunpowder; if not, or if it stills tastes salty, then it needs to be purified by repeating the aforementioned procedures. There are three criteria to test the purity of saltpetre: the acicular crystallization should be very fine; the color should be very bright; the taste should be not at all salty. The extremely white saltpetre that does not shine still contains impurities. If the tongue, having lipped it, could still sense saltiness and acerbity, then the saltpetre is not free of salts and alkali.

Adding egg white and animal gelatin to the saltpetre solution is to make the impurities float and hence easy to be skimmed. The regions of Sichuang, Hubei, He'nan, Shandong, Hebei, Shanxi, Shaanxi, Gansu, and the Northeast all produced saltpetre in ancient China. The above mentioned method has been the traditional way of purifying saltpetre in ancient China, a process both physical and chemical and repeated till very pure saltpetre crystallization (with a purity of 98%) is obtained. Dissolve the crude saltpetre in the water first, and then decoct it until it boils to dissolve most of the saltpetre. The solubility of saltpetre in water increases with the rising of temperature, while that of common salt (sodium chloride, NaCl) in water is largely unaffected by temperature. These properties can be taken advantage of to remove common salt in the crude saltpetre. Adding plant ash water makes calcium salts and magnesium salts precipitate and then be eliminated. Adding egg whites, animal gelatin and raphanus sativus root assists to make other impurities gather and float on the solution surface to be easily removed. The purification of saltpetre is a great achievement in ancient Chinese chemistry.

3. 1. 1. 2   Historical Facts of Early Discovery of Gunpowder Mixture in China

Next, sulfur will be the focus of discussion. It is a solid non-metallic element that the pre-Qin Chinese already knew and used. Sulfur, a chemical element with the symbol S and a molecular weight of 32, is a crisp bright yellow or greenish yellow crystalline solid at room temperature. With a hardness of 1.3 to 2.5, a specific gravity of 2.06, it melts at 119 ℃ and burns at 270 ℃ with a blue flame. It is insoluble in water. Native sulfur can exist in a free state. It is often found near hot springs and volcanic regions, also in the sulfur ore, or as a paragenetic mineral of iron ore or clay. Shanxi, Shaanxi, He'nan, Sichuan,

Hunan have been major sources of sulfur in China. Long being used as a drug in ancient China, sulfur has been compiled into the Han Dynasty *Shennong Bencao Jing* under the entry of *shiliuhuang* (石硫黄, yellow sulfur stone)[1]. Sulphur is also a commonly used alchemist reagent. The Eastern Jin Dynasty alchemist, Ge Hong (葛洪, about 281~341), in *Nei Pian* (《内篇》, *The Inner Chapters*) of *Bao Pu Zi* (《抱朴子》, *Book of the Master Who Embraces Simplicity*, about 324), wrote: "The first-class *danyao* is made of cinnabarite, the second-class, of gold and the third class, of yellow sulfur …. The sulfur stone *danyao*, reddish in color, must be a kind of sulfur. It is found by the cliff and river shores. The soft and wet can be swallowed, while the solid must be crushed up and then taken."

Qiu Jun (丘濬, 1420~1495) of the Ming Dynasty related in Volume 122 of *Daxue Yanyi Bu* (《大学衍义补》, *Supplementation to the Explanations of the Great Learning*, 1488), "Sulfur was imported. Before the Tang Dynasty when the islanders on the sea did not have contact with China, the Chinese people had not had sulfur." Qiu Jun's statement is clearly wrong. The word *liuhuang* (硫黄, sulfur) can be found in *Shennong Bencao Jing* annotated by the medical expert Wu Pu (吴普) of the Three Kingdoms (三国), "*liuhuang* (硫黄, sulfur), also called *shiliuhuang* (石硫黄, yellow sulfur stone), according to the renowned doctors Yi He (医和, the sixth century B.C.) and Bian Que (扁鹊, the fifth century B.C.), is bitter in taste and nontoxic. Its origin is Yiyang (易阳) and Hexi (河西)." In 1983, 193.4 grams of sulfur, 1,130 grams of realgar and 219.5 grams of ochre believed to have been applied pharmaceutically[2] were unearthed from the tomb of Zhao Mei (赵眛, about 162 B.C.~122 B.C., King of Nanyue Kingdom, the Western Han Dynasty), near Yuexiu Park in Guangzhou. The burial period was 122 B.C. or two years later. The author of this book saw in person that the unearthed sulfur is rather pure. As the earliest physical sulfur ever found, it reveals that sulfur has long been applied in medicine and alchemy.

It is worth noting that, saltpetre and sulfur seem to have an indissoluble bond in China since ancient times. Whether in medicine or in alchemy, the two are often co-used. *Taiping Huimin Heji Ju Fang* (《太平惠民和剂局方》, *Prescriptions of the Taiping Huimin Heji Hospital*, 1075~1085) quotes from

---

[1] 顾观光.神农本草经:卷三[M].北京:人民出版社,1956.

[2] 麦英豪,黄展岳.西汉南越王墓:上册[M].北京:文物出版社,1991.

Duxiansheng Fang (杜先生方, *Doctor Du's Prescriptions*) the so-called *Laifu Dan* Prescription (来复丹, Laifu Pellet): pellets as big as beans made from saltpetre (1 *liang*), sulfur (1 *liang*), gypsum vitreum (CaSO₄ • 2H₂O) (1 *liang*), flying squirrel's droppings (Faeces Togopteri) (2 *liang*), green tangerine peel (2 *liang*), tangerine pericarp (2 *liang*), some vinegar and some rice, treating vomiting and diarrhea caused by cholera, and also epigastric pain. Wang Haogu (王好古, 1250 ~ 1310), in explanation of the *Taibai Dan* Prescription (太白丹, Taibai Pellet) and the *Laifu Dan* Prescription in his book *Tangye Bencao* (《汤液本草》, *Materia Medica for Decoctions*, 1298), said, "In both prescriptions, saltpetre is assisted by sulfur." Yan Yonghe (严用和, 1198 ~1273) expounded the *Er Qi Dan* (二气丹, Erqi Pellet) in *Ji Sheng Fang* (《济生方》, *Yan Yonghe's Prescriptions to Save Life*, 1253): "Pulverize some sulfur and some saltpetre of equal weight, heat and stir the mixture in a stone container, add some sticky rice mash, make pellets in the diameter of 6 ~ 9 millimetres; the *Er Qi* Pellet treats the cold in hot summer (伏暑伤冷)". Saltpetre, sulfur and realgar being stir-baked by the herbalists could have caused combustive or explosive accidents sometimes.

Alchemists in ancient China had experienced similar accidents. They often used saltpetre, sulfur and realgar as reagents in making *danyao*. When the three are mixed and heated, it is bound to cause combustion, explosion, equipment damage and even the burning of houses. But since the experiments could not go without these substances, the alchemists took some preventive measures, such as laying the reactor outdoors, burying it in the ground, or keeping it away from residential areas. They widely used the so-called *Fu Huo Fa* (伏火法, method for subduing fire), which has three denotations: (1) adding some ingredients to absorb the fire that might be caused by mingling saltpetre and sulfur/realgar, etc.; (2) adding some ingredients to stabilize and subdue the volatile reagents (e.g., mercury) or the flammable reagents (e.g., sulfur); (3) adding some ingredients to ease the hotness and toxicity of the hot and toxic reagents (e.g., sulfur, realgar, etc.).

However, when the alchemists did the fire subduing experiments, the actual results did not necessarily coincide as expected. They had thought that carbon or carbides may help subduing fire, but unfortunately, these substances could be a curse. If some ingredients were added to change the nature of mercury, sulfur, and realgar, they would no longer be the necessary components to synthesize *danyao*. It is questionable whether the *danyao*, which had undergone fire subduing, actually had its toxicity and hotness reduced. Zhao Yi (赵翼, 1727~1814) recorded in Volume 19 of *Ershier Shi Zhaji* (《二十二史札记》, *Notes on Twenty-two Histories*) that six Emperors of the Tang Dynasty died from *danyao* and the number of civilian victims were countless. These historical facts reveal that the fire subduing method did not achieve the expected results. Nevertheless, the alchemists were still engaged in fire subduing experiments, which reached the climax in the Tang Dynasty. Volume 2 of *Qian Gong Jiageng Zhibao Jicheng* (《铅汞甲庚至宝集成》, *The Collection of Lead and Mercury Prescriptions for Danyao*) prefaced by Zhao Naian (赵耐庵, living around the Tang and Song Dynasties) quotes the *Fu Huo Fan Fa* (伏火矾法, the alum method for subduing fire)[1] from *Taishangshengzu Jindan Mijue* (《太上圣祖金丹秘诀》, *Secret Prescriptions of the Emperor's Danyao*) written by the Tang Dynasty alchemist Qingxuzi (清虚子) in 808:

> Pulverize and mix well 2 liang (1 liang = 37.3 grams) of sulfur, 2 liang of saltpetre, 3.3 qian (1 qian = 3.73 grams) of Aristolochia debilis. Insert the mixture into a pot. Put the pot into a ground pit, the surface of the pot on the same level with the ground. Plunge a mass of burning fire into the pot. Smoke rises. Cover the pot with four or five pieces of wet paper, lay two square bricks onto it and then bury it with soil. Take the pot out when it cools. Thus, subdued sulfur is obtained. Heat the mixture of 3 liang of alum and 2 liang of powdered subdued sulfur in a crucible till it melts. Pour the fluid into a stone container. The fluid will be jade green.

The above account about the *Fu Huo Fan Fa* first of all subdues the flammability of sulfur (Figure 106). The ratio of saltpetre, sulfur and Aristolochia debilis is 46 : 46 : 8. The fire subduing experiment is very dangerous. When the Aristolochia debilis powder meets the flame, it will be carbonized, and then the contents in the pot can become a mixture of the proto-gunpowder. Qingxuzi could have become aware of this, because he put the pot in a pit outdoors and when the pot smoked, he covered it with wet paper, bricks and soil. But after this treatment, the flammability and toxicity of sulfur, failing to be subdued, caused unexpected combustion and explosion when it met saltpetre and carbon.

---

[1]　清虚子. 太上圣祖金丹秘诀[M]//铅汞甲庚至宝集成. 上海：涵芬楼影印本，1926.

The Song Dynasty alchemist masterpiece *Zhu Jia Shenpin Dan Fa* (《诸家神品丹法》, *A Collection of Great Danyao Prescriptions of Former Alchemists*)（Volume 5）（Figure 107）makes a collection of the methods for subduing fire recorded by the alchemists from the Jin Dynasty to the Tang Dynasty. The quoted methods in Volume 5 of *Zhu Jia Shenpin Dan Fa* can be dated back to the periods no later than the Tang Dynasty. The *Fu Huo Liuhuang Fa* （伏火硫黄法，"The Sulfur Method for Subduing Fire"）in the book goes as follows:

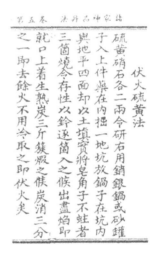

Figure 106　"On *Fu Huo Fan Fa*" (《伏火矾法》, the Alum Method for Subduing Fire, 808), by the Tang Dynasty Alchemist Qingxuzi, taken from Volume 595 of *Dao Zang* (《道藏》, *Taoist Canon*, 1445, a Collection of Taoist Classics)

Figure 107　The *Fu Huo Liuhuang Fa* (伏火硫黄法, the Sulfur Method for Subduing Fire), recorded by the Tang Dynasty Alchemists, kept in Volume 594, *Dao Zang*

Take 2 liang of sulfur and 2 liang of saltpetre, grind them into powder, put the mixture of fine sulfur and saltpetre into an reactor. Lay the reactor into a ground pit, the top of the reactor at the same level with the ground, fill in soil around the reactor. Burn three well-kept gymnocladus chinensis pods into charcoal (not ash), insert them into the reactor one by one (in this way, the charcoal meets saltpetre and sulfur). When the flame in the reactor extinguishes, add 3 jin of charcoal at the upper part of the reactor, calcine the charcoal stack till 1/3 is burnt, remove the charcoal, take out the remaining when it cools. Thus, the flammability of sulfur is subdued.

The above *Fu Huo Liuhuang Fa* recorded by the Tang Dynasty people has similarities with the *Fu Huo Fan Fa* listed by the Tang Dynasty alchemist Qingxuzi. But the *Fu Huo Liuhuang Fa* could be more dangerous, for inside the reactor is actually the mixture of saltpetre, sulfur and charcoal.

Since the alchemists from the Jin Dynasty to the Tang Dynasty were often involved in repeated accidents caused by the fire-subduing experiments in which subduing fire turned into causing fire, some of them began to sum up the lessons to warn the alchemists. *Zhenyuan Miao Dao Yaolüe* (《真元妙道要略》, *Synopsis of Zhenyuan Taoist Theories*) talks about some lessons of failure: "It has happened that burning the mixture of sulfur, realgar, saltpetre and honey, produced flame, burned people's faces or hands and even burned the house out. (有以硫黄、雄黄合硝石并蜜烧之，焰起，烧手面及烬屋舍者。)"❶ (Figure 108) Honey, which, if burnt, turns into charcoal, serves as the source of charcoal, just like the Aristolochia debilis fruits and the gymnocladus chinensis pods. *Zhenyuan Miao Dao Yaolüe* again points out: "Saltpetre in small amount assists the other drugs, while it in large quantities could do harm. It can not be burned with sulfur, realgar or orpiment, as the burning could lead to immediate disastrous accidents. [硝石宜佐诸药，多则败药。生者不可合三黄等烧，(否则)立见祸事。]" It is clearly stated in the book that burning the mixture of saltpetre, sulphur yellow, realgar and orpiment, will immediately lead to explosion. The combustion described in *Zhenyuan Miao Dao Yaolüe* could be something between quick burning and deflagration, and therefore, it should be caused by the mixture of proto-gunpowder. The book was written under the name of Zheng Yin (郑隐, 220～300), a man of the Jin Dynasty. But the accounts in the book mostly date back to the time between the Jin Dynasty and the Tang

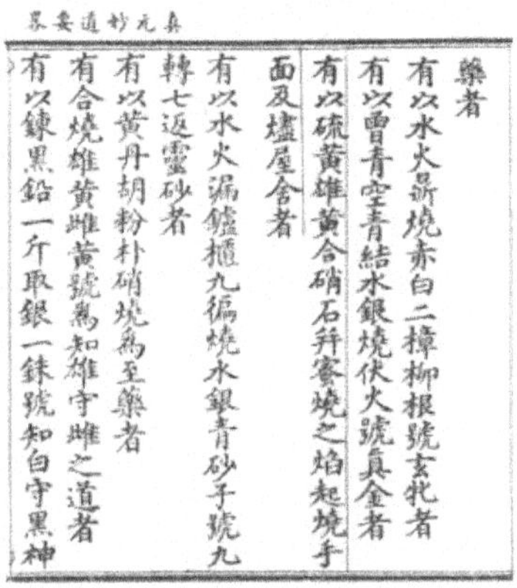
真元妙道要畧
藥者
有以水火鼎燒赤白二樟柳根號玄牝者
有以曾青空青結水銀燒伏火號真金者
有以硫黃雄黃合硝石并蜜燒之焰起燒手
面及燼屋舍者
有以水火漏鑪櫃九徧燒水銀青砂子號九
轉七返靈砂者
有以黃丹胡粉朴硝燒為至藥者
有合燒雄黃雌黃號為知雄守雌之道者
有以鍊黑鉛一斤取銀一銖號知白守黑神

Figure 108 The Accounts on the Combustion of Gunpowder in *Zhenyuan Miao Dao Yaolüe* (《真元妙道要略》, *Synopsis of Zhenyuan Taoist Theories*, from the Ninth Century to the Tenth Century), taken from Volume 596, *Dao Zang*

❶ 真元妙道要略[M]. 上海：涵芬楼影印本，1926.

Dynasty, so the book must have been finished no later than the 10th century[1].

Finally, let's examine the possibility of gunpowder having been invented in countries or regions other than China. Saltpetre (KNO_3) has been named *potassium nitrate* in English and French since the 14th century. *Nitrum* is the Latin term for today's *nitrate*. Before the 14th century, *nitre* appeared in the English Bible; *nitrum* in the Latin Bible; and γιτρον (nitron) in the Greek Bible; all three terms had been derived from the Hebrew word *neter* which means soda or sodium carbonate (Na_2CO_3). Taking *neter*, γιτρον and *nitrum* as saltpetre was an unrealistic conjecture and misunderstanding in history, because they were meant to be soda[2]. Ancient Egyptians used natural soda in the third century B. C., so γιτρον was probably from the Egyptian name *ntrj* of soda. Strabon (63 B. C. ~19 A. D.), an ancient Greek, called the soda lake γιτιδι. Therefore, in the ancient languages of Europe and Egypt, no term has been identified as the name of saltpetre, and consequently, in these regions, people must have not identified saltpetre, nor have they known the existence of such a salt, least of all making purified saltpetre. The first European to talk about the refined saltpetre is a British of the 13th century, Roger Bacon (1214~1292), whose narration of the refined saltpetre was after the introduction of gunpowder technology into Europe. It was after mastering the Chinese gunpowder technique that the Europeans began to apply their ancient name of sodium carbonate to saltpetre.

Al Baytar (1197~1248), the first Arabian who mentioned saltpetre, in his book *Kitāb al-Jāmi fi al-Adwiya al-Mufradi* (1240), called saltpetre "snow of China" (*thalj al-Sīni*)[3], indicating that it was imported from China. But his interpretation of it as the "Asian stone" might not be right, because the term Asian stone (λιθος Ασιos) seen in the Roman scholar Pliny the Elder's (23~79) *Historia Naturalis* (73) refers to the lime of his time. Hassan al-Rammāh (1265~1295), another Arabian who spoke of saltpetre and its purification, quoted many Chinese documents in his book *Kitāb al-Furūsiya wa al-Munāsab al-Harbiya* (around 1285), in which he didn't call saltpetre the "Asian stone",

[1]   NEEDHAM J, et al. Science and Civilization in China: vol. 5[M]. Cambridge: Cambridge University Press, 1986.

[2]   PARTINGTON J R. A History of Greek Fire and Gunpowder[M]. Cambridge: Haffer & Sons, Ltd. , 1960.

[3]   IBNEL BEITHAR. Traité des simples, tom 1. Traduit par Leclere L[M]. Paris: Imprimerie Nationale, 1877.

but used the Arabian local name *bārūd* ❶. According to Yūsuf ibn Ismā'il al-Kutubī, among the people in Iraq, *bārūd* meant the "wall salt" (*mih al-hāyit*), a term rather similar to the one coined about saltpetre by the Chinese Ma Zhi (马志) — ground frost (地霜).

Although India also has natural saltpetre, no fixed term could be found in ancient Sanskrit to mean saltpetre, a proof for the fact that the ancient Indians did not know saltpetre. *Shuraka*, the word in Sanskrit referring to saltpetre, appeared after 1400, and was borrowed from the Persian *shurāj* (saltpetre). There is a record of gunpowder (*agni-cūrna*) prescription in the Sanskrit writing during the beginning of the Mughal Dynasty (1526~1857), but it is far later than the Chinese discovery of gunpowder. Another thing needs to be clarified here. According to *Jinshibu Wujiu Shujue* (《金石簿五九数诀》, *On Metals and Stones Used in Refining Danyao*, about 670), around 664, a Brahman named *Zhifalin* (支法林) arrived in China and claimed that the saltpetre produced in Lingshi County, Fenzhou (汾州灵石县) is not as good as that in Uddiyana, India. Based on the above record, *Zaohua Fu Gong Tu* (《造化伏汞图》, *Illustrations of Subduing Mercury*) written by *Shengxuanzi* (升玄子) in the Song Dynasty, infers that Uddiyana of India also produces saltpetre. Since *Jinshibu Wujin Shujue* was written by an anonymous writer and whether the Brahman named *Zhifalin* recorded in it really existed in history is questionable, then Shengxuanzi's inference that Uddiyana produced saltpetre in the seventh century must be groundless and unconvincing. Japan and Korea, two other ancient countries in East Asia, borrowed herbalism from China, but they did not develop alchemy very well. The purification of saltpetre for the purpose of making gunpowder in both countries began even later than that in the Arabian countries and in Europe.

It can be seen through the above analysis that although natural saltpetre has existed in the countries or regions besides China since ancient times, the residents of these areas failed to recognize this nitrate and its purification for practical purposes, thus blocking the road to the invention of gunpowder. Dr. Joseph Needham ever commented, in this way, it could come to light, the long well-known evidences for the fact that China first invented gunpowder: that it seems obvious that a lack of saltpetre in the west must have been a limiting

---

❶ PARTINGTON J R. A History of Greek Fire and Gunpowder[M]. Cambridge: Haffer & Sons, Ltd. , 1960.

factor for the development of gunpowder there; that the earliest mention of gunpowder was undoubtedly at the end of the 13th century, the time before gunpowder technique was extensively spread in Europe in the 14th century; and that before gunpowder was introduced to the Islamic world and Europe in the 13th century, it had been widely used in China for military purposes. ❶ The same is true for Africa and the rest of Asia. This is why it is impossible for other countries or regions to have made gunpowder earlier than China.

What is the long-established evidence that could make the invention of gunpowder in China come into light (in Needham's words)? Why was gunpowder invented in China? We have given the answer in a systematic way in this section, which can be summarized into the following 5 aspects:

(1) During the fourth century B.C. and the third century B.C., China first discovered saltpetre, named it *xiaoshi* and began to use it in drugs. The tradition has been passed down in the following two thousand years. There is sufficient literature evidence and unearthed physical evidence in this respect.

(2) Saltpetre has been taken orally as an medicine in China since the end of the Warring States Period (战国, the third century B.C.). Its application as oral medicine has been recorded in the medical prescriptions of renowned doctors since the Western Han Dynasty (the second century B.C.). Both facts indicate that the people then must have mastered the purification technique of saltpetre.

(3) Since the Western Han Dynasty (the second century B.C.), the ancient Chinese alchemists used saltpetre as a common reagent in refining *danyao*. They carried out long-term research of its physical and chemical properties. Workshops that manufactured purified saltpetre through the recrystallization method reached a considerable scale in the Han Dynasties.

(4) Saltpetre was compiled as the first-class medicine into the earliest work of pharmacology in China, *Shennong Bencao Jing* (《神农本草经》, *Shennong's Classic of Materia Medica*, finished in the first century B.C.). In the book, it is named *puxiao*, and *xiaoshi* is used to mean sodium sulphate, thus causing confusion in the naming for potassium nitrate and sodium sulphate. Fortunately, Tao Hongjing (陶弘景) distinguished the difference between the two in his scientific experiments and designated the right name for saltpetre in

---

❶   NEEDHAM J, et al. Spagyrical Discovery and Inventions: Apparatus, Theories and Gifts[M]. Cambridge: Cambridge University Press, 1980.

his book *Bencao Jingji Zhu*. Later on，Ma Zhi's（马志）*Kaibao Bencao* completely ended the confusion of the names for potassium nitrate and sodium sulfate.

(5) Since the Jin and Tang Dynasties（the third to ninth centuries），the Chinese alchemists often mixed saltpetre powder with sulfur powder or realgar powder，added some carbonaceous matter，and then heated the mixture in the reactor. As described in *Zhen Yuan Miao Dao Yao Lüe*，the result of these experiments was "flame produced，people's faces or hands burned and the house burned out（焰起，烧手面及烬屋舍者）". In this way，the primitive mixture of gunpowder was invented，which preluded the military application of gunpowder.

So it is logical that gunpowder was invented in China.

3.1.2　Early Firearms and Earliest Military Gunpowder Prescriptions in China

3.1.2.1　Early Firearms in the Northern Song Dynasty Since 10th Century

As mentioned in the previous section，the alchemists in the Tang Dynasty heated the mixture of saltpetre, sulfur and carbonaceous matter in containers in their "fire-subduing" experiments，caused deflagration that burned their bodies and houses，and then accidentally discovered the proto-gunpowder. But causing deflagration and discovering the proto-gunpowder were not the intention of the alchemists who did the experiments to change some ingredient's nature and thus to avoid deflagration. "Subduing fire" was meant to prevent the formation of the gunpowder mixture，but as long as saltpetre，sulfur，and charcoal met，"subduing fire" was turned into setting fire. This was the result of the nature of the three（saltpetre，sulfur，and charcoal），not to be altered by human will. Once recognized，the properties of the three mixed would acquire potential military and practical value，and might be used in fire attacks. The problem with the application of the mixture in fire attacks lay in how the explosion of gunpowder mixture could be artificially controlled and how safety devices could be built so that the explosive force could only be emitted after launching and that the casters were not hurt. Solving the problem would naturally bring about the invention of gunpowder and firearms. The invention of gunpowder，which should be synchronized with its actual application，would be completed only if gunpowder was made into practical combustive and explosive devices.

Technological invention and scientific discovery are two different concepts. Scientific discovery can lead to technological invention，but it is not equal to invention. Similarly, the chemical discovery of gunpowder mixture explosion by the alchemists cannot be equated with the invention of gunpowder technology, just as the discovery of electromagnetic induction in physics cannot be identified with the invention of the electromotor. Nevertheless, it should be acknowledged that the discovery of gunpowder by the Chinese alchemists prompted the invention of gunpowder technology. The discovery of gunpowder explosion is relatively easy, but the invention of gunpowder technology is more difficult, because to consciously make explosive mixtures and firearms require more risky experiments which may cause lost of lives. The technicians and craftsmen making weapons in the Chinese ancient ordnance factories, with courage and exploration, successfully made a batch of the earliest firearms, which was the fruit of their collective effort and wisdom. Therefore, gunpowder, like papermaking, should not be the invention of an individual.

It is non-serious academic research as well as misleading of some previous books to set the invention of gunpowder at a much earlier time and to attribute it to some individual groundlessly. For example, the Ming Dynasty man Luo Qi (罗颀) said in *Wu Yuan* (《物原》, *Origin of Important Inventions*, 15th century), "Ma Jun (马钧) of the Wei Dynasty made *baozhang* (爆仗, firecrackers). Emperor Yang of the Sui Dynasty (隋炀帝) took gunpowder as a gimmick for fun. (魏马钧制爆仗（纸砲），隋炀帝益以火药为杂戏。)"[1] But Fuxuan's (傅玄, 217~278) *Maxiansheng Zhuan* (《马先生传》, *The Biography of Ma Jun*) quoted in and annotated by Volume 29, *Wei Shu* (《魏书》, *History of the Wei Dynasty*) of *Sanguo Zhi* (《三国志》, *Records of the Three Kingdoms*), only keeps a record of Ma Jun (马钧, 207~260) displaying craftsmanship in improving the silk weaving machine, in making the water-powered acrobatic machine and in making the stone launching machine, but not in making firecrackers[2]. The poem of Emperor Yang of the Sui Dynasty, "Ten thousand lanterns hanging on the trees, are colorful and brilliant, as if the trees blossomed countless seven-colored flowers of flame. (灯树千光照，花焰七枝开)"[3], has nothing to do with fireworks. The inspection of the historical documents verifies the incredibility of the contents in Luo Qi's *Wu Yuan*. The

[1]　罗颀. 物原：兵原第十四[M]. 上海：商务印书馆，1937.

[2]　陈寿. 三国志：卷廿九[M]. 裴松之，注. 上海：上海古籍出版社，1986.

[3]　杨广. 正月十五日放通衢建灯夜升南楼诗[M]//丁福保. 全汉三国晋南北朝诗：全隋诗. 北京：中华书局，1959.

book, whose accounts about the origins of many things are mostly without substantial evidence, cannot be a reliable source of citation. Unfortunately, some people in modern times still believe in the misstatements of *Wu Yuan*.

Some modern authors think the alchemist Sun Simiao (孙思邈, 581~682) of the Sui and Tang Dynasties (the seventh century) is the inventor of gunpowder, and ascribes to him *Fu Huo Liu Huang Fa* collected in Volume 5 of *Zhu Jia Shenpin Dan Fa* (《诸家神品丹法》, *A Collection of Great Danyao Prescriptions of Former Alchemists*), a book of the Song Dynasty[1]. The content in *Fu Huo Liu Huang Fa* is indeed related to gunpowder mixture, but whether it can be attributed to Sun Simiao is questionable. The quotation order in the original *Zhu Jia Shenpin Dan Fa* is: *Fu Liuhuang Fa* (《伏硫黄法》, *The Method for Subduing Sulfur*) within *Sunzhenren Dan Jing* (《孙真人丹经》, *The Danyao Prescriptions of Sun Simiao*), *Huangsanguanren Fu Liuhuang Fa* (《黄三官人伏硫黄法》, *Huangsanguanren's Method for Subduing Sulfur*), *Fu Huo Liuhuang Fa*, and finally *Fu Xionghuang Fa* (《伏雄黄法》, *The Method for Subduing Realgar*) within *Gexianweng Dan Jing* (《葛仙翁丹经》, *The Danyao Prescriptions of Ge Hong*). The anonymous *Fu Huo Liuhuang Fa*, does not follow Sun Simiao's *Fu Liuhuang Fa* in the book closely, but lies after *Huangsanguanren Fu Liuhuang Fa*, also having nothing to do with gunpowder). Hence, it is unreasonable to conclude that the anonymous *Fu Huo Liuhuang Fa* was written by Sun Simiao. Nathan Sivin, an American scholar, first noticed this, and Joseph Needham et al. reaffirmed the message in their writings[2]. Besides, *Sunzhenren Dan Jing* was compiled by some Tang Dynasty people from 758 to 760, more than 100 years later than Sun Simiao's living time; in other words, it was not written by Sun Simiao himself. It can be believed that the alchemists only discovered the explosive phenomenon of gunpowder, but did not invent the gunpowder technology. It is obviously wrong to deem Sun Simiao the inventor of gunpowder.

In discussing the invention time of gunpowder, it is necessary to realize that the gunpowder technique must have been a product of highly developed science and technology, and that it required breakthroughs in prior chemistry and military technologies. Gunpowder could not have been invented before the

[1] 冯家昇. 火药的发明和西传[M]. 上海：上海人民出版社,1954.

[2] NEEDHAM J, et al. Spagyrical Discovery and Invention[M]. Cambridge: Cambridge University Press, 1976.

Tang Dynasty, because it was not until the Tang Dynasty that people started to conduct chemical experiments by mixing saltpetre, sulfur and charcoal (or some carbonaceous matter), even if the former medical experts and alchemists, having known much about the nature of saltpetre and sulfur, had mixed them and added other ingredients to make drugs or *danyao*. The early records of proto-gunpowder seen today are mostly on the relevant products of the Tang Dynasty and the Five Dynasties, and included in the writings of the Song Dynasty people. The dangerous chemical experiments involving gunpowder demanded certain operational skills and safety measures, despite of which accidents still happened. It did take some time to experiment and explore in order to get the possible explosion of the proto-gunpowder mixture under control, which decided that the gunpowder technology could only emerge after the Tang Dynasty.

Before the Tang Dynasty, military fire attack techniques generally made use of common incendiary agents, but technological innovations happened in the Five Dynasties: a more violent petroleum product, *menghuoyou* (猛火油, fierce fire oil), replaced the ancient incendiary agents. It was loaded into a metal single-cylinder single-rod double-acting piston force-pump for a liquid to jet flame. *Menghuoyouji* (猛火油机, fierce fire oil machine), a new type of incendiary weapon then, was really fierce and powerful. On the other hand, the ancient *Paoshiji* (抛石机, trebuchet) called *pao* (礮 or 砲) mainly focused on projecting stones and could merely cause sectional mechanical damage. But from the end of the Tang Dynasty (904), the incendiary matters had been projected by *paoshiji*, bringing about the destructive effect of burning and expanding the functions of *Paoshiji*. At the same time crossbow manufacturing techniques were improved to cover an increased range and to shoot incendiary agents. But these two new military technologies were in urgent needs of improvement: *menghuoyouji* need a fast ignition agent, while *paoshiji* and the crossbow need an incendiary agent that had more potent combustion power and was not easy to get extinct. Gunpowder just met these needs. In China, gunpowder appeared after *menghuoyouji*. Gunpowder, without the deficiencies of a liquid such as inconvenience and being unsuitable for long-distance fire attack, was a perfect substitute for the *menghuoyou* (猛火油, fierce fire oil).

There were records of proto-gunpowder in the eighth to the ninth centuries' of Tang Dynasty. The first half of the tenth century (the period between the Tang Dynasty and the Five Dynasties) witnessed the experimental research of military gunpowder. The second half of the tenth century (the time between the Five Dynasties and the Northern Song Dynasty) saw the practical application of gunpowder. The firearms of this period were mostly recorded in the history documents of the Northern Song Dynasty and those after. The Yuan Dynasty historian Wang Yinglin (王应麟, 1223~1296) wrote in Volume 150 of *Yu Hai* (《玉海》, *the Encyclopedia Edited by Wang Yinglin*, 1267):

> *In Nov. 3, 1002, after Shi Pu (石普, 951~1021) claimed that he could launch* huoqiu *(火球, gunpowder package) and* huoyaojian *(火箭 or 火药箭, gunpowder arrow). Emperor Zhenzong of the Song Dynasty (宋真宗) reckoned him to the palace to have a show which was watched by all the ministers. Years before, in March, 969, Feng Jisheng (冯继昇, 930~990) and Yue Yifang (岳义方) presented a show of* huoyaojian *to the Emperor then and after the success of the show they were awarded some silk cloth by the Emperor.*

Wang Tang (王棠), a prime minister of the Southern Kingdom (南院) during the reign of Emperor Daozong of the Liao Dynasty (辽道宗, 1055~1100), echoed the above information in *Yanzaige Zhixin Lu* (《燕在阁知新录》, *A Dictionary of Folk Culture Terms Compiled in Yanzai Pavilion*): "In 969, Feng Jisheng (冯继昇, 930~990) and Yue Yifang (岳义方) presented the method of *huoyaojian* to Emperor Taizong of the Song Dynasty (宋太宗). (宋太宗开宝二年,冯继昇、岳义方上火箭法。)" Tuotuo (脱脱, 1314~1355), a Yuan Dynasty historian, wrote in *Bing Zhi* (《兵志》, *Records of Military Affairs*), Volume 197 of *Song Shi* (《宋史》, *The History of Song Dynasty*, 1345):

> *In 970, Feng Jisheng (冯继昇) and his colleagues (officials of the Ministry of Defense), presented the method to make* Huo Yao Jian, *demonstrated it, and was awarded clothes and silk cloth by the Emperor. In August, 1000, Tang Fu (唐福, 970~1030), the captain of the royal waterborne army presented the well-made* Huo Yao Jian, huoqiu *and* huojili *(火蒺藜, barbed gunpowder package) to the Emperor, and Xiang Wan (项绾), an official of the Ship-making Department, presented ships for sea battles. They were both awarded some money. In 1002, Liu Yongxi (刘永锡) presented* shoupao *(手砲, hand grenade). The Emperor ordered it be mass-manufactured and be applied in the army.*

Song Huiyao Jigao (《宋会要辑稿》, *The Edited Collection of the Song*

Dynasty Official Documents) also records, "In 1000, Tang Fu (唐福, 970～ 1030), the commander of the royal waterborne army presented to the Emperor the well-made gunpowder arrows, gunpowder packages and barbed gunpowder packages."[1] Volume 52 of *Xu Zizhi Tongjian Changbian* (《续资治通鉴长编》, *A Sequel to History as a Mirror*, 1183) by a Song Dynasty man Li Tao (李焘, 1115～1184) relates, "In Nov. 3, 1002, after Shi Pu (石普, 951～1021) claimed that he could launch *huoqiu* and *huoyaojian*, the Emperor reckoned him to the palace to have a show which was watched by all the ministers."[2] According to *Shi Pu Zhuan* (《石普传》, *The Biography of Shi Pu*), Volume 324 of *Song Shi*, Shi Pu (石普, 951～1021), a native of Youzhou (幽州, now Beijing) and famous army commander of the early Song Dynasty, was brave in the battle and knew well the military books, the Taoist philosophy, prognostics, astronomy and the calendar.[3] *Taizu Ji San* (《太祖纪三》, *Biography of Emperor Taizu of the Song Dynasty*, Ⅲ), Volume 4 of *Song Shi* records, "On August 28, 976, the Wuyue Kingdom (吴越国) introduced some soldiers to launch *huoyaojian* for the Emperor". This historical fact shows that *huoyaojian* was available in the Wuyue Kingdom (吴越国, 907～978) located along the southeast coast of ancient China.

The above historical documents demonstrate that in the early stages of the second half of the tenth century (between the Five Dynasties to the Northern Song Dynasty), the Chinese military technologists of the different regions have developed a number of early firearms in order to adapt to the needs of the war. Firstly, Feng Jisheng (冯继昇, 930～990) and Yue Yifang (岳义方), officials of the Ministry of Defense in the capital Bianjing (汴京, now Kaifeng, He'nan Province), in March 969, offered some sample *huoyaojian* and the method of launching to the Emperor, who ordered it be launched and awarded them some clothes and silk as an encouragement for the success of the show. *Huoyaojian* mentioned here, which can be called "gunpowder arrow", refers to a gunpowder package bound to the arrow, and then fired and shot with a bow or crossbow (Figure 109) with a range of 150～200 *bu* (248 m～330 m)[4]. *Huoyaojian* shot by a bow is thus named "bow gunpowder arrow"; and

---

❶ 宋绶. 宋会要辑稿[M]. 北平:国立北平图书馆影印,1936.

❷ 李焘. 续资治通鉴长编[M]. 上海:上海古籍出版社,1986.

❸ 脱脱. 宋史:石普传[M]. 上海:上海古籍出版社,1986.

❹ 吉田光邦. 宋元の军事技术[M]//薮内清. 宋元时代の科学技术史. 京都:中村印刷株式会社,1969.

huoyaojian shot by an crossbow, "crossbow gunpowder arrow". Feng Jisheng
(冯继昇) and Yue Yifang (岳义方) were the inventors of firearms and founders
of gunpowder technology first seen in literature. They invented the gunpowder
arrows in 969, a time just over a thousand years ago. Then, on August 28, 976,
the Wuyue Kingdom (吴越国, 907~978), which surrendered to the Northern
Song Empire, dedicated some sergeants good at shooting gunpowder arrows to
the Northern Song government.

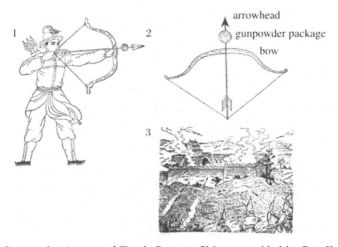

Figure 109   Gunpowder Arrows of Tenth Century China, provided by Pan Jixing (1987)
1. Bow Gunpowder Arrow Launching Method Invented by Feng Jisheng (冯继昇) et al. in 10th Century
2. Components of Bow Gunpowder Arrow   3. Picture of Using Bow Gunpowder Arrows to Attack a City,
Painted by Some Westerner, Taken from Raketové zbraně (1958)

In August 1000, Tang Fu (唐福, 970~1030), the captain of the royal
waterborne army of the imperial guard in the Northern Song Dynasty,
presented some *huoyaojian*, *huoqiu* and *huojili* made by himself to Emperor
Zhenzong of the Song Dynasty and was awarded cash. *Huoqiu* is a gunpowder
package weighing 5 *jin* (2985 grams, about 3 kilograms), which can be
projected through *paoshiji* in a siege. *Danshaopao* (单梢砲, light-weight
catapult) has a range of 60 *bu* (99 metres) and *shuangshaopao* (双梢砲, double-
weight catapult) (Figure 110) covers a range of 80 *bu* (132 metres)❶. *Huojili*
is a gunpowder package with iron blades tied to it and iron thorns hidden inside
(Figure 111). After *huojili* is cast by *paoshiji*, the iron blades fix the package

---

❶   吉田光邦.宋元の軍事技術[M]//薮内清.宋元時代の科学技術史.京都:中村印刷株式
会社,1969.

to the city wall，and the iron thorns darting out after the explosion has supplementary killing power. On November 3，1002，the commander of Jizhou Army and military technologist Shi Pu（石普，951～1021）wrote to Emperor Zhenzong of the Song Dynasty，saying he trained the soldiers to launch *huoqiu* and *huoyaojian*. The Emperor reckoned him to the imperial palace to present a show watched by the Emperor himself and the ministers. In 1002，Liu Yongxi （刘永锡），the head of the Shanxi Ningwu Army，invented the hand grenade. The imperial court ordered it to be made and applied in the border armies.

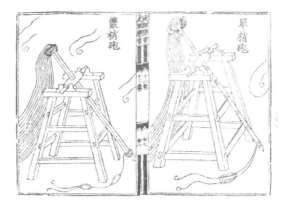

Figure 110   *Danshaopao*（单梢砲，light-weight catapult）and *shuangshaopao* （双梢砲，double-weight catapult）Used to Launch Gunpowder Packages，taken from *Collection of the Most Important Military Techniques*

In ancient China，"礮" or "砲"（pronounced *pao*）meant a catapult for projecting stones. Since the Northern Song Dynasty，the catapults for projecting gunpowder packages were called "火砲"（pronounced *huopao*，denoting catapult throwing gunpowder package）. Since the Ming Dynasty，all firearms that have gunpowder loaded into metal tubes were named "火炮"（pronounced *huopao*，denoting cannon）.

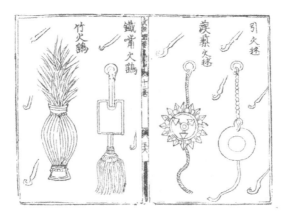

Figure 111   Firearms of *huoqiu*（火球，gunpowder package）and *Huojili*（火蒺藜，barbed gunpowder package）Recorded in *Collection of the Most Important Military Techniques*

After the Northern Song Empire had been established in 960, it had to conquer some independent regimes and reunify the country. Besides, it had to face the military threat of the Liao Dynasty (辽, 916~1125). Therefore, the Northern Song Empire attached great importance to the improvement of their armies' weapons and equipment. In 976, the Northern Song government started two ordnance workshops in its capital city, appointed some government officials as supervisors, and recruited many skilled artisans from all over the country to work there. In 1007, the two ordnance workshops were amalgamated into one, which was divided into eight factories of larger scale in 1023 after the invention of firearms. *Zhu Shi* (《麈史》, *A Comprehensive Record of All Fields*, 1115) by a Northern Song citizen Wang Dechen (王得臣, 1036~1116) in Volume 1 quotes from Song Minqiu's [宋敏求, also named Song Cidao (宋次道), 1019~ 1079] *Dongjing Ji* (《东京记》, *Anecdotes of Dongjing City*, 1040):

> Song Cidao (宋次道) in his Dong Jing Ji (《东京记》, *Anecdotes of Dongjing City*, 1040) wrote, besides the eight ordnance factories, in Dongjing City (the capital then) there was still the city-siege equipment workshop, within which were altogether 11 subdivisions, two of them specializing in the production of gunpowder and menghuoyou (猛火油, fierce fire oil) respectively. Each subdivision has its regulations and operating procedures that are to be remembered by the workers. Leakage of the secret recipes is strictly prohibited.

3.1.2.2 Earliest Military Gunpowder Prescriptions that Appeared in 10th Century

The 10th century gunpowder workshop of Kaifeng State (开封府, also called Dongjing, the capital of the Northern Song Dynasty) was the country's central arsenal of gunpowder and firearms. There should be other provincial ordnance factories as well. The total production scale was large enough to equip an army of up to hundreds of thousands of soldiers. The factories had strict management systems and operating procedures that the craftsmen must keep in mind; it was forbidden to betray the technological secrets. Owing to the confidentiality of the gunpowder technology in the early Northern Song Dynasty, it was more than seventy years later that the technical details were seen in Zeng Gongliang (曾公亮, 999~1078) and Ding Du's (丁度, 990~1053) *Wujing Zongyao* (《武经总要》, *Collection of the Most Important Military Techniques*). Zeng Gongliang, also named Mingzhong, a minister of the Northern Song Dynasty, native of Quanzhou, Fujian Province, *jinshi* (进士, *jinshi* degree holder) of the Tiansheng Years (天圣年间), was once the *zhifu* (知府, mayor or governor) of Kuaiji (会稽), Zhengzhou (郑州) and Kaifeng (开封) respectively. He was appointed to the court as a royal consultant in 1047, and known for his knowledge of the law and regulations, he was appointed the Prime Minister in 1061. By order of Emperor Renzong of the Song Dynasty (宋仁宗), in 1040, he led Ding Du and others to edit *Collection of the Most Important Military Techniques*. The 20-volume book was first published in 1044 and reprinted in 1231. The version we see now is the 1231 edition reprinted in 1505, the first set of which was photocopied in 1959 by the Shanghai Division of Zhonghua Book Company and reprinted by Shanghai Ancient Classics Press in 1988. *Collection of the Most Important Military Techniques* is an authoritative official military encyclopedia of good language and vivid illustrations, prefaced by Emperor Renzong of the Song Dynasty. Since the author took good advantage of the secret archives of the Song government, *Collection of the Most Important Military Techniques* has great historical value. The early gunpowder formula and firearms described in the book are presented below:

Prescription One   *Prescription of the* duyaoyanqiu (毒药烟球, *gunpowder ball producing poisonous smoke*).

The gunpowder ball producing poisonous smoke, weighing 5 jin *(1* jin *in the Song Dynasty* = 16 liang = 596.8 grams*), is composed of sulphur (15* liang*) (1* liang *in the Song Dynasty* = 37.3 grams*), aconitum carmichaeli (5* liang*), saltpetre (30* liang*), croton tiglium (5* liang*), stellera chamaejasme (5* liang*), tung oil (2.5* liang*), vegetable oil (2.5* liang*), charcoal powder (5* liang*), asphalt (2.5* liang*), arsenic (2* liang*), beeswax (1* liang*), bamboo shavings (bambusae caulis im taeniam, 1.1* liang*), hemp shavings (1.1* liang*). All the ingredients are mixed, pounded up, and made into a ball, traversed by a 1.2-*zhang *half-*chi *hemp rope (1* zhang *in the Song Dynasty* = 10 chi = 307.2 centimetres*) functioning as the fuze. Smash and mix old paper (12.5* liang*), hemp bark (1* liang*), asphalt (2.5* liang*), beeswax (2.5* liang*), yellow lead (lead orthoplumbate, Pb_3O_4) (1.1* liang*) and charcoal powder (0.5* jin*). Apply and bound the mixture to the ball. The smelly poisonous ball can cause mouth bleeding and nosebleed, and when projected by* paoshiji, *can cause casualties among the city attacking enemy. All the combustive matters fastened to a gunpowder arrow can be shot by* a bow, a crossbow, or a bed-like crossbow, considering the distance of the target. The detailed instructions can be seen in gongshou (攻守, attacking and defense) and qixie (器械, weapons and instrument). ❶

Prescription Two   Prescription of the gunpowder package thrown by catapults.

Ingredients: sulfur of Jinzhou (晋州, now Linfen, Shanxi Province) (14 liang), impure sulfur (7 liang), saltpetre (40 liang), hemp shavings (1 liang), dry paint (1 liang), yellow arsenic (1 liang), starch (1 liang), bamboo shavings (1 liang), yellow lead (1 liang), beeswax (0.5 liang), vegetable oil (0.1 liang), tung oil (0.5 liang), turpentine (14 liang), animal fat (0.1 liang). Pound the mixture of Jinzhou sulfur, impure sulfur and saltpetre, and then sieve it. Grind the mixture of yellow arsenic, starch and yellow lead into fine powder; pestle the dry paint; heat and stir the hemp shavings and the bamboo shavings for a while till they crush. Have all the powders mixed. Heat the mixture of beeswax, turpentine, vegetable oil, tung oil, and animal fat into paste. Add the paste into the powder and mix them well. Wrap the new mixture in five pieces of paper and then bind the paper ball with hemp ropes. Apply melted turpentine to the ball. Project the ball with paoshiji (抛石机, trebuchet). There is still the method for duyaoyanqiu (毒药烟球, gunpowder ball producing poisonous smoke) kept in Huo Gong Men (火攻门, Ways of Fire Attacks). ❷

Prescription Three   Prescription of jili huoqiu (蒺藜火球, barbed gunpowder package) gunpowder.

Jili huoqiu is a gunpowder package surrounded by three six-tipped blazes, penetrated by a 1-zhang-2-chi (1 zhang = 10 chi = 312 centimetres) hemp rope, coated by paper, with other drugs stuck to its outer layer and eight barbed iron thorns hidden inside. Before throwing it, use a well-heated iron awl to ignite the package. The procedure to prepare the gunpowder package: pulverize the mixture of sulfur (20 liang), saltpetre (40 liang), coarse charcoal powder (5 liang), asphalt (2.5 liang) and dry paint (2.5 liang); crush the bamboo

---

❶　曾公亮. 武经总要：火攻[M]. 上海：中华书局上海编辑所，1959.
❷　曾公亮. 武经总要：守城[M]. 上海：中华书局上海编辑所，1959.

shavings (1.1 liang) *and the hemp shavings* (1.1 liang); *mix and knead the above powders into a gunpowder package; mix and heat tung oil* (2.5 liang), *vegetable oil* (2.5 liang) *and beeswax* (2.5 liang), *and then smear the paste onto the gunpowder package.* ❶

The three gunpowder prescriptions included in *Collection of the Most Important Military Techniques* (Figure 112), which actually reflect the tenth-century gunpowder technology presented by Feng Jisheng, Tang Fu and Shi Pu to the imperial court, are the world's oldest military gunpowder prescriptions in existence. In addition to saltpetre, sulfur, carbon, the prescriptions also include dry paint, asphalt, beeswax, bamboo and hemp shavings, yellow lead, turpentine, and vegetable oil. Some prescriptions even have within some toxic ingredients such as arsenic (As_2O_3), stellera chamaejasmc, aconitum carmichaeli and croton tiglium. For the sake of safety in operation and transportation, 10% of the gunpowder package which could weigh up to 5 *jin* (2984 grams, nearly 3 kilograms), is vegetable oil used to make the gunpowder paste-like. If the proportions of saltpetre, sulfur and carbon are calculated and the other ingredients are omitted, then *duyaoyanqiu* gunpowder contains 30

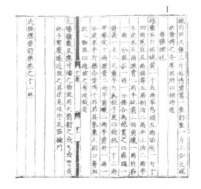

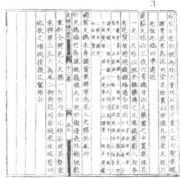

Figure 112   Earliest Military Gunpowder Prescriptions Recorded in *Collection of the Most Important Military Techniques*
1. Prescription of *Duyaoyanqiu*
2. Prescription of Gunpowder Package thrown by Catapults
3. Prescription of *Jili Huoqiu*

---

❶   曾公亮.武经总要：守城[M].上海：中华书局上海编辑所，1959.

liang of saltpetre (60%), 15 *liang* of sulphur (30%) and 5 *liang* of carbon (10%); *Jihihuoqiu* gunpowder has 40 *liang* of saltpetre (61.5%), 20 *liang* of sulphur (30.8%) and 5 *liang* of carbon (7.7%); the gunpowder package thrown by catapults includes 40 *liang* of saltpetre, 21 *liang* of sulfur and 5 *liang* of carbon (estimated, for the weight of carbon used is overlooked in the document) (the ratio of saltpetre to sulfur to carbon is 60.6 : 31.8 : 7.6).

Thus, it can be seen that the average ratio of saltpetre to sulfur to carbon was 60.7 : 30.9 : 8.4 in the tenth century. Its high proportion of sulfur, relatively low ratio of saltpetre and carbon, and paste form make it impossible as a propellant, but only possible as a burning and explosive weapon. The gunpowder package thrown by the bow or the crossbow makes *huoyaojian*; that projected by the catapult forms *huopao*; that containing poisonous drugs is termed *duyaoyanqiu*; that with iron thorns is referred to as *jili huoqiu*. These are the earliest forms of firearms just like the gunpowder arrows, fireballs and *huojili* developed by Feng Jisheng and others. The early gunpowder has two distinctive features: being paste-like, which makes it safe and stable, but leads to slow explosion and slow ignition (it is usually ignited by a hot iron awl piercing through); and being mixed with the traditional incendiary agents such as asphalt, wax, bamboo, and oil, which improves its initial flaming effect, but adds the unnecessary load, and thus plays down the leading role of saltpetre, sulfur and carbon. But still, it is gunpowder in the real sense, only with less powerful effects than the gunpowder that appeared later.

The aforementioned firearms were used in actual combat. *Huo Yao Jian* was employed when Emperor Taizu of the Song Dynasty (宋太祖) conquered the Northern Tang Kingdom (南唐) in 975. Xu Mengsheng (徐梦莘, 1126～1207), in Volume 97 of *Sanchao Beimeng Huibian* (《三朝北盟会编》, *Collections of Materials on the Maintaining of Peace Between the Song Dynasty and the Jin Dynasty*, 1193), quoted from Xia Shaozeng's (夏少曾, 1080～1150) *Chaoye Qianyan* (《朝野佥言》, *Records of the Song and Jin Disputes*, 1120) the record that in 1126 when the Northern Song Empire ended its reign, over 20,000 *huoyaojian* were left. Volume 12 of *Collection of the Most Important Military Techniques* speaks of *huoyaojianbian* (火药鞭箭, gunpowder whip arrow) launched by the springy force of a bamboo pole (Figure 113). The bamboo pole with a gunpowder package fastened to it resembles a javelin. The Jin regime (金政权, 1115～1234) founded by the Nüzhen nationality (女真族) learned the gunpowder technology after it ended the rule of the Northern Song

Dynasty in 1126. Later，in the battles between the Jin Regime and the Southern Song Empire，both sides made use of firearms，such as gunpowder arrows shot by the bow or the crossbow，*jili huopao*（蒺藜火砲，barbed gunpowder package）and *pilihuopao*（霹雳火砲，thunder-bolt missile or rocket-propelled bomb）launched by *paoshiji*. Later，the firearms were improved. In 1161，a Southern Song Dynasty troop leader，Wei Sheng（魏胜，1120～1164）invented *paoche*（砲车，wheeled catapult）during the battle with the Jin troops in Haizhou（now Lianyungang in Jiangsu Province）. The incident was recorded as "The gunpowder packages and stones were launched by *paoche* in the battle to hit targets 200 *bu* away（1 *bu* = 1.65 metres）. Wei Sheng's invention was presented to the court and the court ordered the weapon be manufactured and used in all the troops.（在阵中施火石砲,亦二百步⋯⋯以其制上于朝,诏诸军遵其式造焉）" *Paoche* is a wheeled catapult of high mobility that could throw the gunpowder packages and stones to more than 200 *bu* away.

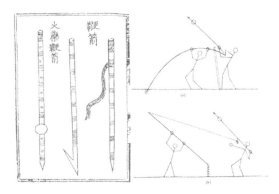

Figure 113   Tenth Century *Huoyaojianbian*（火药鞭箭，gunpowder whip arrow）
in China and Its Launching Technique，taken from Needham（1986）

3.2 Manufacture of Powdered Gunpowder of High Nitrate

3.2.1 Fireworks and Firecrackers in the 10ᵗʰ Century China

Black powder is so sensitive that it may be ignited by any spark. Friction，impact and even heat may trigger explosion.❶The sensitivity of gunpowder to

---

❶ 江洪.烟花炮竹生产与安全［M］.北京:轻工业出版社,1980.

impact is 1 kg/cm^2 to 2 kg/cm^2, that is to say, if a hammer with the weight of 1 kg to 2 kg drops from the height about 1 meter to an area of 1 cm^2, it will make the gunpowder burst and explode. Even if gunpowder is rubbed among wood boards, it will explode. The most sensitive friction surfaces for gunpowder are between iron and iron, iron and stone and lastly stone and stone. In the process of making gunpowder, stone mills have to be used to grind the mixture of saltpeter, sulfur and carbon. What's more, in the process of application and transportation, gunpowder is also vulnerable to heat, strike and other impact. So accidents often occur in arsenals due to improper practices. Then for safety's purpose, arsenal workers added vegetable oil into the mixture of saltpeter, sulfur and carbon to make it paste-like. Since the granularity has been increased, the physical and chemical stability of gunpowder will be improved and it can also become more resistant to the water at the same time. But all the protection measures made it hard to ignite the powder. During the war time, the slow process of igniting the powder was rather inconvenient. Because the gunpowder was made in the paste form for safety, it could only be used as explosive and could not be used as propellant powder, which actually restricted the full use of gunpowder at that time. In the 10th century Northern Song Dynasty, the early gunpowder packages were launched by trebuchet, which were named as *huopao* in Chinese. But they are rather poor in hit ratio and slow in the start-up of explosion. Some other smaller gunpowder packages were fastened onto spears and then projected by bows, which were called *huojian* in Chinese, meaning "fire arrow". Similarly they were not effective enough. All these told us that the techniques of early gunpowder and firearms need to be improved. After more than 50 years of the publication of the book *Collection of the Most Important Military Techniques*, techniques of the manufacture of gunpowder ushered in breakthroughs in development. From the end of the 11th century to the early 12th century, high-saltpeter powdered gunpowder was successfully made, which marked a new chapter in the history of gunpowder and firearms.

Solid powdered gunpowder was first used to make fireworks and firecrackers. Fireworks are made of small amount of powdered gunpowder stuffed inside paper tube connected to fuse, which can be ignited to produce loud sound and shining light to entertain people on special occasions, such as on festivals and other ceremonies. The tradition of lighting fireworks on special days has been kept till now in China. Firecrackers work in a similar way as fireworks do and makes loud bangs when they are lit. Even during opera performances, *huoxi* (火戏, fire show) was practiced by combination of the loud sound of firecrackers and the shining lights of fireworks. Since economy was booming at the late Northern Song Dynasty, gunpowder was used to make small play stuff like fireworks or firecrackers at peace time. The gunpowder was made solid and more saltpeter was added (about 70%) but less sulfur. Emperor Song Huizong, who once indulged in entertainment and enjoyment, let off many fireworks and firecrackers to amuse his imperial concubines during Zhenghe Reign (1111~1118)and Xuanhe Reign (1119~1125). Then royal and wealthy families copied the practice, which propelled the development of powdered gunpowder and as a result, the safety devices got perfected.

Firecracker got its name in an earlier time, but its meaning changed throughout ages. In the ancient time, bamboos were put in the fire at festivals to produce loud noise of cracking, which — it is said — could frighten off the ghosts and bad luck. Quoting from *Sheng Yi Jing* (《神异经》, *Records of the Spiritual and the Strange*), Zong Lin (about 500~563) wrote in his book *Jing Chu Suishi Ji* (《荆楚岁时记》, *Annual Folk Customs of the States of Jing and Chu*): "On the early morning of the first day of the lunar new year, people put bamboos into the fire to crack in their courtyard, the noise of which can drive off mountain spirits and devils." This tradition was kept till the Sui and Tang Dynasties, but after the Northern Song Dynasty people put gunpowder instead into paper tubes to produce the loud noise. That is to say, although the name of "firecracker" (*bao zhu* in Chinese) remained the same, the essence of it changed. Fireworks also named as fire-flowers or fire-flames can be set off by being held in hand or directly into the sky (Figure 114). Different colors of flames or other small particles can be let off from the paper tubes. Fireworks, which can fly into the sky, are called the "flying fire" or the "fire rocket". Gunpowder coated

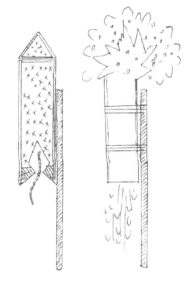

Figure 114   Illustration of Structure of Flying Fire in Song Dynasty, painted by Pan Jixing

with flashlight composition, such as, the yellow lead and the purple powder would be put into the paper tube. After the ignition of the fuse and with the sound of a whistle, the paper tube would fly high into the sky and let off a beautiful display of different colors of fire flames. Flying fire is different from double bang firecracker in that explosives instead of flashlight composition are inserted into the paper tube to produce loud noise of cracking. Both are retroactive lift-off devices.

Sometimes firecrackers can be mixed with fireworks through the connection of fuses. The mixture was put in a big box and placed on a high scaffold used to be called "Fireworks on Frameworks" in the Song Dynasty, but now is called "Framework Fireworks", which can display the double effects of firecrackers and fireworks. Thus the importance of the invention of fuse can be seen through this example. The fuse, once named as powder twist or powder thread, is used to ignite the gunpowder. Twines on the slips of paper around 3 cm wide with boosters added would be twisted together. In order to make the fuse hard and solid, alum and pastry were used to coat it. Fuse is different from other explosives in that it contains less sulfur. Generally speaking, it contains 73% of saltpeter, 4% of sulfur and 23% of charcoal. It has to be solid gunpowder in form. The fuse can safely ignite the gunpowder, control the ignition time, and can also connect series of fuses together to carry out a one-time ignition.

It is said that gunpowder was first used for non-military purposes such as fireworks. But the truth is just the opposite. Gunpowder was first used for military purposes. During peaceful time, it was used for civil purposes, such as firecrackers and fireworks for entertainment. Those gunpowder toys though small in size but well-made for amusement had a certain degree of safety and reliability. The development of gunpowder technology experienced some breakthroughs, many of which were recorded in original documents in the Song Dynasty.

There are quite a lot of historic documents about the fireworks in the Song Dynasty. In the fourth volume of the book *Bencao Yanyi* (《本草衍义》, *Expanded Materia Medica*) written by Kou Zongshi (1071~1149), a herbalist in the Northern Song Dynasty, the usage of saltpetre was recorded as "something able to make fireworks"[1]. This book was finished in 1116 and published in 1119, which proved the fact that in the early reign time of Emperor Song Suizong (1101~1115), solid powdered gunpowder was already used to make

---

[1]　寇宗奭.本草衍义:卷四[M].北京:商务印书馆,1957.

fireworks. Also during the reign of Emperor Song Suizong, Meng Yuanlao who once lived in Kaifeng, the Capital City, described his life at that time in the book *Dongjing Menghua Lu* (《东京梦华录》, *Dreams of the Glories of the Eastern Capital*, 1147). One of the descriptions is about the fireworks — "On the New Year's Eve, the loud noise of the firecrackers could be heard even out of the Royal Palace. All families in the city, rich or poor, stayed up the whole night sitting around the furnace to welcome coming of the New Year, which is called *shousui* (守岁, guarding the Year)."❶ That is to say, then the Royal Palace would set off firecrackers so much that the noise could be heard afar. Also Shi Su recorded the following description in *Kuaiji Zhi* (《会稽志》, *Records of Kuaiji*, 1202): "On the lunar new year's eve, the sound of fire-crackers could be heard everywhere, but sometimes there were people who mixed sulphur with other chemicals to cause even more violent explosions and these were called "bao zhang" in Chinese."❷ They must have been paper firecrackers.

Around the Song and Yuan Dynasties, Zhou Mi (1232~1298) depicted his predecessors' memories about their living in Hangzhou city during years from 1163 to 1189 in his book *Wulin Jiushi* (《武林旧事》, *Customs and Institutions of the Old Capital*). It reads as follows:

> On the lunar New Year's eve, "a variety of lanterns were displayed at the Palace of Qinyan and Minhua. The loud noise of the firecrackers could be heard even out of the Royal Palace. People followed the example of firecrackers' performance since Emperor Huizong's reign, also called Xuanhe Years (1109~1125)."❸

So it is safe to say that the solid gunpowder with more than 70% sulphur has been used to produce fireworks and firecrackers in Kaifeng since Emperor Huizong's reign period (1101~1126).

Since 1126, Hangzhou has been the Capital City in the Southern Song Dynasty. More fireworks were produced at that time. Even more retroactive devices were applied into the fireworks to entertain. Zhou Mi recorded descriptions about firecrackers on the Eve of Spring Festival (1163~1189) at the Royal Palace in Hang Zhou in the third volume of his book *Wu Ling Jiu*

---

❶   孟元老. 东京梦华录：除夕[M]. 北京：中华书局，1982.

❷   施宿. 嘉泰会稽志：节序[Z]. 清嘉庆戊辰年采鞠轩重刻本. 1808.

❸   周密. 武林旧事：元夕[M]. 杭州：西湖书社，1981.

Shi. He wrote:

> There were many firecrackers, some made in the shape of fruits
> or human figures, and between them there were fuses so arranged that
> when you lit one it set off hundreds of others connected with it. ❶

Wu Zimu (1231 ~ 1309), a native of Qiantang in the Southern Song
Dynasty, wrote about similar firecrackers event in Hangzhou in his book named
Mengliang Lu. He wrote:

> On the Eve of Spring Festival, there were stalls selling fire-
> crackers and fireworks on frameworks, and that kind of thing ...
> Inside the palace the firecrackers made a large and glorious noise,
> which could be heard in the streets outside. ❷

The above-mentioned fireworks on frameworks refer to a combination of
fireworks and firecrackers. The fireworks and firecrackers are linked by a long
fuse. Once the fuse ignited, firecrackers would be set off to make a loud noise
and then followed by paper ghost images, paper flower pieces and colored
flames and smoke. In *Jinpingmei Cihua* (《金瓶梅词话》, *The Plum in the
Golden Vase*), a Ming Dynasty novel, the author described in the 12th Chapter
that the protagonist Xi Mengqin displayed fireworks in front of his house to
show off the wealth.

Zhou Mi in the 11th volume of his book *Qi Dong Yeyu* (《齐东野语》, *Rustic
Talks in Eastern Qi*, 1290) mentioned a kind of fireworks called "Ground Rat".
The incident was reported as follows:

> In the first year of Emperor Li Zong's reign, he prepared a feast
> in honor of his mother, the empress dowager, at Qing Yan Dian
> Palace Hall on the Lantern Festival night (the 15th day of the first
> Lunar month). A display of fireworks was given in the courtyard
> then. The "Ground Rat" type of fireworks darted directly towards
> the steps of the throne of the Empress Dowager which scared her so
> much that she stood up in anger, gathered her skirts around her,
> stopped the dinner and left the scene at once. ❸

---

❶ 周密.武林旧事:岁除[M].杭州:西湖书社,1981.

❷ 吴自牧.梦粱录:卷六[M].北京:中国商业出版社,1982.

❸ 周密.齐东野语:御宴烟火[M].北京:中华书局,1983.

In 1419, Persians once saw the "Ground Rat"❶ spinning on the ground at the Lantern Festival in Beijing. This is a kind of spinning fireworks which works with the principle that the air produced by the burning of the gunpowder in the paper tube created a retro-action to propel. If three paper tubes filled with gunpowder are linked, the retro-action power will make the fireworks spin around the axis (Figure 115).

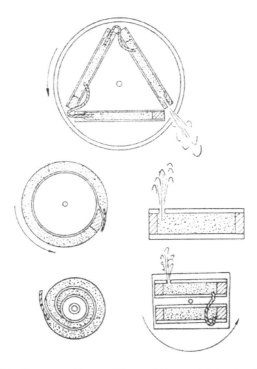

Figure 115   Structure Chart of Ground Rat and Spinning Fireworks

Zhou Mi also talked about his experience of watching performances such as fireworks, firing wheels, comets, etc.,❷ while visiting the West Lake in Hangzhou (1174~1189). Comets are also called "flying fire", which is also a retroactive device. The "flying fire" is made of projectile in the paper tube with some coloring agents added (such as iron powder). When the fuse is ignited, the flame and air will be burst off from the rear of the tube, which propels the tube into the sky by the reaction force. The "flying fire" gives out colorful and

❶   YULE H. Cathay and the Way Thither: Being a Collection of Mediaeval Notices of China[M]. London: Hakluyt Society, 1913.
❷   周密.武林旧事:西湖游幸[M].杭州:西湖书社,1981.

shining lights in the sky, which earns itself the name as "comet". Besides the function as a kind of entertainment and amusement, the "flying fire" can be used as military signal as well. If more explosives are added to the projectile, the "flying fire" will give out a big bang while projected into the sky. They are thus named as "Flying Crackers" or "Double Bang Firecrackers". The "Ascending Wheel" works in the same way as the "Ground Rat", but it flies into the sky instead of revolving on the ground. The "Ascending Wheel" is later called the "Chinese Wheel" after being introduced to the Arabian regions.

The techniques of fireworks which originated in the Northern Song Dynasty possess great theoretical and pragmatic values. Only after the mastery of safe production of the high-saltpetre solid gunpowder, can the full use of it be made and the application scale be expanded. Once the solid gunpowder is ignited in a tube with the forepart open and the rear closed, the air produces powerful spurting energy, which projects flames or other objects to a distant place. But if the gunpowder is ignited in a tube with the forepart closed and the rear open, the air produces reaction thrust, which propels the stuff inside high into the sky. Unconsciously the Newton's second law, that is, the law of action and reaction is applied in the making of reaction levitation devices. Based on these techniques, the above-mentioned entertainment devices are posing potential military values and significance. In the Song Dynasty, the craftsmen in the arsenal and the officers and men who used early firearms developed all kinds of new firearms based on these technological achievements and improved their operational capabilities.

3.2.2　Bombard and Fire Lance in the Southern Song Dynasty

3.2.2.1　Bottle-Shaped Metal Bombard in 1128

Since the late Northern Song Dynasty powdered gunpowder of high nitrate had been used for entertainment as fireworks and firecrackers, people generally observed that gas and flame would be projected out of the paper tube swiftly with great impact, which can be used for fire attack in the military war. When the last Emperor Huizong (1101~1125) of Northern Song Dynasty enjoyed the fireworks in the palace, Jurchen leader Wanyan Aguda became Emperor and established the Jin regime (1115~1234). After eliminating the Liao Dynasty in 1125, they quickened their steps to end the Song Dynasty. In 1125, Jin troops began to attack the Song territory and then Emperor Huizong passed the crown to Prince Qingzong. From 1126, Jin's troops besieged Kaifeng, the capital city of the Northern Song Dynasty, and conquered the city in November, which was called the "Jingkang Coup" in history, representing the fall of the Northern Song Dynasty. In 1127, the younger brother of Emperor Qingzong, Gaozong ascended to the throne in Nanjing (now Shangqiu in Henan Province) and then after moving to the new capital Ling'an (Hangzhou), he started the Southern Song Dynasty (1127~1279). Under the threat of Jin's armies, the patriots and craftsmen in the Song Dynasty developed new types of anti-Jin firearms, which served for military purposes during the Jianyan Years (1127 ~ 1129) and Shaoxing Years (1131~1162).

First of all, it is necessary to point out that the new types of firearms in the early Southern Song Dynasty have common characteristics, that is, the effective working parts of the firearms are long tube shaped. These tubes made of paper, bamboo, wood and metal cylinder barrel, with propellant inside, are designed according to the principle of fireworks. What kind of materials are to be used depends on the willingness of the manufacturers and other specific circumstances. In the Song Dynasty, which was highly advanced in science and technology, the 4 kinds of materials for the tubes could all be easily made. But people tended to think that the metal tubes appeared later than other ones, which is actually not true. The earliest physical image of the existing tubular firearms shows that they are made of metal. This should not be surprising, because the casting technology of the Song Dynasty is highly developed, and there is no difficulty in casting any firearms of any shape.

In the June of 1985, Robin Yates, who is the collaborator of Joseph Needham, once visited the Song Dynasty Buddhist cave-temple at Dazu in Sichuan Province and saw some stone carving images of bottle shaped or pear shaped hand-gun. But the visit was too brief to identify the exact year of the carving. In order to find out more information, in November 21, 1986, Joseph Needham, Lu Gwei-Gjen and the author made another expedition to the cave-temple at Dazu. In the No. 149 cave, we saw some saint images on both sides of Avalokiteshvara and among them there were some figures holding bottle-shaped bombard and hand-gun. Figure 116 is the sketch of the stone carvings, since the photo taken is not clear enough due to the darkness of the cave. The exact year for the carvings is 1128 in the early Southern Song Dynasty (1128). After confirmation, we think that this is the earliest physical presentation of bombard. ❶

---

❶ 鲁桂珍,李约瑟,潘吉星. 铳炮的最早实物形象[M]//潘吉星. 李约瑟集. 天津:天津人民出版社,1998.

Figure 116　Sketch of Figures Holding Bottle-shaped Bombard and Grenade

Through careful observation of the image of the stone carvings and the hand-held things, we exclude the possibility that they are the God of thunder or the God of wind. What they held in hand is clearly firearms and it has been ignited and was shooting out flames. These bottle-shaped objects can not be "*pipa*" (one kind of traditional Chinese musical instrument) either, because a projectile is being shot out of the muzzle of the bottle together with the flames. Theoretically, there should be a wooden handle in the tail in order to hold it. But due to limited space, the carvers could not express it fully, so the firearms were all created as hand-held stuff. The firearms, which is able to shoot off projectiles, can not be made of wood, owing to the overheat of the explosion chamber. It can only be made of metal, and its inner bore should be a long cylinder. But there is a typical bulbous thickening of the metal wall around the explosion chamber to make the whole shape look like a bottle. The shape of the firearms is quite archaic, so is the western early mortar. With the progress of the firearms technology, the bulge part of the metal wall around the explosion chamber is gradually reduced, and the shape of the whole firearms is in a long tube shape. This development vein can be clearly seen from the unearthed cultural relics at home and abroad.

The following inscriptions can be found on the stone wall outside of the No. 149 cave temple:

> The Mayor Ren Zongyi and his wife Ms. Du initiated the building of the Buddhist shrine for the citizens to worship, wishing for peace and prosperity for the country. In April, the second year of Jianyan's reign (1128).

That is to say, the then Mayor Ren Zongyi and his wife Du Huixiu initiated the idea to build the Buddhist cave temple in April 1128. The inscriptions have been recorded in books such as *Dazu Shike Yanjiu* (《大足石刻研究》, *Studies of the Carvings at the Dazu Cave Temple*). ❶

Dazu hand-gun is apparently made of iron or bronze. At that time, metal hand-guns have existed in Sichuan, still part of the Southern Song Empire. Besides the figure holding firearms in hand, there are other 37 figures in the cave holding non-gunpowder weapons such as sword or spear in hand. Why does the scene on the wall look like this? Most probably the donor hoped for the cease of the war or believed that Avalokiteshvara and the armed saints could promise the peace in that place when Jin's troops began to invade cities around Sichuan.

The author has restored the exterior appearance and the interior structure of the metal bombard at Dazu cave temple (Figure 117). As the figure shows, the effective working part for the hand-gun is the long metal gunpowder tube. There is a bulbous thickened metal wall around the explosion chamber, above which there is a vent for fuses to be inserted. The high-saltpeter solid gunpowder is poured into the muzzle by little bronze spoons. There is a socket cast on behind the tube, into which a wooden handle is fitted for handling or grasping. The firearms are usually used by two soldiers for both attacking and defending. In the Song Dynasty, this was also called *huopao*, which is often confused with the bomb projected from a trebuchet. In the study of the history

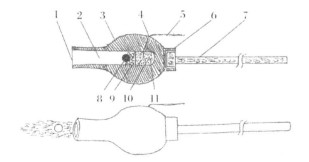

Figure 117   The Exterior Appearance and the Interior Structure
of the Metal Bombard, restored by Pan Jixing

1. Muzzle   2. Bore   3. Metal Wall   4. Vent   5. Fuse   6. Socket   7. Wooden Handle
8. Projectile   9. Gunpowder   10. Combustion Chamber   11. Paper Pad

---

❶   刘长久,胡文和,李永翘. 大足石刻研究[M].成都:四川社会科学院出版社,1985.

of early firearms in China, this kind of mixing of terms often occurs and needs to be clarified everywhere.

In the west, the firearms were called "bombard" sometimes but "hand-gun" at others. Nowadays we use the term *huochong* (火铳, hand-gun) to refer to this kind of metal bottle-shaped firearms. The thundering bomb described in *Wu Bei Zhi* maintained the feature of bottle shape, so they were not the firearms in the Ming Dynasty but before it, bearing influences from the Song Dynasty. On the other hand, they are much larger in size and heavier in weight compared with the Dazu hand-gun.

3.2.2.2　Fire Lance Made in 1132

In the early Southern Song Dynasty, a militarist named Cheng Gui (1072~1141) invented another new firearms in 1132, which was called "fire lance". From the perspective of function, it is a kind of spurting fire tube made of natural bamboo tube. Apparently they were transferred directly from the fireworks, but of a larger scale. Paper tubes for fireworks were replaced by stronger bamboo tubes for the fire lance. It was developed and created by Cheng Gui as an effective anti-enemy firearms in the emergency period of warfare.

According to *Cheng Gui Zhuang* (《陈规传》, *Biography of Cheng Gui*), he was from Anqiu (now Zhucheng in Shandong Province) with a style name as Yuan Ze. In 1132, Li Heng led troops to attack De'an (Now Anlu in Hubei Province). They built a flying bridge to besiege the city. Then "Cheng Gui, with 60 men carrying fire lances, made a sally from the West Gate, and using a fire-fox to assist them, burnt the flying bridges, so that in a short time all were completely destroyed. So Li Heng pulled up his stockades and ran away."❶

According to his *Shou Cheng Lu* (《守城录》, *Accounts of Defending the City*) "twenty fire lances were made from gunpowder and long bamboo tubes, together with several striking lances and spears with hooks on. Two soldiers handled one weapon, which would be used whenever opponents' flying bridges approached the city (Figure 118)".

However, usage of fire lances in warfare was not mentioned until 1132 when Song garrisons used them during the Siege of De'an, in modern-day Anlu, Hubei Province, when fire lance troops led the vanguard in a sortie against the Jin Dynasty (1115~1234). So actually "flame thrower" may serve as a better

---

❶　脱脱. 宋史:陈规传[M]. 上海:上海古籍出版社,1986.

name for "fire lance" by Cheng Gui. The main function of it is to spray flames when it is close to the enemy or enemy facilities, causing burning but not sending out projectiles. Just as the name suggests, the device is a spear-like weapon used to gain a critical shock advantage right at the start of a melee. Since the bamboo tube is quite long and adequate gunpowder can be added, accordingly the barrel is lengthened to protect the safety of soldiers. The total length of the barrel is about 6 *chi* (around 200 cm). If too long, it will be hard to control and easy to break.

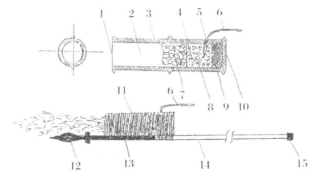

Figure 118   Illustration of Fire Lance in 1132, restored by Pan Jixing
1. Muzzle   2. Bore   3. Metal Wall   4. Combustion Chamber   5. Vent   6. Fuse
7. Paper Pad   8. Gunpowder   9. Mud Layer   10. Bottom Wall   11. Cord
12. Spear Head   13. Iron Sheet   14. Stick   15. Socket

Long and wide phyllostachys pubescens are divided into several segments, membranes of which are made through with the last left covered by clay. A small hole should be drilled at the end of the bamboo tube with clay on it to stuff the fuse. Since bamboo tubes tend to crack when heated, they are strengthened by iron wire and twine wrapped around it. Propellants are stuffed into the bamboo tube, which are tied to the front of a spear. Therefore, fire lances may look like rockets but actually they are not. The bocca of a fire lance is placed at the top of the weapon, but the opposite is true for rockets. According to above-mentioned descriptions, people outlined the shape and structure of fire lance, which provided the possibility for the restoration of firearms.

The image of the fire lance made by Cheng Gui is not untraceable, since the shape of the "bamboo fire lance" described in *Wu Bei Zhi*[1] provided

---

[1]　茅元仪. 武备志: 竹火枪[M]. 沈阳: 辽沈书社, 1989.

references for the understanding of the structure of the fire lance in spite of the fact that the length of the bamboo tube was shorter than what was depicted in the book. The "long bamboo tube" for the making of fire lance mentioned in *Song Shi* did not refer to the whole piece of bamboo tube, since it might be too long to let the flames come out. It actually referred to the thick phyllostachys pubescens tube divided into several segments. According to *Wu Bei Zhi*, in 1215 *Li Quan* in the Song Dynasty once utilized firearms called "Pear Flower Spear" which was similar to Cheng Gui's type. At that time, *Li Quan* always used it in his heroic exploits in Shandong Province, where there was a saying that "Twenty pear flower spears in hand, no rivals in the world,"[1] proving the power of the firearms. But the inner walls of the gunpowder tube might be burnt easily after the fire-spurting, so they had to be updated promptly.

For the convenience of operation, the fire lance was wielded by one soldier and ignited by the other. Once ignited, the flame thrower would ideally eject a stream of flames in the direction of the spearhead and lasted for 3 minutes. [2] If it is used widely, it must be a great threat to the enemy. According to *Shou Cheng Lu* (《守城录》, *Accounts of Defending the City*), Cheng Gui made 20 fire lances but according to *Song Shi* the number is 30. In 1139, fire lance distinguished itself in the Shunchang Victory (now Fuyang in Anhui Province) against the Jin. Metal fire lance made in Sichuan in 1128 is different from the bamboo fire lance made in Hubei in 1132, each having its own merits. The former is solid, reusable and capable of erupting projectile to a distance, but high at cost, while the latter is easy to make and low at cost, but non-reusable. In spite of the differences, both of them are the predecessors of all tube type firearms in the world. According to their structure and the principle of launching, all kinds of firearms with different functions have come into being and were spread abroad, triggering the second technological revolution in the development history of firearms since the 12th Century, the early Southern Song Dynasty. The revolution is characterized by the use of high-saltpeter solid gunpowder as propellants and explosives, which are stuffed into a tube type chamber made of different materials to be ignited with fuse. A strong flame or a lethal projectile or arrow from a combustor can lead to unprecedented

---

[1] 茅元仪.武备志:梨花枪[M].沈阳:辽沈书社,1989.

[2] NEEDHAM J. Science in Traditional China[M]. Cambridge, Massachusetts: Harvard University Press,1981.

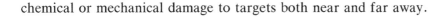

chemical or mechanical damage to targets both near and far away.

3.3   Flame-Spurting Lance, Rocket and Bomb of Hard Shell in the Southern Song Dynasty

3.3.1   Flame-Spurting Lance Made in 1259

After the appearance of the bottle shaped metal bombard and the fire lance, a new type of firearms emerged. It can be literally translated as "flame-spurting lance" in English. Its structure is something between a metal bombard and a fire lance. The following account is included in the 197th Volume of *Song Shi*: "In 1259, Shou Chun Governmental Arsenal (now Shou County in Anhui Province) manufactured flame-spurting lance, which was made of a bamboo tube. It can shot forth a kind of projectile (stone or metal ball). After the flame is over, the projectile will be shot off with loud noise like bombard, which can even be heard 150 paces away."[❶]

According to *Song Shi*, flame-spurting lance was first created by Shouchun Governmental Arsenal in 1259 under the reign of Emperor Lizhong. Although these firearms have already become the object of discussion, its shape and structure are rarely studied and analyzed.

Actually it is not easy to analyze the structure of flame-spurting lance. "Flame spurting" means spurting the fire swiftly, but it is different from fire lance and fire thrower in that it contains bullets in the tube. "*Zike*" in Chinese literally means the eggs in the nest, but here figuratively refers to the stone or metal ball, which can be spurted out from the lance. According to the account in *Song Shi* — "After the flame is over, the projectile will be shot off with loud noise like bombard", this description actually needs some modifications for technological mistakes. Because after all the smoke and flame ends, there will be no projection power and the projectiles can not be shot off automatically. Thus the only possible explanation is the flame-spurting lance shots forth the projectile and flame together. Dr. Joseph Needham noticed it and named "*zike*" in the flame-spurting lance as "co-viative projectile". [❷]

---

❶ 脱脱.宋史:兵志十一(器甲之制)[M].上海:上海古籍出版社,1986.

❷ NEEDHAM J. Guns of Kaifeng-fu: China's development of man's first chemical explosive[J]. The Times Literary Supplement, 1980(4007):39.

In the early Southern Song (1128), metal bombard of bottle shape appeared and then in 1132 "huoqiang" (fire lance or flame thrower) was used for city defence. So the appearance of flame-spurting lance more than 100 years later should not be regarded as the earliest tube-type firearms. People should attach more importance to the image of metal bombard of bottle shape in the Dazu Cave Temple. Actually flame-spurting lance is a modified version or hybrid of flame-spurting lance and fire lance. The range of projectile is definitely greater than the distance of the spurting flame, so the main function is to utilize the power of the projectiles. In other words, it is a tube-type shooting firearms, which can be used to repel the attack or charge into the enemy ranks. This kind of firearms can be held by hand for its lightness. The first three chapters in the book *Wu Bei Zhi* described firearms called "Invincible Bamboo Fire Lance"❶ which may be an advanced version of flame-spurting lance. The account in the book is of great help for the study of the early forms of flame-spurting lance, so in this sense *Wu Bei Zhi* can be viewed as a military encyclopedia not only reflecting the firearms technology in the Ming Dynasty but also in the Song, Jin and Yuan Dynasties.

Chinese early firearms are characterized by the simultaneity of the shooting of projectiles and flames, which is an initial stage in the process of its production and development. Because the initial diameter of the projectile is less than the bore diameter, the front bore in front of the combustion chamber is not fully stuffed. The flame is then ejected from the muzzle through the gap, resulting in the waste of some powder, which in turn affects the initial velocity and the impact force of the projectile. This problem has been gradually solved. Early Western firearms didn't experience the same trouble, because the firearms in the Yuan Dynasty, which was introduced to the West, had already solved the "running fire" problem. So the Westerners were able to go through the primary stage and enter the development stage directly.

Flame-spurting lance was used widely in the folk for its low cost and easiness of making. The British Museum in London has such firearms collection from China (No. 9572). It is made of bamboo tube wrapped with vines. Although it had been used but was still in good shape. Clayton Bredt made the photo public❷ and through analysis, we get the following information about the

---

❶ 茅元仪. 武备志:无敌竹将军砲[M]. 沈阳:辽沈书社,1989.

❷ BREDT C. Fighting for fun and in earnest[J]. Hemisphere, 1977,21(10):9.

flame-spurting lance made by Shouchun Arsenal in 1259.

Long and wide phyllostachys pubescens are cut to a tube about 132~165 cm long, the scapes of which are removed with only the last one left for inserting the wooden handle. An iron sheet circle is put into the tube and then sealed with two inches of clay. A vent is placed above the clay for the fuses. Twines wrap tightly around the bamboo. When in use, gunpowder is put into the tube first and then the stone projectiles (Figure 119). When the enemy is about 200 paces away, the Flame-spurting lance will be ignited. The projectile is shot forth with flames and the noise could be heard more than 150 paces away (around 225 m). A soldier can carry about 10 flame-spurting lances at a time. But after a few times of firing, the firearms need to be renewed and replaced. In war-fares, it is better to use the flame-spurting lance together with the fire arrows. The following figure shows its outline and inner structure.

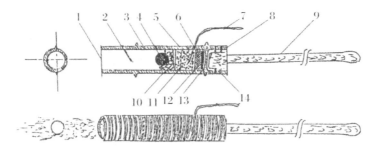

Figure 119   Illustration of Flame-spurting Lance in 1259, restored by Pan Jixing
1. Muzzle   2. Bore   3. Metal Wall   4. Projectile   5. Combustion Chamber   6. Vent   7. Fuse   8. Pit
9. Wooden Handle   10. Paper Pad   11. Gunpowder   12. Mud Layer   13. Iron Pad   14. Iron Nail

3.3.2   Rocket and Rocket-Propelled Bomb

Rockets evolved directly from firecrackers, since rockets and firecrackers both belonged to the retroactive device. The initial launch of the rocket occurred on the battlefield between the Southern Song and Jin in 1161. Rockets fall into two categories: explosive and conflagrant. In the November of 1161, Jin Emperor Wanyan Liang with an army of 400,000 soldiers marched south to the northern coast of Yangtze River to assail the Song and occupy Jian Kang (now Nanjing City). Yu Yunwen, an anti-Jin general fought off the Jin's soldiers by using *haiqiu*, a swift battleship and also thunderclap bombs, which set a good example of defeating enemy troops with a force inferior in number.

As for the original descriptions about the "thunderclap bomb", Yang Wanli (1127~1206), a famous Southern Song poet and a friend of Yu Yunwen, the

Song Admiral, described in his book *Haiqiu Fu Houxu* (《海鳅赋后序》, *Rhasodic Ode to the Sea-Ele Paddle-Wheel Warships*) how the firearms helped Yu to win a great victory over Jurchen Jin forces which were trying to cross the Yangtze River and invade the south. Yang Wanli wrote as follows:

> In the Xinsi Year of the Shaoxing reign-period, the rebels of Liang (Wanyan) came to the north bank of the River and captured the people's boats, intending to cross over. But our fleet hid behind Qi Bao Shan Island ... All of a sudden, the thunderclap bomb were let off. They came dropping down from the air and exploded with a noise like thunder upon meeting the water, the sulphur bursting into flames. The carton case broke and scattered the lime to form a smoky fog which blinded the men and horses so that they could see nothing. Our warships then went forward to attack theirs, and their men and horses were all drowned, so that they were utterly defeated. ❶

The account above has been translated into many occidental and oriental languages, which attracted the attention of many researchers and experts at home and abroad. But the nature of the weapon is not easily determined, for people confuse it with "cannon". In addition, Yang Wanli missed the components of charcoal and saltpetre, which led to an unreasonable explanation of the explosion. Therefore, the problem about what kind of firearms it was could not be solved for a long time. After the research, the author proposed that the thunderclap bomb used by the naval army of the Southern Song Dynasty was actually a kind of bomb launched like a rocket, which can be called "rocket bomb". ❷

There were 2 key points in Yang Wanli's accounts: (1) This firearms were made of paper tube with gunpowder inside. Once ignited, it exploded with a noise like thunder and then the carton case broke and scattered the lime to form a smoky fog. But this explosive effect could not be produced only by lime and sulfur. Saltpeter and charcoal had to be added to achieve the effect. (2) The firearms came dropping down from the air after it had been projected into the sky. The true reason for the explosion was due to the gunpowder inside, not because of "the sulphur upon the water". Propellants were placed at the bottom

---

❶ 杨万里. 诚斋集:海鳅赋后序[M]. 上海:商务印书馆,1936.

❷ Pan Jixing. On the Origin of Rockets[J]. The Exploration of Nature, 1984(3): 173-184.

of the tube while explosives and lime were placed at the top of it. Theoretically, the ratio of saltpetre, sulfur and charcoal in the solid gunpowder should be 70 : 10 : 20.

While Jin Emperor Wangyan Liang was leading the land troops to assault the Southern Song, he designated his Prime Minister Su Baoheng leading the navy to occupy Ling'an (now Hangzhou), the Capital of Song. But they were defeated by the Song Admiral Li Bao (1120~1165) at the Tang Island, Jiaoxi County (now Jiao Zhou Wan), Shandong Province. Admiral Li Bao served under Yue Fei and he fought against Jin's troops with 120 warships and 3000 archers in 1161. What's more, on the way to the North, he rescued the anti-Jin volunteer army led by Wei Sheng trapped in Haizhou (now Lianyungang in Jiangsu Province). He even got in touch with the volunteer armies in Shandong Province. Li Bao's army launched a sudden counter attack at the Tang Island. "He ordered the soldiers to project fire arrows from all directions. The warship would be set on fire with flames and smoke rising immediately upon being shot by the fire arrows. The weapon may get hundreds of ships on fire soon."[1] The majority of Jin's warships were burnt down and the Jin Admiral Su Baoheng escaped on his own making his navy utterly defeated. And the internal contradictions in Jin's troops were worsened resulting in the death of Wanyan Kang killed by his subordinate.

Victories of the two sea battles at Caishi and Tang Island were both won by the Song troops thanks to the use of firearms, which set a good example of defeating enemy troops with a force inferior in number.

After the battle, Bao Hui, the Chief Procurator of Fujian Province, said: "The victory on the sea won by Li Bao was attributed to the use of firearms. Li ordered the soldiers to project fire arrows from all directions. The warship would be set on fire with flames and smoke rising immediately upon being shot by the fire arrows. The weapon may get hundreds of ships on fire soon."[2] Li Bao used fire arrows at the same year as Yu Yunwen used the rocket bombs. Both weapons were supposed to be retroactive devices according to the technical background at that time. They diverged in that the fire arrows used by Li Bao meant to spread fire instead of exploding (Figure 120). It needs to be pointed out that the gunpowder arrows were not completely eliminated after the

---

[1] 脱脱. 宋史：李宝传[M]. 上海：上海古籍出版社，1986.
[2] 包恢. 敝帚稿略：海防寇申省状（宋人集）[Z]. 李氏宜秋馆，1921.

appearance of rocket bombs. They were actually used in conjunction with each other for some time. At the close of the Southern Song Dynasty, the gunpowder arrows then began to withdraw from the historical stage.

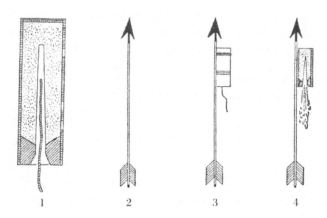

Figure 120   Illustration of Rocket in the Southern Song Dynasty, painted by Pan Jixing

Jin's troops were defeated by the rocket bombs attacks in many wars during the Southern Song, so they were determined to master the firearms. In April 1232, Mongol Emperor Wo Kuotai sent Su Butai to lead troops to attack Kaifeng, the capital of Jin. They set around 100 catapults to project boulders into the city, day and night so that the city was almost buried under the boulders. Then they even projected gunpowder packages into the city. General Chizhan Hexi (about 1180~1232) of the Jin Dynasty also fought back with the same firearms.

> The defenders had at their disposal the thunder-crash bombs which consisted of gunpowder put into an iron container. And they also had the flying-fire spears which were filled with gunpowder and when ignited, the flames shot forwards for a distance of more than 10 paces, so that no one dared to come near. These thunder-crash bombs and flying-fire spears (lance) were the only two weapons that the Mongol soldiers were really scared of. ❶

So Su Butai had to retreat due to the threat of the firearms. In May 1233, Jin General Pucha Guannu (around 1188 ~ 1233) of Guide Prefecture (now Shangqiu in Henan Province) led a troop of 450 soldiers to attack Mongol General Temoutai's camp at Wangjiasi.

---

❶　脱脱.金史:蒲察官奴传[M].上海:上海古籍出版社,1986.

Timid at first, the Jin troop began the attack at about two or three o'clock in the morning. General Pucha divided the troops into several groups of about 50 or 70 soldiers each. They held the flying-fire spears to attack the camp from the front and rear. The Mongol troop could not resist the attack and lost the battle completely. More than 3500 Mongol soldiers were drowned and the enemy camp was all burned down. The flying-fire spears was once used by the Jin troops in the defence of Bianjing (now Kaifeng in He'nan Province). Now they are reused in the battlefield. ❶

The flying fire lance (Figure 121) was introduced from the Southern Song, the recorded documents of which have been translated into many eastern and western languages. But scholars and researchers diverged on the idea that what kind of firearms it really was. There are mainly two opinions: some people think that they were rockets❷ while others think they were just fire lance or flame thrower❸ once used by Cheng Gui in 1132.

Figure 121   Flying Fire Lance Used by Jin's Troops in 1232

The common point of the two opinions is that the gunpowder tube is tied to a spear with a spear head. According to the accounts of *Jin Shi* (《金史》, *History of the Jin Dynasty*), "flames would be spurted out forward to a distance of about 3 meters", which refers to the longest distance the flames from the gunpowder tube might spurt to. But the point is from which direction the flames spurt out, from the direction of the spearhead or the opposite, which will finally determine the nature of the firearms, a fire lance or a rocket?

Fire lance is a melee weapon while confronted with the enemies. Since it is usually held by hand, the effective range is very short (about 3 meters). In the above mentioned battle, General Chizhan Hexi (about 1180~1232) of Jin

---

❶   脱脱. 金史: 蒲察官奴传[M]. 上海: 上海古籍出版社, 1986.

❷   LALANE L. Recherches sur le Feu Gregeois et sur l'Introduction de Poudre a Canon en Europe[M]. Paris: J. Corread, 1845.

❸   冯家昇. 火药的发明和西传[M]. 2版. 上海: 上海人民出版社, 1978.

defended Kaifeng Prefecture by using flying fire lances to fight back the Mongol soldiers. At that time, the city wall of Kaifeng was over 5 feet high and there was also a moat outside the city. If the flying fire lance was just the common fire lance or flame thrower, how can the soldiers standing at the city wall spurt out the fire by holding it in hand and scare off the Mongol soldiers outside of the city? In fact, the range of fire lance was only about 3 meters, how could it hurt the enemies before it burned out? Therefore, the only reasonable explanation is that Jin soldiers were using the flying fire lance, which actually should be called the rocket.

According to *Jin Shi*, the gunpowder tube was about 2 chi (about 70 cm) and more explosives could be stuffed in. The range was over 500 paces (about $450 \sim 500$ meters maximum) which was feasible for a rocket. The rockets launched from the city walls damaged the Mongolia soldiers in a devastating way. Jin learned the application of rockets from the Jin-Song wars and then triumphed over the overwhelming Mongol soldiers in two important battles. Suppose the firearms that Jin soldiers used to defeat the foes were fire lance, the gunpowder tube had to be more than 16 meters long to spurt the flames out to the soldiers outside of the city wall, which was actually unfeasible.

Therefore we agree to the judgment made by some French or German firearms historians that the flying fire lance is actually rocket. (Figure 124) The fire lance looked like a rocket, but different from it. The bocca of a fire lance is upward while rocket works in the opposite way. "Flames would be spurted out forward to a distance of about 3 meters" explains the function of a fire lance. But a flying fire lance can dart out to a much farther distance, which distinguished itself from the common fire lance. There is a kind of reaction fireworks called "the flying fire" in the Song Dynasty. The flying fire lance is launched out also by the reaction, so they are similar both in their operating principle and their names.

3.3.3 Grenades and Bomb of Hard Shell

During the $10^{th} \sim 11^{th}$ centuries in the Northern Song Dynasty, people made the ball-like gunpowder packages from solid gunpowder, linens, paper and ropes, which were called "the fire ball". The gunpowder packages would be thrown out by trebuchets to explode. This is the early form of bombard. The gunpowder packages were supposed to be lit by hot iron cones and projected by catapults to attack. But the weakness was that they had to be operated by many people, which restricted the speed and the flexibility of the weapon.

Sometimes things like barbs, broken porcelain or lime were added into the gunpowder package to increase extra damaging effect. Generally speaking, these were two main explosive firearms in the Song Dynasty.

Since the 12th century and the appearance of high-saltpetre solid gunpowder and fuse in the early Southern Song Dynasty, the soft gunpowder package has been transformed to the hard package one.

The solid powder had been placed in a hard shell container, which would be ignited by the fuse to produce greater chemical and mechanical damages after the explosion. This new kind of bombard gradually improved and its explosive power was strengthened as a result.

Lu You (1125~1210), a poet of the Southern Song Dynasty, described the Zhong Xiang and Yang Mo Up-rise in his book *Laoxuean Biji* (《老学庵笔记》, *Notes from the Hall of Learned Old Age*):

> The official armies made lime bombs with poison, lime and barbs stuffed in fragile pots. After the bombs targeted and struck the rebels' ships in the battle, the scattered lime formed a smoky fog which blinded the men and horses so that they could see nothing. The rebels wanted to make similar weapons, but they didn't know how to, so that they were utterly defeated. ❶

From 1133 to 1134, the lime bomb made by pots served multiple functions as explosive bombs and smoke bombs, etc. In 1134 when Jin troops assaulted Haozhou (now Fengyang in Anhui Province), the Song troops still used the lime bombs or the jars of lime projected from the trebuchet to fight back.

At the same time, the technology of grenades and bombs also developed in the Jin Dynasty. Undoubtedly it was introduced from the Song. Yuan Haowen, a scholar of the Jin Dynasty, recorded an interesting story in the second volume of his book *Xu Yijian Zhi* (《续夷坚志》, *More Strange Stories from Yijian*, 1225). It reads as follows:

> In the Beizheng Village of Yangqu (now Dingxiang in Shanxi Province), a man called Tieli made a living by hunting foxes. One day in 1189, he used a pigeon as the bait to set a trap near the tomb north of the river. He climbed up into a tree and carried a pot of

---

❶   陆游. 老学庵笔记:卷一[M].北京:中华书局,1979.

gunpowder at his waist. When the foxes came under the tree, he lit the fuse and threw the pot down. It burst with loud noise and scared all the foxes, who finally rushed into the trap he had prepared for them.

In 1189, Tieli in Shanxi made the grenade by pot (Figure 122), which was then ignited by fuse. That is to say, the explosive used was actually solid gunpowder. This kind of grenade can be definitely used in war-fares. The Jin Dynasty ceramic barbed grenade, which was excavated near Dalian City, is now possessed by Lvshun Museum of Liaoning Province. Barbs are added to produce more devastating bits and fragments after explosion.

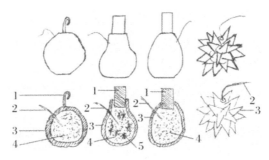

Figure 122   Illustration of Grenade Produced in Jin Dynasty, painted by Pan Jixing
1. Handle 2. Fuse 3. Shell 4. Gunpowder 5. Lime

In order to promote the power of bombs and grenades, iron was used to make the shell. Big bombard with iron shell named as iron bombard, which was to be projected out by trebuchet. However, it was not cannon in its real sense. From *Xinsi Qi Ji Lu* (《辛巳泣蕲录》, *The Sorrowful Records of the Siege of Jizhou in the Xinsi Year*, 1230) written by Zhao Yujun, we learned that in Xinsi Years (1221)Jin soldiers attacked Jizhou (now Jichun in Hubei Province) by using iron bombards, which were projected off by trebuchets. The Song soldiers hit by the bombards were all killed. Zhao Yu depicted it as: "Shaped like a gourd with a small muzzle, the bombard made of cast iron about 2 inches thick sounded like thunder when exploding."

In 1231, Mongol soldiers crossed the Yellow River to assault Jin's Hezhong Prefecture (now Yongji, Shanxi Province). Jin General Wanyan Eke was escaping after the defeat and was pursued by Mongol soldiers. The pursuit was so hot that Jin soldiers were not able to retreat. So just at that critical moment, Jin soldiers launched the iron bombard named "Sky-Shaking Thunder" to break through. *Zhentianlei* (震天雷, thundering bomb), as a kind of iron bombard, was used in the marine battle for the first time in history while in Europe it was first used in 1469 in the French battlefield.

According to *Chi Zhan He Xi Zhuan* (《赤盏合喜传》, *Biography of Chi Zhan He Xi*), the 113ᵗʰ volume of *Jin Shi*, in the March of 1232, Mongol General Subutai led the troop to siege Kaifeng. More than one hundred trebuchets were erected to project huge stones into the city day and night. Explosives bombs were also used to attack. "Among the weapons of the defenders there was *zhentianlei*. It consisted of gunpowder put into an iron container. Upon ignition, there was a great explosion and the noise sounded like thunder, audible for over a hundred *li* (about 500 m) and the vegetation was scorched by the heat over an area of more than half *mu* (about 330 m²). When lit, even iron armour could be pierced through."

Liu Qi, a scholar of the Jin realm, he wrote the following in the 11 volume of his book *Gui Qian Zhi* (《归潜志》, *On Returning to a Life of Obscurity*, 1235):

> *In the march of 1232, the army of the Northerners (the Mongols) then sieged the city of Kaifeng. The attack became more and fiercer, so that the trebuchet stones flew through the air like rain. But the defenders of the city had heaven-shaking thunder-crash bomb to fight back. As a result, hundreds of soldiers in the Northern troops were burned to death by the fires caused by the explosion.*

There are still physical remains of the heaven-shaking thunder-crash bomb used by the Jin troops to assault the Mongols in 1232 in the Ming Dynasty. He Mengchun (1474~1536) of the Ming Dynasty once wrote the following in his book *Yu Dong Xulu Zhaichao* (《馀冬绪录摘钞》, *Selected Excerpts from the 'Later Winter Talks'*):

> *On the city wall of Xian, I saw some old iron bombs, which was known as the heaven-shaking thunder-crash bomb in previous times. In shape they looked like two bowls that could be joined together to form a ball, at the top of which there was a hole only the size of a finger. These weapons are not used by the army now, but I think that they were one kind of the firearms used by the Jin people when defending Kaifeng.*

Judging from the descriptions above, we know that *zhentianlei* is actually a kind of bomb, which is made of cast iron in the shape looking like two bowls that can be joined together to form a ball. Several kilograms of explosives are filled in. Upon ignition, there comes a great noise, and the explosion will reach an area of about half *mu* (484 m²). It is said that the sound can be heard in dozens of miles. This kind of powerful bomb can be projected by a trebuchet or be thrown by hand. Because of the powerful force of the explosion, the shattered iron bits can even pierce through iron armor, which scares the Mongols most.

According to the thirty-nine volume of *Jinding Jiankang Zhi* (《景定建康志》, *Local Chronicles of the Capital Jiankang*, 1261) by Zhou Yinghe (1213~1280), Ma Guangzu served as the official in charge of military affairs in Jiankang (now Nanjing City) and he ordered to expand the production of military weapons. In 1259, still in Jiankang (now Nanjing City) "more than 300,000 items of weapons were produced and renovated" among which there were 63,754 items of firearms. To be specific, 4 iron bomb-shells, 5 kilograms each; 7 iron bomb-shells, 3.5 kilograms each; 100 iron bomb-shells, 3 kilograms each; 13,104 iron bomb-shells, 2.5 kilograms each; 22,044 iron bomb-shells, 1.5 kilograms each; 1,000 fire arrows; 333 fire lances; 333 fire barbs; 100 thunder-clap bomb-shells.

Iron-shell bombs can be classified into 5 different scales according to their weight: 10 *jin* (5 kg), 7 *jin* (3.5 kg), 6 *jin* (3 kg), 5 *jin* (2.5 kg) and 3 *jin* (1.5 kg), among which the last two are the most common. The structure of those is similar to *zhentianlei* in the Jin Dynasty. In other words, *zhentianlei* is the largest of the same kind. The so-called *huojili* in the Southern Song Dynasty is actually earthen grenade. The "thunder-clap bomb" refers to a barbed ball stuffed with gunpowder, earth-made or iron-made. Since it is in fact a projectile from a trebuchet, *huopao* is not a proper name for it.

3.4 Development of Gunpowder and Firearms During the Yuan and Ming Dynasties

3.4.1 Hand-gun in the Yuan Dynasty

Mongols rose and developed in the distant northern deserts, but it soon spread into the Central Plains. In the war with Jin and the Southern Song, it also mastered the techniques of gunpowder and firearms. After the elimination of the Jin Dynasty in 1234, all the military factories, craftsmen and firearms under the rule of Jin were now all owned by Mongol. Later it occupied some important cities in the Southern Song Dynasty, so the gunpowder and firearms technology in Mongol combined the comprehensive results of both the Southern Song and Jin Dynasties.

Zhou Mi (1232~1298) at the end of the late Song and the early Yuan Dynasties noted an arsenal explosion incident in *Kuixin Zashi* (《癸辛杂识·前集》, *Miscellanea Information From Kuixin Street*, 1290). It reads as follows:

> *The then Prime Minister Zhao Nanzhong reared four tigers at his private house in Liyang, and kept them within a palisade near the gunpowder arsenal. One day, while the gunpowder was being dried, a fire broke out and a terrible explosion followed. The noise sounded like thunder-clap; the ground was shattered; many houses collapsed and the four tigers were killed instantly.*

The above accounts in the book tell us that in the early Yuan Dynasty, Ali who is the Governor of Jiangsu and Zhejiang Province, once ruled in Yangzhou (1276~1279) and he converted the private house of the late Southern Song Dynasty Prime Minister Zhao Kui (the style name Nan Zhong, 1186~1266) in Liyang (now northern Jiangsu Province) to an arsenal to produce gunpowder and firearms. Four tigers were reared around to guard the arsenal. The great explosion at Yangzhou Arsenal in 1280 was still more terrible. At first all the workers were Han people in the south, who knew very well about the firearms techniques, but after a short time all the jobs were given to the Mongols. Unfortunately, these men knew little about the handling of gunpowder. One day while sulphur was being ground fine, it suddenly burst into flame. Then the stored fire lances caught fire, and flashed hither and thither like frightened snakes. At first the workers thought it was funny, laughing and joking, but very soon the fire got into the bomb store and then came the noise like a volcanic eruption and the howling of a storm at sea. The whole city was terrified, thinking that an army was approaching, and panic soon spread among the people, who did not know what had happened and what to do next. Even at a distance of a hundred *li* (about 50,000 meters) tiles shook and houses trembled. Alarms of fire were given, but the troops were held strictly to discipline. The disturbance lasted a whole day and night. After order had been restored and inspection was made, and it was found that a hundred of the guards had been blown to pieces; beams and pillars of the house had been cleft asunder or carried away by the force of the explosion to a distance of over ten *li* (about 5,000 meters). The smooth ground was scooped into craters and trenches more than ten feet deep. Above two hundred families living in the neighborhood were victims of this unexpected disaster. This is the most serious gunpowder explosion accident ever recorded in the historical documents in the Yuan Dynasty.

The above two arsenal explosion accidents are avoidable. After the Mongols learned the painful lessons, they realized that in the production of gunpowder, the rules of safety operation must be strictly observed. Many of the explosions in history were caused by sparks, and the sparks were usually caused by workers who violate the rules of operation. There is a lack of fireproof and explosion-proof facilities inside the workshop, and the workers are not skilled in techniques, which all account for the accident. In ancient safety operations, the tools for production of gunpowder are restricted to wood, copper or stone, and iron is strictly prohibited. Since no fire was allowed to be used indoors, much work was done in the daytime.

There are two methods of operation: "raw mix" and "ripe mix". The former method will grind and sieve the charcoal, sulphur and saltpetre respectively, and then mix them according to the proportion. Easy to operate, this method is usually used in ancient times. But small saltpetre particles are hard to be mixed with charcoal and sulphur. So for the second method, sulphur and saltpetre or charcoal and sulphur would be mixed first and then the mixture would be ground and sieved to be blended with the third ingredient. Although the method is complicated, the three ingredients can be evenly and finely mixed. Therefore what methods are used generally depends on the specific circumstances. 8%～10% of water or alcohol will be added to the mixture of charcoal, sulphur and saltpetre to make it paste-like. Then the solid gunpowder will be ground, dried, crushed and sieved.

After the gunpowder is made, it will be put into the large porcelain tank and sealed for storage. In general, gunpowder will not be immediately loaded into firearms, so as to avoid the occurrence of deliquescence. The distance between the gunpowder storehouse and the pharmacy should not be too close, and there is a firewall apart in the middle. The arsenal should be located in the suburbs of the city but not close to the residential area. The craftsman must fulfill their duties with concentration to make the powder clean, fine, even and solid.

All the arsenals in the Yuan Dynasty adopted the above security measures, and the quality of the gunpowder was better than that in the Jin and Song Dynasties. During 1161 and 1232 of the Song and Jin Dynasties, the propellants contained about 70% of the saltpetre, 7%~8% of the sulfur and 20% of the charcoal, while the powder of the Yuan Dynasty contained about 72% of saltpetre, 7% of sulfur and about 21% of charcoal. According to some Chinese records and documents, an Arabian named Hassan al Rammh (1265~1298) from Mongol Yili Khanate territory wrote about "Chinese fire arrows" with the recipe of 73% saltpetre, 8% sulfur and 19% charcoal; "flying fire" 71% saltpetre, 7% sulfur and 22% charcoal respectively. The powder formula is very similar to the Yuan Dynasty one. In the Yuan Dynasty this kind of gunpowder is used to make a varieties of new firearms invented since the Southern Song Dynasty, which are modified and improved according to the features of battles and warfares. Till the thirteenth century Westward Marching Movement, the firearms technology has been spread to Arabia and Europe.

As mentioned earlier, the earliest real image of bottle shaped metal gun was found in Dazu Cave Temple in 1128, the early Southern Song Dynasty. This kind of weapon with gunpowder placed in metal tubes can project projectiles to kill. It can be held by hand or set up on a scaffold to fire and it is both durable and powerful, possessing a major strategic significance of military tactics. This technique is assumed to develop in the Southern Song Dynasty, but unfortunately little is known about the development history in this period. The reason may attribute to the mixed and confusing use of different firearms names.

Name of the weapon in the Dazu cave temple was not indicated, but it may be called *huopao* (hand-gun) at that time. The ancients are fond of the word *pao*. The stone projectiles from the trebuchet are called *pao* or the stone bomb; the soft-shell explosive packages from the trebuchet are called *huopao* (incendiary bombs); the iron-shell bombs from the trebuchet are called *tiehuopao* (explosive bombs of cast iron); even grenades are named as *Pao* as well. What is more, the retroactive device like rocket was called *pili pao*. The Chinese character *pao* is abused seriously, so the mixed and confusing use of different firearms terms put scholars and researchers in trouble for it is hard for them to identify different types of weapons in historic documents and books.

There was an epic siege of Xiangyang (now Xiangfan in Hubei Province) by the Mongols. Two gallant Song officers Zhang Shun and Zhang Gui led a troop of 3,000 soldiers with 100 paddle-boat warships and organized a relief convoy which successfully re-provisioned the city. They loaded the paddle-boat warships with weapons like fire lances, bombards, charcoals, arrows, axes, etc. The troop fought violently against the Mongols on the river and finally succeeded in re-provisioning the city.

Here the above-mentioned fire lance was actually the flame-spurting lance created in 1132, while the bombard was similar to *chong* (gun) in 1128, which was sort of shooting weapon. The techniques of these weapons were mastered by Mongols soon after. The metal gun or bombard fit the Mongol troops very well for the weapons were easy to be taken along and even mounted on the horse for the cavalrymen to battle since they often journeyed a long way from the north to the Southern Song territory.

From *Li Ting Zhuan* (《李庭传》, *Biography of Li Ting*) we learn that in 1287 a rebellion was initiated by a Mongol prince Naiyan, which was put down by Li Ting, a commander in the Yuan service. During this campaign gunpowder weapons were much used. The text says: "Li Ting personally led a detachment of 10 brave soldiers holding *huopao* (portable bombard or hand-gun), and in a night attack penetrated the enemy's camp. Then they let off *pao* (bomb), which caused great damage and such confusion that the enemy soldiers attacked and killed each other, escaping in all directions."

There is a further statement concerning another warfare just in the following year, 1288. It reads as follows:

> Li Ting chose gun-soldiers, concealing those who bore huopao (hand-gun) on their backs; then at night he crossed the river, moved upstream, and fired off the weapons. This threw all the enemy's horses and men into great confusion ... at last he won a great victory.

Joseph Needham concluded like this: "Here we have such explicit statement that hand-guns or portable bombards must have been involved rather than grenades or small bombs." We totally agree with him to his point.

Li Ting with the style name Lao Shan, came from an old Jurchen family called Pu Cha. After settling down in Shandong Province at the late Jin Dynasty, he changed his family name to Li and joined the Mongol Army in 1269, when Liu Guojie, a native of Shandong Province, joined the army as well. Liu also came from a Jurchen family named Wu Gu Lun. Liu Guojie once led the troop to attack Xiangfan and got hurt in the thigh by the bombard in the battle. Then he banded the wound and fought again. The "bombard" used by the Song Army in the 1269 Xiangfan Battle was supposed to be the metal hand-gun. In recent years, the metal hand-guns used during the Yuan Dynasty have been unearthed.

In the July of 1970, three pieces of charger accessories, one small bronze tripod, one copper mirror and one bronze hand-gun (Figure 123) were excavated at Banlachengzi Village on the bank of the Ash River in Acheng district, in modern Heilongjiang Province. The bronze hand-gun without inscription weighed 3.55 kg and spanned 340 mm long with the inner diameter of muzzle being 26 mm. The socket of the hand-gun was 165 mm long and designed hollow for the installation of the handle. The gunpowder chamber was provided with a touch hole. Generally the bronze hand-gun was roughly made with an uneven surface and some irregularity in the cylindrical shape. Archaeologists concluded that the casting time of the hand-gun was not later than year 1290 after comparing it with similar objects used in 1332 and 1351. This is the earliest metal hand-gun existing in the world. Evidently the Acheng bronze hand-gun evolved from the Dazu metal hand-gun in 1128. For the latter there was a bulge over the explosion chamber, but with the progress of application, the bulge became smaller and the barrel lengthened. Therefore, the construction of the hand-gun changed from the bottle-shaped or pear-shaped to the long cylindrical one, which reflected the improvement of casting technology. The Acheng bronze hand-gun had a fine shape and a reasonable organization.

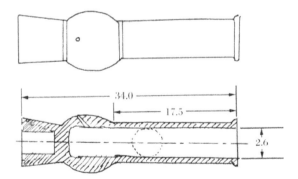

Figure 123   Illustration of Bronze Hand-gun in 1290, painted by Pan Jixing (unit: cm)

Another Yuan Dynasty bronze hand-gun (Figure 124) was excavated in a construction site in Xi'an City, Shanxi Province in the August of 1974. With a full length of 36.5 cm, it weighed 1.78 kg. The barrel length and the muzzle diameter was 14 cm and 23 mm respectively. There were six reinforcement hoops around the hand-gun from the muzzle to the socket. There was still some black gunpowder about 10 g~15 g left over in the barrel. Compared with the hand-guns in early Ming Dynasty (the later half of the 14th century), the surface of the Yuan hand-guns were rough and the thickness of the inner walls was uneven. In a word, they were not as delicate as the later similar objects. So the reporter concluded that the existing time for the unearthed hand-gun in Xi'an was somewhat between the Acheng hand-gun period and the early Ming period, which is the end of the 13th century and the early 14th century.

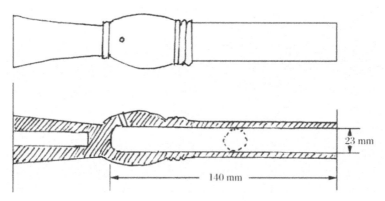

Figure 124　Bronze Hand-gun in Yuan Dynasty, painted by Pan Jixing

The judgement is acceptable but it is possible to make the time more specific. We proposed that it was made between 1297 and 1307, during the Chengzong reign-period. Although the left-over gunpowder was tested, the result (the ratio for saltpeter, sulfur, charcoal as 60% : 20% : 20%) was not convincing for it was not a reasonable recipe for propellant powder. The problem may be attributed to the process of the test or the loss of some saltpeter in the hand-gun.

National Historical Museum in Beijing possessed the Yuan Dynasty bronze hand-gun with the date inscription. Its full length, muzzle diameter, socket diameter, barrel diameter and weight is 35.3 cm, 10.5 cm, 7.7 cm, 8 cm and 6.94 kg respectively. On both sides of the hand-gun socket, there was a square hole, 2 cm in diameter. The inscription on the middle part of hand-gun reads as follows: "Made on the 14th day of the second month of the third year of the Zhishun reign-period. Border-pacifying anti-bandit forces, No. 300, Mashan." Zhishun was the title of Emperor Yuanzhong's reign, so the casting time for the hand-gun was on March 11th, 1332. It was used by the border-pacifying anti-bandit forces which were stationed at a place called Mashan. Due to its large bore diameter and trumpet-shaped muzzle, the hand-gun was capable of projecting big stones (Figure 125). Wang Rong noticed that two square holes' center and the gun barrel axis were in the same plane, so he thought the two holes were used for the iron bolts to go through. The iron bolts, which served as trunnions, could connect the gun body with the carpenter's bench.

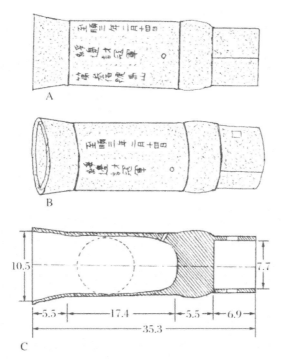

Figure 125   Illustration of Bronze Hand-gun in 1332, painted by Pan Jixing (unit: cm)

A. Front View   B. Side View   C. Cross-section

The restored device made according to the above-mentioned principle was widely accepted and quoted at home and abroad. Mr. Wang Rong also said that the socket diameter of the bronze hand-gun in 1332 was 77 mm. Although a wooden handle might be fitted into the socket, it was actually too big for the hands. Therefore, it should be mounted on a bench-like frame to shoot and wedges be inserted to adjust the angle. There is still room for discussion for his idea. For example, if the two holes were used as trunnions for the iron bolts to go through, the shape of the two holes should be round instead of square. What is more, the trunnions were supposed to be placed near the middle of the hand-gun body instead of the rear part. With the length of 35.3 cm and the weight of 6.94 kg, the hand-gun, just as Figure 126 shows, can be held in hand. People at that time would not choose to fix the portable and flexible hand-gun onto the clumsy carpenter's bench.

Figure 126   Operation Chart of Bronze Hand-gun in 1332, painted by Pan Jixing

After analyzing the weapon's inner structure, size and weight, we hold the opinion that the two square holes were used for installing wooden handle. Since the barrel diameter was large and one part of the handle was as wide as 7.7 cm, a thinner wooden stick was supposed to be connected to it for the easy handle of the firearms. This weapon was actually hand-gun instead of cannon. Its structure can be improved due to its short barrel and trumpet-shaped muzzle. Mr. Shi Zhilian from the National Historical Museum told me that this hand-gun was allocated and transferred from the Capital Historical and Construction Museum by Beijing Cultural Relics Management Committee in 1959 when the National Historical Museum was originally founded. The Capital Historical and Construction Museum got the hand-gun from a temple in Fangshan County.

The Capital Historical and Construction Museum is the predecessor of the Capital Museum. With a clear source, the hand-gun had been appraised by cultural relics experts before it was collected by the two museums. Some people thought it was only a fake one from Liulichang Antique Market even without actually seeing it. This is groundless. The Rotunda Museum of Artillery at Woolwich, London possessed an iron hand-gun of the 14th century excavated in China. With a full length of 47.5 cm and the muzzle diameter 10.5 cm, the hand-gun's shape and inner structure is similar to the year 1332 bronze hand-gun, which proves the existence of such weapons in the Yuan Dynasty.

The Yuan Dynasty bronze hand-gun excavated in Zhangjiakou, Hebei Province in 1961, which is now possessed by Hebei Provincial Museum, resembles the bronze hand-gun made in the 3rd year of the Zhi Shun reign-period. But the trumpet-shaped muzzle of the former was even wider, the bore size being 12 cm and the full length being 38.5 cm. So in the early Ming Dynasty this type of hand-gun was labeled as *dawankou hand-gun* (bowl-shaped hand-gun). For reasons already given, there is no doubt of their authenticity.

We have even restored this type of hand-gun. Basically they are handled by hands and, if possible, can be mounted on simple frames to shoot.

The cast bronze hand-gun in 1351 (Figure 127), the 11th year under the reign of Zhi Zheng, which is now possessed by the Military Museum of the Chinese People, was delicately cast. Its full length, muzzle diameter, barrel length and weight are 43.5 cm, 3 cm, 28.9 cm and 4750 g (4.75 kg) respectively. The 11.4 cm long socket has 2 holes about 3 mm~4 mm on both sides, which are made for fixing the wooden handle by nails.

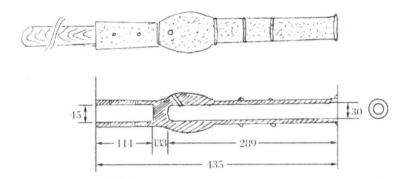

Figure 127   Illustration of Bronze Hand-gun in 1351, painted by Pan Jixing (unit: cm)

Mr. Wang Rong restored it quite well, but he did not mark the length of the wooden handle. We proposed it was about 130 cm and 173 cm in total for the hand-gun. There were altogether 6 reinforcement hoops around the hand-gun. Inscriptions were marked, which read in Chinese "*She Chuan Baizha // Sheng Dong Jiutian*" (Shooting Far and Thundering Loud) in the front part, "*Tian Shan // Sheng Fei*" in the middle, and "*Zhizheng // Xinmao*" in the rear. "*She Chuan Baizha // Sheng Dong Jiutian*" in Chinese means the firearms when ignited will make a loud noise like thunder and the projectiles from the hand-gun are powerful and penetrating.

3.4.2   Hand-gun, Cannon and Arrow Launched from Hand-gun in the Ming Dynasty

The firearms of the Ming Dynasty were developed on the basis of inheriting the Yuan Dynasty. The first Emperor of the Ming Dynasty, Zhu Yuanzhang was supposed to give special thanks to the use of firearms in the process of overthrowing the Mongol rule, quelling separatist forces in different regions and finally unifying the whole China. From *Bao Yue Lu* (《保越录》, *The Defending of Yue City*) written by Xu Mianzhi (1320~1380) we learned that cannon and handguns were used by both sides, firing not only stone balls but iron ones too.

From the third volume of *Ming Shi Jishi Benmo* (《明史纪事本末》, *The Accounts and Records of the Ming Dynasty*, 1658) by Gu Yingtai (1620～1690), we know that in the April of the 23rd year of Zhizheng-reign period, Cheng Youliang, one of the provincial rulers who resisted the rise of the Ming, built a huge warship about 3 meters high with 600 thousand soldiers in his claim to siege Nanchang. Deng Yu (1337～1377) designated by Zhu Yuanzhang was to guard Fuzhou Gate. He defended Nanchang successfully with handguns.

Although this kind of firearms is still called *pao* in Chinese, it in fact falls into the category of cannon. So if we read ancient Chinese books about firearms, we'd better understand the real meaning of the firearms terms according to their functions instead of their pronunciations.

3.4.3 Rocket, Multiple Rockets Launcher, Two-stage Rocket and Rocket Flight

Rockets originated in the Song and Jin Dynasties developed further in the Yuan and Ming Dynasties. After the emergence of rockets in 1161 in the early Southern Song period, the gunpowder arrows, which were gunpowder-tipped incendiary arrows fired from crossbows or longbows and were first used in the early Northern Song period (the 10th century), were not totally replaced. These two coexisted for a certain period. What's more, at that time the gunpowder arrow was fired from a crossbow or longbow once for a time. But after the Yuan Dynasty, the gunpowder arrows were eliminated by the rockets, and all the flights over the battlefield were retro-active rockets.

Due to the limit amount of gunpowder that each rocket arrow could carry, a swarm of arrows aimed at a certain target would definitely increase the effectiveness of the weapon. The Chinese rocket designers connected several single gunpowder arrows with a fuse and then put them into one launcher. After the fuse had been ignited, the rocket arrows were launched in a cluster. This practice, which surely multiplied the power and effectiveness of the weapon, was actually inspired by the firecracker working principle in the early Southern Song period. During the Second World War, the Nazi German troops were greatly terrified by the Katyusha rocket, which was in fact made according to the principle cluster rockets.

The ancient Chinese cluster rockets were made and used in the battlefield in the middle and late Southern Song Dynasty (1225 ~ 1235). Mongols, who fought with the Song at that time, soon masted the techniques of making this kind of weapon. The second Mongol Emperor Wokuotai launched the second westward march from 1235 to 1244 and the gunners in the army then carried the cluster rockets along (Figure 128). According to the accounts in Historia Polonica by the 15th century Polish historian Jan Dlugosz (1415~1480), in the 1241 Liegnitz battle in Poland Mongol armies once launched the cluster rockets against the Poland and the Deutsche allied forces. Walenly Sabisch (1577 ~ 1657) created the series of pictures reflecting the battle in 1640, according to the descriptions of Jan Dlugosz and the frescoes in the monastery near Liegnitz.

Figure 128   Illustration of the Ancient Chinese Cluster Rockets in 1241

According to the reports by Polish scholar Wladislaw Geisler, the rocket-launcher depicted in Sabisch pictures was decorated with dragon heads. Sabisch also drew deep conical recesses in the fuel which played the role of nozzles. The mention of the balance-weight and the cluster of rocket-arrows in the dragon-head rocket tube very significant. So Wladislaw Geisler drew the conclusion that this was the so-called cluster rocket in the modern sense.

The rocket-arrows made in the reign of Southern Song Lizhong-reign period were used by the Mongol troops in the European battlefield in 1241. But they were first used domestically in China till Ming and Qin Dynasties. The conical rocket-launcher described in *Huo Long Jing* (《火龙经》, *The Fire Dragon Manual*), a book recording the firearms technology development during the Yuan and Ming period (the 14th century), was actually kind of cluster rocket.

Later the rocket-arrows evolved into different varieties, many of which were mainly used in the battlefield. According to the 126 and 127 chapters of the book *Wu Bei Zhi* there were many types of rocket-arrows, such as "five-tigers-spring-from-a-cave" rocket-arrow (range: 500 *bu*, around 800 meters); seven-fold tube arrow (7 arrows); fire-crossbow meteoric rocket-arrow (10 arrows); rocket-arrow firing basket (17 ~ 20 arrows); long-serpent enemy-destroying rocket-arrow (30 arrows; range: 200 paces, around 320 meters); wasp's net rocket-arrow (32 arrows; range: 300 paces, around 480 meters); leopard-pack-unexpectedly-scattering rocket-arrow (40 arrows; range: 400 paces, around 640 meters); forty-nine simultaneously-fired rocket-arrow (49 arrows) and " pack-of-100-tigers-running-together rocket-arrow" (100 arrows; range: 300 paces, around 480 meters).

Figure 129   Illustration of Flying Crow with Magic Fire

Volume 131 of *Wu Bei Zhi* keeps the description about "*shenhuo feiya*", the "flying crow with magic fire" (Figure 129) and the passage runs as follows:

The body of the bird is made of fine bamboo laths (or reeds) forming an elongated basketwork. It has paper glued over to strengthen it and it is filled with explosive gunpowder. All is sealed up with more paper, with head and tail fixed on before and behind, and the two wings nailed firmly on both sides, so that it looks just like a flying crow. Under each wing there are two slanting rockets. the four-fold branching fuse about a foot long connected with the rockets, is put through a hole drilled on the back of the crow. When in use, the main fuse is lit first. The winged rocket-bomb will burn the enemy camp in the land battle and destroy the warships in the naval battle.

The earlier time "fire bird" in Europe which is the same as the "flying crow with magic fire", is supposed to be introduced from China. The biggest achievement of ancient Chinese rocket technology is the invention of the multiple stage rocket. When ancient people succeeded in projecting explosives into the air by rocket devices, they might naturally think about the idea that rocket could be launched off by another one in order to fly to a further distance, which forms the two-stage rocket as a sort of multiple stage rocket. *Wu Bei Zhi* recorded two kinds of two-stage rockets: fire dragon rocket flying on the water (Figure 130) and sand thrower flying in the air.

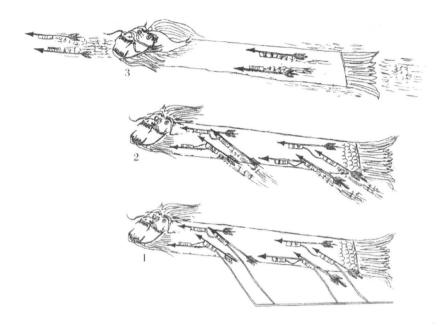

Figure 130   Illustration of Fire Dragon Rocket Flying on the Water, painted by Pan Jixing
1. Before Launch   2. During Flight   3. After Launch

Fire dragon rocket flying on the water was also mentioned in the military treatise *Huolong Jing* published prior to *Wu Bei Zhi*, so Needham thought the rocket might be made in the early 14th century, that is to say, the early and middle Ming Dynasty (14th~15th century) witnessed the emergence of two-stage rocket in China. Fire dragon rocket flying on the water were used exclusively in naval battles.

A tube of 5-feet-long bamboo is taken, the septa removed, and the nodes scraped smooth with an iron knife. A piece of wood is carved into the shape of a dragon's head and fitted on at the front while a wooden dragon tail is made for the rear end. The mouth of the dragon must be facing upwards and in the belly there are several rocket-arrows connected by one fuse. Beneath the dragon head on both sides there are two big rocket-tubes. Their fuses and orifices should face downwards. The fuses of rocket arrows within the belly lead out from the head of the dragon so as to be connected with the front ends of the outside rocket-tubes. The fuses of the four rockets are twisted into a single one. Upon lighting it will fly over the water as far as 2 or 3 *li* (about 1 km). Upon lighting the first-stage rocket, that is the big rocket tube, will fly over the water, looking like a flying dragon coming out of the water. When the gunpowder about 3.58 kg in weight is nearly all finished, the rocket-arrows within the belly of the dragon is ignited and then fly forth, destroying the enemy and his ships.

Sand thrower flying in the air (Figure 131) is made of two rocket tubes with explosives inside, but with two orifices opposite in direction. One rocket tube is connected with the explosive tube by a fuse, the mouth of which must be facing downwards, while the other rocket tube is also connected with the explosive tube by a fuse, the mouth of which must be facing upwards. Then the two rocket tubes and the explosive tube are attached to a bamboo stick with twine or swimming bladder. The whole device is 7 feet long, the explosive tube is 7 inches long (22 cm), and the diameter is 2.2 cm. Once ignited the rocket tube with the downward mouth will fly to the enemy field. Upon explosion the sand inside the explosive tube will scatter and blur the sight of the enemies. Afterwards the other rocket tube will be ignited and boosted back to the original launching place (Figure 132), which confuses the enemies a lot.

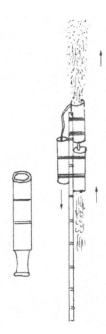

Figure 131　Sand Thrower Flying in the Air

Actually this weapon is a kind of two-stage reciprocating rocket, which is characterized by the fact that the second stage rocket moves in the opposite direction of the first

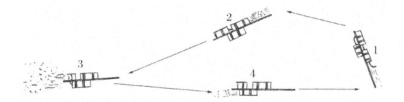

turn of the fifteenth century, if records are correct, was an official who experimented with rockets. Let's give Wan-Hoo credit for being the first to try to use rockets as a means of transportation. Wan-Hoo first secured two large kites, arranged them side by side and fixed a chair to a framework between them. On the frame he attached 47 of the largest rockets he could buy. When all was

Figure 132　Two-stage Reciprocating Rocket

1. One-stage Rocket Launch　　2. One-stage Flight

3. One-stage Rocket igniting Two-stage Rocket

4. Two-stage Rocket Returning to the Original Launch Position

stage one. So it is easy to see that technicians in the Ming Dynasty even possessed the thought and technology of rocket recovery after the launch.

According to a copy of one ancient Chinese book circulated to the United States, in the 14th ~ 15th Century a man named Wan Hoo in the early Ming Dynasty made a bold experiment. He first secured two large kites, arranged them side by side and fixed a chair to a framework between them. On the frame he attached 47 of the largest rockets he could buy. When all was arranged, Wan Hoo sat down on the chair and commanded his assistants to stand by with torches (Figure 133). At a signal these assistants ran up and applied their torches to 47 rockets. There was a roar and blast of flame. Wan Hoo, the experimenter, disappeared in a burst of flame and smoke. The first attempt at a rocket flight was not a success. American

arranged Wan-Hoo sat down on the chair and commanded his coolies to stand by with torches. At a signal these assistants ran up and applied their torches to all 47 rockets. There was a roar and blast of flame. Wan-Hoo, the experimenter, disappeared in a burst of flame and smoke. The first attempt at rocket flight was not a success

Figure 133   Wan Hoo and His Rocket Vehicle

rocket scientist Herbert S. Zim gave Wan Hoo the credit for being "the first to try to use rockets as a means of transportation." Russian rocket scientists V. I. Feodos'ev and G. B. Siniarev once wrote as follows: "Chinese people are not only the inventor of rocket, but the fantasies of the first attempt to use a solid fuel rocket to carry people into the air."

Although the manned rocket flight experiment could not be realized under the conditions at that time, it was of revolutionary significance. As a result, the American astronaut took a rocket vehicle to the moon and named a ring-shaped mountain on the back of the moon with the name of Wan Hoo.

The German scholar Wiley Lee (Willy Ley) commented on Wan Hoo as follows: "The knowledgeable and adventurous Chinese official invented and experimented a rocket airplane in around 1500, but sacrificed himself quite spectacularly". W. R. Maxwell, a British rocket historian, spelled the name as Wan Hoo instead of Wan Hoo, considering his experiment in 1500 was a significant event in the early rocket history.

When Zim made the historical documents public, he didn't mention the title of the Chinese book. Therefore it is almost impossible to identify Wan Hoo, which is only the transliteration of the name. Someone suggested that Wan Hoo stood for a military rank, which was pronounced similarly. But Wan Hoo, as a military rank in the Yuan Dynasty, was later eliminated and revised as commander-in-chief instead in the early Ming Dynasty. So it can't be an official title in the military. According to historical documents, American painter MacDonald (James MacDonald) also drew a schematic diagram attached to Kim's book (Figure 132). It should be said that there are some technical problems in the diagram, for example, the shape of the kites are not right and they are not supposed to be held in hand. So another diagram should be painted.

After the three Dynasties of Song, Jin and Yuan, China's rocket technology reached its peak in the Ming Dynasty. American chemistry historian Tenny L. Davis (1890~1949) wrote: "Before the 17th century Chinese people's creation talent has made the rocket apply to a variety of special purposes, and the rocket with high explosives represents useful strategical techniques even in modern times."

Needham called the two-stage rocket in the Ming Dynasty "proto Apollo rocket". He commented as follows: "Many further development of great interest followed during the Ming and Qin Dynasties. First of all, there were large two-stage rockets, surprising as it may seem, reminiscent of the Apollo spacecraft in which propulsion rockets were ignited in two successive stages, releasing automatically toward the end of the trajectory a swarm of rocket-propelled arrows to harass the enemy's troop concentrations. Rockets were provided with wings and endowed with a bird-like shape, in early attempts to give some aerodynamic stability to the rocket flight. There were also multiple rocket-arrow launchers in which one fuse would ignite as many as fifty projectiles; later these were mounted on wheelbarrows so that whole batteries could be trundled into action positions like the regular artillery of a still later date."

3.4.4　Bomb, Time Bomb, Mine and Submarine Mine

The iron bomb-shell used in the Song and Jin Dynasties developed further in the Yuan and Ming Dynasties to be applied in the battlefield. In 1273, Kublai Khan led a troop of 30 thousand people marched east to Japan and battled with 100 thousand Japanese soldiers in Hakata Bay (now SetoNaikai). A witness of the war called Takezaki Suenaga described what he had experienced to the author of the book *Menggu Xilai Huici* (《蒙古袭来绘词》, *Illustrated Narrative of the Mongol Invasion of Japan*, 1293). It reads as follows:

Mongol soldiers sent iron bomb-shell flying against us, which made our side dizzy and confused. Our soldiers were frightened out of their wits by the thundering explosions; their eyes were blinded, their ears deafened, so that they could hardly distinguish east from the west.

Observed from the illustrations, the iron bomb-shell looked like a ball made of two semicircular bowls. They were aimed for close combat. Both *Huolong Jing* and *Wu Bei Zhi* kept a record of different kinds of iron bomb-shell at that time, but termed it as "*pao*" inappropriately, such as "*zhuanfeng shenghuo liuxing pao* (magic-fire meteoric bomb that goes against the wind)". According to *Qing Shi Lu · Taizu Shi Lu* (《清实录·太祖实录》, *The Records of Emperor Taizu* in *the Records of the Qing Dynasty*, 1636), Nurhachi led troops to assault Ningyuan (now Xingcheng in Liaoning Province) and the city was held by its gallant commanding general Yuan Chonghuan. The siege was finally raised after a spirited defence by using iron bombs.

A land or sea mine is an explosive device concealed under the ground or the water and designed to destroy or disable enemy targets. But still they were named as *pao* in ancient China. Needham thinks that something more of the nature of a land-mine was available in the Southern Song Dynasty in the 13th century but specific descriptions of mines were not found until the Ming Dynasty. The word "mine" first appeared in *Huolong Jing*, for example, *wudi dilei pao* (无敌地雷炮, invincible ground-thunder mine), which was also mentioned in Volume 134 of the book *Wu Bei Zhi*. *Wudi dilei pao* made of cast iron is spherical in shape (Figure 134). Gunpowder is put into the sphere

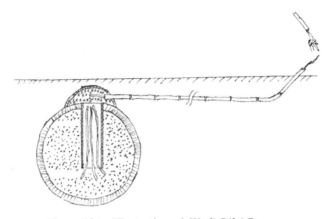

Figure 134   Illustration of *Wudi Dilei Pao*

through a hole on the top and a wooden tube with 3 fuses inside is inserted as well. The fuse is connected to a firing-device which ignites them when disturbed. The mines are buried in places where the enemy is expected to come. When the enemy is induced to enter the minefield, the firing device lights the fuse and then detonates the mine. Sometimes people dig pits and bury several dozen mines in the ground. All mines are connected by fuses through the gunpowder fire-ducts.

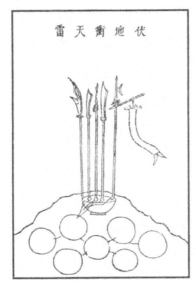

Figure 135　Illustration of *Fudi Chongtianlei Pao*

Volume 134 of *Wu Bei Zhi* describes another mine called *"fudi chongtianlei pao"* (underground sky-soaring thunder mine) (Figure 135). The mines are placed three feet underground on the way where the enemy is expected to come. The fuses are led to a point below a bowl containing a slow-burning incandescent material called *huozhong*. Lances or pikes with long handles are set up vertically above the bowl. Since the mine and the bowl are both buried underground, the enemy can only notice the lances and pikes above-ground. When the enemy comes to take out the weapons, the bowl is upset and the mine fuses ignited.

Zhang Yin (1637～1708) once cited from Qu Rushuo's *Bin Lüe Zuanwen* (《兵略纂闻》, *Classified Compendium of Things Seen and Heard on Military Matters*) in the 213 volume of his own book *Yuan Jian Lei Han* (《渊鉴类函》, *Royal Classified Records*) which was completed in 1677 and published in 1710. The script reads like this:

> *Zeng Xian served at the border and made land mines. The mines with explosives inside are buried several meters under the ground and covered with stone and sand. Slow-burning incandescent material is concealed below, which can endure for even months. Since they are triggered by the enemy passing by and upon ignition the stone fragments from the explosion burst and kill people, the enemy troops are scared and amazed.*

According to *Huolong Jing* and *Wu Bei Zhi*, *shuidi longwang pao* (水底龙王炮, submarine dragon king) is a kind of floating time sea mine. In Volume

133 of *Wu Bei Zhi*, some accounts go as follows:

> The sea mine is made of wrought iron and carried on a wooden
> board. The firing mechanism consists of a floated incense-stick (joss-
> stick) which lights the fuse when it burns down. (Maybe two hours or
> six hours. The time can be accurately determined by the length of the
> incense.) The mine is enclosed in niu pao (牛炮, an ox-bladder).
> But without air the glowing of the joss-stick would of course go out,
> so the container is connected with the mine by a long piece of goat's
> intestine through which passes the fuse. At the upper end of joss-stick
> in the container is kept floating by an arrangement of goose and wild-
> duck feathers, so that it moves up and down with the ripples of the
> water. In the naval battle, the mine is sent downstream towards the
> enemy's ships and when the joss-stick has burnt down to the fuse,
> there is a great explosion, which will shatter the bottom of the ship
> and the enemy will get drowned or caught.

Some information is omitted or missed in the above descriptions, and the
illustrations are not accurately drawn, so it is not easy to get the main points
immediately. The Qing Dynasty version of *Huolong Jing* is similar to this one,
that proves the fact that the book has not been carefully proofread since the
early Ming Dynasty. *Wu Bei Zhi* quotes from it, but still lacks in proofreading,
which makes it rather difficult to read and understand.

Li Chongzhou made some restoration research on the sea mine — *shuidi
longwang pao* in 1985 referring to books like *Tiangong Kai Wu* to make people
better understand the working mechanism of it. The submarine-dragon-king
mine made of iron and spherical in shape weighs 4~6 *jin* (about 2.4 kg to 3.6
kg) and is loaded with gunpowder (5.2 L to 10.4 L). One fuse tube is inserted
into the hole on the top of the sphere and the end of the fuse is connected with
the gunpowder while the front part with the joss-stick (Figure 136). The mine
must be contained in an ox-bladder to escape from being soaked underwater.
But without air the glowing of the joss-stick would of course go out, so the
container is connected with the mine by a long piece of goat's intestine through
which passes the fuse. The ox-bladder is fixed on the board with two stones
fastened underneath to keep the whole device submerged all the time. After
learning the parking place of the enemy's ships, they send the sea mine
downstream on a dark night. The burning speed of the joss-stick will determine
the explosion time and the distance of floating. At the same time, the weight of

the stones will decide the depth to which the total device goes. Through careful calculation, the sea mine explodes just at the time when it arrives at the bottom of the enemy ship. After the explosion, enemies are drowned or injured so that it provides a good opportunity for the coming attack. "That is where the subtlety lies." Obviously, it is an exquisite device full of creativity.

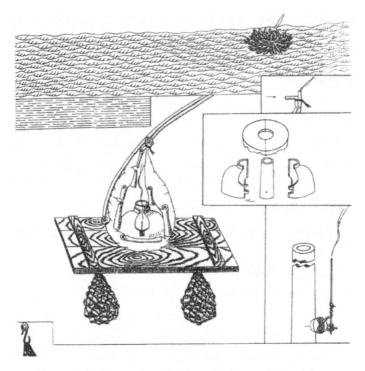

Figure 136　Illustration of Submarine-Dragon-King Mine

Qi Jiguang (1527~1587), the militarist of the Ming Dynasty made "self-trespassing steel wheel mine" in 1580, the eighth year of Wanli's reign. The device contains a wooden box with all the fuses put inside. At the bottom of the box, explosives and steel wheel are concealed with flint beside. When the enemy ventures on to ground containing the device, the steel wheel will rotate to light the fuse by the sparks from the flints igniting the tinder. Loud explosion confuses and frightens the enemy a lot.

This type of mine works according to the steel and flint firing mechanism. The firing device is touched off by the enemy, one of the "self-trespassing" type. Volume 134 of *Wu Bei Zhi* by Mao Yuanyi describes a kind of explosive mine (*zhapao*) with the same firing mechanism. It reads as follows:

The explosive mine is made of cast iron about the size of a rice bowl, *hollow inside with gunpowder rammed into it*. *A small bamboo tube is inserted for the fuse to pass through*. *Outside*, *a long fuse runs through a fire-duct* ("huocao"). *Choose a place where the enemy will have to pass through*, *dig pits and bury several dozen such mines under the ground*. *All the mines are connected by fuses through the gunpowder fire-ducts*, *and all originate from a steel wheel*. *This must be well concealed from the enemy*. *On triggering the firing device the mines will explode*, *sending pieces of iron flying in all directions and shooting up flames towards the sky*.

Xu Zonggan (1796~1866) of the Qing Dynasty, the writer of *Bing Jian* (《兵鉴》), once cited from the book *Huo Gong Da* (《火攻答》, *Answers to Questions on Fire-weapons and Firearms*) by Wang Minghe (1550~1610) in the Ming Dynasty (1550~1610). It reads:

No matter whether big or small, *the land mines apply bamboo tubes for fuses to run through*. *Some people use steel wheel to ignite and trigger the explosion*. *Once the mine is trespassed*, *the fuses will be lit immediately and series of mines will explode at the same time*.

It also refers to the steel wheel firing mechanism. Therefore, it is obvious to see that this way of firing the land mine has been applauded by most Chinese militarists since the 16th century and has been effectively used in real battles.

What is the steel wheel firing mechanism? *Wu Bei Zhi* gives an elaborate account of it with illustrations for explanation as well. There are 3 kinds of steel wheel firing devices introduced. Although they are used on different occasions but they are all created by the same mechanism. A running thread will control the steel wheel and make it rotate when pulled. As a result, the wheel produces sparks by rubbing against the flints, thus lighting the fuses and setting off the mines. The outside long fuse is either controlled by the gunner or triggered by the enemy unintentionally. In conclusion, it is ready to be set off at a touch.

Volume 134 of *Wu Bei Zhi* introduced the steel wheel firing device like this:

The container for the steel wheel is often made of elm or locust tree. *Two steel wheels are connected with the flints*. *Steel wheels are intersecting with a rotatable iron axle*. *A cord is wound around the axle*, *which is attached to a weight at one end*. *The device is kept in position by a pin*. *The whole device is buried underground covered with grass and tree branches. When the pin is removed by an enemy tripping over a thread attached to the pin or the gunner pulls the thread, the pin will release the weight to make the wheels rotate. Then the steel rubs the flint to produce sparks. Finally the fuse is ignited and explosions come. It works like magic and always performs well.*

Flint is a hard, sedimentary cryptocrystalline form of the mineral quartz, categorized as a variety of chert. It occurs chiefly as nodules and masses in sedimentary rocks, such as chalks and limestones. Inside the nodule, flint is usually dark gray, black, green, white or brown in colour, and often has a glassy or waxy appearance. When struck against steel, a flint edge produces sparks. The hard flint edge shaves off a particle of the steel that exposes iron, which reacts with oxygen from the atmosphere and can ignite the proper tinder. Prior to the wide availability of steel, rocks of pyrite would be used along with the flint, in a similar (but more time-consuming) way. People in ancient China use flint for practicing traditional fire-starting skills in exquisite and subtle devices. The steel wheel firing device is made of a rectangular wooden box, one flint fixed on each upper side of the box. Two steel wheels are connected with the flints above respectively (Figure 137). Steel wheels are intersecting with a rotatable iron axle. A cord is wound around the axle, which is attached to a weight at one end. The device is kept in position by a pin. When the pin is removed by an enemy tripping over a thread attached to the pin or the gunner

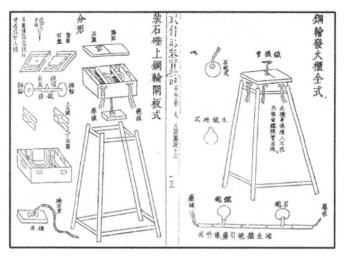

Figure 137　The Steel Wheel Firing Device

pulls the thread, the pin will release the weight to make the wheels rotate. As a result, the wheel produces sparks by rubbing against the flints, thus lighting the fuses and setting off the mines.

In 1973, Liu Xianzhou restored the three kinds of steel wheel firing devices introduced in the book *Wu Bei Zhi* one by one (Figure 138). Some people supposed that this kind of weight-drive flint-and-steel firing device was introduced to China by Jesuit missionaries. But Needham does not agree to the point. The device did not exist in Europe until 1573, so China was about a hundred years ahead of it. But Mao Yuanyi traced the origin of the device back to Zhuge Liang's time in the third century, which is actually groundless.

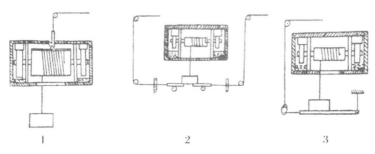

Figure 138　Illustration of Three Kinds of Steel Wheel Firing Devices

All in all, land mines, sea mines and time bombs developed after the Ming Dynasty apply altogether 3 kinds of firing devices: burning-down of incense (joss-stick), slow-burning incandescent material (*huozhong*) and flint-and-steel firing mechanism, the last of which is the most advanced and exquisite. All these devices are native to China.

3.5　Spread of China's Gunpowder Technique in Europe

3.5.1　Western Expedition of the Mongol Army Leading to Westward Spread of Gunpowder and Firearms

Fire attack was universally used by the ancient people in warfare. As aforementioned, "sea fire" ($\pi\tilde{\upsilon}\rho$ $\theta\alpha\lambda\acute{\alpha}\sigma\sigma\iota o\nu$, *pûr thalássion*), composed of sulfur, pine charcoal, asphalt and hemp dust, was used by the Greeks from the fifth century B.C. to the fourth century B.C. and was hence known as Greek fire by later generations. After the seventh century, the Byzantine Empire improved "sea fire" by adding petroleum, resins, and lime, by pushing the fluid fuel with a pump, and by spurting fire through a tube. The best firearms available to medieval Europeans, though, Greek fire could not rival gunpowder weapons, because Greek fire, even improved, was unable to produce destructive explosions and could only be used at close range. Moreover, Greek fire was generally used in naval battles and much less used on land. If used on land, it was thrown by hand or by some mechanical ejection devices. Naturally, troops armed with Greek fire and cold weapons could hardly resist the cavalry and infantry equipped with firearms.

While the Europeans had mastered Chinese papermaking technology since the 12th century, Chinese gunpowder technology was introduced to Europe after the 13th century either directly with the western expedition of the Mongol army or indirectly by the Arabs. Having acceded to the throne, Ögödei, Emperor Taizong of the Yuan Dynasty (元太宗窝阔台, 1229~1241), in order to fight against the repeated attacks from the Turkic tribes in Kiptchak in Northeast Europe, sent Batu (拔都, 1209~1209), Subotai (速不台, 1170~1248) and a troop of 150,000 soldiers to wage the second westward expedition (1236~1242)[1]. Since the expedition was joined by Prince Batu (拔都), Prince Haitu

(海都), Prince Möngke (蒙哥) and Prince Baida'er (拜答儿), the eldest sons of Juchi (术赤), Ögödei (窝阔台), Tölüi (拖雷), and Chahetai (察合台) respectively, also the grandsons of Genghis Khan, Emperor Taizu of the Yuan Dynasty (元太祖成吉思汗, 1162~1227), each leading a troop, the second western expedition is known as the western expedition of the eldest sons. All four troops consisted of regiments of gunners, armed with huochong (Figure 139), rockets, flame-

Figure 139 *Huo Chong* (火铳, hand-gun) used by the Mongol Cavalry, by Pan Jixing (2001)

spurting lances, bombs (actually *huopao*) and other firearms to assist the cavalry siege and large-scale field warfare. In all the battles, Batu was the commander in chief aided by General Subotai (速不台). In the autumn of 1236, the Mongol army breached the capital city of Bulgares (now south of Kazan,

---

❶ 张星烺. 中西交通史料汇编:第二册[M]. 北平:京城印书局,1030.

Russia) on the Volga River. In the spring of 1237, it advanced into Kiptchak. The Kiptchak army retreated to Russia and joined the Russian army to resist the Mongol army. Subotai led his army northward in pursuit of the Kiptchak army, smashed it and occupied the area along the Volga River. After the Mongol army destroyed Kiptchak, it entered the Russian heartland and in the spring of 1238, it captured the Russian city Moscow with the aid of cannons.

In 1239, Batu's army pushed deep into the south of Russia, seized Novgorod, and pressed on towards Kiev, then the capital of Russia. In 1240, Batu ordered *Huo Pao* be set up around the city, and started a fierce siege of the city. Then, Kiev was occupied and Russia was conquered. In the spring of 1241, the troop led by Baidaer and Wulianghetai (兀良合台) went southward from Russia and invaded Bular (now Poland). The troop led by Haitu and Subotai entered Magyars (now Hungary), where the Mongol army used rockets at the battle of Sejo and occupied Budapest at the end of 1241❶. In February, 1241, the Baidaer forces crossed the Vistula river in the icy waters, arrived in Krakow of Poland, moved forward to Silesia, crossed the Odra River, and then attacked Breslau (Wroslaw), the capital of the Silesia King Henry Ⅱ. Henry retreated to Legnica, assembling altogether 30,000 soldiers of the Polish army and the Germanic army to fight the Mongol army.

Legnica is in the southwest of today's Poland, near Germany. On April 9, 1241, the Mongolian army commanded by Baidaer was engaged in a fierce battle in the great Wahlstatt Plain near Legnica, with the allied forces of the Polish army and the Germanic army. The 15th-century Polish historian Jan Dlugosz (1415~1480) in *History of Poland* (*Historia Polonica*, 12 vols., 1470 ~1480) related that the Mongol army used rockets in the battle of Legnica in 1241 (Figure 128). According to the oil paintings and descriptive texts finished by the Polish architect Walenty Sebisch (1577~1657) in 1640 in Wroclaw (Breslau), inside the rocket's cartridge column is a deep conical recess acting as a nozzle, and the cartridge is attached to a rod. The polish rocket historian Wladyslaw Geisler said, at a monastery near the old battlefields of Legnica is kept an old painting of the precise Mongolian military rocket, the original of Sebisch's paintings of the Legnica Campaign.

The Mongolian military rocket described in the Sebisch paintings, is a

---

❶   VON BRAUN W. ORDWAY F I. History of Rocketry and Space Travel[M]. London-New York: Crowell Co., 1966.

cluster in a barrel, i. e., multiple rockets could be launched at a time. There is also a dragon head on the launch barrel, so the rocket is called "the Chinese dragon belching fire" in Poland. It bears some resemblance with *huolong jian* (火笼箭, rocket-arrow firing basket), *changshe pozhen Jian* (长蛇破阵箭, long-serpent enemy-destroying rocket-arrow) and *yiwofeng huojian* (一窝蜂火箭, wasp's net rocket-arrow) as recorded in Volumes 126 & 127 of *Wu Bei Zhi*, for it can launch 30 rockets per barrel and reach a range of 200 to 300 *bu* (330 metres~495 metres). Because of the superiority of the Mongol army in number and equipment, the allied forces of the Polish army and the Germanic army were soon defeated. The Mongols went on to cross the Danube River and to attack Bohemia (now Czech) and Austria. Only when the obituary of Ögödei Khan arrived did they return to the east. Batu was then honored to found Kiptchac Khanate (1243~1480) in the conquered area, with Sarai (now in Astrakhan, Russia) as its capital which was garrisoned by the Mongol army.

Although the western expedition of the Mongol army brought disasters to the invaded countries, objectively speaking, it also opened up the once blocked land passage between the east and the west, through which messengers, businessmen, academics, craftsmen of both China and Europe could pay mutual visits frequently in the 13th ~ 14th centuries, thus enabling the cultural and technological exchanges between China and Europe. In the first half of the 13th century, the Europeans experienced the power of gunpowder from China in person and on their own soil. Mongolian troops equipped with gunpowder weapons continued to be stationed in Kiptchac Khanate, a place near them. They must have made every effort to explore the technology of manufacturing firearms, which could lead to the direct introduction of Chinese gunpowder technology into Europe. Besides, having acceded to the throne, Möngke, Emperor Xianzong of the Yuan Dynasty (元宪宗蒙哥, 1208~1259) sent his brother Hülegü Khan (旭列兀) to lead the army into the third western expedition (1253~1259). In 1258, Hülegü's troop took Baghdad, capital of the Arab Empire with rockets and *huopao*, destroyed the Abbasid Dynasty, and built there Il Khanate (1260~1353) with Tabriz in Persia as its capital. In this way, the Chinese gunpowder technology was introduced into the Arabian world. Then Europe could also obtain technical information on gunpowder through the Arabs.

3.5.2 European Pioneers Introducing Gunpowder Knowledge

The first European to record gunpowder is Roger Bacon, a famous 13th-century British scientist who studied and wrote at Oxford University and joined the Franciscan Order. He arrived at the University of Paris around 1236 to teach Aristotelian philosophy, then stayed in Italy for a while, and returned from France to Oxford around 1251. He was knowledgeable, versatile in Greek, Latin, French, Italian, Hebrew and Arabic. His major works, *Greater Work* (*Opus Majus*), *Lesser Work* (*Opus Minus*), and *Third Work* (*Opus Tertium*), were written in Latin from 1266 to 1267[1]. *Opus Majus* is his masterpiece, including his main ideas. *Opus Minus* is the supplement of the former, and *Opus Tertium* is the supplement of *Opus Minus*. It is difficult to decide on the exact period when his works were written, because he often rewrote the same work many times and used the same material repeatedly in different works.

Roger Bacon mentioned gunpowder in *Opus Majus*, *Opus Tertium* and his books on alchemy. In the 1267 *Opus Majus*, he referred to Greek fire, pointing out that it contained petroleum to ignite things, the flame being water-resistant. He introduced the new incendiary matter (his original text is Latin; J. R. Partington provided the English translation):

> *Quædam vero auditum perturbant in tantum, guod si subito et de nocte et artificio sufficienti fierent, nec posset civitas nec exercitus sustinere. Nullus tonitrui fragor posset talibus comparari. Quædam tantum terrorem visui incutiunt, quod coruscationes nubium longe minus et sine comparatione perturbant; quibus operibus Gideon in castris Midianitarum consimilia æstimatur fuisse operatus.*

> *Certain inventions disturb the hearing to such a degree that if they are set off suddenly at night with sufficient skill, neither city nor army can endure them. No clap of thunder can compare with such noises. Some of these strike such terror to the sight that the coruscations of the clouds disturb it incomparably less. Gideon is thought to have employed inventions similar to these in the camp of the Midianites.* [2]

Gideon Jerubbaal, the leader of the Jewish in *The Book of Judges* of *The*

[1] SARTON G. Introduction to the History of Science[M]. Baltimore: Williams & Wilkins Co., 1931.

[2] PARTINGTON J R. A History of Greek Fire and Gunpowder[M]. Cambridge: Heffer & Sons, Ltd., 1960.

Bible, led 300 people to defeat the Midianites and freed his people from oppression. Bacon quoted the allusion in the subjunctive mood, meaning that if Gideon had mastered this weapon, it would have been applied to the fight against the Midianites. After the above passage, Bacon went on to give the following statement:

> *Et experimentum hujus rei capimus ex hoc ludicro puerili, quod fit in multis mundi partibus, scilicet ut instrumento facto ad quantitatem pollicis humani ex violentia illius salis, qui sal petrae vocatur, tam horribilis sonus nascitur in ruptura tam modicae rei, scilicet modici pergameni, quod fortis tonitrui sentiatur excedere rugitum, et coruscat ionem maximam sui luminis jubar excedit.*
>
> *We have an example of these things in that children's toy which is made in many parts of the world, viz. an instrument made as large as the human thumb. From the force of the salt called saltpetre so horrible a sound is produced by the bursting of so small a thing, viz. a small piece of parchment, that we perceive it exceeds the roar of strong thunder and the flash exceeds the greatest brilliancy of the lighting.*

The German gunpowder historian S. J. Von Romocki believed this new invention described by Bacon is an explosion (sprengträftigen) device❶. But in the above text, Bacon only mentioned saltpetre, but not sulphur or charcoal. We might as well agree with Romocki's judgment that the phenomenon described by Bacon is related to gunpowder explosion. But the thumb-sized children's toy was actually the firecrackers played by children in the 12th century in the Southern Song Dynasty, 100 years earlier than Bacon's time. Song Renwang (宋人王, 1091~1161), in his *Za Zuanxu* (《杂纂续》, *A Sequel to the Collection of Assorted Affairs*) listed many things that made people happy and fearful, "children playing firecrackers" included [seen Volume 76, *Shuo Fu* (《说郛》, *Collection of Unofficial Literature*)]. This inevitably makes people think that in the era of Bacon Chinese fireworks and firecrackers had been exported to some places in the west as entertainment. At least, the Chinese in Kiptchac Khanate and Il Khanate celebrated their festivals with fireworks, which were spread to Western Europe, and then known by Bacon. He again spoke of

---

❶ VON ROMOCKI S J. Geschichte der Explosive stoffe[M]. Berlin: Oppenheim, 1895.

firecrackers in *Opus Tertium* （1267）, in the same tone as in *Opus Majus*. But somewhere else he wrote differently (the following is J. R. Partington's translation):

> *By the flash and combustion of fires, and by the horror of sounds, wonders can be wrought, and at any distance that we wish — so that a man can hardly protect himself or endure it. There is a child's toy of sound and fire made in various parts of the world with powder of saltpetre, sulphur and charcoal of hardwood. This powder is enclosed in an instrument of parchment the size of a finger, and since this can make such a noise that it seriously distresses the ears of men, especially if one is taken unawares, and the terrible flash is also very alarming, if an instrument of large size were used, no one could stand the terror of the noise and flash. If the instrument were made of solid material the violence of the explosion would be much greater (quod si fieret instrumentum de solidis corporibus, tunc longe major fieret violentia).* ❶

Bacon's *De Arte Chymiae Scripta* in Latin, first published in 1603 in Frankfurt, Germany, includes *De naturis metallorum in ratione alkimica et artificiali transformatione (Breve breviarium* for short) written by himself, which mentions that saltpetre grows on some stone and blazes once meeting charcoal. It points out that when nitrite is purified, it is dissolved in water and filtered to form a long, white, bright, needle-like crystalloid. It also says that nitrite can ooze out of the earthenware, "because I saw it in the experiment"❷. The recrystallization method to purify saltpetre mentioned by Bacon corresponds with the account in the Chinese doctor Ma Zhi's （马志, about 935 ～1004） *Kaibao Bencao*: "sweep the raw nitrite, pour water over it, decoct the solution and then needle-like matter will be formed （扫取（硝石）以水淋汁后,乃煎炼而成,状如钗脚（针状））". But that nitrite grows on some kind of stone was claimed by the Arabs.

To sum up, Bacon did introduce the knowledge of nitrite, gunpowder and firecrackers, which he repeatedly wrote in several books. His description of firecrackers in *Opus Majus* shows that this kind of gunpowder products had

---

❶ PARTINGTON J R. A History of Greek Fire and Gunpowder[M]. Cambridge: Heffer & Sons, Ltd. , 1960.

❷ SARTON G. Introduction to the History of Science[M]. Baltimore: Williams & Wilkins Co. , 1931.

been seen somewhere before his time, because he stressed that this children toy "is made in many parts of the world" ("quod fit in multis mundi partibus"). The regions that had made gunpowder and gunpowder products in Bacon's words surely include China where gunpowder was made more than three hundred years earlier than his time. Gunpowder knowledge never started from Bacon, but he was observant and incisive enough to introduce it to the European readers at a much earlier time in history.

Where did Bacon's gunpowder knowledge come from? Given the historical background, it is obvious that China was the origin in a direct way or indirect way in Yuan Dynasty. He could have heard it from travellers during his stay in France and Italy, then the European continent's cultural hub. The two western expeditions of the Mongol army from 1236 to 1242 and from 1253 to 1258 both involved the use of firearms in Europe and its surrounding areas. The Europeans were startled by the deafening artillery and the fire-breathing weapons on the battlefields. The fear was still lingering in the era of Bacon. Afraid to be attacked again, many times, the Popes and the Kings sent ambassadors to the palace of the Mongolian khans. Apparently, the visiting of the missions was of political, religious and commercial purposes, but detecting the technical secrets of firearms could not be ruled out. In 1245, Pope Innocent Ⅳ (1243～1254) sent from France Jean Plano de Carpini (1182～1252), the former archbishop of the Germanic and Spanish parishes and Italian Francis monk, to Mongolian Yuan, accompanied by a Polish named Benedict and some Austrian businessmen. They travelled eastward, passed Kiptchac Khanate's capital Salai, arrived in Kharakorum in 1246, met Güyüg, Emperor Dingzong of the Yuan Dynasty (元定宗贵由), and returned to Lyon, France in 1247. Plano de Carpini wrote *A Brief History of Mongolia* to record what he saw and heard. ❶

From 1247 to 1248, the Pope sent three members of the Dominican Order, Ascelin, Simon de St. Quentin and Guiscard de Cremona to shuttle between China and Europe. In 1248 and 1249, King Louis Ⅸ of France (1214～1270) sent a French Dominican André Longjumeau and others to Kharakorum. But because of the death of Emperor Dingzong of the Yuan Dynasty, the delegation was unable to fulfill its mission. In 1252, King Louis Ⅸ again sent a French Franciscan missionary, Guillaume de Rubrouck (1215～1270) to Mongolia,

---

❶ ROCKHILL W W. The Journey of Rubruck to the Eastern Parts of the World[M]. London: Hakluyt Society, 1900.

accompanied by an Italian missionary named Bartolomeo de Cremona. They met Batu while passing Kiptchac Khanate, arrived in Kharakorum at the end of 1253, visited Möngke, Emperor Xianzong of the Yuan Dynasty (元宪宗蒙哥) in the beginning of 1254, stayed in Yuan Empire for a few months and returned to France in 1255. Rubrouck wrote his travel to the east into *Itinerarium Fratris Wilhelmi de Rubruk de Ordine Fratrum Minorum*, *Anno Gratiae* 1253 *as Partes Orientales* (hereinafter referred to as *Itinerarium ad Orientales*). He presented the book to Louis Ⅸ, for which he was allowed to settle in Paris by the King. There, Bacon got to know him❶. Bacon had read his *Itinerarium ad Orientales* and mentioned him in *Opus Majus*❷.

In *Itinerarium ad Orientales*, Rubrouck said that during his visit in Kharakorum, Mongolia, he was acquainted with some Teutons, Russians, Frenchmen, Englishmen and Hungarians who worked there, including a goldsmith, Guillaume Boucher, born in Paris and then making gold and silver ware for the khan, his wife born in Lorraine; a French woman, Paquette de Metz, married to a Russian architect and settled in Kharakorum; and an Englishman, Basil, who knew many languages. These skilled Europeans were discovered in the western expedition of the Mongol army and then brought back to Kharakorum. As can be seen in the above information, 30 years earlier than the time Bacon recorded gunpowder, several batches of European missionaries had been on exchange trips between China and Europe. They were likely to have brought back to the continent the gunpowder knowledge appealing most to the Europeans. Then, it could have attracted Bacon's attention. Although these travellers did not write reports on the subject, they could have had oral presentations. Bacon's stay in France and travel in Italy allowed him the chance to meet the missionaries who had just returned from China, Rubrouck included (he is seen in written documents). Since Rubrouck had contact with the technologists working in China who were from France, Germany, Russia and other countries, most probably he was one of those who had brought the gunpowder knowledge back to Europe. Besides, he had publicized in his travelogues the issue and the printing techniques of paper money in the Yuan Dynasty.

---

❶   CORDIER H. Histoire Generale de la Chine[M]. Paris: Geuthner, 1920.

❷   DAWSON C. The Mongol Mission: Narratives and Letters of the Franciscan Missionaries in Mongolia and China in the 13th and 14th Centuries[M]. London: Sheed & Ward, 1955.

Bacon's narrative of gunpowder and firecrackers (children's toy) makes people assume that he had such samples at hand. The firearms historians V. Foley & K. Perry[1], Frank H. Winter and J. Needham concluded that Bacon's missionary friends who returned from their 1245~1255 trips to China had given him a pack of small Chinese firecrackers as samples for his research and also provided him verbal instructions, which Bacon really followed to do his experiments. Bacon, on the other hand, acquired the knowledge of saltpetre and its purification from the early Arab records, especially the account of the China snow (thalj al-Sīni) in the book *Kitbāb al-Jami fi al-Adwiya al-Mufradi* written in Arabic by Ibn al-Baytar (1197~1248) in 1248. Bacon knew Arabic and the Arab culture, read many Arabic writings of science and technology, and was well-acquainted with the German translator Hermann Alemann who had translated many Arab works into Latin[2].

Another early European who introduced the Chinese gunpowder knowledge was Roger Bacon's contemporary, a German named Saint Albertus Magnus (1200~1280). Both Albertus and Bacon were known as the most learned men in medieval Europe. Albertus was born in an aristocratic family in Lauingen, Bavaria, pursued his education in Padova, Italy during his early years, joined the Dominican Order there in 1229, and later taught in the churches all over Germany. In 1245 ~ 1248, he studied in the University of Paris and was awarded Doctor of Divinity. From 1248 to 1254, he taught in Cologne, Germany. In 1254, he was made provincial of the Dominican Order. In 1256, he visited Paris again. From 1260 to 1262, he was the bishop of Regensburg, Germany[3]. With more than 300 works involving various disciplines, he was proclaimed "Doctor Universalis" or "Saint", which shows his great influence in Europe. Not knowing Arabic, he had read many Latin translations of Arabic works. Besides, he was devoted to collecting new knowledge of diverse origins.

Among Albertus's works, the one that interests us most is *De Mirabilibus Mundi* (*The Wonders of the World*), which writes about gunpowder and flying

---

[1] FOLEY V, PERRY K. In Defence of Liber Ignium: Arab Alchemy, Roger Bacon, and the Introduction of Gunpowder into the West[J]. Journal for the History of Arabic Science, 1979, 3(2): 207.

[2] SARTON G. Introduction to the History of Science[M]. Baltimore: Williams & Wilkins Co. , 1931.

[3] PARTINGTON J R. A History of Greek Fire and Gunpowder[M]. Cambridge: Heffer & Sons, Ltd. , 1960.

fire. It is hard to decide on exact time of writing for the book. The earliest extant manuscript that dates back to the 13th century is kept in Wolfenbuttel Library. The version in St. Mark of Venice dates back to the 14th century; those in Paris and Florence are from the 15th century. The Latin version of this book was first published in Venice in 1472. Partington translated some chapters into English, and the complete English version was published in Oxford in 1973. Now, we quote a relevant paragraph on gunpowder in the book (the Latin text) and present Partington's translation of it as follows:

> *Ignis volans. Accipe libran unam sulphuris, libras duas carbonum salis, libras sex salis petrosi, quae tria sabtilissime terantur in lapide marmorei, postes aliquid posterius ad libitum in tunica de papyro volanti, vel tonitruum faciente, ponatur. Tunica ad volandum debet esse longa, gracilis, pulvere illo optime plena, ad faciendum vero tonitruum longer grossa et semiplena.*

> *Flying fire* (*Ignis volans*). *Take one pound of sulphur, two pounds of willow charcoal, and six pounds of saltpetre, which three things grind finely on a marble stone. Then put as much as you wish into a paper case to make flying fire or thunder. The case for flying fire should be long and thin and well-filled with powder, that for making thunder short and thick and half-filled with powder.* ❶

The above account of Albertus was almost entirely quoted from Recipe 13 of the Latin translation of *Book on Fire for Burning Enemies* (*Liber Ignium ad Comburendos Hostes*), a book on the art of war written by the Arabs. According to Hime H., the book was completed around 1225~1250❷. This helps the decision of the writing time of Albertus's *De Mirabilibus Mundi*: it must have been written in the second half of the 13th century. Here we transcribe the original Latin text and provide Partington's translation of Recipe 13, *Book on Fire for Burning Enemies*:

> *Secundus modus ignis volatilis hoc modo conficitur. R. Acc. 1.*
> *I sulfuris vivi, 1. Ⅱ carbonum tiliae vel cilie, Ⅵ 1. salis petrosi,*

❶　BEST M R, Brightman F H. The Book of Secrets of Albertus Magnus, of The Virtues of Herbs, Stones and Certain Bests, also a Book of The marvels of the World[M]. Oxford: Clarendon Press, 1973.

❷　HIME H. Gunpowder and Ammunition: Their Origin and Progress[M]. London: Longmans Green, 1904.

quae tria substilissime terantur in lapide marmoreo. Postea pulverem ad libitum in tunica reponatis volatili vel tonitruum facientem. Nota tunica ad volandum debet esse gracilis et longa et cum praedicto pulvere optime conculato repleta. Tunica vero tonitruum faciens debet esse brevis et grossa et praedicto pulvere semiplena et ab utraque parte fortissime filo ferreo bene ligata.

The second kind of flying fire is made in this way. Take 1 lb. of native sulphur, 2 lb. of linden or willow charcoal, 6 lb. of saltpetre, which three things are very finely powdered on a marble slab. Then put as much powder as desired into a case to make flying fire or thunder. Note. — The case for flying fire should be narrow and long and filled with well-pressed powder. The case for making thunder should be short and thick and half filled with the said powder and at each end strongly bound with iron wire.

Having compared the paragraph on gunpowder in *De Mirabilibus Mundi* and Recipe 13, *Book on Fire for Burning Enemies*, we can see that they are the same in a word-for-word way, except that in his book, Albertus did not quote the first and last sentences of Recipe 13 of *Book on Fire for Burning Enemies*. It is evident that Albertus's knowledge of gunpowder came from the Arab war books, which can be exemplified by other transcripts of the Arab war books found in *De Mirabilibus Mundi*. Hime H. believed the flying fire prescription was actually referring to rockets and firecrackers, the latter being the children's toys in Bacon's words. Johann Beckmann (1739 ~ 1811), in his *History of Inventions* (*Beiträge zur Geschichte der Erfindungen*, 1780~1805), thought that the Arabian *Book on Fire for Burning Enemies*, the German Albertus's *De Mirabilibus Mundi* and the British Bacon's works are very similar in their narrations on gunpowder and that the three's gunpowder messages must have the same source[1]. According to the French expert Joseph Toussaint Reinaud (1795~1867) and Ildephone Favé, their common source was China, and from China gunpowder technology was introduced to Arab and Europe around 1250 with the westward expedition of the Mongol army[2].

---

[1] BECKMANN J. History of Inventions[Z]. London: H. G. Bohn, 1846.

[2] REINAUD J T, FAVÉ I. De Feu Grégeois, de feux de guerre, et des origines de la poudre à canon chez les Arabes, Persans et les Chinois[J]. Journal Asiatique, 1849 (14): 316.

Feng Jiasheng （冯家昇，1911～1970）concluded that "the Mongols used gunpowder in conquering Europe, but they didn't introduce it into Europe", arguing that the Mongols would not disclose their military secrets about gunpowder to others, that the Europeans, overwhelmed on the battlefield and scattered to escape, did not have a chance to know the secret, and that the Europeans, poorly educated, did not know a thing about gunpowder❶. We believe that the above reasons and assertions cannot be justified. The Chinese gunpowder prescriptions and methods of making firearms were no longer secrets since they were made public in the Northern Song Dynasty in 1044 when *Collection of the Most Important Military Techniques* was published. Saltpetre was sold overseas during the Song and Yuan Dynasties. In the Yuan Dynasty, firearms were exported despite of the government's ban. The Mongolian Kiptchac Khanate and Il Khanate, adjacent to Europe, manufactured gunpowder and firearms on their own to provide military supplies. The Mongolian Yuan also called some Arabs and Europeans to work in all industries, including the military industry. For instance, two muslin gunners 'Ala al-Din and Ismā'il, once did a meritorious military service in 1274 in the Mongol army. So the Arabs and the Europeans had plenty of opportunities to learn Chinese gunpowder technology. At the same time, the 13th-century Europeans cannot be deemed poorly educated as a whole.

In short, 30 years after the Mongols had used firearms in the European battlefields during their western expeditions, induced by the Mongol western expedition, some incisive European scholars wrote firearms into their books as the latest inventions or novelties. The Mongolian Yuan firearms, developed on the basis of the Song and Jin firearms, were much improved in powder preparation, structure and performance than before. Metal materials were more widely applied in the manufacture of firearms. As a result, the bombs, hand-guns, catapults throwing gunpowder packages, and flame-spurting lances with which the Mongol army was equipped were very powerful. The cluster rockets had increased burning force and wider range. These advanced firearms, when applied on the battlefield, would surely attract the attention of the Europeans and lure them to imitate, for they found it urgent to grasp the firearms technology. The secret of advanced firearms couldn't be monopolized by the Chinese for long. In the 1260s and the 1270s, the Europeans learned the

---

❶   冯家昇. 火药的发明和西传[M]. 2版. 上海：上海人民出版社，1978.

Chinese gunpowder knowledge, quickened their experimental pace to copy the firearms used by the Mongolian army, and in the 14th century put firearms into use gradually (Figure 140).

Figure 140　A 15th-century European Sketch of Gunpowder Combustion Testing, collected by Vienna Hofmuseum (Drawn by Pan Jixing in 1987 on the Basis of Its Photograph)

3.5.3　Early Firearms in Europe

The earliest extant European firearms pictures are two bombard illustrations in the manuscript of *De Nobilitatibus, Sapientiis et Prudentis Regum* by Walter de Milamete (1295~1357), kept in Bodleian Library, Oxford University. Walter de Milamete was the penitentiary of King Edward Ⅲ (1312 ~1377) and church pastor of Cornwall. The manuscript itself does not mention the firearms, but the color illustrations of the bombards appear on the back of Page 44 of the manuscript: a bottle-shaped or pear-shaped bombard placed on the table; a knight lighting the vent field with hot blazing iron bars; in the front of the bombard, an arrow shooting into the entrance to the castle passage; the shooter in a protective cap, without a shawl armor, bare-footed, and brown-faced (Figure 141). Having looked at the original painting, Partington thought that the shooter looks like a Moor or a North African Muslim. The bombard in the other illustration has a similar shape, with the gunman, and three other people standing by, possibly assistants helping with cleaning up the bombard-bore, charging, and putting on arrows.

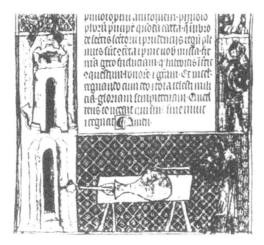

Figure 141　Bottle-shaped Bombard in the 1327 Milamete Manuscript,
taken from Partington (1960)

Undoubtedly, the bombards drawn in the Milamete's manuscript should be made of metal. It needs to be noted that the early bombards were small, only a span long, and not heavy. They could be held in hand if having handles or tied to wooden poles. Painting them as set on the table seems technically inaccurate. In the manuscript they were not drawn in the correct proportion, so they appeared too large. Besides, the handle was omitted. The German researcher Bernhard Rathgen thought the picture might have been borrowed from some books like *Feuerwerkbuch* from the 14th century in Germany, which, like China's *Huolong Jing*, has many copies. He also pointed out that there should be a round board in the bore of the bombard to fire the arrow out[1], which is equivalent to *musongzi* (木送子, wooden sender) in the Chinese hand-guns. Given that the original figure does not present its internal structure, we made a technical recovery (Figure 142) and drew the handle. Later, the European bombard began to grow larger and heavier and must be placed on the wooden carriage. Its evolving trend was the same with that in China.

The 1327 bombard was shaped like a bottle, but the bore was a long tube. Needham suspected that such kind of bombard could have originated from China, because his collaborator Robin Yates found some stone carvings of

---

[1]　RATHGEN B. Das Geschütz im Mittelalter[J]. Quellenkritische Untersuchungen von Bernhard Rathgen, 1928(1):124-125.

bottle-shaped firearms while visiting Dazu Grottoes of the Song Dynasty in Sichuan in June, 1985. But Yates did not have time to figure out the exact age of the stone carvings. In order to find it out, on November 21, 1986, Joseph Needham, Lu Gwei-Djen, and the author of this book had the site visit at the Dazu Grottoes in Sichuan, China. We saw in Cave No. 149 one god statue holding a blazing hand-held bombard, another holding a bottle-shaped firearms firing a cannonball (Figure 116), both of them standing by the statue of Avalokites' vara (观自在如意轮菩萨) built in 1128 (the Southern Song Dynasty). We believe that the Southern-Song-Dynasty hand-gun of 1128 could be the ancestor of the 1327 European bombard❶. The bottle-shaped bombard should be used with the handle or on the shelf, but because of the limited space in the grottoes, the stonemasons had to carve them as hand-held.

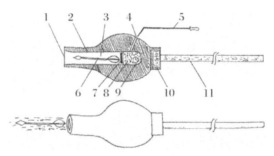

Figure 142　Restored Picture of Earliest Bombard in Europe of 1327,
drawn by Pan Jixing (2000)
1. Muzzle　2. Bore surface　3. Bore　4. Vent　5. Iron cone　6. Arrow　7. Baffle
8. Combustion chamber　9. Gunpowder　10. Breech　11. Wooden handle

As the combustion chamber could overheat, there was a thick layer of metal around it, making the whole weapon look like a bottle, which was a common feature of the early bombards in China and those in Europe. The thick outer wall of the combustion chamber helped it to withstand the pressure, prolong its life, and ensure the safety of the gunner. China was ahead of Europe in developing bottle-shaped metal bombards for nearly 200 years, a reasonable interval for Chinese firearms to spread westward. The western expedition of the Mongol army accelerated the process of transmission, for the firearms the Mongols used in the European battlefield could reach Europe within 30 or 40 years. In his 1979 speech at the University of London, Joseph Needham proposed that the bottle-shaped bombard in the 1327 Milamete

---

❶　潘吉星. 李约瑟集[M]. 天津:天津人民出版社,1998.

manuscript originated in China. His assertion was confirmed by physical data in 1986.

As gunpowder performance improved and firearms technology advanced, the spherical bulges on the metal tubular firearms grew smaller and smaller till a standard tube came into being, enabling the large firearms to be laid on wooden gallows and the small ones to be held by hand or mounted on horses for the cavalry. The above evolution was completed during the Mongol Yuan (the 13th century) and used for actual combat. In the 14th-century Germany, France, Italy, Britain, and other European countries, the barrels of the copper bombards were no longer bottle-shaped, but long cylindrical tubes that could shoot pellets made of stone, iron or lead. The small bombards were held by hand; the larger ones placed and ignited on the shelf (Figures 143 & 144).

Figure 143   Illustration of Early Flame-spurting Lance in Latin Manuscript of 1396, National Library of France in Paris, taken from Reinaud et Favé (1845)

Figure 144   Illustration of Hand-gun in a 14th-century Latin Manuscript, taken from Hogg (1980)

New changes in firearms technology in the Chinese Mongol Yuan subsequently appeared in some European literature in the 14th century. An example is the Latin copy of *War Fortifications* (*Bellifortis*) written on a parchment board kept in Göttingen University Library of Germany, believed to have been finished between 1395 and 1405❶, authored by the German military engineer Conrad Kyeser von Eystädt (1366 ~ 1405). The book talks about military gunpowder arrows, fireworks, rockets, bombs and hand-guns, the knowledge of which came from the Arabic scripts whose information was in

---

❶   SARTON G. Introduction to the History of Science[M]. Baltimore: Williams & Wilkins Co., 1947.

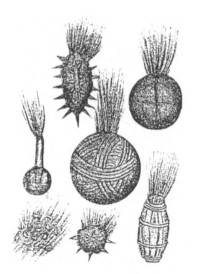

Figure 145 Bombs in Kyeser's Manuscript, taken from Partington (1960)

turn from China. The figures in the illustrations of *War Fortifications* are dressed in Arab costumes. The gunpowder recipes are quoted from *Book on Fire for Burning Enemies*. The recipes contain such assistants as arsenic, realgar and lime, being consistent with the Chinese formulas. The "flying dragon" has the fire cartridge tied to the guide pole with ropes, and the "flying bird" is similar to the Chinese *Shenhuo Feiya* (神火飞鸦, flying crow with magic fire). The bomb in the book (Figure 145) resembles the Chinese hard shell bombs with tipped thorns and *zhentianlei* (震天雷, thundering bomb). S. J. Von Romocki believed that Kyeser had obtained a lot of useful material for his book while traveling in Italy.

Italy, as the cradle of Renaissance and the seat of the Roman Pope, is one of the European countries that had acquired gunpowder knowledge earlier than the rest, because in the $13^{th} \sim 14^{th}$ centuries, it was close to the Muslim Mameluke Kingdom (1250~1517) in North Africa and Il Khanate in Western Asia, the merchants from Venice, Genoa and Fiorentina were keen to trade with China, and the Pope had sent envoys to China for many times. The word *rocket* first appeared as "rochette" in Italian. According to the 18th-century Italian historian Ludovico Antonio Muratori's (1672~1750) research of the ancient Italian manuscripts, the Republics of Genoa and Venice broke into a fierce battle near the fort at Chioggia Island from 1379 to 1380 for maritime trade monopoly. In the battle, rockets were launched ("igne imissio cum rochetis"). Western firearms historians believe that the rockets used in the battle of Chioggia from 1379 to 1380 are the earliest records of rockets made in Europe. Chioggia was off the Adriatic coast in northern Italy, south of Venice.

Today the word *rocket* in English, *roquette* (*or fusée*) in French, *rokete* in German, *raketa* in Russian and ロケット in Japanese, were derived from the Italian *rochette*. Some authors think that the military technician Guido da Vigevano (1280 ~ 1350) in Pavia, Italy mentioned a weapon named *una rochette* (rocket) in his book *Handbook for a Crusader* finished in 1330, and that rockets were also used in the Forli Campaign in 1281 near Emilia, south of Chioggia. If so, the initial use of rockets in Italy or Europe could be traced back to the late 13th century. Either way, it has been widely acknowledged by scholars that the early European combat rockets first emerged in Italy.

European pyrotechnics also first appeared in Italy, the firework products of Florence and Siena being the most famous. Many places regularly presented large-scale firework shows. From the 14th century to the end of the 17th century, Italy was leading the way in the manufacture of fireworks. In the libraries of Munich and Weimar, Germany, was kept the 1420 Latin copy of *Bellicorum Instrumentorum Liber*, written in secret languages, but well illustrated, and authored by a technician in Venetia, Italy named Giovannic da Fantana (1395 ~1455). The book referred to rockets, bowl-muzzled hand-guns, bombs and fire birds. In 1463, a Roman, Roberto Valturio (1413~1482) wrote about the purification of saltpeter, gunpowder manufacture, bombs, fireworks, flying fire, and fire lances, etc., in his *De re Militari Libri* XII in Latin.

In the 19th century, the French orientalist Joseph Toussaint Reinaud (1795~1867) and the artillery colonel Ildephonse Favé reported that Bibliothèque Royale had a 1396 Latin script (MS Bib. Roy., 7 239) which includes the painting of a cavalry holding a fire lance spurting flame[1] (Figure 143). To ensure safety, the rider wears fire protection clothing and armour and the horse also wears a protecting cover. This is the historical testimony of the spread westward of China's fire lance. Having mastered the technology of fire lance from the Southern Song, the Mongols changed bamboo tubes into metal cylinders and used the improved firearms in their westward expeditions in the 13th century, thus bringing them to the west.

With the two expeditions in the 13th century of the Mongol army, and due to their use of various firearms in the Arab and European battlefields, Chinese gunpowder and firearms technology quickly spread in the west. In the 14th century, Europe learned the Chinese technology and imitated the Chinese sample firearms to make their own hand-guns, rockets, flame-spurting lances, grenades, bombs and fireworks. On this basis, with the development of economy and technology in Europe, the Europeans stepped forward from the

---

[1]   REINAUD J T, Favé. Histoires de l'Artillerie, Du Feu Grégeois, des Feux de Guerre et des Origines de la Poudre à Canon, d'apres des Texts Nouveaux[M]. Paris: J. Dumaine, 1845.

Figure 146　Rocket from the 16th-century Haas' Manuscript, taken from Pan Jixing (1987)

imitation stage to the stage of self-development in the $15^{th} \sim 16^{th}$ centuries, thus improving the existing firearms. For example, the German rocket technician Conrad Haas who had worked in Sibiu Arsenal from 1529 to 1569, in his manuscript written in old German, proposed that launching a rocket of several rocket tubes tied onto a single shaft should be able to increase the range and the thrust (Figure 146), coinciding with the ideas of a Chinese named Wan Hoo (万虎) who lived earlier than him. Haas also proposed to make two-stage rockets and three-stage rockets[1]. Although the Chinese had put similar ideas into practice earlier than him, he was still the first European to have such ideas. The aforementioned Italian Giovannic da Fantana, in *Bellicorum Instrumentorum Liber* (1420), introduced the jet cart, a four-wheeled vehicle pushed by the reaction force of rockets.

After the mid-15th century, European copper bombards, like those in China, gradually grew larger, with longer barrels and wider calibers, and thus could shoot very heavy projectiles, functioning as good siege weapons and evolving towards cannons. As the bombards' weight was increased, they were usually placed on carriers. Large bombards are still in existence now, such as the Paris Bombard (1404), with a length of 3.65 metres, a diameter of 39 centimetres and a weight of 4,597 kilograms; the Belgium Ghent Bombard (1450~1452), with a length of 4.96 metres, a diameter of 62 centimetres, and a weight of 340 kilograms. Such bombards are equivalent to "*jiangjun pao*" (将军炮, general bombard) in the Ming Dynasty. It takes some time to clear the barrel, recharge the cartridge and reload the cartridge after firing. In the second half of the 14th century, a new type of breech-loading cannon was developed in Europe, and it could be made of cast iron (Figure 147). It was introduced to the Ming Dynasty in the 16th century, known as *folangji* (佛朗机)

❶　CARAFOLI E, NITA M. Romanian Rocketry in the 16th Century[J]. Essays on the History of Rocketry and Astronautics, 1972(1): 3-8.

by the Chinese and called Frankish culverin by the Europeans (the name may be inappropriate).

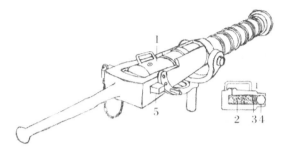

Figure 147   The Breech-loading *folangji* (佛朗机, Frankish Culverin) appearing in Europe in the Second Half of the 14th Century, taken from Needham (1986)

1. Charging chamber   2. Gunpowder chamber   3. Springy plug   4. Cannonball   5. Wooden or iron bolt

Folangji is actually a breech-loading cannon invented by the Europeans. It basically consists of two parts: the cannon barrel or the main body of a cannon, including, in the front, the chase with reinforce rings, in the middle, trunnions for setting the cannon on the carriage to adjust the shooting angle, at the tail, an empty slot, and in the end, the knob; the charging chamber or sub gun, a short and handled cylinder loaded with gunpowder and projectiles, with the vent in the upper part. The charging chamber, which just fits into the empty slot in the cannon barrel, is packed there and then fixed by the bolt. There is the front sight in the front of the cannon body for aiming. In non-battle time, people could prepare as many charging chambers well packed with powder and projectile as possible so that they are ready to replace the used charging chambers. The combustion chamber of the Frankish culverin is mobile rather than fixed, thus shortening the preparation time required for firing. The Metropolitan Museum of Art in New York exhibits the Frankish culverin used on a Spanish sea ship in 1475 (Figure 147). It has a length of 94 centimetres, a weight of 50 kilograms, and a calibre of 11.7 centimetres. Volume 122 of *Wu Bei Zhi* introduced *folangji*.

Chapter 4　Invention, Development and Spread of China's Compass

4.1　Ancient Methods for Direction Determination Before Compass Invention

4.1.1　Method for Measuring Sun's Shadow Length by Using Sundial Since the Warring States Period

A compass, composed by a magnetic needle and a direction dial, is an instrument used to show direction for navigation and orientation, which is one of the Four Great Inventions of ancient Chinese science and technology. Its invention is not a sudden result, but a great achievement of nearly one thousand years' practices and unceasing explorations of how to determine orientation by Chinese people in the Warring States Period or the 5th century B.C., which is closely linked to the evolution of ancient Chinese direction culture. Compass is a location device, which is made by using the property of the lodestone's directivity and polarity. Before it, ancient people resorted to non-magnetic astronomic methods to orient directions. The magnetic needle of a compass is a man-made magnetized one, but before it, ancient Chinese adopted natural lodestone to make the earliest instrument to give directions, which was called *sinan* (司南, south-pointer), the predecessor of compass. From the history of Chinese direction culture, we can conclude that the compass evolution by and large experienced 3 stages from astronomic direction determination, to *sinan*, and finally to the real compass, which just reflected the gradual deepening of understanding of nature of ancient Chinese, and the constant improvement of orientation technology. Before the invention of compass is studied, its historical process of evolution should be disclosed firstly.

Since ancient times, people have felt it is necessary to orient the right directions of the north, south, east, and west in their long-term life experiences and productive activities. The Works *Zhouli · Tianguan · Zhongzai Diyi* (《周礼·天官·冢宰第一》, Chapter One of *Zhongzai Prime Minister*, *Celestial*

Offices in the Rites of the Zhou Dynasty）, which was compiled in the Warring States Period (5th century B.C. ～3rd century B.C.), wrote about this. "When the king establishes the capital, he should figure out directions and determine location of the sacred shrine and the palace. He should also demarcate territories between the capital and the suburb, then decide different government posts to set examples for these civilians. So the king appoints Zhongzai（冢宰）as Prime Minister, and lets him lead his subordinates to help govern the country and make every class keep its place and perform its duty. （惟王建国,辨方正位,体国经野,设官分职,以为民极。乃立天官冢宰,使帅其属而掌邦治。）" Thus, direction determination and territory demarcation are significant events in the establishment of rulership for a country. From ancient times to the present, for Chinese people, the south direction is the symbol of dignity and excellency, so for an Emperor or vassal Kings, when they meet with their courtiers or ministers, they often sit facing the south. *Yi · Shuogua*（《易·说卦》, *Explaining the Hexagrams in The Book of Change*）says: "These sages sit facing the south, governing the whole country. （圣人南面而听天下,响（向）明而治。）"

Therefore, direction determination and layouts for capitals, palaces and all kinds of urban architectures embody the idea of facing the south, even mausoleums face south, let alone official mansions and the houses for common people. For a long time, this idea of direction interpenetrates all aspects of daily life in Chinese architectural history. Besides, before a number of social activities, directions need to be determined first, such as observation of horoscope, sacrificial rites, marching and fighting, mining, land and water travelling, demarcation of administrative districts, mapping and charting, etc.

Ancient people originally determined direction by using astronomical calculation. They often adopted two simple methods to reach this purpose. They respectively measured the sun's shadow length by using the sundial as well as orienting direction by identifying the position of Polaris, which could be echoed in the record of Confucianism in *Zhouli · Dongguan · Kaogong Ji*（《周礼·冬官·考工记》, *The Rites of the Zhou Dynasty · Winter Offices · Records on the Examination of Craftsmanship*）. It says: "In the daytime, (the officials) refer to the shadow of the sun at noon while identifying the orientation of the Polaris at night, thus they can calculate the length of a day. （昼参诸日中之景（影）,夜考之极星,以正朝夕。）" *Yi zheng zhaoxi* （以正朝夕）means more specifically that once they first oriented the right directions of the east and the west, naturally they can measure the right directions of the south and the north. Ancient people discovered long ago that the length and direction of a tree shadow would change regularly with the time accordingly, so they took advantage of this effect to make the oldest astronomic instrument, namely the real Chinese sundial, also called gnomon. Take an upright wooden pole or a stone pillar, which is called *nie* (臬), *nie* (槷) and *bi* (髀) by ancient Chinese people, and then observe its shadow under the sun to determine its orientation. This is because the rising and setting of the sun, and the path of its movement in the daytime is basically symmetrical, relative to meridian line of the observing site. So by observing the direction of sunrise and sunset, we can identify the orientation of four points (east and west, south and north).

The sundial is simple, but very unusual, because it shares multiple functions, such as determination of direction, mark of the time and measurement of the solar terms, of which the earliest application is the determination of direction. In the Banpo Neolithic Village (半坡村) located in the city of Xi'an, Shaanxi Province, experts in astronomy found that all the gates of these 46 well-preserved house bases about 6,000 years ago face the south. In *Dadunzi Site* (大墩子遗址) located in Pizhou City, Jiangsu Province, Tombs Complex of the same age follows the north-south direction. This illustrates that our forerunners in the primitive society have grasped the methods of how to orient directions. Besides the method of Polaris orientation, the other way is setting up an upright wooden pole to measure the length of its shadow. ❶ One of the Chinese idioms *li gan jian ying* (立竿见影) just originated from it. Even some present primitive tribes, for example, tribe members in Kalimantan Island of Indonesia, still adopt wooden poles to measure the length of the sun shadow to tell directions, and the strategy is the same as our Chinese ancestors did.

Records can still be found in Pre-Qin Classics about how ancient Chinese measured the length of the sun shadow by a sundial to determine direction. In *The Book of Songs*, the first collection of poems in China, compiled in the Spring and Autumn period (春秋时期,770～476 B.C.), there are some events recorded, which could be traced back to the Shang and Zhou Dynasties (商周, 1700 B.C. ～221 B.C.). According to the record in *Shijing · Daya · Gongliu* (《诗经·大雅·公刘》, *The Book of Songs · Major Odes · Gongliu Chapter*),

---

❶ 薄树人. 中国天文学史[M]. 北京:科学出版社,1981.

the Duke of Zhou (周公旦) assisted King Cheng of Zhou (周成王, 1042 B.C. ~ 1021 B.C.) to rule the country when the king just ascended to the throne. To help the young king understand difficulties and hardship about his ancestors in the establishment of the Kingdom of Zhou, the Duke of Zhou recounted his achievements, who later founded the Zhou dynasty. The Duke of Liu (1700 B.C. ~ 1600 B.C.), who was the great-grandson of the Lord of Millet (后稷, the ancestor deity of the Zhou dynasty), was dispelled by the ruler of the Xia Dynasty (夏, 2070 B.C. ~1600 B.C.). Therefore, he chose to leave and led his clans to set up a new country in Bin area (豳, in the central Shaanxi Province, the present south-west of Xunyi City). The Duke of Zhou said: "Great Duke of Liu, you are so honest and tolerant, you help exploit a large square of fields, and you determine the direction by measuring the length of the sun shadow, even climbing to the hill for it. (笃公刘,既溥既长,既景乃冈。)" Zhu Xi (朱熹, 1130~1200), a Song Dynasty Confucian scholar, added the annotations to *ji ying nai gang* (既景乃冈). He said: "ying (景) was just the sun shadow, which showed that the Duke of Liu measured it to identify four points of direction. Gang (冈) was a means to climb high for better observing and orienting the length of the sun's shadow. (景,考日景以正四方也。冈,登高以望也。)"❶ This is the earliest documentary evidence about setting up an upright pole to measure the length of the sun's shadow.

Before the Western Zhou Dynasty, the capital of the late Yin Dynasty (1600 B. C. ~ 1046 B. C.) was located in present Xiaotun in Anyang City, Henan Province. Further excavations of 56 base addresses of buildings in the site revealed that all these buildings were in order and well-organized according to certain planning. Many base addresses were found facing the right south-north or the east-west direction when checked by modern magnetoresistive compass. The layouts of some compounds were similar to the "courtyard" mode of *siheyuan* (四合院) in Beijing. Besides the Main Hall (正房), there were East Wing Room (东厢房) and West Wing Room (西厢房). In Liang Sicheng's treatise, these names are lower-case form and italic. This showed that people in the Yin Dynasty had mastered the skills of orientation technology, and identification of direction could be reached by the method of setting up an upright pole to measure the length of the sun's shadow. We still get evidence from inscriptions on bones or tortoise shells of the Shang Dynasty, such as the

---

❶   朱熹. 诗经集传:大雅(公刘)[M].北京:中国书店,1985.

character *zhong* (中) and the phrase *li zhong* (立中), both of which mean setting up a pole to measure directions and the four seasons. Some characters share multiple connotations, for example, the character *nie* (臬) in the inscriptions on bones or tortoise shells means setting up a wooden pole as an arrow target. It also means setting up a pole to measure the sun's shadow. For another example, one sentence from the oracle inscriptions of the Shang Dynasty says: "On the Yiyou day, the official responsible for divination asks: 'Can we leave for *zhengfang* (征方) by following *si nie* (司臬)? '" Here, *nie* (臬) means the person who took charge of it, namely, the one who set up a pole to measure the sun's shadow. Hence, it may be inferred that at the time of the Duke of Liu, the statement is credible as to the identification of directions by setting a pole to measure the sun's shadow in *The Book of Songs*.

How did ancient Chinese people determine directions by measuring the sun's shadow with a gnomon? *Zhouli • Dongguan • Kaogong Ji • Jiangren* (《周礼·冬官·考工记·匠人》, *The Rites of the Zhou Dynasty • Winter Offices • Records on the Examination of Craftsmanship*) says: "When these craftsmen construct the capital, they try to keep the foundation base straight and level by the way of suspending a right plummet and referring to water surface. By means of this, they can keep the pillar vertical to the base through observation of the sun's shadow. First, they make a sundial, then observe its shadow under the sum, and affirm the shadow of sunrise and sunset respectively. In the day they refer to the sun's shadow, while at night they turn to the orientation of Polaris, thus they can calculate the length of a day. (匠人建国,水地以悬,置槷以悬,眡 (shì,同视)以景(影)。为规,识日出之景与日入之景。昼参诸日中之景,夜考之极星,以正朝夕。)" We need further explanation about the quotation. Craftsmen means technical officials who are responsible for the construction of the capital. Before any construction, they should orient the position of the building to be constructed. "水地以悬,置槷以悬" means that we should correct the base surface by referring to water surface, thus making it level and horizontal. If not, the cast shadow of the sun would produce error. Then set up a wooden pole (8 feet long), which is called *biao* (表) or *niebiao* (臬表) in Chinese. Next suspend a string with a plummet, to check whether the pole is perpendicular to the base surface, and then observe the sun's shadow of the pole. "为规,识日出之景与日入之景" means: first take Pole A as a center, draw a circle with the length of the sunrise shadow and sunset shadow as radius (Figure 148), which are of equal length. The linking line *BB'* between the vertex of the sunrise shadow B and the vertex of the sunset shadow B' is the forward directions of the east and the west, while the joining line of the bisecting point M of the line BB' and the center A is the forward directions of the south and the north.

However, the intersection points of the shadows of sunrise and sunset with the circle cannot be easily precisely established as a standard. So, according to *Zhou Li • Dong Guan • Kaogong Ji • Jiang Ren*（周礼·冬官·考工记·匠人, *The Rites of the Zhou Dynasty • Winter Offices • Records on the Examination of Craftsmanship*）, these craftsmen must observe the orientations

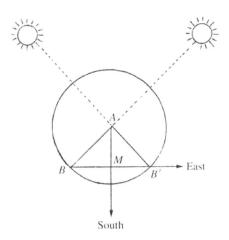

Figure 148   Schematic of Direction Determination by Raising Poles Mentioned in *Kaogong Ji*, from Bo Shuren

of the south and the north by the Polaris at night to correct the directions of the east and the west by measuring the sun's shadow in the day. Later, several improved methods about measuring the sun's shadow through a sundial appeared. We can find evidence in a famous treatise *Huainanzi • Tianwen Xun* (120 B.C.)（淮南子·天文训, *Patterns of Heaven* in *The Masters of Huainan*）, which was compiled by the court of King of Huainan Liu An (179 B.C. ~122 B.C.)（淮南王刘安）and his followers. The chapter says: "First, set up a pole in the morning, take the east as an observation point. Then hold another pole behind the previous pole about 10 feet away, and take it as a fixed reference of observation. When the sun rises to its northeast angle, adjust the second pole, and keep the first and the second poles and the sun on one straight line; at this time, the shadows of the two poles overlap and project to the south-west direction; then fix the second pole. At the sunset, set up the third pole located in the south relative to the first pole, use the second pole as a fixed reference of observation. When the sun lies between north of the second and the third pole and when the overlapped shadows of them point to south-east, fix the the third pole, thus the forward east direction could be measured. Namely, the midpoint of the joining line of the first pole and the third one is right opposite to the second pole. Thus, joining them, people could obtain the direction of the east and the west."❶

One outstanding feature about the proposed method in *Huainanzi · Tianwen Xun* (120 B.C.) (淮南子，*The Masters of Huainan*) lies in that people use a fixed pole and two movable poles to construct an equation, and calculate the sun's shadow by using these poles. First stand a fixed pole A (Figure 149), then fix a movable pole B 10 feet away from the east of Pole A. When the sun begins to rise, observe these two sundials from west to east (north-east), namely, from the fixed sundial to the movable sundial, thus making Sundial *A*, Sundial *B* and the sun *S* on the same line. Then wait for the sun to set. Still fix a movable sundial *B′* 10 feet away from the east of *A*, observe these two sundials from east to west (north-west), and keep Sundial *A*, Sundial *B′* and the sun *S* on the same line. Thus we can get $AB = AB'$. The joining line of midpoint M of the line *BB′* and *A* is just the forward east and west directions, while *BB′* is the forward south and north directions.

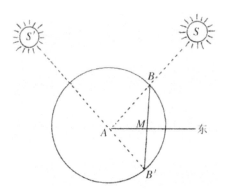

Figure 149 Schematic of Direction Determination by Raising Poles mentioned in *Huainanzi* (淮南子，*The Masters of Huainan*), from Bo Shuren (1981)

Both *Records on the Examination of Craftsmanship* and *Huainanzi*, following the same principle, use the two equi-length lines of the directions of the poles' shadows at the sunrise and the sunset to orient four forward directions of the east, the west, the south and the north, but all the observations and measurements were carried out outside doors. However, in certain situations, this method should be adjusted, because people could not fix poles to measure the shadows at any moment, such as marching and fighting, water transportation and astronomy observation, etc. So a miniature instrument of direction measurement was invented by applying the above principle, which was called Sundial (晷仪). According to *Han Shu · Lüli Zhi Shang* (83) (《汉书 83 · 律历志上》, *Treatise on Rhythm and the Calendar of History of the Western Han Dynasty*, Volume 21), in the first year of the Han dynasty (104 B.C.), Emperor Wu ordered officials such as Sun Qing (孙卿), Hu Sui (壶遂) and Sima

---

❶ 淮南子：天文训[M]．杭州：浙江人民出版社，1984．

Qian who was *Taishi Ling* (太史令, Court Astrologer) to go to court, planning to compile a Chinese calendar. They were to stand a sundial so that the nation could orient directions of the east and the west, and to set up an ancient Chinese timer *louke* (漏刻) by which to calculate the general distance from twenty-eight lunar mansions to the four forward directions (east, west, south and north) in the country. Finally, based on the above practice and analysis, these important calendar elements were determined, including lunations, equinox and solstice of four seasons, the law of the lunisolar moving and crescents and the oppositions for the Grand Inception of system. (议造汉历。乃定东西,立晷仪。下漏刻,以追二十八宿相距于四方。举终以定朔晦、分至、躔离、弦望。)[1] Here, *ding dongxi* (定东西, orient the directions of the east and the west) shares the same meaning as *zheng zhaoxi* (正朝夕, orient the direction of the east and the west) in *Kaogong Ji* (《考工记》, *Records on the Examination of Craftsmanship*), and both mean the orientation of four directions of the east, the west, the south and the north. To reach this purpose, it was necessary to stand a sundial, because it could not only determine time but also orient directions, like a gnomon.

In 1897 (the 23th year in the period of Guangxu in the Qing Dynasty), in the Togtoh city of Inner Mongolia was unearthed a stone scale sundial plate in the early western Han Dynasty (200 B. C.), which was collected by Duang Fang, and is now well preserved in the Museum of Chinese History in Beijing. The plate is square (Figure 150), 27.5 cm × 27.4 cm, and 3.5 cm thick. On the plate, there are 3 circles of different diameters. Ex-circle diameter is 23.2 cm × 23.6 cm. There are seal characters, with a size of 0.4 cm ～

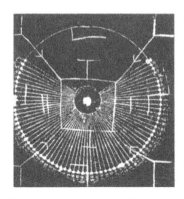

Figure 150 Gnomon (200 B.C.)
Unearthed in Togtoh City, 1877,
from Li Jiancheng (1978)

0.6 cm. There is a circular hole in the center, 0.65 cm its diameter, not penetrated, for standing a gnomon, which has fallen off and disappeared. Outside the inner circle, 69 radiation lines were carved, and the corresponding excircle had numbers from 1 to 69, but with no hints of orientation or other characters. Mr. Li Jiancheng, historian of astronomy, has made a special study

---

[1] 班固. 汉书:律历志上[M].上海:上海古籍出版社,1986.

of it, and confirmed that it is an instrument for determination of directions. Although there are no carved directions, people can still get right directions by observing from the front surface, because the up and down sides of the square in the middle means the directions of the east and the west, while the right and the left ones point to the south and the north directions, which is self-evident.

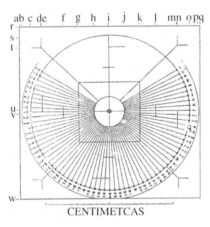

Figure 151　Sundial of the West Han Dynasty, collected in Royal Ontario Museum in Canada, from White (1938)

In 1932, the similar plate of a sundial was extracted from the tomb in Jincun of Luoyang City in Henan Province. It has no pedestal for standing a pole, which is now collected in Royal Ontario Museum in Canada. ❶ The plate, basically square, is 28. 4 cm long, 27. 5 cm wide, and 3 cm thick, with a circular hole in the center, 0. 65 cm in diameter and 1. 5 cm in depth, not penetrated. (Figure 151) On the plate, there are two circles and one incomplete circle. Between the inner circle and outer circle are carved 69 radiation lines. At the intersecting points of the radiation lines and the outer circle, there are small circular holes about 0. 16 cm in depth. Holes are marked with numbers from 1 ～ 69 (in seal characters). All radiation lines are equant. The circumference is divided into one hundred parts, and the angle of every two adjacent lines is 3. 6525°. A square is carved between the two circles, four angles of which have corresponding extended lines of diagonal lines, thus dividing the circumference equally into four equal parts and eight equal parts. Therefore, the scales and the words of the above two sundials on the plates are completely identical. The only difference is that the latter is more exquisite than the former. The time of their creation should be of the same period, and they share the same function.

The uses of a sundial mentioned above have been proven in one of the famous treatises of the same age, *Zhoubi Suanjing* (100 B. C.) (《周髀算经》, *The Book of Ancient China's Astronomy and Mathematics*). The book says: "At the sunrise, set up a sundial to mark its position and then re-mark its position

---

❶ WHITE W C, MILIMAN P M. An Ancient Chinese Sundial[J]. Journal of the Royal Astronomical Society of Canada, 1938(32):417.

when the sun is set. Then the joining line of the both ends would be perpendicular to the joining line of the sundial and the midpoint of the both ends, which just points to the forward direction of the east and the west, and the midpoint to the point of the sundial is just the forward direction of the south and the north. (以日始出，立表而识其晷，日入复识其晷。晷之两端相直者，正东西也。中折之指表者，正南北也。)" Hence, the method mentioned is essentially the same as the one proposed in *Records on the Examination of Craftsmanship*, with the only change of a horizontal plane into a small level plate on which are carved circles, the possible lines of the sun's shadows of the poles at different seasons, and the marked numbers of the intersection points between the lines of the sunrise's shadows and the sunset's shadow. So long as one stands a sundial in the center of the plate, keep the sundial level, observe the intersection points of the shadow of sunrise and sunset and mark them, then the joining line of the intersection points would be the directions of the east and the west, while the joining line of the midpoint of intersection points and the center of the circle is the forward directions of the south and the north. In this way, when weather is fine, people can carry it to any place, such as on a ship or on a wagon, to determine directions. If people used a wooden material to make the plate, it would be smaller, thus the gnomon could become a portable instrument to measure directions, which was more convenient.

4.1.2 Ancient Method for Determining Directions by Pole Star Observation

From *Zhouli • Dongguan • Kaogong Ji • Jiangren* (《周礼 • 冬官 • 考工记 • 匠人》, *The Rites of the Zhou Dynasty • Winter Offices • Records on the Examination of Craftsmanship*), we can conclude that ancient people also referred to the pole star at night besides resorting to the sun's shadow in the day. Here the pole star means Polaris. When people in the northern hemisphere (where China is located in) watch the sky, almost every star in the northern region of the sky on the horizon would be within the sweep of sight. In diurnal motion of stars in the northern region of the sky, only the polestar lies in the area of northern celestial pole of the celestial sphere, with a myriad of stars revolving round it. So the polestar, also called Kochab (帝星, the Emperor Star) or Polaris (北辰, the North Star), is the azimuth star most easily seen, which is endowed with unrivaled supremacy by ancient people. One quotation from Confucius (551 B.C. ~479 B.C.) in the Chapter *Lunyu • Weizheng* (《论

语·为政》，*The Confucian Analects · Exercising Government*）says："He who rules by moral force is like the polestar, which remains in its place while all the lesser stars do homage to it.（为政以德，譬如北辰，居其所而众星拱之）". *Shiji · Tianguan Shu*（《史记·天官书》，*Records of the Historian · Tianguan Astronomy*）also says："The north pole star stands in the middle palace（heaven's location around the north star）. It is the brightest, in which *Taiyi*（太一，First Great One or the Ultimate One）resides.（中居天极星，其一明者，太一常居也）" In fact, because of precession of the equinoxes, the polestar does not always remain constant. In the Zhou Dynasty（1046 B. C.～256 B. C.）and the Qin Dynasty（221 B. C.～206 B. C.）, northern celestial pole was moving to the ursa minor（β）in the constellation. Since the Sui and Tang Dynasties, the star α Gouchen（勾陈，Curved Array）in the constellation had become the pole star, which was called Gouchen First（勾陈一，α UMi）, roughly 1. 0 degree across from the northern celestial pole. Around year 2095, α UMi（勾陈一）will be closer to the celestial pole（15″）.

People can lock the pole star through identifying the stars（Constellation Ursa Major）. The ursa major is called *Beidouqixing*［北斗七星，Big Dipper, so called because it is shaped like a measuring tool for grain *dou*（斗）or *shao*（杓）in China］, an asterism consisting of the seven brightest stars of the constellation Ursa Major in the celestial pole. Their names are as follows：（1）α UMa（天枢，Dubhe）；（2）β UMa（天璇，Merak）；（3）γ UMa（天玑，Phecda）；（4）δ UMa（天权，Megrez）；（5）ε UMa（玉衡，Alioth）；（6）ζ UMa（开阳，Mizar）；（7）η UMa（摇光，Alkaid）. Among them, the former four stars in（1）～（4）in the big dipper were called *doukui*（斗魁，bucket）, also known as *xuanji*（璇玑）, while the latter three were called *doubing*（斗柄，bucket handle）or *yuheng*（玉衡，the handle of the dipper）. The pole star is the most identifiable in the sky, if the connecting line of β UMa and α UMa were extended along the β→α direction forward five times longer, the pole star would be located（Figure 152）. The Big Dipper is of great significance for marking the

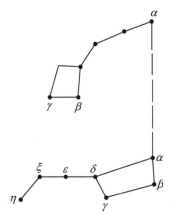

Figure 152　Sketch Map for Locating Pointers from Polaris, from *Ci Hai*（辞海，*Chinese Word Ocean Dictionary*）

true north and learning the constellations, so both stars, α UMa (天枢, Dubhe) and β UMa (天璇, Merak) are known as Pointers.

In the daytime, people could determine directions by measuring the sun's shadows through a sundial, while at night, people who wanted to identify orientations only relied on the Polaris and the pointers. When sailing on the sea, they would become disoriented without the stars to guide them. The Chapter *Qisu* of *Huainanzi* (《淮南子·齐俗训》, *Placing Customs on a Par* in *Huainanzi*) says: "When on board a boat or ship, those who lose their sense of directions and can not tell east from west will be perplexed, but as soon as they see the pole star and the Big Dipper, they will come to their senses. [夫乘舟而惑者,不知东西,见斗极则寤(wù,悟)矣。]"❶ During the Eastern Jin Dynasty of China, Faxian, a Chinese Buddhist monk (337~422), pilgrimaged from China to India by land at the time of the 3ʳᵈ year of Long'an Emperor for Buddhist scriptures, returned to the country by sea, and then reached Qing Province (Present Yangzhou City, Jiangsu Province) (412). He said in the chapter *Fuhai Donggui of his Foguo Ji* (《佛国记·浮海东归》, *Floating on the Sea for East Return* in *A Record of Buddhist Kingdoms*): "The sea was so vast and boundless that we could not recognize the directions. We only relied on the sun, the moon, the polestar and pointers to move forward. When it rained, all these directions would be distorted by wind and we were disoriented. When the night fell, we could only see the great waves ... we did not know the exact directions. We could not tell east from west until the sky cleared up so that we could adjust our directions and move on to right directions. (大海弥漫无边,不识东西,唯望日月、星宿而进。若阴雨时,为逐风去,亦无准。当夜暗时,但见大浪相搏……不知那向。至天晴已,乃知东西,还复望正而进。)"❷

Prior to the introduction of the compass, celestial navigation was a key element for safety of sailing on the sea and reaching preassigned locations. Even after its invention, people still adopted the method of celestial navigation as a reference. In ancient time, when people went for a short voyage near the mainland, they only determined the directions by watching the sun and other stars. However, if they embarked on a long voyage far away from the mainland, it was not enough to know the sailing direction, because there was still the possibility to deviate in the navigation, so apart from the sailing

---

❶　淮南子:齐俗训[M].杭州:浙江人民出版社,1984.

❷　法显.法显传[M].章巽,校注.上海:上海人民出版社,1985.

direction, the information on the location of a ship in the sea was also needed. Ever since the ancient times, helmsmen or seamen had found an interesting phenomenon: the farther they voyaged from north to south, the smaller the stellar ring in the northern region of the sky, and the lower the height of the North Star above ground level (AGL). Therefore, people could orient the geographical latitude of ships in the sea according to AGL of the North Star and its position. Height above ground level (AGL) is a height measured with respect to the underlying ground surface, which mainly means the angle of elevation or depression angle between the horizontal line and the line to the celestial body, which can be measured through observation in different places. Thus, celestial navigation has changed from the qualitative stage to the quantitative stage. It is an inevitable tendency for its development, just like the changes from the qualitative stage of determining directions of east and west by measuring the sun's shadow through a sundial in pre-Qin time, to the quantitative stage of using a gnomon with a plate on which number values are carved in Qin and Han dynasties.

As for when Chinese began the time of quantitative navigation by measuring the height above ground level (AGL), it still needs further studies. But one thing is certain that Chinese people began it at least in the 27th year of Kaiyuan era of Emperor Xuan in the Tang dynasty (724).

The Chinese astronomer Sen Yixing (僧一行, 683 ~ 727) and his participator Nangong Yue designed the instruments, which were widely used across the country to calculate quickly the degrees between the height of the North Star above ground level from the north pole by measuring its squint angle. [❶] According to the interpretation given by Bo Shuren (薄树人) and Chen Meidong (陈美东), historians of astronomy, [❷] the reversed protractor is just one kind of semicircular goniometer (Figure 153), with a support of a rotation axis in the center, through which the protractor can pivot around the axis for measuring degrees. For observations, first make the plumb line through the circle center coincide with the 90° division line of the goniometer. Then turn the goniometer and aim at the North Star along the direction of the diameter, thus the degree of the plumb line can be read. The angle between this degree

---

❶   刘昫. 旧唐书：天文志上[M]. 上海：上海古籍出版社，1986.
❷   薄树人. 中国天文学简史[M]. 天津：天津科学技术出版社，1978.
　　陈美东. 一行传[M]//杜石然. 中国古代科学家传记：上集. 北京：科学出版社，1992.

and the 90° division line would be the
height of the North Star above ground
level（AGL）or height altitude，namely
the geographic latitude of the observation
point. This simple method is convenient
for determining the positions of ships.

With the backing of the
government，Yi Xing and Nangong Yue
measured the heights of the North Star
above ground level（AGL）in 13 areas，
which extended from Tiele at 55° north
latitude （ present southwest area of
Ulaanbaatar，Mongolia）to Linyi at 17°

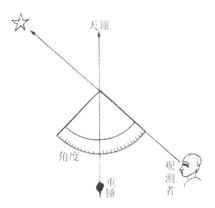

Figure 153   Sketch Map of Reversed
Protractor（覆矩仪）by Pan Jixing（2002）

north latitude（present south-central Vietnam）. In this process，they found that
altitude difference of the heights was 1° for every 129. 22 km（351 *li* 80 *feet* in
ancient China）. Therefore，according to the calculation，the length of meridian
line could be measured：the arc length about 1° was 166. 14 km，a research result
which no one in the world had discovered at that time. The surveying team also
observed in the Beibu gulf the star of Old Man or Old Man of South pole（α
Car，Carina，*chuandizuo* in Chinese，meaning "the bottom of boat
constellation"）at 20° north latitude in the southern sky，which was rarely ever
seen in central China.

Jia Dan（贾耽，730 ～ 805），one famous Chinese scholar official and
geographer recorded 7 common sea trade routes for communications with
foreigners in his works *Huanghua Si Da Ji*（《皇华四达记》，*Sea Routes to the
Overseas Countries for the Empire* ）. Among these routes，there was one
maritime Silk Road，starting from Guangzhou Port，with stops in present
Vietnam，Malaya，Indonesia，then across the Indian Ocean，through Sri Lanka
and India towards Arabian Peninsula in the Persian Gulf，and finally to eastern
Africa. ❶ Chinese merchant vessels must resort to quantitative methods of
celestial navigation to orient the directions and locations on the sea in such an
oceangoing voyage. Moreover，the simplest way is to measure the height of the
North Star or the azimuth star in South Pole above ground level（AGL）in
different places. Unfortunately，the early information on navigation had not

❶   欧阳修. 新唐书：地理志［M］. 上海：上海古籍出版社，1986.

been preserved, and even in the work *Huanghua Sida Ji*, there were left only some pieces of early navigational information in *Xin Tang Shu • Dili Zhi* [《新唐书·地理志》(1061), *Geography*, Volume 43 in *The New Book of Tang*]. The great development in navigation technology of the Age of Sail in the Song and Yuan Dynasties was undoubtedly based on the navigation skills in the Tang Dynasty. Therefore, Doctor Joseph Needham thought that the quantitative navigation time, which was marked by celestial navigation, began in the Tang Dynasty.

4.1.3 *Qian Xing Shu* Used for Celestial Navigation in Ancient China

Besides the use of the Reversed Protractor to measure the altitude above the horizon by Yixing and Nangong Yue in the Tang Dynasty, there was an incredible way adopted to reach the same purpose, namely measuring the polar altitude by using the guiding-star stretch-boards (牵星板). This kind of celestial navigation skills was called the *Qian Xing Shu* (star guiding art). Yan Dunjie (1917~1988), a well-known scientist, made great efforts in this respect, which enabled us to understand more details about it. He quoted the words from *Jie-an Laoren Manbi • Zhoubi Suan Chi* (《戒庵老人漫笔 • 周髀算尺》, *The Arithmetical Instrument of the Gnomon and the Circular Paths of Heave* (Volume 1), in 1590, *Rambles from the Old Man of Jie-an Study*) written by Li Xu (1505~1592): "Ma Huaide from Suzhou City has collected one pair of guiding star stretching-out boards, which have 12 small chips made from ebony. Their lengths are various from small to large, of which the longest is over 7 *cun* (寸, a Chinese unit of length, about 1/3 decimetre). They were marked as one *zhi* (指, one kind of units about scale), two *zhi*, up to 12 *zhi*. All these data have been carved in the chips for measuring the height, just like the rulers to measure length." "Besides, there was a piece of ivory, two *cun* long, with four angles missing, on which there were equal parts of semi-*zhi*, semi-*jiao* (角), one-*jiao*, tri-*jiao* and so on, and they were reversed in opposite directions. This was so called *Zhoubi Suan Chi* (《周髀算尺》, *The Arithmetical Instrument of the Gnomon and the Circular Paths of Heave*) (Figure 154)."

The boards, which are made of diospyros ebenum, can locate directions and geographic latitudes of ships in the sea. They consist of 12 square boards of different sizes, with an increase of 2 cm for each board. The smallest board is 2 cm×2 cm, while the largest board is 24 cm×24 cm (about 7.6 *cun* in the units of *chi* in the Ming Dynasty). The smallest one is called one *zhi*, so the boards

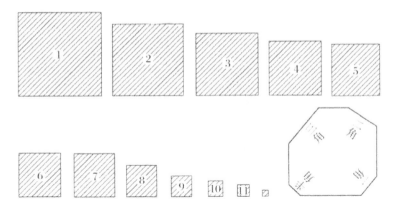

Figure 154   Guiding-Star Stretch Boards, from Yan Dunjie (1966)

vary between 1 and 12 *zhi*. A rope is fastened through the centers of these boards, and the rope's length is the distance from the eye to the hand holding the boards, equivalent to the length of one stretching arm. When in use, the left hand holds the boards, the right one pulls the rope and makes it straight (Figure 155). When the top of the board aims at the North Star, and the bottom of the board is flush with the horizontal line, the horizontal altitude between the place and the North Star can

Figure 155   Drawing of Operation of a Guiding-Star Stretch-Board, painted by Pan Jixing (2001)

be probably measured. The unit under *zhi* is called *jiao*, which can be calculated with an ivory board of 6 cm×6 cm. The four corners of the board are cut in different sizes, indicating half *jiao*, one *jiao*, two *jiao* and three *jiao*. One *zhi* equals four *jiao* or one *jiao* is 1/4 *zhi*. If the top and the bottom of one board are not asymmetric through observation, then it should be replaced by another one, thus 12 chips or ivory boards can be changed in turn. After the polar altitude is given, the geographic latitude can be calculated by Gougu theorem (勾股之理, Pythagorean proposition).

Some related data from *Wu Bei Zhi* written by Mao Yuanyi (茅元仪, 1570~1637) in the Ming Dynasty showed that Zheng He (1371~1435), the famous navigator, in his last great voyage in 1430 (宣德五年, the 5th year of the Xuande period，Ming Dynasty），also referred to celestial navigation to locate directions besides compass. In *Zheng He's Navigation Map* (《郑和航海图》，1422~1430)，his helmsmen often used the units of *zhi* and *jiao* to measure the height of the stars above ground level in different places. This is another example of application of *qianxing* technology. *Shunfeng Xiangsong* (《顺风相送》，*Voyage with a Tail Wind* or *Wind Both Hands*)，one contemporaneous book with the map，also described the similar navigation technology，the collection of which is now housed in the Bodleian Library of the University of Oxford. Annotated Works of the book by Xiangda（1900~1966）was published by Zhonghua Publishing House in 1961. According to the exponential difference and latitude difference of the North Star at the observation site，professor Yan Dunjie estimated that one *zhi* equals a value between $1°34'~1°36'$. He also compared the latitude of the site with the the height of the the North Star above ground level（AGL），and figured out the average modification value of $4°54'$，because the North Star is not in the area of the pole. Therefore，the sum of the polygonal number of the North Star and the modification value is the geographic latitude，which can be further proved by Zheng He's Navigation Map（《自宝船厂开船从龙江出水直抵外国诸番图》），originally named as *Map of Navigating to the Western Oceans，beginning from Boalong Factory，then through Longjiang River to the Sea*).

For instance，in the map，*Kezhi Guo*（柯枝国，Cochin，a major port city on the south-west coast of India by the Arabian Sea and the Laccadive Sea）was marked 3 *zhi* and one *jiao* with reference to the North Star，equivalent to $5°12'$. With the modification value $4°54'$ taken into consideration，then the site was situated at about $10°06'$ north latitude. Guli Guo（古里国，present Calicut City，it is a metropolitan city in the state of Kerala in southern India on the Malabar Coast）was marked 4 *zhi*，equivalent to $6°24'$. After the addition of the modification value $4°54'$，the latitude of the place is $11°18'$. Related instances such as these were extremely common in the map. Together with *Zheng He's Navigation Map*，there are four *Qianxing Maps Across the Sea* (《过洋牵星图》). Figure 156 was just the *Qianxing Map from Sri Lanka to Samdura* (《锡兰山回苏门答剌过洋牵星图》)，which described the voyage from present Sri Lanka to Samudra Islands of Indonesia，namely，about $10°$ between the equator and the north latitude，where the North Star could not be seen，so it was replaced by the azimuth stars in the south pole. Professor Yan Wen thought that Huagai Star（华盖星，Canopy of the Emperor）was ursa minor β and γ（β UMi and γ UMi）. Moreover，Denglonggu Star（灯笼骨星，the Southern Cross or Crux）was α Car or Old-man Star called by Chinese. Northwestern Busi Xing （西北布司星，β Geminorum/Pollux）was α auriga （4 *zhi*，equivalent to 6°24′.）. Southwestern Busi Xing （西南布司星，α Canis Minrois/ Procyon）was α Scorpius. Nanmen Shuangxing（南门双星，Alpha Centauri，binary stars）was α grus and β grus （15 *zhi*，equivalent to 24°）. Zhinüxing（织女星，vega）was α lyrae. The North Star in Sri Lanka was only one *zhi* （1°35′），which could not be seen any more when ships moved further to the south.

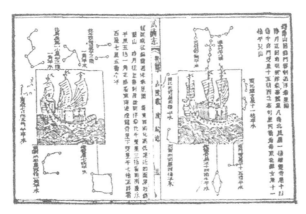

Figure 156   Map Drawn by Using a Guiding-Star Stretch-Board in
Zheng He's Navigation Map（《郑和航海图》，1422~1430），contained in *Wu Bei Zhi*

A research team of astronomical navigation，which was composed of 11 scientists from different research institutes （such as the department of geography in South China Normal University （SCNU），the National Astronomical Observatories of the Chinese Academy of Sciences （NAOC），Guangzhou shipyard international Company Limited （GSI） and Shanghai Maritime University （SMU）），has systematically collected these historical materials of ancient Chinese astronomical navigation and surveyed the traditional technology of civilian celestial navigation. When studying the *Qian Xing Shu* （guiding star stretching-out art），they estimated the unit of *zhi* equivalent to 1.9° or 1°54′. They pointed out that Mawangdui Silk Texts *Wu Xing Zhan* （《五星占》，*Divination by Five Stars* or *Prognostics of the Five Stars*，170 B.C.） unearthed in the 3rd Graves of the West Han dynasty of Mawangdui Tombs in Changsha City of Hunan Province still adopted *zhi* as the angular unit. The book read in part："When the moon meets Venus （Great White Planet），and the moon appears to the south of the Venus，Yang Guo （阳国，the central state） should resort to arms. When the moon appears to the north of

Venus, then Yin Guo (阴国, Noncentral countries) would use military forces ... supported by the army. They divine in the army that 3 *zhi* of the star would make people worry about the safety of the capital, and 2 *zhi* of the star ... " The team further gave more details in the book *Kaiyuan Zhan Jing* (《开元占 经》, *The Divination Classics of Kaiyuan era in the Tang Dynasty*, 729). The book mentioned with citations from *Wuxian Zhan* (《巫咸占》, *Divination Given by Diviner Wuxian*) in the Han Dynasty that the maximal angular distance of the south and the north between Venus and the moon was 5 *zhi*, thus people could calculate that the maximal declination difference between Venus and the moon was 9.4°. By this inference, the team concluded that one *zhi* equals 1.9° or 1°54′. They thought that the the method of measurement in *qianxingshu* (guiding star stretching-out art) Chinese astronomical navigation history was employed very early, which could be dated back to the Warring States Period and the Han Dynasty.

What is noteworthy is that the set of *Qian Xing Ban* (牵星板, guiding-star stretch-boards) was collected by Ma Huaide, which had been narrated above through the introduction of Li Xu in the Ming Dynasty. Ma Huaide (1015~ 1065), also known as Dezhi, from Xiangfu County in Kaifeng City (开封祥符) in the Northern Song Dynasty, joined the military, and later served as an inspector for East Lu (东路巡检) recommended by Fan Zhongyan (989~1052) and Han Qi. He made unusual contributions by defeating Xixia forces (西夏, the Xixia regime, 1038~1227) repeatedly, and was promoted as *a deputy of jiedushi in Jingnan Army* (静难军节度观察留后) at the reign of Emperor Yingzong (1064), and moved to Suzhou after his retirement. What made Joseph Needham regrettable was that the achievement of this military officer of the 11th century in celestial navigations has not been recorded in *Chouren Zhuan* [《畴人传》, *Biographies of Mathematics and Astronomers*, edited by Ruan Yuan (1764~1849) in the Qing Dynasty], neither in any biographies for these philosophers. After all, the *Qian Xing Shu* (guiding star stretching-out art) of the Northern Song Dynasty (960~1127) had been carried down to the early Ming Dynasty by going through the Yuan Dynasty. In visiting the boatmen in Wenchang county of Hainan province, the team learned that these boatmen had their own ways to measure the height of the stars. These people often observed the height of the stars by holding a graduation ruler with upright arms, and the bottom of the ruler should be flush with the sea level. They also got some related material *Zhenlu Bu* (《针路簿》, *Reference of Navigation in the Sea*) collected from a sea fleet in Xiamen of Fujian Province. In the chapter *The Method to Measure the Height of Meridian* (定子午高低法) of the book, it said that the height of Merdian in Luzon in the Philippines was 5 *cun* and 6 *fen*, which meant that the height of the North Star above ground level (AGL) in Luzon was 5 *cun* and 6 *fen*. This was another method by which to observe the location of the positioning star by a graduation ruler instead of *qianxinban* (guiding star stretching-out boards), however both methods were based on the same principle. The only difference lay in their units of measurement, but these units could be fully cross-referenced by interconversion. Moreover, the application of the method must be time-honored.

According to the research report given by Han Zhenhua[1], a star-surveying ruler was found in the 13rd cabin for helmsmen in one seagoing vessel of the Song Dynasty unearthed in Xiamen of Fujian Province in 1973. It was made of bamboo; 20.7 cm long; 2.3 cm wide. Half of the ruler was divided into five units, 1 *cun* for every unit following the regulations of a small-size ruler in the Tang Dynasty. The large size of a ruler (31.1 cm) equals 1 *chi* and 2 *cun* of the small- size ruler, so here every unit would be 2.6 cm. The other half was not divided for the convenience of holding it in hand and its length was similar to 3 units or 3 *cun* (7.8 cm). Mr. Han thought, when in use, the ruler would be perpendicular to the arm, the top of the ruler aims at the measured star and the bottom is tangent to sea horizon. The length of the arm is about 20 cm. The ratio between the length of the arm and half of the ruler approximately amounts to the ration between the ruler and the height measured by gnomon (5 : 1). Its calculation result is that one *cun* equals about $2°50'$ on average. He still thought that the star surveying ruler had different specifications and types, such as 12 units, 10 units, and 20 units, etc.

In general, we agree with Mr. Han on his conclusion and related judgment, but we want to give some complementary discussions and comments. We think that there are 10 small units for *fen* in the big unit of *cun*, otherwise we cannot read the text "the height of Merdian in Luzon in the Philippines was 5 *cun* and 6 *fen* (吕宋子午高五寸六分)". This kind of small units was finely carved, which could not be read because it had been buried in the ground for a long time. At the time, measurement data could not simply be read if the ruler

[1]　韩振华. 中国古代航海用的量天尺[J]. 文物集刊,1980(2).

孙光圻. 中国古代航海史[M]. 北京:海洋出版社,1989.

was perpendicular to the arm, with its top end aiming at the measured star and its bottom end tangent to the sea surface. We deem that only by making the top end aim at the measured star and observing which graduation of the ruler is tangent to sea surface could they read the measurement data. Another method is: first set up a movable pole perpendicular to the ruler, which can move between different graduations of the ruler; then straighten the left arm and hold the half of the ruler without graduation; the right hand moves the movable pole; when the top of the pole aims at the star and the bottom is tangent to the sea surface, the measurement data is given. This is just the Cross-Staff Surveying Instrument that Joseph Needham has mentioned in his work or so-called Jacob's Staff in the western world.

It is generally acknowledged that the Cross Staff Surveying Instrument mentioned by Jews Levi ben Gerson (1288 ~ 1394) was used in celestial navigation in the West after 1321, but this conclusion is unreliable. Because the earliest possible age for the surveying instrument was 1571, there were no reliable documents to prove that the western people used the cross staff surveying instrument before the 16th century. ❶ Chinese used the instrument for navigation at least during the 11th century in the Song Dynasty. *Song Huiyao Jigao* recorded that in the 3rd year of Emperor Jianyan (1129), Lin Zhiping, the supervisory censor, began to take charge of the national defense of Yangtze River and coastal regions. He asked that all ships of Fujian and Gunagdong should be armed with *wangdou* (望斗, the dipper observer), cannons, rockets and other fire-protection devices. Here the mentioned *wangdou* is obviously an instrument for celestial navigation through observing the height of the North Star above ground level (AGL). When combined with the star surveying ruler (量天尺) unearthed in a sea vessel of the Song Dynasty in 1973, these materials proved the conclusion that the cross staff surveying instrument had been invented and used in the 11th century in China, 300 years earlier before Europe learned its secret as it had claimed. ❷ The time difference could prove that the western surveying instrument came from China. The measuring instrument had its origin very early in China. *Wangshan* (望山), the sighting device with

❶ BEAUJOUAN G, POULLE E. Les Origines de la Navigation Astronomique Aux 14e et 15e Siècles[M]//Proceedings of the First International Colloquium of Maritime History. Paris, 1956.

❷ NEEDHAM J, et al. Civil Engineering and Nautics[M]. Cambridge: Cambridge University Press, 1971.

graduation in ancient crossbows just operated with the same principle. So, the Song Dynasty used the cross height measuring instrument *Liang Tian Chi* （量天 尺, the star surveying ruler） as celestial navigator, which was only the lower boundary of the time of its use.

Joseph Needham also thought there were reasons to believe that Chinese helmsmen continued to use the cross staff surveying instrument as the type of the star surveying ruler in the Song Dynasty to navigate in the sea, while keeping using *Qianxingban* （guiding star stretching-out boards） for navigation, one of which Ma Huaide had collected. It should be noted that the movable pole in the star surveying ruler unearthed in Quanzhou city fell off long ago under the ground, with only a bamboo ruler left behind. Interestingly enough, some Arabians who sailed in the Indian Ocean also used guiding star stretching-out boards for navigation. It was called *kamal*. Every set of it had nine square boards on which there were knotted ropes. ❶ Each rope shared a standard length, as long as an arm. By using it, they observed the stars to calculate the angle between the horizon line and the star. The unit of the angle was *isba*, similar to *zhi* in China. One *isba* was approximately equal to $1°36'25''$, and $1/8$ *isba* was called *zam*. It is obvious that the technology of Arabian celestial navigation was the same as Chinese *Qian Xing Shu*, thus a question arises: which one came first? As previously mentioned, Chinese helmsmen used the guiding-star stretch-boards and the star surveying rulers for cross-staff surveying instrument to observe the height of the azimuth stars above the ground level and location in the sea at least in the 11th century in the Northern Song Dynasty, which could be confirmed by unearthed objects as well as documentary records. So far, there is no evidence to prove that the sailors of Arab and India have used instruments to measure the height by observing azimuth stars 1300 years ago. It means that the Arabian technology of guiding star stretch board has been obtained from China.

Europeans knew *Qian Xing Shu* （guiding star stretching-out art） through the nautical book *Mahit* in Arabian, which was written by one Turkish soldier Reis ibn Husain in 1553. The book was based on the collected data in India, mainly drawn from the book written by Arabian Sulaiman Al Mahri （1480～ 1550） and the book *The Book of Benefits on the Principles of the Science of*

---

❶   PRINSEP J. Note on the Nautical Instrument of the Arabs[J]. Journal of the Royal Asiatic Society of Bengal, 1836(5):784.

Navigation (*Kitāb al-Fawā'id fi usul'ilm al-Bahr wa'l-Qawa'id*) written by Shihāb Al-Dīn Aḥ mad Ibn Mājid (1436~1500) in 1475. Shihāb Al-Dīn Aḥ mad Ibn Mājid also participated in the sailing expedition hosted by Portuguese Vasco da Gama (1460~1524) as Arabian navigator. Later Portuguese used *Qian Xing Shu* for a period of time, and translated *isba* into the language of Portuguese as *polegada*, and considered it as a unit of measurement. Thus it can be seen that the *Qian Xing Shu* spread to Europe for celestial navigation through the Arabian media.

4.2   Invention of *Sinanyi* (**South-Pointer**), **Predecessor of Compass**

4.2.1   Discovery of Polarity of Lodestone and Manufacture of *Sinanyi*

Although ancient people determined direction by observing the Sun's Shadows in the day or the stars at night, these methods had their weak points. In the event of bad weather, people could not see the sun in the day or the stars at night, therefore, this kind of method could not work, because it was not the all-weather technology of navigation and orientation. When meeting wild weather in the sea, ships had to follow the waves. The sailors had to wait and adjusted their directions when it turned fine again, thus they became passive in navigation. So did traveling or marching on land or in the desert. The use of the late-model directional device made by applying magnetic methods could overcome the limitations of celestial orientation. Because the earth is a magnificent magnetic substance, there is magnetic field in the earth and near its terrestrial space. If a lodestone were made in the form of a strip and in a state of free rotation, then in the effect of the planet's magnetic field, the two ends of magnetic bar always point to the north and the south when the rotation stops. With that mechanism, the pointing device thus constructed can operate in all weather conditions, because the earth magnetic field will not be affected by the weather.

The magnetic pointing device does not rely on any heavenly body, but on the earth magnetic field to give directions at any time in any place, which help people to change from celestial orientation by observing heavenly bodies to magnetic orientation. The earliest magnetic pointing device was *sinanyi* made by natural permanent magnet bar, which appeared in the late period of the

Warring States （in the 3rd century B. C.） and further developed in the Han Dynasty. The invention of *sinanyi* was of historical significance because it was a new-type navigational device with magnetism，totally different from the principle of astronomical orientation，which could work in any weather conditions，give directions promptly and was convenient to operate simply and portably. When technological improvement was applied to *sinanyi*，then the world's first compass came into being，so *sinanyi* was the predecessor of the compass. It was a significant reform on the ideas of human development as well as a giant leap for humankind in learning about nature，when people transfered the objects on which pointing devices had relied from the sun and stars to the earth essential for human existence. Hence，there was an alternative for orientation determination.

The principle of *sinan* was based on magnetism，in which the ancient Chinese were experts in. As early as the Warring States periods，the magnetic force of magnetite was discovered and recorded in some books，such as in *Classic of the North Mountains* in *Classic of Mountains and Seas* （《山海经·北山经》，5th century B.C. to 4th century B.C.），which said that in Guanci Mountain there were a lot of lodestones. ❶ In another work，*Dishu* of *Guanzi* （《管子·地数》，the 4th century B.C.，*Guanzi · On Administering Financial Transactions According to the Geographical Conditions*），in Guan Zhong's name （管仲，720 B.C. ~ 645 B.C.，a chancellor and reformer of the State of Qi during the Spring and Autumn Period of Chinese history，a scholar of the Jixia Academ），said that "When lodestone appears on the surface of a mountain，there must be a copper mine inside. （上有慈石者，其下有铜金。）"❷ While they found the lodestone，they also discovered its magnetic nature. *Lüshi Chun Qiu* （吕氏春秋，300 B.C. ~ 235 B.C.，*The Spring and Autumn of Lü Buwei*，Vol. 9） written by followers of the Qin Dynasty Chancellor Lü Buwei，said that "A lodestone can attract iron for there is an attractive force in it. （慈石召铁，或引之也。）"❸Gao You （175~225），a scholar of the Han Dynasty annotated："The lodestone is the mother of iron. Because of its existence，it can attract iron. A stone with no magnetic property cannot attract any iron. ［（慈）石，铁之母也。以有慈石，故能引其子。石之不慈者，亦不能引也。］"

❶ 袁珂. 山海经校注：卷三[M]. 上海：上海古籍出版社，1980.
❷ 管仲. 管子：地数篇[M]. 杭州：浙江人民出版社，1984.
❸ 高诱. 吕氏春秋：季秋（精通）[M]. 上海：上海古籍出版社，1989.

As we all know, Fe_3O_4 is the main ingredient of natural lodestone, and it can attract such iron-group material as iron, nickel and cobalt etc. Its magnetism comes from its internal charge movement. Ancient people found that the stone could attract iron, just like a kind mother summoning her sons to come together, so the stone was called *cishi* (慈石, loving stone). Based on the connotation of the character 慈 (*ci*), another character 磁 (*ci*) was coined, which was a homophone for 慈 (*ci*) in Chinese. An ancient Chinese text *Huainanzi · Lanming Xun* (《淮南子 · 览冥训》, 120 B. C., *Huainanzi · Peering into the Obscure*), which was edited by followers of the king of Huainan Liu An in the Han Dynasty, also discussed the property of a lodestone of attracting iron, and gave a preliminary theoretical explanation, which was rare in ancient books from western countries. It says: "However, if one tries to use a lodestone to attract a tile in the way it attracts iron, this is difficult. It is natural that the myriad things cannot be differentiated in terms of importance. With regard to the phenomena of *yangsui* (the sun-igniter made of brass speculum of concave form) producing fire with the help of sunlight, magnet drawing iron, crabs ruining lacquer, and sunflowers turning towards the sun, even people of excellent intelligence cannot explain." The book pointed out that the lodestone can attract iron but not a tile because substances possess generality and diversity as wells, thus making it so. Lodestone and iron share some general character, so they can attract each other, while there is a difference between a lodestone and a tile, so it is difficult to draw together. If the talents do not research these phenomena, they still cannot understand why they happen. This is the theoretical enlightenment brought to us by the book *Huainanzi*.

Basic Annals of Emperor Qinshihuang (《秦始皇纪》) in *Records of the Historian* (《史记》) written by Sima Qian recorded that the Emperor constructed the massive E-Pang Palace (阿房宫) in the 35th year of Emperor Qinshihuang. Later, Zhang Shoujie (750~820) gave annotations of the paragraph by citing the proof from the works *Sanfu Jiushi* (《三辅旧事》, *Reminiscence in Chang'an*) written by Wei Shi of the Jin Dynasty. He said: "E-Pang Palace is 3 *li* (1 li = 500 meters) long and 500 *paces* wide. In addition, it can accommodate over 10, 000 people. In the building are also built 12 bronze human figures in front of the palace, the gate of which is made of lodestone." Quotation in the *Sanfu Huangtu* [《三辅黄图》, *Yellow* (*i. e.*, *imperial*) *Maps of the Three Metropolitan Areas* or *Description of Palace Buildings in* (*the Han capital*) *Chang'an*], which was written in the 4th century, also talks about it. "The front hall was constructed with Magnolia liliflora as roof beams, and lodestone as the gate, thus these people who carried blades or swords can not get in." The Records of the West Capital in the *Taiping Yulan* (《太平御览》, 983, *Imperial Readings of the Taiping Era*) also claimed that the gate of E-Pang Palace was made of lodestone, so those who carry swords could not get in. Another proof given by Pan Yue in his *Xizheng Fu* (《西征赋》, *Verse of Marching West*) said that the palace constructed by the Emperor of the Han Dynasty was rare and smart. Because the gate was made of lodestone and the roof beam of lily magnolia,❶ it can further indicate that the attraction property of a lodestone has been employed in practical applications. Since *Shennong Bencao Jing*, the lodestone had been listed as a medicine in the Chinese herbal work as well as an object of study for these alchemists and pharmacologists.

Chinese people not only found magnetism of a lodestone, but also its directivity and polarity, by means of which they invented the instrument pointing direction called *sinan*. Han Fei (280 B. C. ~233 B. C.), an influential political philosopher of the Warring States Period, who belonged to the "Chinese Legalist" school, claimed in the sixth chapter *Having Regulations: A Memorial* of *Han Feizi* (《韩非子》, 255 B. C., *The Complete Works of Han Feizi*) that a country should be run in accordance with law. If a king curbed all his officials by law, he would not be bullied by his subordinates. The king should master the statecraft, namely how to rule his people. Then, Han Fei continued:

> Indeed, the ministers would trespass against the sovereign in the court as in the undulating terrain. Step by step, they would make the king unable to stay true to the mission until he loses his direction by turning from east to west and is not conscious of the change. To guard against such misleadings, the early kings set up sinanyi (the south-pointing needle) to ascertain the directions of sunrise and sunset.

Since the change from the Zhou Dynasty to the Qin Dynasty, there were specific regulations to make monarch-subject relationship clear, and to maintain it with morality as well as by laws. There were a lot of specifications to ensure the supremacy of the rulers in their actions, costumes, mansion houses and

---

❶　李昉. 太平御览：居处部(门)[M]. 北京：中华书局，1960.

ranking in the court. Take the position of an Emperor in the imperial court as an example, generally speaking, the Emperor sits north facing the south, and his ministers stand on western and eastern sides. When they have state affairs to report or submit memorials to the Emperor, they should face the north, thus the monarch and subjects are all in their respective positions. The paragraph above taken from *Han Feizi* mainly means that if a nation lost its law system and is in disorder, then the subjects would trespass against the sovereign step by step, just like cutting down the land from the kingdom progressively, which would then lead to the disorder of monarch-subject relationship. If the sovereign did not scent this out, then the country would be in danger. So the early kings set up *sinan* (the south-pointing needle) to ascertain the orientation of the west and the east so as to control the ministers' behaviors, since *sinan* can correct directions, because it is an instrument to give directions. But here in the citation, it also serves as ritual vessel standing in front of the imperial palace to warn the subjects not to break the law of the country. The monarch governs his country by laws, ensures that the laws are strictly observed and exert influence on his subjects. In that way, the whole county would be naturally in the heyday of peace.

Sinanyi, which is based on the lodestone's directivity and polarity, has been applied and developed from the Warring States Period, via the Dynasties of Han and Jin, to the Northern and Southern Dynasties (420～589). Zhang Heng (78～139) in his *Dongjing Fu* (《东京(洛阳)赋》, 107, *Eastern Metropolis Verse*) said: "I was so shallow and ignorant, because I often took wrongs for granted, so I got lost. Now I am so fortunate to meet you to guide me as a south-pointer. (鄙哉予乎,习非而遂迷也,幸见指南于吾子。)"[1] That is to say, people realized the truth after getting astray in the wrong direction and retracted from it. It is a delicious yet subtle pun, which means the wrong life path as well as getting lost on the way. In *Wudu Fu* (《吴都赋》, 281, *Verse of Metropolis of Wu Kingdom*) written by Zuo Si (250～305) of the Jin Dynasty, he said that cavalrymen of the spearhead were driving along the road and the south-pointer gave the direction they wanted. Here, the involved *zhi nan* means the instrument that gives the right direction of marching. There was a similar expression about it in *Strategy* of *Guiguzi* written in the 3rd century (《鬼谷子》, *The Sage of Ghost Valley*). It says: "When craftsmen in the state of

---

[1]　张衡.东京赋[M]//萧统.文选.上海:上海古籍出版社,1986.

Zheng get into the mountain to exploit jade, they are certain to drive a carriage carrying *sinan*, so as not to be misled in the mountain." In the sentence above, "*zhi*" (之) in the phrase "*zai sinan zhi che*" (载司南之车) is a polysemic character, which means "于" (*yu*, in, or on) here, thus the phrase means carrying the south-pointer on the carriage. "*Che*" (车) means the carriage that transports stone material. It was not the later south pointing carriage which was made by gear train with automatic clutch. There are some incorrect translations in the English version of the book, which should be translated as: When the people of Zheng go out to collect jade, they must carry a south-pointer on their carriage so as not to lose their way. The author of *Guiguzi* was called *Guiguzi* in the late Warring States period, which would be the master of Su Qin and Zhang Yi according to the legend. But he was not recorded in *Treatise on Literature* of *History of the Former Han*. Only later in *Treatises on Classic Works* of *History of the Sui Dynasty* (《隋书·经籍志》), he was listed in the School of Vertical and Horizontal Alliances. *Guiguzi* was annotated by Huangfu Mi (215~282), and the extant edition was the one in the 3rd century.

4.2.2 Shape and Uses of South-Pointer

Ancient people have mentioned many times about the south-pointer's function of orientation, but less was known in historical records about its shape and uses. Only in *Shi Ying* (《是应》, *Auguries Verified*) of *Lun Heng* (论衡, 83, *Balanced Discourses*, Vol. 17), did the famous ideologist Wang Chong (27~97) in the Eastern Han Dynasty (25~220) make an introduction as described below:

> When put the scoop handle of the south-pointer on a flat surface, its handle would always point south.

According to the description given by Wang Chong, the shape of a south-pointer is similar to a spoon. "*Di* (柢)" means the end of the spoon, namely the spoon handle. The sentence above means that the spoon handle points south. The reason for such a shape is due to big dipper (Ursa Major), because ancient people observed the North Star through it when orienting directions. If these seven stars of the constellation Ursa major were connected with a line, then the shape of these stars would be similar to the shape of a spoon, also called *doushao* (斗勺), consisting of a *dou* (斗, a kind of grain container) and a *shao* (勺, a spoon). Since the Warring States Period and the Qin and the Han Dynasties, people have correlated the North Star with the south-pointer. The former is in the sky, while the latter is on the earth, but both of them can give directions. The aim of setting up the south-pointer in front of the great palace is to ascertain the relationship between the sovereign and ministers, and the North Star can be embedded into the jade article with the shape of a spoon. Both of them are also the symbol of monarchy in some sense. According to *The Biography of Wang Mang* of *History of the Western Han Dynasty* (《汉书·王莽传》), after Wang Mang (43 B.C. ~23 A.D.) proclaimed himself as Emperor, "he ordered to cast a *weidou* (威斗) vessel in the southern suburb of the capital Chang'an in August in the 4th year of Tianfeng (17 A.D.). *Weidou* was made in minerals of five colors (copper of five colors), shaped like the northern Dipper. It was 2.5 *chi* (尺, a traditional unit of length, equal to 1/3 meter) in length (57.6 cm), in the hope that its magical properties would deter a rise of military activity. Once the vessel was made, he ordered his *siming* (司命, the Director of Fate, who is tasked with calculating all mortal life-spans.) to carry it on his shoulder. When Wang Mang went out of the great palace, his followers waited upon him at his side." According to the *Bao Pu Zi* (《抱朴子》, *Book of the Master who Embraces Simplicity*), five minerals should include realgar (雄黄), cinnabar (丹砂), orpiment (雌黄), alum (矾石) and laminar malachite (曾青) showing different colors of yellow, red, white and green, which symbolized *Wuxing* (五行, the five elements: metal, wood, water, fire and earth), probably used to decorate the bronze ware *weidou*. Accompanied by his ministers and warriors, he ordered his official in charge of business of *wuxing* to carry *weidou* on his shoulder, and assumed himself as the star of the celestial pole on the earth. By performing such a ceremony of divination (combined with Taoist belief), Wang Mang wanted to bring all the people to his knees by its magical properties.

In the Northern and Southern Dynasties, *weidou* was unearthed in Nanjing city, which had been well preserved until the Song Dynasty. According to the description of *Yeke Congshu* (《野客丛书》, 1210, *Collection by A Villager*, Vol. 13) written by Wang Mao (1151 ~ 1515), *weidou*, which Han Yu collected, was made in the reign of Wang Mang, with an inscription of *the 4th year of Tianfeng* on it. Its shape was like a *shao* (勺, a spoon), 1.3 *chi* long (39.4 cm; the Song Dynasty length unit), weighing 3 *jin* (a unit for measuring weight equal to 10 *liang* in the Song Dynasty) and 9 *liang* (about 920 g). ❶ *Weidou*

---

❶ 王楙. 野客丛书:新莽威斗[M].扬州:广陵古籍刻印社,1984.

itself was not the instrument to give direction. Only the south-pointer shared such a function. However, according to the description by Wang Chong, its shape was similar to a *shaodou*. *Weidou* was made of bronze, while the south-pointer was made of lodestone. They were different in quality as well as in property.

As for its use, Wang Chong only mentioned that people only "put the scoop handle of the south-pointer on a flat surface". He did not give specific descriptions about its use. Then, what does *di*（地）mean? How can it give direction when put on a flat surface? It has been a problem without any explanation for a long time. Later Wang Zhenduo（王振铎，1912～1992），a well-known historian of science, in his article《司南、指南针与罗经盘》(*The South-pointer, Compass and Fengshui Compass*) successfully solved the puzzle. He also performed the recovery research for *sinanyi*. Based on his study, *di* in *Lun Heng*（《论衡》，*Balanced Discourses*) was not the land but a square earth plate of diviner's board for divination, also called *shi*（an instrument for divination in ancient China）. ❶ *Rizhe Liezhuan*（《日者列传》，*The Biographies of the Soothsayers*) of *The Records of the Grand Historian* said: "These days, when diviners perform divination, they must follow the law of nature, get imago through recognition of changes of the four seasons, and conform to the principle of righteousness, thus distinguishing various methods of divination and determining the hexagram. Then turn *shi* and practice divination formally, explain gains and losses of this secular world and foretell good or ill luck and success or failure." Sima Zhen（司马贞，679～732）of the Tang Dynasty gave annotation and said: "In the original work, *shi*（式）is *shi*（杖），*xuan*（旋）means turning or rotating. The circle above symbolizes the heaven, and the square below means the earth. When using it, turn the law of the heaven and match the time of the earth, thus the method is called *xuanshi*（旋式，turning *shi*）." So *shi* is composed of two parts: one is the square earth plate; the other the round heaven board with the latter above the former, but both can turn and share the same axis.

According to *The Biography of Wang Mang* of *History of the Former Han*, until the fourth year of Dihuang of the Xin Dynasty, Liu Xiu（刘秀，6 B.C.～57 A.D., Emperor Guangwu of Han, the founder of the Later Han or Eastern Han）marched to Chang'an to attack Wang Mang and defeated him:

---

❶　王振铎. 司南、指南针与罗经盘: 上[J]. 中国考古学报，1948（3）: 119-260.

Fire reached the Hall of Chengming, where the daughter of Wang Mang lived. Wang Mang hurried to escape from the fire to the front Hall of Xuanshi, but the fire sneaked its way out and still followed him. The palace attendants and maids or ladies cried what they had to do. At this time, Wang Mang could not save them, but wore the dark blue and red clothing, carrying the seal of the Emperor with silk ribbon and holding the dagger of Yu Shun, with his official in charge of astronomy to take shi *ahead to measure the time. The official turned the handle of* shi *to measure direction, while Wang Mang seated himself facing the direction that the handle pointed, and said, "Now that the heaven blessed me virtue and chose me as the Great, then what can these armies of the Han Dynasty do to me?"*

Just at this time, the army rushed to the palace, and Wang Mang was killed. Therefore, something can be inferred from this record that before the death of Wang Mang, he used *shi* for divination and seated himself facing the south as indicated by the south-pointer.

Of the shape and use of ancient *shi* for divination, many documentary records of the Han Dynasty had explicit and specific explanations, such as in the *Treatise on Literature* of the *History of the Former Han*, *Classics of Shi* of different versions since the Warring States Period was recorded. In 1925, in one of the unearthed cultural relics from the tomb of Wang Xu in the Lelang Prefecture, which was established by the government of the Han Dynasty at that time in the Korean Peninsula, the faculty of letters of Tokyo University found a set of diviner's boards made of wood smoothly painted, with a square earth plate and a round heaven plate. The board was damaged, and the Japanese scholar Tazawa Kingo made a recovery of it. The square earth plate and the round heaven plate were all made of wood. The outer was painted black, red and yellow. The diameter of the heaven plate was 3 *cun* (13. 5 cm), and 5 mm thick. On the surface of the board there were six red *circles* (Figure 157): in the first circle, there was the image of Big Dipper; in the second, the gods of 12 months; the third, blank; the fourth, 24 directions composed by eight *Tiangan* (heavenly stems, 甲乙丙丁庚辛壬癸), twelve *Dizhi* (Earthly Branches, 子丑寅卯辰巳午未申酉戌亥) and four *Gua* (Diagram, 乾艮巽坤), of which, four forward directions were divided by *zi*, *wu*, *mao* and *you* (子午卯酉), and four dimensions matched with *ji*, *ji*, *wu* and *wu* (己己戊戊); the fifth and the sixth were divided by four black lines corresponding to *ji*, *ji*, *wu* and *wu*（己己戊戊）. Every circle had four lines respectively.

The earth plate is a square of 20.5 cm × 20.5 cm，and 5 mm in thickness，with quadruple grid squares and four diagrams（乾艮巽坤）in four opposite angles of each grid. Four diagrams are inscribed in the grids of the directions：the direction of *mao* （East）is *zhen* diagram；the direction of *you* （West）is *dui* diagram；the direction of *wu* （South）is *li* diagram and the direction of *zi* （North）is *kan* diagram. There are also four constellations in the grids：*fang* （房，Room，π Sco），*mao*（昴，Hairy Head，17 Tau），*xing*（星，Star，α Hya）and *xu* （虚，Emptiness，β Aqr） etc.

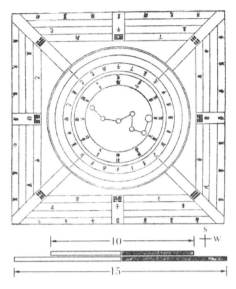

Figure 157　Replica of *shi* of the Eastern Han Dynasty （25~220）unearthed in Lelang Relics in 1925，by Tazawa Kingo （unit：cm）

Quadruple grids are from outside to inside. The first lists twenty-eight lunar mansions，seven for each side. The second is blank，and the third one is inscribed with *zi*，*wu*，*mao* and *you* （子午卯酉），which forms 12 *dizhi* （地支，12 earthly branches）by matching *haichou* （亥丑），*siwei* （巳未），*yinchen* （寅辰）and *shenxu* （申戌）. The fourth square is the innermost one inscribed with eight heavenly stems. The direction of *mao* has *jia* and *yi* （甲乙），the direction of *you geng* and *xin* （庚辛），the direction of *wu bing* and *ding* （丙丁），and the direction of *zi ren* and *kui* （壬癸）. The words are black-painted in small seal scripts，and the red line is used as columns. The blank of the outer tangent line in the innermost part of the heaven board is painted with red lacquer，and the North Star is painted red in the plate. One of the north stars is pierced through the center of a circle，used for putting an axis in the center of a square earth plate，thus the two boards conform to each other and the heaven plate could be turned.

From March to April in 1972，a set of *shi* was unearthed in the 62th tomb of the Han Dynasty in the Mozizui Village in Wuwei County in Gansu Province. ❶

---

❶　甘肃省博物馆.武威磨嘴子三座汉墓发掘简报[J].文物，1972（12）：9-19.

The tomb belonged to the joint burial tomb of a couple, which could be proved by two articles on the lacquer ear cups of the first year of Suihe (绥和元年, 8 B.C.), and its time was proved to be the end of the Western Han Dynasty. The main body of *shi* was wooden with painted lacquer, dark brown. The diameter of the circle heaven plate is 5.6 cm, 0.2 cm thick for each side, and its center is 1 cm thick. The square earth plate is 9 cm × 9 cm with the center pierced, which could connect with the center of the heaven plate with a bamboo axis. The heaven board could be turned, and the innermost is inscribed with the north stars with bamboo pearls. The fifth star is the center, and a fine line connects these stars in series. The second circle is incised with 12 moon gods. The outermost circle is inscribed anticlockwise with twenty-eight lunar mansions (Figure 158). There are two levels of inscribed characters in the earth plate: in the inner one, the eight heavenly stems and 12 earthly branches are incised in seal characters, arranged clockwise. There are 20 words in all. The four characters 子午卯酉 (*zi wu mao mou*) are inscribed in the grid, and its bottom is inlaid with bamboo pearls. Twenty-eight lunar mansions are inscribed on the outer periphery, seven constellations at each side with the arrangement similar to the heaven board. There are four radiating doublets in the centre of the earth plate, joining the four angles, and each angle is inlaid with one big and two small bamboo pearls, which are used to mean four diagrams of *qian*, *kun*, *xun* and *gen* (乾坤巽艮). Small dots are inscribed above the characters meaning the graduation. The present heaven board shares over 150 scales, while the earth plate 182, which represents that one ecliptic is 365 1/4 degrees.

Besides Lelang and Wuwei, another *shi* was unearthed in the tomb of the first Lord Xia Hou Yin (汝阴侯夏侯婴) of the Western Han Dynasty in Fuyang, Anhui Province. Its time was during the reign of Emperor Wen of Han (文帝, the fifth Emperor of the Han dynasty), which was the earliest *shi* in the history. Besides the painted wooden *shi*, there were ones made of bronze. The four-door Mirror rubbings of the Han Dyansty recorded by a specialist in epigraphy Liu Xinyuan in the Qing Dynasty, was a real earth plate of *shi* of the Han Dynasty, 14 cm long, and 13.7 cm wide, with a round slot in the center. The longest diameter of the incurvated part of the board was 8.7 cm, and the shortest diameter 8.4 cm. Tangent to the grid of the earth plate (Figure 159) in the blank of four angles there were 3 round vacant and solid buttons. The earth plate had ternary grids, the innermost incised with eight heavenly stems (干), the second grid 12 earthly branches, and the outermost twenty-eight lunar mansions. The four diagrams of *qian*，*kun*，*xun* and *gen*（乾坤巽艮）had no images of the hexagrams provided，each marked as Xutian Gate（戌天门），Jiru Gate（己入门），Wuchu Gate（戊出门）and Jigui Gate（己鬼门）. The 12 earthly branches were decorated with the four spirits（Dragon，White Tiger，Suzaku，Basaltic beast），which represented four directions of the east，the west，the south and the north.

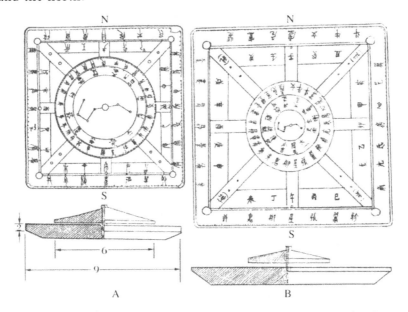

Figure 158   Painted *Shi* of the End of the Western Han Dynasty unearthed in Wuwei，Gansu Province（A），and Interpretation（B），from *Wenwu*（*Antiquity Journal*）（unit：cm）

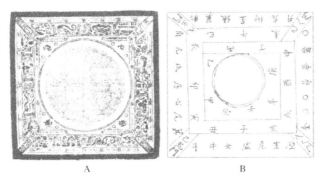

Figure 159   The Four-door Mirror Rubbings of the Han Dynasty recorded by Liu Xinyuan of the Qing Dynasty（A），and Interpretation（B）

From Figure 159，it can be seen that though four diagrams，eight heavenly stems and 12 earthly branches are located at different positions in the bronze plate recorded by Liu Xinyuan，people could still find that the surface of the board is divided into 24 equal parts by connecting the characters on the plate to the center of the heaven board with dotted lines (Figure 160). Then, if the dotted lines continue to be extended, they would be intersected with the painted circle outside the plate. Then it can be inferred that each orientation difference is 15° (Figure 161). The order mechanism of 24 orientations about *shi* of the Han Dynasty is the basis not only of the later south-pointer, but also of the later compass. Namely, the surface of the earth plate is the surface of the early south-pointer. When the painted image of the North Star in the heaven board was replaced by a spoon of magnetism, *shi* turned into the south-pointer. This is the main idea contained in *Lun Heng (Balanced Discourses)* written by Wang Chong, which laid a good foundation for the recovery of *sinanyi*.

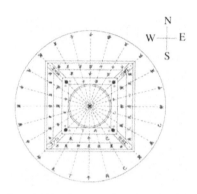

Figure 160 Permutation Graph of 24
Directions, the Han Dynasty,
painted by Pan Jixing (2001)

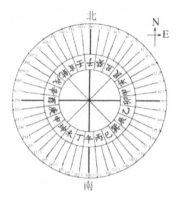

Figure 161 Comparison Diagram of
360° about *Shi* of Ancient China,
24 Directions of Compass and Modern
Compass, painted by Pan Jixing (2001)

According to the experimental research of the recovery by Wang Zhenduo, the spoon of the south-pointer was made by natural lodestone, which was chosen and polished according to the direction of the earth magnetic field. If this kind of lodestone spoon were put on a wooden earth plate, then its directivity and polarity would be hard to display because of huge frictional resistance, but if put on the polished surface of a bronze-made earth plate, the handle would always point to the south when the spoon's rotation stops. ❶ Figure 162 is a Recovery Picture of the South-pointer by Mr. Wang Zhenduo.

The recovery model of the south-pointer made by Mr. Wang Zhenduo was often put on display at home and abroad. But today it seems that his

---

❶ 王振铎. 司南、指南针与罗经盘：上[J]. 中国考古学报,1948(3).

reconstruction project is worth more discussion for improvement. His practices are still open to discussion that the earth plate of the south-pointer which he reconstructed was the same as the earth plate in the *shi* for divination, and the magnetic spoon should be polished rigidly as the shape of a real spoon for tableware. It is well-known that the south-pointer is used for giving directions instead of divination, so its earth plate should be compact, and it should not copy from the earth plate for divination without any change. In other words, at least the outermost should be dispelled. It is enough only to preserve 24

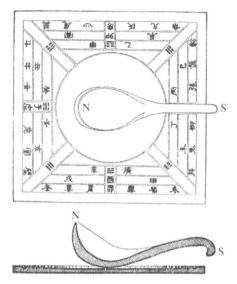

Figure 162   *Sinan* of the Han Dynasty and its Restoration Image of the Earth Plate, from Wang Zhenduo (1948)

directions composed by eight heavenly stems, 12 earthy branches and four diagrams. From the compass of *Fengshui* to navigational compass, the difference lies in that the content of geomancy is deleted while the orientation knowledge is retained. Therefore, the south-pointer, which evolved from *shi*, should follow the same principle.

As for the shape of a magnetic handle, Mr. Wang polished the lodestone bar as the shape of a soup ladle, with a large body (13.3 cm long) and a handle (47% of the whole length). Though it can point south, technically speaking, it is not a practically plausible shape. *Piao Fu* (《瓢赋》, *Verse on Gourd Ladle*) written by Wei Zhao of the Tang Dynasty, which was recorded in *Quan Tang Wen* (《全唐文》, Vol. 439, *Complete Prose of the Tang Dynasty*), discussed its uses, and said: "if it were taken as a toy, then the south-pointer made of gourd ladle would be better. Some people would value it because of its smallness, while some others think that its preciousness is the soul of its simplicity. (充玩好, 校司南以为可。有以小为贵, 有以约为珍。)" No matter whether we consider it from its function of pleasure or practicability, the magnetic handle should be small and the earth plate should be simple, both of which should be considered in making the south-pointer when using a gourd, because it conforms to the manufacturing principle of the ancient south-pointer. It is easy to be made and turned when it is small in shape, short in its handle and light in weight; and the content of the earth plate is concise and practical, thus the instrument could be made small and exquisite. The body of the magnetic spoon should also conform to several principles: even distribution for its magnetic domain, steady center of gravity, and easy rotation without tilt.

Therefore, when polishing the lodestone bar, people should follow the principle of self-righting doll, stabilize its center of gravity, and make its swinging radius less than the radius of the center plate, and the shape should be a hollow hemi-elliptical spherical body with a short handle, just like a practical gourd ladle instead of a real spoon. The ladle is 3 cm, the radius of center plate 7 cm, the earth plate 10 cm × 10 cm and 8 cm thick, with the center of the heaven board polished. The earth plate becomes tiny because some layers are dismissed. According to such kind of reconstruction, the south-pointer can be made into a mock-up, easy and portable for use and more mirrors its ancient shape. Now our reconstruction of it is presented as in Figure 163, which is not yet confirmed by authoritative researchers whether it is reliable or appropriate.

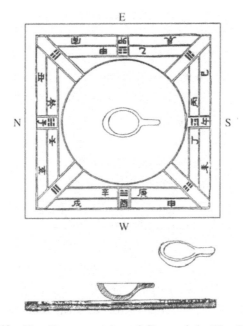

Figure 163　New Reconstruction of *Sinan* of the Han Dynasty,
by Pan Jixing (2002), in contrast with Figure 162

4.2.3　Technical Improvement of South-Pointer During the Jin and Tang Dynasties

It should be pointed out that, though the south-pointer made of natural lodestone shares the function of giving direction, its weak magnetism would affect the instrument's sensibility because it is roughly polished. Another factor which led to its reduction of its sensibility was friction resistance of the lodestone bar on the bronze board. Secondly, the lodestone in the center could not be fixed in a specific position, so people had to use hands to put it in a proper location and then turn it with hand. Therefore, it is a manual device. It can not read the orientations automatically, and neither can it be used in jolting situations. Therefore, people make efforts to change its shape and uses so as to improve its performance since the Han Dynasty. There are many factors which resulted in the improvement of *sinanyi* in many fields, such as in *kanyushu* (堪舆术, Chinese Geomancy, the study of the heaven and the earth), overseas trade, massive measurement for the undulating terrain and orientation, overseas navigational needs, etc. The invention of compass is the unambiguous result of a series of improvement. However, from the south-pointer to the compass, it experiences several technology transitions, mainly from the Dynasties of Wei and Jin, the Northern and Southern Dynasties (420~589) to the Tang Dynasty. There were few literatures to record the phenomenon of magnetic attraction of a lodestone before, but since the Han Dynasty, more literatures had been found to record it, which was essential for the invention of compass. Wang Chong pointed out in the chapter *Luan Long* (《乱龙》, *Disorder Dragon*) of his *Lun Heng*:

> *The shells of hawksbills can attract small objects through friction, while a lodestone can attract an iron needle, because both are real, which cannot be replaced by other similar materials. Even though these things are similar, they still cannot attract something. Why? Because their nature of* qi (气, *a vital force forming part of any living thing*) *is different, so they cannot affect each other.*

Here, *dunmou* (顿牟) is a hawksbill. Its shell is a nonconductor, and it can produce electrostatic energy by friction, which can attract mustard. A lodestone can attract an iron needle instead of a bronze needle. From Wang Chong's view, it is because of the different properties of two metallic materials. He attributed this to the affection of *qi*, which can be understood now as magnetic induction. Guo Pu of the Jin Dynasty also pointed out in his prose *Cishi Zan* (《磁石赞》, *Ode to the Lodestone*) that a lodestone could attract an iron needle while the shell of a hawksbill could attract mustard (through friction). Tao Hongjing (陶弘景, 456~536) of the Liang Dynasty also talked about it in his works *Bencao Jingji Zhu*. He said: "Nowadays, the lodestones originate in the south of China. They can attract iron needles by suspending. The best one can suspend 3, 4 or 5 iron needles one by one." In the fourth volume *Jade of the Xin Xiu Bencao* (659) compiled by Su Jing of the Tang Dynasty, the lodestone was recorded, "which shares another name *xuanshi* (玄石, black jade), or *chushi* (处石, the stone giving directions). It comes from mountains or valleys or the back of *Ci* Mountain. If there were iron ore near the lodestone, then it would be located in the south of the iron. It can be mined at any time ... nowadays, it is found in the south. The good one can suspend an iron needle. If it can suspend 3, 4 or 5 needles one by one, it is the best one."❶

Here *chu* (处) means orientation or position. Because such stone can determine direction, it is called *chushi*, namely, load-stone or lodestone about 1000 yeas later in the western world. The lodestone can attract iron needles, and the needles can become affected, thus becoming magnetic, which can also attract another 3 or 5 needles. The magnetized needles once get the property of magnetism, they can share properties of the lodestone's directivity and polarity. Especially after being rubbed with the lodestone, a needle has made its chaotic magnetic domain into regular arrangement, thus the needle becomes one magnetic substance.

Needle-like magnetic substance is more reasonable than those polished spoon-like natural lodestones in shape, because the former is more regular and even in magnetic domain, so its sensibility of polarity becomes stronger. That is to say, people from the the Jin Dynasty, the Northern and Southern Dynasties, and the Tang Dynasty once noticed its properties of magnetization and attraction, and they certainly found the the magnetized needle's directivity and polarity. Then the spoon-like natural lodestone on the south-pointer was replaced by a magnetized needle, thus the improvement of magnetic substance in material and shape was achieved. Meanwhile, the square earth plate must be replaced by a round direction dial so as to fit the orientation needs of a magnetic

❶　苏敬. 新修本草：玉石部[M]. 上海：上海科学技术出版社, 1978.

needle. But the magnetic needle cannot be put on the earth plate directly as the magnetic spoon, because the increase of the contact plane between the needle and the plate would strengthen frictional resistance so that the magnetic needle could not rotate on the plate smoothly and give the right direction. If we want to make the magnetic needle pivot freely on the plate to reach the aim of determining direction, we should take some measures to reduce drag and increase its performance. As we have learned from the historical literature, ancient people achieved this by employing three methods: the first one is to suspend the magnetic needle with a silk thread over the earth plate in a place with no wind so that it would give direction when it stops turning (Figure 164). This method was tried out before the Tang Dynasty, but it was found that there was a limit that the needle was easy to be affected by the ambient air. Therefore, to prevent its swing, ancient people made refinements in its shape by manufacturing it into the sheet of a tadpole or a fish, which was inherited continuously by the later dynasties.

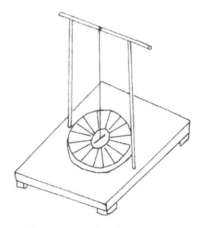

Figure 164   The Compass Map of Suspending A Needle, painted by Pan Jixing (2001)

Cui Bao (255~320) of the Jin Dynasty in his *Gujin Zhu* (《古今注》, 300, *Annotations of Ancient and Modern Objects*) wrote: "A tadpole is the immature form of frogs and toads, called *xuanzhen* (悬针, a suspending needle) or *xuanyu* (悬鱼, a black fish)."

Here, *xuanzhen* is a pun, which means the magnetic needle attracting other iron needles as well as suspending on the earth plate to determine directions. Therefore, in the western historic literatures, it should be translated into *suspending needle*. It is inappropriate for Joseph Needham to call it *mysterious needle*. "玄" (*xuan*) means black, the color shared by the lodestone, the magnetic needle and the tadpole, so the lodestone is called black stone, and the magnetic needle black needle. Ma Gao of the later Tang Dynasty in the period of the Five Dynasties had some relevant recordings in his works *Zhonghua Gujin Zhu* (《中华古今注》, 924, *Annotations of Chinese Ancient and Modern Objects*), similar to the book *Gujin Zhu*. Later Chen Zhensun (1183~

1261) quoted the similar content in *Zhongxing Guange Shumu* (《中兴馆阁书目》, *Catalogue Compiled by Academic Writers in the Zhongxing Government Institution*) in his catalogic book *Zhizhai Shulu Jieti* (《直斋书录解题》, 1176, *Explanation and Remarks on the List of Books from the Studio of Straightness*). He juxtaposed *Gujin Zhu* with *Zhonghua Gujin Zhu* to make the right estimation, and the extant edition of *Gujin Zhu* is stilled regarded as the original work of Cui Bao. The method of employing a suspending magnetic needle to match the plate of orientation can really help point to the south, which was used in the Jin Dynasty and the Northern and Southern Dynasties, thus it is acknowledged that the compass experienced the process from the spoon to the needle as Joseph Needham had said in his work. This is the key step from the south-pointer to the compass. It is not unique. Heinrich Julius Klaproth (1783~1835), German sinologist, told us that the earliest name for a compass in Europe was *Calamita*, which just means tadpole. ❶ This kind of tadpole-shaped or fish-looking compass can still be found in *Sphaerd Mundi* (published in 1485) written by Italian Erhard Ratdolt (1442~1528). In 1777, the French Academy of Sciences offered a reward for the best design of the compass, and Charles Augustin de Coulum (1753~1806), French physician, was awared for his method of suspending the needle with a silk thread.

The biggest disadvantage of the suspending needle is that it cannot escape the negative impact of airflow around it on its stability, thus, it cannot determine the right direction quickly, and is completely unable to be employed in jolting situations. Its greatest contribution to history was that the magnetic spoon was replaced by the magnetic needle, completing the technical improvement in the shape of magnetic substance. The next challenge was how to make it freely turn when a magnetic needle was suspended above the earth plate.

The second method for its improvement is to turn the flat surface of the heaven board to a concave surface, with water in it so that the magnetic needle could rotate on the surface. When rotation stops, the needle could point to the direction on the surrounding graduation. The purpose of adopting such a method is to reduce the impact of airflow around it. This is the shape of the wet compass recorded in the literature of the Northern Song Dynasty in the earlier

---

❶ KLAPROTH H J. Lettre M. le Baron Alexander de Humboldt sur l Invention de la Boussole[M]. Paris: Dondey Dupré, 1834.

part of the 11th century. Then did these reformers before the Northern Song really manufacture the wet compass? The answer is yes. The evidence suggests that the makers of the Tang Dynasty have taken a decisive step to manufacture compass for geomancy.

In these works on *fengshui* (geomancy) at the later stage of the Tang Dynasty, the earliest records of magnetic declination have been mentioned. Meanwhile, to wipe out its error, people set up a unit for readjustment of orientations on an azimuth plate. The south-pointing device, by observing magnetic declination, is not the traditional sense of the south-pointer, because there were no records of magnetic declination in the time of popularity of *sinan*. The invention of wet compass solves the problem of how to put the magnetic needle on the earth plate, which is a revolutionary change for the better.

The third method to make the magnetic needle freely turn on the azimuth plate is to support the needle on the plate by a pivot, which was the way applied in the Song Dynasty. In this sense, both methods, suspending a needle and floating a needle, are proved to be used by the reformers of the south-pointer before the Song Dynasty, and the latter is proved to be more feasible and practical. Further discussion will be extended in detail in the next section.

However, these reformers from the Jin Dynasty, the Northern and Southern Dynasties to the Tang Dynasty, and these early manufacturers of suspending needles and floating needles did not record their manufacturing skills and secrets in their works. Another possibility is that such kinds of works were missing or lost gradually. Most of the relevant works cannot be found in these historic literatures, such as *Yiwen Zhi* (《艺文志》, *Treatise on Literature*) of *Jiu Tang Shu* (《旧唐书》, *Treatise on Literature* of *The Old Book of Tang*), corresponding parts in *The New Book of the Tang Dynasty* (*New History of the Tang*) and in *The History of Song Dynasty* respectively. However, related traces can be obtained from some citations of different authors since the Song Dynasty. Because the technical achievement of the early compass can still be found in the works of the Jin Dynasty, the Northern and Southern Dynasties and the Tang Dynasty, we can trace out its manufacturing technique by means of modern scientific knowledge. On the other hand, the scholars of the Song Dynasty also filled in the blank with the information on the early compass technology and made full details public. The Northern Song Dynasty (960 ～ 1126) was at the height of prosperity and cultural development, especially in the explosive growth of magnetics knowledge, which experienced a gradual accumulation. It was just triggered by the suspending needle and the floating needle before the Song Dynasty. People made technological improvement on the south-pointer in the period of the Jin Dynasty and the Northern and Southern Dynasties, while at the later stage of the Tang Dynasty, people completed the transition from the south-pointer to magnetic compass, which was shared and promoted by the people of the Northern Song. This is probably the clear clue of the development course of the compass.

The above-mentioned transition first happened in China, mainly because the theory and practice of magnetism in China was once in the leading position in the world for a long time. As we all know, natural magnetites are found all over the world. Though almost every nation in ancient time had already found it, not every one had an equal chance in making it into the instruments to point to the south, so there was a difference in the time line for its manufacture. Socrates (464 B. C. ∼399 B. C.) of ancient Greece pointed that the lodestone could attract iron hoop. Pliny the Elder, Ancient Roman scholar, once talked in his book *Historia Naturalis* about the lodestone's application by using an arched roof so as to suspend the cast-iron statue of the late Kaiserin in the air in Arsinoe Temple, Alexander City. [1] Although the attraction of the lodestone had been discovered by Europeans for a long time, they knew nothing about its directivity and polarity until the end of 12th century, thus they could not invent magnetic pointing device. Their invention of a magnetic pointing device was over 1000 years behind that of China, because people in China made *sinanyi* in the 3rd century B. C.. In fact, 300 years earlier before English Alexander Neckam (1157 ∼ 1217), in his *De Naturis Rerum* [2], first mentioned the lodestone's directivity and magnetic induction in Europe in 1190, Chinese people had tried to solve the problem of magnetic declination, and made the early magnetoresistive compass.

Arabians' records about the lodestone first appeared in the book *Tawq al-Hamāmà* written by Ali ibn Ahmad ibn Hazm (994∼1064), who was born in Persia in the 11th century. In the book, he only talked about the attraction of the lodestone. There was a plot in the book which described a story of a spoony

---

[1] CAJORI F. 物理学史[M]. 戴念祖,译. 呼和浩特:内蒙古人民出版社,1981.

[2] BROMEHEAD C E N. Alexander Neckam on the Compass Needle[J]. Geographical Journal, 1944(104):63.

man running after a girl. The man compared himself as a sheet iron, and
compared the girl to a lodestone, and he was attracted deeply by the lover. ❶
Before the 11th century, Arabians did not understand the lodestone's directivity
and polarity. In 1232 (the fifth year of Shaoding during the reign of Emperor
Lizong, the Southern Song Dynasty), Muhammad al-Awfi (1202 ~ 1257)
mentioned navigational compass in his Persian work *Jami al-Hikayat*, which
was similar to Chinese wet compass, ❷ but 300 years later than Chinese. In
addition, in this field, Indians were not earlier than Europeans and Arabians.

From the above analysis, we can conclude that an intermediate link is
necessary from the discovery of the lodestone's directivity and polarity to the
manufacture of a south-pointer. However, other civilized countries and regions
than China lacked this necessary intermediate link for the past one thousand
years from BCE to the 12th century. This proves that the device of south-
pointing originated from China, and was spread to every corner of the world.
More details will be given in the next chapter.

Incidentally, ancient Indians in Mexico in the Central America carved the
natural magnetite ores into human images, animals or articles of daily use,
which called close attention from some western archaeologists when some were
unearthed during the time from 1966 to 1976. These archaeologists found that
there was a magnetic field near some statues, and that some bulging parts such
as claws or nose shared magnetic polarity, which is natural because they were
decided by properties of the material itself. Ancient Chinese, Europeans, and
Indians all made magnetite ores into practical stuffs, which shared the similar
physical properties, but it cannot prove that such kind of materials must be a
south-pointing device. Someone picked a magnet bar out from these relics
unearthed in Olmec of Veracrus State, Mexico, and then by means of it they
came to a conclusion that ancient Indians invented the south-pointer. This kind
of deduction may be too hasty. ❸ The history of Chinese south-pointer was
insufficiently understood by those western researchers holding such an opinion,
and they could not measure accurately the time of these unearthed objects in

---

❶   ARBERRY A J. The Ring of the Love Translated from the Tanq al Ham m of Ali ibn
     Ahmad ibn Hazm[M]. London: Luzac, 1953.
❷   WIEDEMANN E. Zur Geschichte des Kompasses bei den Arabern[J].
     Verhandlungen der Deutschen Physikalischen Gesellschaft, 1907, 9(24):764.
❸   CARLSON J. Lodestone Compass: China or Olmec Primacy[M]. Science, 1975
     (189): 753-760.

Mexico. They only compared it with Chinese ones in a rush, so their conclusion was not reliable. The polarity parts of other unearthed statues were all fixed and could not be turned, so it was evident that they were impossible to point south.

For all of these reasons, China further developed its knowledge of magnetism beginning from the Jin Dynasty, the Northern and Southern Dynasties to the Tang Dynasty (4 century ～10 century). People had done some technological improvement on *sinanyi* since the Warring States Period (5th century B. C. to 3rd century B. C.) and the Qin and the Han Dynasties. They used the artificial magnet shaped like a fish or a tadpole to replace polished spoon-like natural lodestone according to the principle of friction method of transferring magnetism, so as to increase the sensitivity of its pointing polarity and correspondingly altered the way to put a magnetic body on the graduation plate. Moreover, the graduation plate, namely the earth plate, changed from a square-shaped one to a disk-shaped one. In the Northern Song Dynasty, some scholars recorded its manufacture method and process in their works coincidently, which proved that its advent and application preceded what scholars had identified in their works. Therefore, the invention of the device went back to some time before the Northern Song Dynasty. Actually, the magnetic compass emerged in the late Tang Dynasty, especially *fengshui luopan* or geomantic compass.

4.3　Invention and Early Development of Compass

4.3.1　Invention of Water-Floating Compass in Late Tang Dynasty

The south pointing needle is also called magnetic compass or "罗盘针" (*luopan zhen*) in Chinese, "compass" in English, "compasso" in Italian, "Kompass" in German, "compas" in Spanish and "boussole" in French, which is composed of two parts: turnable magnetic needle and graduated circle. Though the new-type device of orientation originated from *sinanyi*, it was totally different from the traditional south-pointer, so the improvement of it was a fundamental transformation, which can be regarded as an invention. People always confused the compass with the south-pointing carriage in history, which also confounded researchers in the historic study of compass. In the past 50 years, scholars both at home and abroad finally distinguished them through in-depth studies. The south-pointing carriage or *sinan che* （司 南 车） is a mechanical device of gear train with automatic clutch, and can keep a fixed orientation in motion, which is totally different from the compass. The carriage originated from the Western Han Dynasty,❶ but some scholars after the Jin Dynasty dated it back to the Yellow Emperor （黄帝，Huangdi） era in legendary ancient time or the Western Zhou Dynasty （1122 B. C. ~ 771 A. D.）.❷ However, there was no historical evidence to prove it. The manufacture of the south-pointing carriage was one of the significant achievements in ancient mechanical engineering, but it shares no connection with the discussed theme in the book, because it is beyond the scope of this article.

The earliest developmental form of compass is wet compass employed by these geomancers of the late Tang Dynasty, mainly used for determining directions on land. Geomancy or *Fengshui* is a Chinese philosophical system of harmonizing everyone with the surrounding environment. It has a long history in China, mainly applied in choosing so-called geomantic treasure land or good occasions such as the best terrain, orientation and time for constructing mansion houses, palaces, temples, cities or graveyards, for making them auspicious, ensuring that the family-line carry on, avoiding omens, and for all those that the ancients valued most. However, geomancers often determine the orientation by observing the undulating terrain of mountains and waters, and some of the practices may be unfounded, but not all. They must have contributed to the compass for its improvement, development of magnetoresistive compass and the knowledge of magnetics. The early geomancers kept their manufacture methods to themselves; so there were few related works left down for the present studies. The present data about compass are most after the Song Dynasty, some of which were the well preserved knowledge of earlier generations, especially those about the earlier records of compass, so people cannot doubt or deny all these records in the works of geomancy.

The right attitude is that we should, on one side, distance ourselves from the elements of superstition in these books of geomancy, while on the other hand, we should dig out the traditional scientific essence. Joseph Needham set an example for us in his work. He found implicit records about magnetic

---

❶　刘仙洲.中国机械工程发明史[M].北京:科学出版社,1962.
❷　虞喜.志林新书[M]//李昉.太平御览.北京:中华书局,1960.

declination in the geomancy book *Guanshi Dili Zhimeng* (《管氏地理指蒙》, *Master Guan's Geomantic Instructor*), which was written in the Tang Dynasty (in the middle of the 9th century). According to *Supplements to Treatise on Literature* of *The History of Song Dynasty*, it recorded such following information. "Two books of *Master Guan's Geomantic Instructor* are the works in Guan Lu's name (a famous geomancer in the period of three kingdoms), practically compiled and annotated by Xiao Ji in the Sui Dynasty, Yuan Tiangang and Li Chunfeng in the Tang Dynasty, and Wang Ji in the Song Dynasty, but we don't know its real bookmakers." Guan Lu (管辂, 209~256), also named Gongming, was from Pingyuan under Wei Kingdom in the Three Kingdoms Period, who liked to observe stars in the sky since young. When he was an adult, he was good at divination and physiognomy, and later he was called as the adjutant for literature cause. In the early years of Zhengyuan reign, he was appointed as *shaofucheng* (少府丞, the director for guiding chamberlain for the Palace Revenues). Before long, he passed away at the age of 50 (254 A. D.). ❶ The contents of *Guanshi Dili Zhimeng* (*Master Guan's Geomantic Instructor*) showed that it was in the name of Guan Lu by a person of the Tang Dynasty. It was a compilation of geomancy books of previous dynasties and the dynasty of his day. Wang Ji (990~1050), who wrote annotations for the book, was a geomancer in the early Song Dynasty, thus it can prove that the book was written in the late Tang Dynasty (the 9th century). *Yiwen Zhi* (《艺文志》, *Treatise on Literature* of *New History of Tang*) recorded two volumes about *Guanshi Zhilüe* (《管氏指略》, *Master Guan's Brief Doctrine*), which was the same book with *Master Guan's Geomantic Instructor*. It stated in the book:

> The lodestone shares the properties of a mother, while the needle is made of an iron bar, and the lodestone and the needle echo each other in the field just like a mother and her child, so they have mutual induction. The needle shares the properties of a lodestone after attraction, so it was completely like a lodestone. The body of a needle is light and right, and its pointing direction is upright, because it is induced by so-called qi. However, it is set on the earth plate, and the direction of the needle may produce the deviation. Its two ends should point south and north, but it points partially to east

---

❶ 陈寿. 三国志：魏书（管辂传）[M]. 上海：上海古籍出版社，1986.

and west. If the precession of the equinoxes along ekliptik were taken into consideration, it would make this difference, thus the phenomenon of deviation can be understood ... the needle points south and north, merely because the mother loves her son.

Some key points in the book are very clear. The first is that the needle is unavoidably influenced by the lodestone so it has magnetic property and could point the polarity. Also, because the body of the needle is light and straight, its pointing directions are accurate. The second is that there is a magnetized needle set up in these instruments for direction-finding. The third lies in its deviation sometimes. The magnetic needle can determine the directions of the south and the north. However, sometimes its pointing directions are not the forward directions of the prime meridian (south and north) or the constellation of forward south and the Emptiness of forward north. It would point to the location of south by west (丁位, the orientation of *ding*) and the location of north by east (癸位, the orientation of *kui*) (Figure 161). The deviation is just 15°. The testing location might be in the city Chang'an, at approximately 108°57′ east longitude, and 34°16′ north latitude. Because of employing a magnetic needle instead of a spoon-like natural lodestone, the compass' sensitivity of the geographic cardinal points is greatly improved, thus its deviation in pointing directions can be found. Even today, magnetic declination is the angle on the horizontal plane between magnetic north (the direction the north end of a compass needle points, corresponding to the direction of the Earth's magnetic field lines) and true north (the direction along a meridian towards the geographic North Pole). This angle varies depending on positions on the Earth's surface, and changes over time. The cause lies in that the geomagnetic pole does not coincide with the geographical poles, and it is also influenced by solar radiation and cosmic rays etc., so that the magnetic field of the earth varies with the time. In fact, it has nothing to do with the precession of the equinoxes.

The annotator of *Master Guan's Geomantic Instructor*, Wang Ji, also named Zhaoqing, was a geomancer in the early Song Dynasty and came of a well-known family background with knowledge. His grandfather Wang Chuna (914~981) and his father Wang Xiyuan (961~1018) were good at astrology and divination, and both held the positions of Directorate of Astronomy and Calendar, which was recorded in the related biographies of *The History of Song Dynasty*. Wang Ji was born in the first year of Chunhua era (纯化元年) under the Emperor Taizong of Song (990). Later, he followed his ancestors' footsteps to be an official in charge of astrology and calendar. He mentioned the phenomenon of magnetic declination in his poem *Zhenfa Shi* (《针法诗》, the poem about setting methods of the needle) in the 8th year of Tiansheng era under the Emperor Renzong of the Song Dynasty: The needle's pointing directions can be clearly found between the Emptiness and the Rooftop (two of the seven mansions of the north).

> *We should work out the degree of* wu (午, *the seventh of 12 mundane branches) whose center is in the 3rd calibrated unit of the Extended Net* (张宿) *in the seven mansions of the south on the plate.*
>
> *The right directions of* kan (坎, *meaning water, one of 64 hexagrams) and* li (离, *meaning flame, one of 64 hexagrams) are hard for people to tell.*
>
> *If we make a little mistake, thus the measurement results will be inaccurate.*

From the title of the poem, people can conclude that the poem is mainly a short introduction of how to use the magnetic needle to determine directions and make it fit different magnetic declinations in various places. In the poem, *xu* (虚) means the Emptiness in the right north of the seven mansions, equivalent to the position of *zi* (子, the first of 12 mundane branches) in the compass. *Wei* (危) means the Rooftop to the right of the Emptiness, similar to the position of *ren* (壬, the ninth of the ten Heavenly Stems) in the compass. *Zhang* (张) means the Extended Net in the south of the seven mansions towards the eastern Extended Net, similar to the location of *bing* (丙, the fourth of the ten Heavenly Stems). *Kan* (坎) is one of the 64 hexagrams, located in the right north of *zi*. Wang Ji thought in the poem that the needle should have pointed to the right north and the south, but it actually pointed north to west or south to east. That was the result of observation in Kaifeng, the capital of the Northern Song, which was located at and 34°52′ north latitude and 114°38′ east longitude, different from Chang'an City. Mr. Wang Zhenduo did not investigate the life experience of Wang Ji, and said he was the person of the Southern Song Dynasty who lived in the 12th century, and then concluded that *fengshui luopan* would not be earlier than the Southern Song Dynasty. Now it is time for us to revise it.

According to the study given by Joseph Needham, *fengshui luopan* appeared not later than the Southern Song Dynasty as it was often supposed to be. In fact, its origin can be traced back to the Tang Dynasty. Since the late Tang Dynasty and the early Song Dynasty, geomancers found that the needle did not always point the right directions of the south and the north. To rectify it, they set up additional needles for correction of direction on the plate of the compass according to different localities, which was the origins of *zhengzhen* （正针, the correction needle）, *fengzhen* （缝针, the seam needle） and *zhongzhen* （中针, the central needle）. In the early of the Qing Dynasty, *Luojing Zhinan Bowu Ji* （《罗经指南拨雾集》, *The Guidance for Using A Luopan Compass*） written by the famous geomancer Ye Tai in the 31st year during the rulership of Emperor Kangxi （1692）, was a more popular book of geomancy. It was annotated by his contemporary Wu Tianhong, and renamed as *Luo Jing Jie* （《罗经解》, *How to Decipher A Luopan Compass*）, and published in the 32nd year of Emperor Kangxi （1693）. Based on the historical records about these three needles, it wrote:

> The main ideas of the above passage mean that the central needle and the seam needle are not set without any foundation, because there is some truth behind. There is so-called qi to show its images in the world. Luojing compass is mysterious and abstruse, and revered Qiu got it from Taiyi Immortal. There is one correction needle, three needle plate rings by golden divisions （fenjin, 分金） signifying Heaven, Earth and Human. The earth plate ring follows the correction needle for allocating the coming mountain ranges and the orientations, which is well-known to geomancers. In the golden divisions, the position of zi （子, the first of 12 mundane branches） is the north by northeast, and the position of wu （午, the seventh of 12 mundane branches） is the south by southwest, so revered Yang added the seam needle for allocating the water ways and determining the locations of golden divisions. In the heaven plate ring, zi （子） is a little south-west and wu （午） a little south-east, so revered Lai added the central needle for allocating hills and determining the right locations in the heaven plate ring. Thus, such a setup is reasonably sound and can determine orientation by different image-directions.

If we put aside some geomancy elements in the above quotation, as far as the three needles reflecting magnetic declination are concerned, the correction needle shows astronomical north and south points, while the other two needles of the seam needle and the central needle show the magnetic north and south points. The seam needle expresses the correction orientation, which means that the needle position is clockwise by 7.5° from north by east of the celestial azimuth or west by south of the celestial azimuth, while the central needle means the correction orientation that the needle position is clockwise by 7.5° from north by west of the celestial azimuth or south by east of the celestial azimuth. There are many concentric circles painted on the existing *fengshui luopan* and recorded in some related works. The number of circles can reach over 40. Among them there are 3 circles marking positions of 24 bearings (Figure 165), and the rest of circles are about fengshui. The blank center circle is so-called *tianchi* (天池, Heavenly Pond), which is employed for setting the magnetic needles. The innermost circle is the earth plate ring, in which there is the correction needle of Qiu the Revered (丘公), which was named after Qiu Yanhan (丘延翰, 688～752), a geomancer in the Tang Dynasty. He set the correction needle in the 18th year of Kaiyuan era, and arranged original 24 orientations on four sides of the square earth plate of *sinanyi* in the earth plate ring of the compass averagely. The seam needle of Yang the Revered is set in the outermost heaven plate ring, which was added by the geomancer Yang Yunsong (杨筠松, 839～903) in the first year of Guangming era (880) at the end of the Tang Dynasty, and the needle also gives directions for the magnetic guiding device (so-called compass). When compared with the celestial azimuth, it is north by east or south by west. Because its points of the compass lie in the

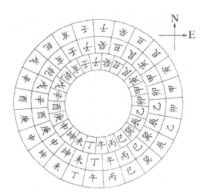

Figure 165　Three Needles Map Marked on the Plate of *Fengshui Luopan*,
from *Luo Jing Jie* (《罗经解》, *How to Decipher A Luopan Compass*)

1. The innermost layer is *zhengzhen* (正针, the correction needle), indicating the astronomical bearings;
2. The middle one is *zhongzhen* (中针, the central needle), indicating the magnetic bearings found in the 12th century, north by west; 3. The outermost layer is *fengzhen* (缝针, the seam needle), indicating the magnetic bearings found in the 9th century, north by east

clipper seams of the correction needle positions, it is also called the seam needle. In the central plate ring, namely the Human plate, is the central needle of Lai the Revered, who was the geomancer Lai Wenjun (赖文俊,1106~1172) in the Northern Song Dynasty. He added it in the 10th year of Shaoxing era, which was used to adjust orientation based on his finding that the direction given by the magnetic needle deviates to north by west when compared with the celestial azimuth in the 12th century. Thus, the needle was arranged between the earth plate ring and the heaven plate ring or between the correction needle and the seam needle, so it was called the central needle. Therefore, *fengshui luopan* in the Dynasties of Ming and Qing was the finalized design based on the achievements of geomancy in the Dynasties of Tang and Song. Thus, it can be seen that since the Tang Dynasty magnetic declination had been discovered. Generally speaking, the south end of magnetic meridian in the east of the meridian line is called east of true north, while in the west it is called west of true north. By convention, declination is positive when magnetic north is east of true north, while negative when it is in the west. Therefore, there are 3 different needle positions marked on the plate of the compass adapted for employing the same compass in various times and places. It is obvious that *zhengzhen* (正针, the correction needle) of Qiu the Revered was found the earliest; the next *fengzhen* (缝针, the seam needle) of Yang the Revered, and the latest is *zhongzhen* (中针, the central needle) of Lai the Revered.

The theory of three needles is the general doctrine of geomancy, proved by the above-mentioned works. It is consistent with the compass plate of the Ming and Qing Dynasties. Moreover, it was further confirmed by the present historical researches. Mr. Qiu, Yang and Zhang played a major role in the historic development of compass, and their achievements can be found in some precious historical records. Some researchers thought that the three-needle theory was not at all convincing because it was just in the name of Qiu the Revered, Yang the Revered and Zhang the Revered, so it was unavoidably arbitrary. Another early book of geomancy related to the three-needle theory is *Jiutianxuannü Qingnang Haijiao Jing* (《九天玄女青囊海角经》, *The Geomantic Manual Passed On by A Female Immortal in the Heavenly Palace*), which is worthy of our attention. Some important historic records are connected with this book, such as *Haijiao Jing* (《海角经》, *Sea Horn Classic*) written by Chisongzi (赤松子) and *Tianya Haijiao Jing* (《天涯海角经》, *Sea Horn Classic over the World*) given by Li Lin (李麟). Both were recorded in *Treatise on*

Literature of *History of the Song Dynasty*. Besides, *Jiutianxuannü Miaofa* (《九天玄女妙法》, *Secrets from A Female Immortal in the Heavenly Palace*) and *Qingnang Xuannü Zhijue* (《青囊玄女指诀》, *Secrets of Fingers According to Cerulean Satchel Given by A Female Immortal in the Heavenly Palace*) were also related to it, both of which were recorded in *Song Shi* · *Yiwen Zhi Fupian* (《宋史·艺文志附篇》, the *Extra Chapter of Treatise on Literature* of *The History of the Song Dynasty*). This book has two prefaces written by Guo Pu of the West Jin Dynasty as well as Zhang Shiyuan (张士元) of the Yuan Dynasty, but its original author is untraceable. According to the analysis of its content, it has collected and compiled some related data of different times from the Han Dynasty, the Jin Dynasty, via the Tang Dynasty to the Five Dynasties, and its final completion time should be between 9th and 10th centuries. The content of magnetic declination should be newly added at the completion of this book, in particular. Figure 166 was inserted in the book, describing so-called *Fuzhen Fangqi Tu* (浮针方气图, *Map of Direction Determination of Earthly Branches by Floating Needle*). Later, the book and the figure were included in the part of *Proclivities* (《艺术典》, including crafts, divination, games, medicine) in the so-called the imperial encyclopedia of the Qing Dynasty *Gujin Tushu Jicheng* (《古今图书集成》, *The Collection of Ancient and Modern Books*). There is a rather remarkable account of the functions of a compass as follows:

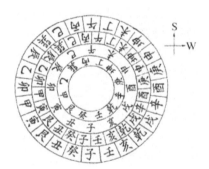

Figure 166　Map of Direction Determination of Earthly Branches by Floating Needle in *The Geomantic Manual Passed on by a Female Immortal in the Heavenly Palace* of Late Tang Dynasty (the 10th century)

In the beginning, heavenly stems and earthly branches decide the directions. Later, the designed copper board is carved with all 24 directions on the plate, thus the heaven plate means heavenly stems, while the earth plate stands for earthly branches. The up-right direction which can receive water conforms to the heaven, while the dragon vein which can gather sands conforms to the earth. Nowadays when people make divination, they use the correction needle to confirm the dragon vein on the heaven plate ring, and use the seam needle to stand for divination on the earth plate ring. The round is obedient to the heaven, the square is conforming to the earth, and both are used to show the laws of the earth.

Here the correction needle and the seam needle are mentioned but not the central needle, which just proves that the books were written earlier than the Northern Song. The mentioned *gelong*（格龙）means the dragon vein on the earth, which can distinguish nobleness from lowliness of these corresponding heaven stars. Some person dated the book back to the Southern Song Dynasty, but it was insufficient. The book and related illustrations provided us examples about the plate image of earlier geomantic compass, thus supplementing the inadequacy of rare object specimen.

Its value lies in that Chinese geomancers replaced the magnetic spoon with the magnetic needle in the 9th and the 10th centuries. Meanwhile, 24 bearings plate of the magnetic substance finished the transition from the square shape to the round shape, thus it shared the form of the later magnetic compass. The emergence of the word *fuzhen*（浮针, floating needle）showed that the earlier magnetic needle was floating on the round water slot, namely the so-called *Tian Chi*. This kind of geomantic compass could be used as a navigational device if simplified slightly by taking out some rings for observing fengshui. Oceangoing ships in the Tang Dynasty often crossed back and forth between Southeast Asia and the Indian Ocean, straight to the Persian Gulf and the east coast of Africa. We cannot say that it had nothing to do with the compass.

It was normal in the past to assume that the magnetic declination was first put forward by Shen Kuo（1031～1095）, a scientist of the Northern Song Dynasty, in his *Mengxi Bitan*（《梦溪笔谈》, *Dream Pool Essays*）, and that its finding was attributed to him. There are plenty of people who still hold such a view these days, but Shen Kuo himself thought that geomancers before him found it. He said："These geomancers use magnetic stone to rub the needle, thus the needle can point south, but regularly eastward, not completely south."[1] We should, first of all, recognize that Shen Kuo accumulated rich

---

❶　沈括.梦溪笔谈:杂志一[M].北京:文物出版社,1975.

experiences and was profound in the theory of magnetics, but the discovery of the magnetic declination would be postponed to too late a time if we attributed the finding of magnetic declination to him in the 3rd year of Yuanyou period in the reign of Emperor Zhezong of the Northern Song (1088). The above historical materials are enough to prove that geomancers of the Tang Dynasty in the 9th century employed wet compass to observe magnetic declination and adjust azimuth by means of it. Yang Yunsong, a famous geomancer in the Tang Dynasty, passed the knowledge down, which was the common origin of magnetics in the Northern Song. When people of the Northern Song Dynasty talked about magnetic declination, they did not regard it as a novelty, because it existed for a long time in previous dynasties.

4.3.2 Construction and Restoration of Wet Compass of the Northern Song Dynasty

The great development of compass in the Song Dynasty is achieved based upon its existing technology since the Tang Dynasty. We can still find overviews of compass of the Tang Dynasty from the pens of people in the Song Dynasty, which supplement the compass corpus and its historical documents. With the help of Ding Du, Zeng Gongliang, the minister of the Northern Song Dynasty, led his team in compiling a large comprehensive military book *Collection of the Most Important Military Techniques*. They began the work from the fourth year of Qingli period in the reign of Emperor Renzong by referring to a collection of books and related archives in the imperial storehouse. The book was divided into two parts, 20 volumes for each part, which was completed in the 7th of Qingli period. Zeng Gongliang wrote in the 15th volume in the first part of the book when talking about the troop marching:

When bad weather was encountered, it became gray, darkness gathered and the troop could not distinguish the directions. Thereupon, they made old horses to move forward, let them recognize the way and offer guidance. Or they used the south-pointing carriage or the south-pointing fish to guide them. The fabrication methods of the south-pointing carriage failed to be handed down. However, the method of making the south-pointing fish is to cut out a thin sheet of metal into an iron sheet with its head and tail shaped like a fish, 2 cun long (6 cm) and 5 fen wide (1.5 cm). Then it was put in the charcoal fire for smithing until completely red-

hot （known today as thermoremanence）. Then a plier was used to clamp the head of the thin sheet out of the fire, and its tail was made to direct to the position of zi （子, the first of 12 mundane branches, the direction of north）. Afterwards it was dipped in a basin of water and its tail was made to slant down for several fun （分）. Then it was put in an iron box. When using it, put it in a bowl of water, and let it float on the water, aligning itself to the south.

Zeng Gongliang here introduced the shape and structure of a magnetoresistive compass and how it was prepared, which was used exclusively for determination of directions, not for *fengshui*.

The method to make artificial magnetic substance based on the theory of thermos-remanence can also be realized by employing natural magnet to convey magnetic influence. The key point of the method is: After heating the iron sheet, people must make sudden cooling of it in the direction of earth magnetic lines of force, thus it can realign the magnetic domain regularly. This is the secret as described in the citation: let its tail direct to the position of zi （子, the direction of north）, and then dip it in the basin of water （"以尾正对子位，蘸于水盆中"）. There was a cold working procedure （quenching） in steel-making in ancient China. Therefore, it was possible that this kind of method would enlighten these craftsmen for making a south-pointing fish, so it would not appear queer for people in the Song Dynasty to make artificial magnet by employing this method. It may be convenient to tell the north from the south by cutting the thin iron sheet into the one of a fish shape, its head pointing south and its tail north. The head and the tail should be turning up a little for floating on the water surface by means of surface tension, and its contact plane with water should be reduced as much as possible. We thought that if the head and tail of the fish pointed to the position of the prime meridian, which was mentioned in the book *Collection of the Most Important Military Techniques*, a device similar to the round direction dial in a compass should be used as one of the accessories. The bowl of water with the fish in is just the so-called *tianchi* on the plate. The magnetic fish does not necessarily assume the shape of a real fish. Neither does the bowl take on the form of a real bowl. A real fish shape does harm the homogeneous distribution of the magnetic domain, and a real bowl, with a big rim and a small bottom, would be hard to keep equilibrium once filled with water, thus it would affect the fish's rotation. All these elements should be taken into consideration in its restoration.

When describing the copy of the south-pointing fish, Mr. Wang Zhenduo was curbed by its literal meaning mentioned in the book *Collection of the Most Important Military Techniques*. He painted it as the shape of a real fish with fish eyes and fish scales, which was really an action to gild the lily. There was no need to do that, and it could not be done in this way. When the fish floated on the bowl of water, since there would not be any azimuth plate around the bowl, how could the bearings of the prime meridian be read? That is to say, this kind of restoration contradicts the traditional manufacturing pattern of the south-pointing device. Though his restored map was cited largely in some related domestic and overseas data, it was still a far cry from the physical truth. So we renew its map (Figure 167). Its azimuth plate is a wooden lacquer-coated round plate with 24 directions around it. In the middle there is a cylindrical bronze-made *tianchi*. When it is filled with water, then the south-pointing fish is floating on the water. When not in use, the south-pointing fish would be put in a sealed container, namely, in an iron-made box to form closed magnetic circuit against losing its magnetism, or put along a certain direction beside the natural lodestone continuing to keep it magnetized. When in use, it should be kept in a windless place. Therefore, the described south-pointing fish in *Collection of the Most Important Military Techniques* is based on the *fengshui luopan* of the Tang Dynasty and is renovated by removing its geomancy elements.

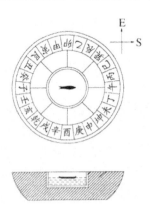

Figure 167 The Restored Map of The South-pointing Fish (Wet Compass) in *Collection of the Most Important Military Techniques* (1044), restored by Pan Jixing

After Zeng Gongliang, Shen Kuo (1031~1095), a famous scientist of the Northern Song, also talked about the compass in the 24th volume of *Mengxi Bitan* (《梦溪笔谈》, *Dream Pool Essays*):

> The geomancers often use the magnetic lodestone to rub the needle peak, thus it can point to the south, but often slightly eastward, not completely southward. The needle would sway on the water surface when floating. It can be done by the hands or on the edge of the bowl. Its rotating is fast and flexible, therefore easy to

fall because it is slippery. It is better to suspend it with a line. The method is like this: first get a single cocoon silk, and glue it with a small piece of wax of about the size of a mustard seed in the middle of the needle, suspend it in a windless place, thus the pinpoint would always point south. But the pinpoint may point north by rubbing it with a natural lodestone. In my house I have both of the needles point south or north.

Shen Kuo said that the geomancers adopted another method to make artificial magnetic needle by rubbing the pinpoint of the needle with a natural lodestone, totally different from the method given by Zeng Gongliang, who resorted to the quenching method by heating the thin iron sheet (thermo remanence). The geomancy compass described by Shen Kuo might well be similar to *diluo* (地螺 or 地罗) mentioned in the book *Yinhualu* (《因话录》, *Tales of Repayment and Retribution*) written in the 16th year of Chunxi period (1189) by Zeng Sanyi (曾三异, 1164～1240) in the Southern Song Dynasty. Zeng Sanyi said:

Diluo (地螺 or 地罗) might well use zhengzhen (正针, the correction needle) of the prime meridian, or fengzhen (缝针, the seam needle) between the prime meridian and the bingwu line. When determining the south-north direction of the heaven and the earth, people should employ the correction needle. Some would think that the regions south of the Yangtze River were out of the way, which were hard to determine directions only by resorting to the prime meridian, so the bingren (丙壬) line should be drawn as a reference. In the ancient time, people calculated the sun shadow in Luoyang city, because it lay in the middle of the heaven and the earth. However, the county outside of Luoyang City would be a little out of the way, hard to determine orientation by the prime meridian, and to some extent, this viewpoint had some truth in it.

Shen Kuo thought that the south pointed by geomancy compass is not really forward south of the terrestrial meridian, but slightly south by east. Zeng Sanyi (1164～1240) said directly that there was an 7.5° magnetic declination between the correction needle of the prime meridian and the seam needle of the *bingren* (丙壬) line, which was obvious in southeastern China, so the seam needle of *bingren* line was needed as a reference. He also listed four methods of testing the south-pointing attribution of a magnetic needle at home. The first method is that after rubbing the iron needle with a magnetic lodestone, float it on the water surface of a bowl, then the pinpoint would point south, the needle head north, however, the water would sway it because of its floating on the surface. The second or the third one is to put the magnetic needle on the thumbnail or at the edge of the bowl, then turn it fast, but it is easy to fall off because of slippage. The fourth one is to glue a new thin thread by a small piece of wax with a size of a mustard seed (2 mm in diameter) in the needle waist, and then suspend it in a windless place, thus the pinpoint would point south, which is more effective.

It appeared that Shen Kuo has verified these four methods. The second and the third one can only test the magnetic needle's directivity and polarity, which cannot be applied to make a pointing-south device, because its center of gravity is not steady and it is easy to fall from the thumbnail or the edge of the bowl. The first method would be the most desirable if we tried to float the needle on the water surface effectively. The fourth method is employed by previous generations, more sensitive than *sinan*. However, it became old-fashioned in the Song Dynasty because new invention of a compass had come into being, but it still has the theoretical significance for present related research. What is really practicable is wet compass. However, Shen Kuo did not give any description about how it could float on the water surface steadily and smoothly. Shen Kuo was a Chinese polymathic scientist of the Song Dynasty in the 11th century, excelling in magnetics and the principles of compass through experimental approaches. He became one of forerunners of Pierre de Maricourt (1224~1279), a French scientist. To increase the buoyancy of the magnetic needle on the water surface, Kou Zongshi (寇宗奭, 1071~1149), a famous herbalist of the Northern Song, mentioned another method in the fifth volume of *Bencao Yanyi*, which said:

> If we use the magnetic lodestone to rub the needle peak, the needle can point south, but often slightly eastward, not completely forward south. The method is as follows: first get a single cocoon silk, and glue it with a small piece of wax of about a half size of a mustard seed in the middle of the needle, then suspend it in a windless place, thus the pinpoint would always point south. The needle can cross through segment (s) of rushes, thus it can point south when floating on the water surface, but to the bing position (丙

位，*the third of Chinese Heavenly Stems)，a little south by east*.

Here，Kou Zongshi added six characters "以针横贯灯心（make the needle across one segment of rushes)" to speak out the knack of the needle's floating on the water surface. Juncus effusus (one kind of rushes) is a perennial herbaceous flowering plant species in the family Juncaceae. The stems of Juncus effusus are smooth cylinders with light pith filling. It grows in large clumps about 1 metre tall and 1.5 mm~4 mm in diameter at the water's edge along streams and ditches. Its rush pith can be used to light oil lamps，so it is called *dengxincao* (rushes). First，cut Juncus effusus into several segments，then use the magnetic needle to cross them one by one，make it float on the water，thus the needle can float on the water steadily and smoothly without turning excessively because of addition of buoyancy and torque resistance produced by Juncus effusus. In fact，the traditional wet compass is made by such a method to suspend the magnetic needle. *Bencao Yanyi* written by Kou Zongshi (1071~1149) was completed in the 6th year of Zhenghe period (1116) in the reign of Emperor Hui in the Northern Song Dynasty，and was published in the first year of Xuanhe period (1119). ❶ The author demonstrated the knowledge of magnetic declination through the term of the needle position，which could supplement records given by Shen Kuo，and meanwhile，it also explained that the magnetic needle in the compass was set on the *tianchi* in the plate of compass just by employing the method he described in the book.

Cheng Qi (程棨，1245~1295) of the Southern Song Dynasty claimed it in his book *Sanliuxuan Zaji* (《三柳轩杂记》，*Miscellanea from Sanliuxuan Study*). He said：

> *Believers of the Yin-yang School determined the directions of the south and the north by making the magnetic needle，when there is the principle about the prime meridian and the bingren (丙壬) line. According to my reasoning analysis，we can use a natural lodestone to rub the pinpoint of the needle，thus the needle becomes magnetic，and it can point south. However，it often points slightly eastward，not completely forward south. The method is like this：first get a single cocoon silk，and glue it with a small piece of wax of about a half size of a mustard seed in the middle of the needle，suspend it in a*

❶   尚志钧，林乾良，郑金生.历代中药文献精华[M].北京:科学技术文献出版社,1989.

windless place, *thus the needle often points south*. *The needle can cross through segments of rushes*, *thus it can point south when floating on the water surface*, *but to the* bing *position*（丙位，*the third of Chinese Heavenly Stems*）, *a little south by east*.

The paragraph just restated the methods given by Shen Kuo and Kou Zong shi, and there was nothing innovative about the wet compass. Cheng Qi only interpreted *Fang jia*（方家，geomancers）as *Yin-yang Jia*（阴阳家，believers of the *Yin-yang* School）, which can prove our previous judgment. Here, he changed *hengguan dengxin*（横贯灯心，cross through a segment of rushes）as *jiguan dengxin*（积贯灯心，cross through segments of rushes）. It illustrates that rushes' traverse is not a segment of rush stem, but several ones. If people do not determine the magnetic heading beforehand when rubbing an iron needle with a magnetic lodestone, then two ends of the needle would generate confusion, as a result, it would not be uncommon that sometimes the pinpoint would point south, and sometimes north. According to the description of *Mengxi Bitan* and *Bencao Yanyi*, we can restore such kind of south-pointing devices（Figure 168）, which is specially used to give directions, namely the wet compass.

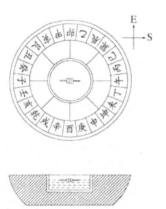

Figure 168   The Recorded Restored Map of Wet Compass in *Mengxi Bitan*, & *Bencao Yanyi*, restored by Pan Jixing（2001）

From the above description, we can find that there are at least two kinds of compasses with only minor differences, which are water-floating compass and wet compass used to travel on the land or navigate in sea voyage. The main difference lies in their production methods and shapes of magnetic needles. One method is to make a thin iron sheet with a shape of a fish into an artificial magnetic substance on the principle of thermoremanent magnetization（TRM）; the other is to make artificial magnetic substance by rubbing an iron needle with a natural lodestone. Compasses made by both methods have been applied in practice and have spread abroad. When the magnetic substance is not in use, it should be preserved in a box for keeping its magnetic property. When in use, first fill the *tianchi* on the compass with water, then put the magnetic fish or needle on the water. This kind of compass is evolved from *fengshui luopan* of the Tang Dynasty, which has been narrated in the previous chapters. The rush that helps to float the magnetic needle on the water later was replaced by sickle feather, which was used for the magnetic needle to cross through and to be more water-proof than rush. ❶ As previously mentioned, wet compass was evolved from *fengshui luopan* of the Tang Dynasty, and became popular in the Northern Song. The compass body is wooden-bodied and lacquer-coated, and *tianchi* is made of bronze, or both can be made by bronze, which should depend on the practical requirement. This kind of wet compass had been used up to the Ming and the Qing Dynasties (14 century~19 century), and such compass for navigation retained the birthmark of wet compass since Tang and Song Dynasties, which reflected technological heredity.

4.3.3　Dry Compass Invented in the Southern Song Dynasty

Besides wet compass, in the 12th century, people in the Song Dynasty designed another device which was fixed on the azimuth plate. The device was used to support the magnetic needle with a copper nail to rotate freely on pivots, therefore the compass did not need any water for determining directions. This is so-called dry compass. Wang Zhenduo pointed out: "Before the Ming Dynasty, all the compasses were the magnetized sewing needles which turned with the help of the buoyancy of water. There was not any other shape and structure of compass beyond that." He thought that Chinese dry compass developed just under the influence of European compass of such kind and the western scholars held the similar view. However, this kind of view should be revised, for material data through recent archaeological excavation suggested that Chinese invented dry compass and applied it to practical requirement in the early of the Southern Song Dynasty (1127~1279), at least 200 years earlier than Europeans. Because people at that time were accustomed to using wet compasses, it, thus, covered the fact of dry compass application. Therefore, the view is absolutely inappropriate that dry compass did not exist before the Ming Dynasty and that it developed in the Ming Dynasty just because of western technological influence. So the origin of dry compass should be rewritten.

In May, 1985, a tomb of the Song Dynasty was excavated in Wenquan township of Linchuan County in Jiangxi Province, from which over 70 historical

---

❶　王振铎. 司南、指南针与罗经盘：下[J]. 中国考古学报，1951(5)：101.

A　　　　B

Figure 169　The Porcelain Figurine *Zhang Xianren* (张仙人, Great Immortal with Surname Zhang) Holds a Dry Compass, unearthed in Linchuan (A) and Sketched Drawing of It (B) by Pan Jixing (2001)

relics were unearthed. The owner of the tomb was Zhu Jinan (朱济南, 1140～1197) from the tombstone record. During his lifetime, he was appointed the grand master of the fifth level for the imperial court, the prefect of Shaowu prefecture in Fujian area, similar to a magistrate of a prefecture, and he was buried in the fourth year of the Qingyuan period under the reign of Emperor Qingzong (1198). What is noteworthy is that there was a porcelain figurine found in the tomb, 22.2 cm tall, with written words "张仙人" (*Zhang Xianren*, Great immortal with surname Zhang) on the bottom (Figure 169). The figurine wore a long gown with an obvious right lapel, his left hand holding a compass in front of the right chest, and his right hand grasping the left sleeve. Here, *Zhang Xianren* was Zhang Jingwen (张景文), also recorded in *Dahan Yuanling Mizang Jing* (《大汉原陵秘葬经》, *Secret Canon of Burial for Chinese*), namely the 19th volume of *Yongle Encyclopedia*.

The compass held by the porcelain figurine Zhang Xianren has a graduation of 16 bearings, the magnetic needle in the middle of the compass differs from the water-floating magnetic needle, and there is a circular hole in the rhombus center, which suggests the supporting structure of a shaft. Geomancer Fan Yibin, under the reign of Emperor Qianlong of the Qing Dynasty, pointed out in the part of his book *Luo Jing Jing Yijie · Zhenshuo* (《罗经精一解·针说》, *Deciphering Secrets of A Luopan Compass · On the Magnetic Needle*): "The dry compass originated from Jiangxi Province, became popular in the previous dynasty, and people used it to determine the pivots of the south and the north." Therefore, in the 12th century, Chinese people had the device of dry compass, which originated in Jiangxi province. There are not any characters in the bearings' graduation in the unearthed porcelain figurine of the Southern Song Dynasty. According to the related studies, we can speculate that these bearings should include eight Heavenly Stems, four earthly branches (*zi*, *wu*, *mao* and *you*, 子午卯酉) and four diagrams (*qian*, *kun*, *xun* and *gen*, 乾坤巽艮). The compass was believed to be a dry compass to determine directions, supported by a copper nail, from which the later dry compass originated. Now we provide our restored map as follows (Figure 170).

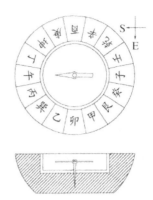

The water compass and the dry compass were most widely used in the Song Dynasty, and even they could be employed as a part of magic tricks by using its operating principle. Chen Yuanjing of the Song Dynasty talked about the south-pointing fish and the south-pointing tortoise in the 10th volume of *Shilin Guangji* (《事林广记》,

Figure 170　Restored Map of Porcelain Figurine Unearthed in Tomb of Southern Song Dynasty (the 12th century), restored by Pan Jixing

Comprehensive Notes on Myriad Things), the earliest edition of which was the chiseled edition from the edition of the second year of Taiding period in the Yuan Dynasty (1325). The book was included in *Song Shi · Yiwen Zhi* (《宋史·艺文志》, *Treatise on Literature* of *The History of Song Dynasty*). According to a systematic textual research given by Joseph Needham, the book was compiled during the Shaoxing period at the beginning of the Southern Song Dynasty (1135～1150). The present edition of the book was collected in the Institute for Human Sciences of Kyoto University in the 12th year of the Genroku period in Japan, which was the chiseled edition followed by the definitive edition of *Yuantai*. When talking about the south-pointing fish, it wrote:

> *Carvea fish from a piece of wood with a size of a thumb (6 cm), open a hole in its belly proper to put a bar of magnetic lodestone in, and then use wax to fill in the gaps. Put half of the needle into the fish from the mouth, let it float on the water, and then the needle would point south automatically. If you use your hand to pluck the needle, it would stay the same as before soon.*

The secret of the magic is as follows: carve a piece of wood into one of a fish body about 6 cm size, open a hole in its belly, put in a magnet bar, then fill in the gaps with wax. Insert a half of the needle into the fish through the mouth, half-hidden, and half exposed. Here, *gouru* (钩入) means "stick the needle in". But in the restored map made by Wang Zhenduo, the exposed part was bent upward (Figure 171-A), which was his interpretation about *gouru* (钩入), but it was not in line with the original meaning, and did not work very well technically. If the natural lodestone were made a little thicker, and then were put under the belly of the wooden fish, the center of gravity would be transferred downward, thus it would undermine its floating on the water. In fact, the magnetic lodestone should be made fine and small, and the magnetic needle should be straight, and the lodestone should be put on the superior area in the belly, so that it can be in a semi-submersible state. Then if the exposed needle were pulled by hand, the fish would turn automatically and would stop at the south-north direction. Pull it again, and it can turn again as before. Note that the needle should not be exposed too much and the action of pulling the needle should be seen by the audience so that it will produce the wrong impression that the wooden fish can turn automatically, because audience cannot find the secret that there is a lodestone hidden in the belly of the fish. Now we provide our improved restored map (Figure 171-B) here for a comparison with the one made by Wang Zhenduo (Figure 171-A).

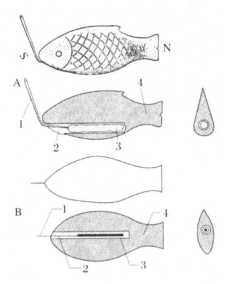

Figure 171　The Restored Map of South-pointing Fish,
One Magic Prop Recorded in Shilin Guangji
A. Restored by Wang Zhenduo (1951)　B. Improved by Pan Jixing (2001)
1. The iron needle　2. Wax　3. The magnetic lodestone　4. A wooden fish

In the 10th volume of *Shilin Guangji*, it said:

　　Carve a piece of wood into a wooden tortoise in the way that a wooden fish is made, but stick the needle into the hole at the end of the tortoise. Then, in the wooden base board, set up a bamboo nail whose size is like the head of a chopstick. The head of the nail should point to the small hollow of the belly for supporting the tortoise. If people pull the needle, it will always point south, and the needle should be put at the end of the tortoise.

Its working secret of the magic prop (the south-pointing tortoise) is as follows: First, cut a piece of wood into a shape of a tortoise, open a hole in the belly, put a magnet bar in it; Then, fill the gaps with wax. Stick half of the needle into the hole at the end of the body and expose the other half in the way that a south-pointing fish is made. The difference is that we should set up a bamboo nail in the wooden baseboard, long as the head of a chopstick (about 4 cm), sharp on the head and its sharp end upward. We can set the needle at the sharp-pointed groove below the belly precisely. Insert it in. Pull the exposed part of the needle with hand, and the tortoise would pivot on the nail. When it stops, the head and the tail would point south and north respectively (Figure 172). Its principle is similar to the support pattern of a dry compass. The intention of two tricks is to make the dead things into living ones, not for demonstrating how they point south. The magic tricks introduced by Chen Yuanjing suggested that before his time people of the Song Dynasty had

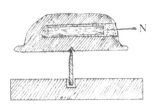

Figure 172   The Restored Map of South-pointing Tortoise, One Magic Prop Reconstructed by Wang Zhenduo (1951)

known dry compass's construction principle and its uses. If people had known nothing about it, then magicians could not have displayed his magic trick by using the turtle shell to deceive audience. The time of the previously mentioned dry compass unearthed in Jiangxi is speculated to be the middle period of the Southern Song Dynasty, thus it can prove the related records in the book *Shilin Guangji* compiled in the the early Southern Song Dynasty. In other words, the time of dry compass's origination can go back to the era between the Northern Song Dynasty and the Southern Song Dynasty or to the early 12th century (1119 ~1140).

In brief, there are two kinds of compasses in the Song Dynasty: one is dry compass; the other wet compass. The wet one is easy to make and convenient for use, however, its magnetic body with a shape of a needle or a fish floats unsteadily on the water; while the dry compass' manufacture is finer by using magnetic needle, so it becomes steady when pivoting. Both have their own advantages respectively.

4.3.4 Wet-and-Dry Compasses in the Ming and Qing Dynasties

Fengshui luopan and Navigational compass in the Dynasties of Yuan, Ming and Qing still adopted the similar production system of the Song Dynasty, so they did not change too much in their shape and structure. From the surviving examples of compasses, we can find that there are more wet compasses and most of their magnetic substances are shaped like needles. Besides lacquer-coated wooden-bodied body, there are some made through bronze-cast methods. For example, the navigational one of the Ming Dynasty which Wang Zhenduo purchased in Beijing (Figure 173) was cast by bronze. The compass is 1.2 cm high and 8 cm in diameter. There are 24 directions cast on the plate, some characters such as *xun* (巽), *qian* (乾), etc. are carved into simplified Chinese characters for convenience of its manufacture. In the inner circle there are eight hexagrams and images of the eight diagrams (八卦), each hexagram including 3 bearings, expressing 8 bearings practically, and there are boundary columns between bearings. However, some navigational compasses have only one circle including 24 bearings. The bottom of the compass is a little inwardly converging in a shape of a saucer. There is a neat line cast in the middle of the bottom, marking the right needle position on the water. Few compasses before the Ming Dynasty were left, but it does not mean that such kind of compasses was never used. Evidence is to be discovered in the future.

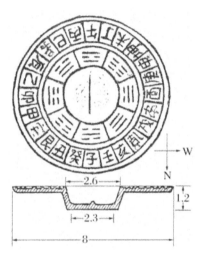

Figure 173 Structural Map of Wet Compass with *Zhengzhen* (正针, the correction needle) in the Ming Dynasty, from Wang Zhenduo (1951), Length Unit: cm

Just as described before, the compass devices, which were introduced to the west, included dry ones and wet ones. However, when westerners made some refit and improvement in actual use, they had copied them, especially the dry compasses. Marine fleet from Portugal and Netherlands came to China and Japan by means of dry compasses since the Ming Dynasty, and brought the improved dry compasses to China directly or via Japan. During the rule of Longqing Emperor of the Ming Dynasty, Li Yuheng (李豫亨) recorded the following anecdote in the seventh volume of his medical book *Tuipeng Wuyu* (《推篷寤语》, *Somniloquy in the Study of Tuipeng*): "Recently the states of *Wu* (Jiangsu), *Yue* (Zhejiang), *Min* (Fujian) and *Guang* (Guangdong) experienced aggressions from Japanese pirates. They often used dry compasses tied at the sterns of their fleets for navigation. When these compasses were obtained by the people of *Wu*, they began to make them by imitating their use. The needle is made by rubbing the lodestone. Once its magnetism is extremely much, it would not work effectively, not as precise as a wet compass (can do)."❶ Wang Dahai (王大海, 1740~1810) of the Qing Dynasty in his *Haidao Yizhi* (《海岛逸志》, *Anecdotes on the Sea*) wrote: "Netherlanders do use south pointing carriage (compass) without a needle. They always adopt one small metal iron for navigation. The two ends of the iron are sharp and pointed and its middle is a little broad, just like a shuttle. In its center, there is a small concave, where a nail is standing to support the iron, rotating just like an umbrella. On the surface is the carved word *Netherlands*, and it has 16 bearings on the plate."

The above two materials talked about European compasses, which had attracted the attention of Chinese people because the compasses had improved slightly over previous Chinese compasses, so they had learned to copy them. At first, the function of original replica was not as convenient as that of old-fashioned wet compass, but they soon found the gist to make it. Since the early Qing Dynasty, communications between Chinese and westerners had deepened, and the Chinese and Western-style dry compasses became popular for navigation. In the 58th year during Kangxi rulership (1719), Emperor Kangxi sent Xu Baoguang (徐葆光, 1682~1725) to Ryukyu. The fleet of ships just used such kind of compasses for navigation (Figure 174-A), which can be found in Xu's works *Zhongshan Chuanxin Lu* (《中山传信录》, *Records about Journey from China to Ryukyu*).❷ A Chinese marine compass of the early Qing Dynasty,

---

❶ 李豫亨. 推篷寤语[M]//裘庆元. 三三医书：第一集. 杭州：三三医社,1924.

❷ 徐葆光.中山传信录[M]//王锡祺. 小方壶斋舆地丛钞：第十帙.上海：著易堂,1897.

which is now kept at the Museum of the History of Science in Florence in Italy，belongs to the same type (Figure 174-B). The diameter of the plate is 9 cm. It can be concluded that in the Yuan Dynasty Chinese ancient gunpowder and firearms technology had spread to the west. After the westerners improved them，these inventions came back to China again in the Ming Dynasty in the form of *folangi* (佛郎机) and *musket* (乌铳) and so on. It could attribute to the history of undeveloped technology and science in China，which fell far behind the West since the middle period of the Ming Dynasty.

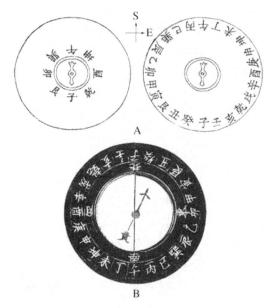

Figure 174　Dry Compass for Navigation under Rule of Emperor Kangxi
in Qing Dynasty (in the early 18th century)
A. Record in *Zhongshan Chuanxin Lu*
B. Collection in Museum of History of Science in Florence，from Needham (1962)

4.4　Application of Compass in Navigation

4.4.1　Mariner's Compass and Navigational Diagram in the Song Dynasty

Since the invention of compass at the end of the Tang Dynasty，the compass was mainly used in determining bearings of *fengshui*，and sometimes it was applied for marine navigation as well. However，the related historical materials are scattered or lost，so more information about its uses in navigation needs to be further discovered and studied. Since the Song Dynasty, there had been a gradual increase in the records about compass's application to marine navigation, the earliest of which was in *Pingzhou Ketan* (《萍洲可谈》, *Pingzhou Table Talks*) written by Zhu Yu (朱彧, 1075~1140). Zhu Yu (朱彧) was a Chinese writer and historian of the Song Dynasty (960~1279), also known as Zhu Wuhuo from Wucheng (the present Wuxing County in Zhejiang Province). His father Zhu Fu (朱服, 1051~1121) was a maritime trade commissioner in Guangzhou Bureau for Foreign Shipping. He was promoted to be the governor of the Bureau from the second year of *Yuanfu* (元符) during the rule of Emperor Zhezong to the first year of Chongning under the rule of Huizong Emperor, mainly responsible for overseas trades and affairs. In the book *Zhizhai Shulu Jieti* (《直斋书录解题》, *Explanation and Remarks on the List of Books from the Studio of Straightness*, 1176), Zhu Yu came to know his father's story and experience, so he wrote his own book in the first year of Xuanhe (1119) during the rule of Huizong Emperor. In the second volume of the book, it said:

> *According to the rules issued by the court, a large vessel can load hundreds of people, while a small one can carry over one hundred people. So the court in most cases appoints these great merchants as the head of a ship* (纲首) *or vice-generals* (副纲首) *or the officials in charge of general affairs* (杂事), *then grants them the certificates with red seal, and allows them to punish those lawless people on shipboard by flogging them with bamboo stick. If anyone died on board, his personal property would be confiscated. These navigators are familiar with geography and have learned how to observe stars at night and sun in the daytime. If unfavorable weather conditions were met, they would resort to the compass, or they would get sea mud with a cord hook more than ten* zhang (丈, *one Chinese length unit*) *long, and smell it, thus they can judge where they have been.*

Jialing (甲令) in the above citation means the rules issued by the court. *Biography of Wu Rui* in the 34th volume of the *History of the Western Han Dynasty* (《汉书》) claimed: "The rise of Changsha King Wu Rui originated from his sticking to truth and righteousness, so his blood lineage can be passed through many generations, and he will not die without descendants. No matter who were his legitimate male offsprings or baseborn ones, they all depended on this gist: sticking to the rule of the court and being absolutely faithful to Emperors."

Among other things, the citation talks about the navigators, who should be familiar with astronomy and geography. If they met bad weather or at night, they should learn to determine directions with a compass, or they use other ways for navigation guidance. However, some people took *Jialing* (甲令, government regulation) for a foreign name *Kling*, and pointed out that the man who used a compass was an Arabian, which is obviously unreliable. In fact, Japanese sinologist Kuwabara Jistuzo (1870~1931) has corrected the error in his treatises. ❶ In the citation, it mainly talks about overseas trade vessels and compasses on board in the early 12th century and here navigators or helmsmen were just Chinese people of the northern Song Dynasty.

In the fifth year of Xuanhe during the reign of Emperor Huizong, Xu Jing who had been sent to Corea as a diplomat wrote in the 34th volume of his *Xuanhe Fengshi Gaoli Tujing* (《宣和奉使高丽图经》, *Illustrated Account of An Official Mission to Korea During the Xuanhe Reign*):

> When the vessels ploughed through the sea area of Penglai Mountain, the seawater became clear and blue just like glass, and waves became stronger. There were big stones appearing indistinctly in the sea called submerged reef. If a vessel struck the rock, it would get grounded or sink. Therefore, the navigators always fear it. On the afternoon this day, the south wind became stronger, and people set up small sails, which were used to resist against the wind or resort to waves to push the vessels forward. They feared that the big sails could not resist the wind, so they set up small sails, and made them sail through the wind. This night people could not stop the vessels in the sea, so they had to move forward by means of stars to determine directions. In bad weather, it became dark and gloomy. Then navigators must resort to compass to direct them. When the night fell, people lit torches and came into the vessels to deal with the approaching of midnight. At this time, the wind shifted to the northwest, and the situation was urgent. Though we had lowered the sails, the vessels were ceaselessly shaking, and basins and jars were toppled and fell. The whole crew on the vessels feared and

---

❶ 桑原骘藏. 唐宋元时代中西通商史[M]. 冯攸, 译. 上海: 商务印书馆, 1930.

shuddered. At dawn, the sea became less tough, people felt calm and continued to set up sails to move forward.

So, the conclusion can be found from the above record that the seagoing vessels, which adopted compass to navigate, belonged to Chinese fleet of ships, not to foreign ships, thus this reconfirmed that the Chinese employment of navigational compasses was more than one century earlier than westerners'. ❶ What is noteworthy is that Xu Jing explicitly pointed out that the use of the floating needle of a compass (指南浮针) could determine south and north directions, which is actually the wet compass we have talked about in the previous section (Figure 171). In the same volume, it also recorded that the diplomatic corps and their entourage started from Mingzhou Harbor (present Ningbo in Zhejiang Province) in eight ships, of which two were great vessels called *Shenzhou* (神舟, Sacred Vessel), and six were guests' vessels called *Kezhou* (客舟, Guests' ships). The guests' vessels were built in Fujian and Zhejiang, over 10 *zhang* long (about 30 m), 2 *zhang* wide (about 6.1 m), 3 *zhang* high (about 9.3 m), and each can accommodate 2,000 *hu* (斛) of food (132,000 *sheng*, Chinese units of capicity and volume) and 60 people, however, the size of *Shenzhou* is three times greater than *Kezhou*. "When the night fell, people lit torches and came into the vessels for the midnight." It might mean the navigators gave right direction by means of compass, and then these seamen bore torches as a signal for leading other ships to move in the same direction.

Zhao Rushi (赵汝适, 1195 ~ 1260), the seventh from the Emperor Taizong, who had attained his *Jinshi* degree, talked about his sea voyage near the seawaters of Hainan Island in the second volume of his book *Zhu Fan Zhi* (《诸蕃志》, *The Barbarian Countries*). He said: "In the east of Hainan Island, land sand extends thousands of miles, stone bed can be seen everywhere. The sea is boundless, melting into the sky at the horizon. The ships coming and going must resort to compass for safe passage, so people in ships observe and guard their ships cautiously. They know that an error by a hairbreadth would lead to destruction of ships."

The first two records above both mentioned that when navigators determined directions on the sea, they should adopt astronomical orientation as

---

❶　EDKINS J. Note on the Magnetic Compass in China[J]. China Review, 1889(18): 197.

well as magnetic orientation to make them complement each other. The compass is commonly used in bad situations such as in dark weather or when the sun or stars cannot be seen in the sky. However, Zhao Rushi of the southern Song Dynasty pointed out that "The ships coming and going must resort to compass for safe passage, so people in ships observe and guard their ships cautiously." It means that, at this time, at least according to what Zhao Rushi saw and heard, coming and going ships should sail in accordance with the determined directions measured by compasses. In addition, they should have people accountable for safety of ships. These specially-assigned people should watch ships cautiously. When any possible problem arises, they should report to helmsmen to steer rightly, because life and death of the whole crews depend on right observations of navigators and no mistakes should be made. All these records prove that navigation technology of magnetic compass has come to maturity. Besides, the citation did not mention nationality of the vessels, but the citation involved experiences in Hainan Island of China, and it also mentioned the seagoing vessels from Quanzhou of Fujian or other harbors, which were related to Chinese vessels instead of foreign ones.

Wu Zimu (吴自牧, 1231~1301), a writer of the southern Song Dynasty, came from Hangzhou, talked about seagoing vessels of Zhejiang in the 12th volume of his *Meng Liang Lu*:

Zhejiang river is an important waterway for vessels to reach inland rivers or sail to the sea, and these trade ships are of different sizes. The great one can carry loads up to 5,000 *liao* (each *liao* is about 60 kg), or accommodate five or six hundred people. The middle-sized one can carry loads from 1,000 *liao* up to 2,000 *liao*, or can accommodate two or three hundred people .... When vessels sailed into the ocean, the sea turned vast and boundless, and the situation was really treacherous ... When eyesight became obscured by wind and rain, the vessels could move forward only by compass, which the navigators took charge of. They must not make any error by a hairbreadth, because life and death of the whole crews depend on their right observations. I often met these great businessmen, who told me about these detailed and thorough anecdotes. However, the seawaters near the reef were shallow. If a vessel struck the rocks, the vessel must be destroyed. So they relied on compass. If there was a slight error, all the people on board would fall into Davy Jones's locker.

So-called *zhenpan* (针盘) or *nanzhen* (南针, namely, compass) mentioned by Wu Zimu both mean navigational compasses with a plate of 24 bearings.

Huozhang（火长）originally referred to junior officers in the military system of the Tang Dynasty. Xin Tang Shu • Bing Zhi（《新唐书 • 兵志》, *Records of Military Affairs* in the *New Book of Tang*) shows: "A group of 50 people forms a team（队）, the person in charge of which is called *zheng*（正）. A group of 10 people forms a *huo*（火）, whose leader is called *zhang*（长）." Later *huozhang* （火长）was used to call navigators in the vessels, or called *zhoushi*（舟师）. The book pointed out that when navigators began to navigate, they must pay close attention to the marking of the exact bearing given by a compass. Sometimes they should adjust their courses and routes properly whenever necessary according to changes of marine meteorology and specific situations around the vessel.

It should be noted that ancient navigators must refer to nautical chart to determine their courses and routes. They would paint and mark the landform, hydrologic condition, coastal geographic names, islands and their bearings as well as natural and humanities landscape around the waters of their passage. When a compass is used for navigation, people should mark the needle position, so the nautical chart is also called *Zhenbu*（《针簿》, *The Book of Needle Positions*), painted by experience of navigators, which is of very practical value in navigation on the sea. Since the Tang Dynasty, China has been a great country in ocean shipping. In fact, during the Song Dynasty, the nautical chart directed by navigational compass was created by navigators, which was an important technical invention at that time. Li Tao（李焘, 1115~1184）of the Song Dynasty wrote it in the 54th volume of his historical writing *Xu Zizhi Tongjian Changbian*. In the sixth year（1003）of Xianping period during the reign of Emperor Zhenzong, magistrates from Guangzhou submitted *Haiwai Zhu Fan Tu*（《海外诸蕃图》, *The Map of Overseas Barbarian Countries*）to the court. It was just the navigational diagram used by Chinese ships in their oceangoing voyages, mainly including the courses from Guangzhou Harbor, via Southeast Asia and the Indian Ocean, through the Persian Gulf to the area of the Red Sea, and it was so-called maritime charts by later westerners. Also Xu Jing（徐兢）talked about it in the 34th volume of his book *Xuanhe Fengshi Gaoli Tujing*. He said: "We marked cautiously every island and islet that our great vessels（神舟）went by in the book, thus making a map for our course." Here the map was just the nautical chart made by navigational compass. *Shenzhou* （神舟）means the giant flagship used for envoys sent by the Song Dynasty to travel by. *Shan*（苫）means the small well-vegetated island islet. The book was named *Tujing* (《图经》, *Map Guides*), to which these important maps must be attached. However, when the book began to be published in 1167, all these maps were deleted. Even the duplicate collected in Xu Jing's mansion disappeared without any trace when he died, which is really a pity.

Zhao Rushi also mentioned the fact in *Zhu Fan Zhi* (《诸蕃志》, *The Barbarian Countries*) that people should read the maps of these overseas countries when they navigate their seagoing vessels, which might be the same with *Haiwai Zhu Fan Tu* submitted to the court by the Guangzhou Bureau for Foreign Shipping in 1003. In the sixth year of Xianchun (1270), during the reign of Emperor Duzong at the end of the Southern Song Dynasty, Jin Yuxiang, who was a scholar from Lanxi county in Zhejiang Province, gave his advice to the court to fight against the Mongolian invaders. In consideration of the armies of the Song Dynasty having no good strategy to fight back when the invaders besieged Xiangfan (a city in present Hubei Province), he suggested the following methods: "I present this strategy so as to hold down the enemies and attack their weak points. I wish that the court could send massive forces to attack them behind enemy lines, namely, to attack Yanji area (the present area of Beijing), thus the enemy troops besieging Xiangfan would retreat naturally without being attacked. I will state the courses which our seagoing vessels will go by in detail. All the information about prefectures and counties, the great oceans and islands, difficulties in marching, and even traffic and distance could be referred to in these well-documented maps. However, the government of the Song Dynasty did not accept it finally. Later, Zhu Xuan (朱瑄) and Zhang Qingxian (张清献) offered advice and suggestions to the court to describe how convenient the sea transportation was, and these courses which they went by were little different from the descriptions of the foregoing statements given by Jin Yuxiang, and later people all admired Jin's accuracy in maps invariably."[1] The map which Jin Yuxiang submitted to the court in 1270 was just the precise nautical chart, which introduced places in detail from Zhejiang coastal waters, northward to the west coast of the Bohai Sea. When compared to the similar one used in the early Yuan period, it can be found that they were almost indistinguishable. Moreover, the nautical charts used by Arabians and European helmsmen could be seen during the latter half of the 13th century, which was behind Chinese ones for more than two centuries.

---

[1] 宋濂. 元史:金履祥传[M]. 上海:上海古籍出版社,1986.

4.4.2　Mariner's Compass and Navigational Diagram in the Yuan Dynasty

The Yuan Dynasty inherited the mechanism of manufacturing navigational compass from the Song Dynasty and further developed it. Since its conquest of the west in the 13th century, Mongolia set up four great khanates in the central Asia, West Asia and Eastern Europe. Besides road link between these khanates and the Mongolian government in Beijing, they have stayed in touch also by sea route. The business vessels of the Yuan Dynasty often came and went for the purpose of economic exchanges in various regions, such as Southeast Asia, India, Perse, Japan and Korea, etc. Oceangoing voyage mainly depends on direction-determination by compass as well as the nautical chart of needle positions. Zhou Daguan (周达观, 1270～1348) of the Yuan Dynasty once recorded such kind of story in his *Zhenla Fengtu Ji* (《真腊风土记》, *Geography and Customs of Zhenla Cambodia*). Zhou Daguan, also called Zhou Caoting, was a native of Yongjia County in Zhejiang Province. In June of the first year of Yuanzhen period (1295) during the reign of Emperor Chengzong Emperor Temür Khan, he was one of the official delegates sent to Cambodia for the diplomatic mission. On February 20th 1296, he set sail from Wenzhou Harbor and arrived at the capital of Cambodia in July. He stayed there for eleven months, left in July of the first year of Dade period (1297) and reached Siming Harbor (now Ningbo) in August. He wrote the book *Zhenla Fengtu Ji* based on his experiences, although the exact date of the book's completion is uncertain. Here *Zhenla* (真腊) was the ancient name of Cambodia. He wrote in the book:

> *Since we set sail from Wenzhou Harbor, we moved according to the direction of* dingwei (丁未) *in the plate given by the needle of the compass, passed the ports of Fuzhou, Guangzhou and Hainan, then through the Seven-islands Sea* (七洲洋) *and the sea off the Central Vietnam coast* (交趾洋), *and at last arrived at Champa* (占城, *now Qui Nhon*). *Then we resumed our trip downwind from Champa* (占城, *now Qui Nhon*), *sailed half a month before we reached the province of Ba Ria* (真蒲, *a place in present-day southeastern Vietnam*), *and entered into Cambodia. Then we continued to move from Ba Ria according to the direction of* kunshen (坤申) *on the plate given by the needle of the compass, through Poulo Condore Sea* (昆仑洋), *and into its harbor. There were dozens of harbors. However, we could only harbor at the forth one. Others were not deep enough*

for our vessels.

Seven-islands Sea（七洲洋）is located in today's sea area near the northeast of Hainan Island, while the sea off the Central Vietnam coast（交趾洋）belongs to the present waters along the east in north Vietnam. Champa is in the middle of Vietnam, and Ba Ria is near Vūng Tàu of Vietnam. In 1296, Chinese diplomatic corps of the Yuan government set sail from Wenzhou Harbor in Zhejiang Province to the south of Cambodia. Its course was as follows: they sail down south and then pass through Taiwan Strait under the guidance of a compass. Here, *xing dingwei zhen*（行丁未针）means that they moved according to the direction of *dingwei*（丁未）given by the needle of the compass in the plate. Namely, they followed the direction between the positions of *ding*（丁）and *wei*（未）directed by the needle of a compass in the plate of 24 bearings (Figure 237), or S. S. W. 202°30′, or due south by west with the angle of direction at 22.5°, passed through waters east of Hainan Island, sailed southwest to the harbor of the central Vietnam and reached Ba Ria in the south Vietnam. Then they continued to move from Ba Ria, following the direction displayed between the positions of *kun*（坤）and *shen*（申）on the plate, or S. 52°30′ W. (South by west 52 1/2°). Then they traversed southern waters of Indo-China Peninsula, kept moving according to the direction between *xu*（戌）and *qian*（乾）on the plate, or W. W. N. 307°30′ or W. 37 1/2 °N, and finally arrived in the coastwise waters in the south Cambodia. The navigational chart was mapped by practical sailing experiences, and these navigators could follow the chart and the direction displayed by a navigation compass, and adjust their courses at any time by directing their helmsmen, thus they could reach the destination and return to the starting point safely and successfully.

Haidao Zhinan Tu（《海道指南图》, *Guidelines of Nautical Chart*）of the Yuan Dynasty, recorded in the book *Haidao Jing*（《海道经》, *Classic of Nautical Chart*）, was edited by an unknown person of the Ming Dynasty. It included the nautical chart from the lower reaches of Yangtze River to the Northern Sea. Actually, it was a nautical chart about sea transport of grain to the capital Beijing in the early Yuan Dynasty and was used for navigation. The chart listed all the harbors along rivers and coasts, islands and anchor places, and marked related directions displayed by a compass, such as due east, due south, due west, due north and southwest and so on. It also included some tips about reefs in some places which people should escape from in their voyage. The chart is the existing earlier nautical chart called *Dayuan Haiyun Ji*（《大元海

运记》, *Shipping Records of the Yuan Dynasty*）, which was edited by the palace eunuchs in the imperial storehouse in the Tianli period (1328~1329) during the reign of Emperor Wenzong in the Yuan Dynasty. It said: "The broad ocean is boundless, people could not determine directions in temperamental weather, and they only predict it by looking up into sky. So ship masters often recruit these experienced helmsmen (navigators) with fat salary, and let them manage such kind of business."❶ During Emperor Shun（元顺帝）of the Yuan dynasty (1333~1368), Wang Dayuan（汪大渊, about 1311~1370）, who traveled with sea vessels to Southeast Asia and the Indian Ocean, wrote in his book *Daoyi Zhilüe*（《岛夷志略》,*Brief Records of the Island Barbarians*）:

> At a glance, the sea is boundless. However, there must be reefs hidden somewhere. Who can get their exact positions? If we escaped from them, our vessel would be lucky; if we hit it, we would be in danger. So the needle of the prime meridian is a safety tool on which people's lives in the vessel depend. But for these sharp-sighted helmsmen, our vessel would have been sunk. Why! The pleasing land, you might as well not head for again, how could people take great wind and waves in the sea as a route?

Here, the mentioned needle of the prime meridian（子午针）, on which the whole crew' lives depend, just means the so-called navigational compass. Wang Dayuan spoke of it when talking of *wanli shitang*（万里石塘）, the surrounding waters of the Xisha Islands in the south of Hainan Island. In this maritime space, there are a lot of reefs, and all ships in their voyage must keep away from them. He also spoke of Poulo Condore Sea（昆仑洋）or waters around Poulo Condore Island in the south of Vietnam. He said: "All vessels, which want to sail to West Ocean, must pass the place. They can go through it for seven days if downwind. A proverb says: in the north, there are seven lands; in the south there is Poulo Condore. If the needle were missing or the rudder uncontrollable, people as well as vessels could not survive."

4.4.3　Zheng He's Diagram of Navigation Course Set by Needle in the Ming Dynasty and Dry Compass of the Qing Dynasty

In the Ming and Qing Dynasties, there were more compasses used as

---

❶　元天历中官撰. 大元海运记[M]//罗振玉. 雪堂丛刻,1915.

navigational devices. From the first year of Yongle period during the reign of Emperor Chengzong in the Ming Dynasty, Zheng He（郑和，1371～1435）served in the highest post as Grand Director, who led a large fleet of 27,000 people to conduct seven sea voyages on behalf of the Emperor, trading and collecting tribute in the eastern Pacific and Indian Oceans. He accomplished a great feat in the world maritime history. He commanded over the seventh and final expedition into the "Western Ocean" (Indian Ocean) from the 6th year of Xuande period to the 8th one during the reign of Emperor Xuanzong (1431～1433). He began his voyage from Nanjing, and the farthest place he arrived at was Kenya in eastern Africa. When the world-famous voyager Zheng He sailed to Indian Ocean, he always carried these navigational diagrams every time, which included navigational map along the way, the full ranges of his marine fleet, and related geography, etc. Of the things he carried, one used by Zheng He in his seventh westward voyage, was recorded in the 24th volume of *Wu Bei Zhi* written by Mao Yuanyi（茅元仪，1570～1637）. *Map of Starting from the Ship Yard through the Long Jiang Guan, to Arrive at Foreign Soils*, is also called *Zheng He's Navigation Map*（《郑和航海图》）for short. Its drawing idea had been inherited and developed from *Haidao Zhinan Tu*（《海道指南图》, *Guidelines of Nautical Chart*）of the Yuan Dynasty. The original copy was scroll binding, collected in the imperial storehouse as a file. Mao Yuanyi changed its form into one complete chart of 24 pages. Of these pages, 20 pages were navigational diagrams, 2 were guiding star stretching-out maps across the oceans, one is the preface given by Mao Yuanyi, and one was blank. The diagram was complete, which was a representative of ancient Chinese navigational diagrams. Later a scholar named Xiang Da (1900～1966) took it out from *Wu Bei Zhi*. After checking and making annotation for it, he then published it as a book of related researches for the sake of convenience.

Zheng He's Navigation Map is also called "*Mao Kun Map*", which is actually a series of maps published in the military encyclopedia *Wu Bei Zhi* in the late-Ming Dynasty. It is lined up from right to left. In this map, the course from Nanjing to today's Sumatera（苏门答腊）in Indonesia is marked by the needle position, not using stars to determine directions. The voyages along the ship routes from the west of Bras Island（龙涎屿，the northwest of Sumatera）to Sri Lanka, and then from the northwest sea area, or across the Indian Ocean to Arabian Peninsula and northeast Africa depend mainly on the needle position for direction determination, and is complemented by the star position. There are 56 needle positions marked on the map to record the course from Taichang Habor in Jiangsu province to Hormuz (now Strait of Hormuz in Iran), and 53 positions marked in the backtracking from the strait to Taichang. In addition to the listed names of places and the marked needle position for navigation, the map also gives detailed information of mountains and rivers following the line of the sea, harbors, cities, islands and beaches. In some places, the map uses temples, pagodas and bridges as land markers. There are more than 500 recorded names of places on the map, of which foreign geographical names take up 3/5 of them. From this map, more details for navigation by the diagram can be found, which can help people understand how ancient Chinese navigators used it in their voyages. For example, for the course from Sumatera to Sri Lanka, people can find the following information in the map:

> *Start from Sumatera, guided by the direction of* qianxu (乾戌) *on the plate given by the needle of the compass, sail about 12* geng (更, *one geng is about 33. 6 km), and you reach Bras Is.. Then start from Bras Is., guided by the direction of Xinxu directed by the needle of the compass, sail for 10* geng, *and then you could set eyes on Nicobar Island. Then, guided by the direction of* danxin (丹辛), *continue to go for 30* geng. *Then, the vessels, directed by the needle of the compass, change course to the direction of* xinyou (辛酉), *travel about 50* geng, *and the vessels could see Ceylan Hill.*

The above mentioned "*qianxuzhen*" [乾戌针, the direction of *qianxu* (乾戌)in the plate given by the needle of the compass] means W. 52. 5°N. (W. b. N. 52. 5°) or W. W. N. 307. 5°. *Danxinzhen* (丹辛针) means the direction of due *xin* (辛) or W. 15°N. (W. b. N. 15°) N. 285°. *Geng* (night watches) is originally employed as a traditional timing unit for Chinese from dusk to dawn, which is similar to 2 hours. Here, *geng* means a leg in one *geng* (2 hours). Huang Shengceng (黄省曾, 1490~1540) of the Ming Dynasty wrote in the first volume of his book *Xiyang Chaogong Dianlu Jiaozhu* (《西洋朝贡典录校注》, *The Collation of Records of the Tributes Paid by the Western Countries*): "In accordance with practice in sea voyage, 60 *li* constitutes one *geng*, and navigators often use *tuo* (托, a stone or steel roller tied with a rope to test the depth of waters by striking soundings, 1 *tuo* is similar to present five feet) to avoid submerged reefs, and choose the sea route by referring to positions directed by a compass." Thus here, one *geng* = 60 *li* = 33. 6 km. (According to the system of length unit in the Ming Dynasty, 1 *li* = 0. 56 km). In the citation, *chuan ping longxianyu*（船平龙涎屿）means the fleet of vessels passed by Bras Is. *Xinyou zhen*（辛酉针）is W. 15°N. (W.b. N. 15°) or the position of W. N. 285°. Great Nicobar Island is the main island of Andaman and Nicobar Islands in the northeast Indian Ocean. Therefore, after the above explanation, the main meaning of the cited paragraph in *Zheng He's Navigation Map* is mainly as follows：

> *Zheng He's fleet arrived at Sumatera on 12 September in the 6ᵗʰ year of Xuande period（1431/9/12）. Then, they started again on 2 November（Figure 239）, sailed over 403.2 km along the needle position of 52.5°（W.b. N.）, and passed by the northwest of Sumatera and Pulau Beras Island in the Indian Ocean. They continued their voyage by the 22.5°（W.b. N.）direction of the needle position for about 336 km and on 14 November in 1431 they arrived at Andaman and Nicobar Islands in the northeast of the Indian Ocean. There, they adjusted their needle position to the direction of 15° west by north（W. b. N. 277.5°）, kept sailing about 1680 km, and finally reached Sri Lanka Island on 28 November in 1431.*

From the above, we can see that Zheng He's fleet reached Sumatera in Indonesia on 12 September in 1431, and stayed there for 50 days. On 2 November, they sailed northwest. During this voyage, they kept on adjusting their directions and charting their course. They passed through Pulau Beras Island, Andaman and Nicobar Islands, and then reached Sri Lanka, thus completing the whole course sailing across the Indian Ocean. The whole course was 3,427.2 km or 6,210 *li*（里, one of Chinese units of measurement, one *li* equals approximately 500 meters）, which took them 20 days and nights, about 127 km（227 *li*）for each day. Ensuring that the fleet would sail by the predetermined route safely and accurately so as to arrive at the scheduled positions, navigators on the vessels should look over the nautical chart all the time, and find right needle positions by referring to a compass. Then they made helmsmen chart their courses. Sailors should hoist sails or clew sails down according to wind direction, and night watchmen should calculate the course by means of the real Chinese sundial, water clock and *gengxiang*（更香, a specially-made incense for timing）and other timing tools. Because shipping lane was curving, they needed to change their needle positions and shift the helm at all times. If they found they were off the course, they should correct it promptly. All the men should stick to their posts day and night and coordinate their respective work.

During 1405～1433，Gong Zhen（巩珍，1376～1440），as a follower of Zheng He's fleet sailing to the west ocean，said in his own preface of the book *Xiyang Fanguo Zhi*（《西洋番国志》，*Record of Barbarian States in the Western Ocean*）："We have been living on the sea for three years during the whole voyage. The sea is boundless，vast and hazy. At the end of the sea，the sky is conjuncted with the waters. When we look around，everything is far away. The sea extends as far as the eye can reach and there is not any shelter for us. Only by observing the rising and setting of the sun and the moon，navigators can determine directions. The measure of height of the stars and the distance all depend on a compass. In making the compass，navigators often use a piece of cut wood as a plate，and mark characters of heavenly stems and earthly branches on it. Then they float a needle on the water in its hollow center，thus navigating vessels by the directions which the needle points. Such kind of work will never stop，round the clock and month after month. The islands in the sea often vary in shapes. We have found that they are either in front of us，or on the right side or the left side of our vessels. We must chart our courses to try to arrive at them. We must calculate the result from the beginning of *geng*（更，a Chinese timing unit）without making any slip so that we can reach our destinations. At the beginning，when planning to sail through Fujian，Guangdong and Zhejiang，we choose some experienced boatmen as maritime technicians，who often go to sea，and they are also called navigators（船师）in our vessels. We give the guide book of the compass and nautical charts to the head of navigators，who is responsible for navigation with full authority in the voyage. Thus，the duty of this work is great and its correspondent responsibility is inherent，which we should take seriously."

We can see from the above lines that the compass used by Zheng He's fleet was kind of water-floating navigational compass with 24 bearings. These navigators were often experienced boatmen chosen from Fujian，Guangdong and Zhejiang and other coastal provinces，and then they were handed the guidebook of the compass and nautical charts，which were prerequisite for their work. *Tushi*（图式，nautical charts）should be a kind of charts as recorded in *Map of Starting From the Ship Yard Through the Long Jiang Guan*，*to Arrive at Foreign Soils* in *Wu Bei Zhi*. *Zhen Jing*（《针经》，*The Guidance of Compass*）was similar to *Haidao Zhen Jing*（《海道针经》，*Classic of Compass in Sea Routes*）. There were two kinds of *Haidao Zhen Jing*（《海道针经》，*Classic of*

Compass in Sea Routes) collected in the Oxford University Library in Britain: one was *Shunfeng Xiangsong* (《顺风相送》, *Voyage with a Tail Wind*); the other was *Zhinan Zhengfa* (《指南正法》, *The True Art of Pointing South*). Later, Xiang Da (向达) made collation and annotation for both, which were published by the Zhonghua Publishing House. The published one mainly introduced various aspects of ancient Chinese seafaring technology. Besides compass, the book also explained ancient *qianxingshu* (guiding star stretching-out art), marine meteorology and seawater measurement, etc. Some related data about the compass and nautical charts can also be found in *Dong-xi Yang Kao* (《东西洋考》,1618, *Studies on the Eastern and Western Oceans*) written by Zhang Xie (张燮, 1574~1640), mainly in the 9th volume *Zhoushi Kao* (舟师考, *Studies of Navigators*), which reflected marine traffic situation in the late Ming Dynasty.

As previously mentioned, dry compass, which mainly used a pivot to support the magnetic needle in the graduated circle, was invented and applied in the Song Dynasty. However, navigators since the Song and the Yuan Dynasties preferred to use wet compasses, thus the invention of dry compass had not been developed. The same is even true of the Ming Dynasty. It was only a matter of habit. On the other hand, the supporting structure of pivot for dry compass needed to be perfected. With the introduction of Chinese compass to the west, dry compass had been greatly valued and obtained large-scale development in Europe, which was then re-introduced back to China. The dry compass, when spread overseas, was immature, but once it was reintroduced back to China, the situation became different. In the Dynasties of Ming and Qing, Chinese learned how to manufacture improved dry compasses for better use. Since the early Qing Dynasty, dry compasses were used in navigation. In the 58th year of Emperor Kangxi (1719), under the Emperor's order, Xu Baoguang (徐葆光, 1682~1725) was sent to Ryukyu on a diplomatic mission. To record this trip, he wrote his sketch book *Zhongshan Chuanxin Lu* (《中山传信录》, *Records about Journey from China to Ryukyu*). The book talked about *fengzhou* (封舟, government-running ship) that he traveled by and the dry compass used in the vessels. Zhou Huang (周煌, 1707 ~ 1784), another official of the Qing Dynasty, with *Jinshi*'s parentage, was also sent to Ryukyu on a diplomatic mission in the 22th year (1757) of Qianlong Emperor. He wrote *Liuqiu Guozhi Lüe* (《琉球国志略》, 1757, *A Brief History of Ryukyu*), in which there were some similar records about compass. There was *zhenfang* (针房, compass room) in the government-running vessel that Chinese envoy took, in which was the place where navigators worked (Figure 175), which was located near the wheelhouse in the stern. Both books talked about the numbers of *geng* calculated by hourglass, and its nautical charm of compass was the same as that in the Ming Dynasty.

Figure 175   *Fengzhou* (封舟), Ocean Vessel of Qing Dynasty; *Zhenfang* (针房, compass room) in the Stern, recorded in *Zhongshan Chuanxin Lu*

4.5   Spread of China's Compass in Europe

4.5.1   Origin of Compass in Europe

Although history proved that scholars of ancient Greece and Roman had long known the attraction of a lodestone, Europeans did not know its directivity and polarity for a long time. The earliest compass of Europe was made by virtue of the technology from China. Many centuries have passed. Some people in Europe have forgotten this history. They refuse to acknowledge that Europe lagged behind China in magnetics before the Renaissance and China's influence on them in this field. They even claimed that there were no important inventions in other parts of the world than in Europe. One of the representatives was William Whewell (1794~1866), an English historian of science of the 19th century. When speaking of the development of magnetics, he ignored arrogantly the achievements made by the ancient Chinese, which, he claimed, had nothing to do with the development of European science at all events. ❶ Nevertheless, he or his contemporaries could not give any historical facts of the progress which Europe had made in the knowledge of directivity and polarity of the lodestone. In fact, when European knew nothing about its polarity, Chinese people had begun to study the knowledge of magnetic declination, and applied azimuth correction skill to the magnetoresistive compass. Because of China's introduction of technology about compass and related magnetics knowledge to Europe, Europeans could develop its magnetics. How can we say that it has nothing to do with the development of science in Europe?

What is more, more than 100 years after the publication of the English version of *Pingzhou Ketan* written by Zhu Yu (朱彧), in which were recorded the reliable historical materials for the use of compass in navigation in the northern Song Dynasty, some people still asserted in 1950 that the compass was absolutely the western invention. ❷ It was astonishing that such an assertion was issued by some so-called science historians. Others even thought that the compass was invented by an Italian named Flavio Gioja in 1300. All these viewpoints were based on scant information, which contradicted the revealing historical facts by scholars both at home and abroad for the recent hundred years or so. Contrary to these extreme views, another scholar admitted that Europeans only found the attraction of a lodestone, but its polarity was first found by Chinese. Based on such kind of discovery, Chinese invented the compass, which was brought to Europe by the world-famous Italian traveler Marco Polo (1254 ~ 1324) from China. But this view set the time for the compasses to spread to the west too late, because more than 100 years before Marco Polo returned to Venice in 1292, Europeans had used compass in their daily life. What is more, Marco Polo did not mention compass in his book *The Travels of Marco Polo*.

Others pre-dated the birth of compass in Europe. They thought that before the 12th century Scandinavians or Vikings obtained the compass from Arabia via Russia and used it for navigation, ❸ but later research proved that the cited

---

❶   WHEWELL W. History of Inductive Science[M]. London: Parker, 1847.

❷   FORBES R. Man the Maker: A History of Technology and Engineering[M]. New York: Schuman, 1950.

❸   WINTER H. Die Nautik der Wikinger und ihre Bedeutung für die Entwicklung der europ ischen Seefahrt[J]. Hansische Geschichtsbl tter, 1937(62):173.

historical materials were not the original records, but the distorted ones. ❶ So far, the earliest European record about compass could be found in *De Naturis Rerum* (*On the Natures of Things*), which was written by Alexander Neckam (1157~1217), an encyclopedic writer of the 12th century in Latin in 1190. Neckam was born in St. Albans. He had been to Paris for furthering his study, and returned to his country in 1186. Later he was an English scholar, teacher, theologian and abbot of Cirencester Abbey from 1213 until his death. His *De Naturis Rerum* (*On the Natures of Things*) was a popular five-volume micropedia at his time, ❷ about scientific information, in which one paragraph in the 2nd volume says below❸:

> *When sailors sailed on the sea, if they met a cloudy day on which they could not see sunshine, or the night at which the whole world was shrouded in darkness and they could not determine directions of navigation, they would turn to the needle and the lodestone, and make them contact each other. At this time, the needle would turn on the plate. When the turning stopped, the needle always pointed north.* ❹

His *De Naturis Rerum* (*On the Natures of Things*) had Thomas Wright's edition (1836) in Latin. In 1945 C. E. N. Bromehead had translated some parts about compass into English. Based on the English version, they were re-translated into Chinese. The original book was printed with no detailed publication year. According to the related researches, after the textual investigations, the time of publication was proved to be in 1190, which was widely accepted over the years by scholars worldwide. Here what Neckam talked about was affirmatively the navigational compass, and the adopted needle was artificial magnetic substance, made through friction between the iron needle and the lodestone, but he did not further introduce its shape,

❶ VON LIPPMANN E O. Geschichte der Magnet Nadel bis zur Erfindung des Kompasses[J]. Quellen und Studien zur Geschichte der Naturwissenschaft und der Medizin, 1933(3):1.

❷ SARTON G. Introduction to the History of Science[M]. Baltimore: Williams & Wilkins Co., 1931.

❸ NECKAM A. Alexander Neckam De Naturis Rerum[M]. London: Her Majesty Stationery Office, 1863.

❹ BROMEHEAD C E. Alexander Neckam on the compass needle[J]. Geographical Journal, 1944(104):63.

structure and construction. When Neckam spoke of compass, he did not emphasize that it was a novelty, which confirmed that the compass was familiar to some people before him, but before the 12th century, European could not grasp the skills of compass. It was possible only in the 12th century.

After Neckam, one French Jews Berakya ben Natronal ha-Adanim also talked about compass. He immigrated to Britain at the end of the 12th century, and had lived in Oxford since 1194. He was perfect in Arabian, and had translated some French literature into Hebrew. Later he compiled a book called *Koah ha-Adanim* in about 1195. In this Hebrew literary works, he narrated properties about 73 kinds of stones, among which were the lodestone and the compass. ❶ According to the research made by K. W. A. Schück, shortly after 1,200 years, compass was used in mining work near Massa in the north of Italy, which reminded people of the Chinese people from Zheng state exploiting jade in remote mountains by using *sinan* several hundred years ago. This demonstrated that European not only used compass in their sea voyage, but also on land to give direction. So, Samuel Purchas (1575~1626), an English travel writer in the 17th century, named compass as "lead stone" in his work *Pilgrimes* (1625), and he thought that it was brought into Italy from Mangiover three hundred years ago, which was now called China. The word *Mangi* originated from *The Travels of Marco Polo* (1299), which was the appellation which Mongolia's army used to call people in the south Yangzi River area after they defeated the southern Song Dynasty (1127~1279).

4.5.2 Influence of Chinese Technique upon European Mariner's Compass

By the early 13th century, compass had been widely used in Europe. Even Guiot de Provins (1148~1218), one French poet at that time, talked about the lodestone and the compass in his satirical verses. Guiot de Provins, also spelled Guyot, was born in the town of Provins in the Champagne area (Seine Maritime) in the southeast outskirts of Paris. He traveled widely, even visiting Germany, Greece, Constantinople and Jerusalem. Later, he settled in the south of France. In 1205, in his later life, he wrote a long satirical poem (2,691 lines) in French with the title *La Bible*, which was used to criticize the Church

---

❶ SARTON G. Introduction to the History of Science[M]. Baltimore: Williams & Wilkins Co. , 1931.

authorities. The whole poem was just a mirror of social life of that time. The poem pointed out that the magnetic needle used by these navigators was made by friction between an iron needle and a lodestone, and its function of guiding the navigation was better than that of the Polaris. In 1834, Heinrich Julius Klaproth (1783~1835), one German orientalist, discussed the compass in his letter written to Baron Alexander von Humboldt (1767 ~ 1859). He first reported the poem written by Guiot de Provins. ❶ In 1840, A Jal, a French marine archaeologist, paid greater attention to the poem by Guiot de Provins. ❷ When Joseph Needham studied the history of compass, he quoted the original poem in ancient French. Because it is hard for today's readers to understand, we try to interpret it here as follows:

> Our pope is like the polestar,
> Stands high above in the sky,
> Sailors on the sea can see him clearly,
> All ships, coming and going at sea,
> Navigate by the polestar,
> Along the right direction to sail.
> Though other stars are moving,
> He stands there still,
> So it is called the North Star.
> Now there is a strange skill grasped by sailors,
> They take black lodestone,
> Rub it with an iron needle to show its magical power,
> Then thread it through a wheat-straw,
> Put it to float on the water,
> Then it would point to the North Star,
> So our navigation would never go wrong,
> And sailors become more confident.
> When the sea is in darkness,
> People cannot see the moon or stars,
> Sailors would immediately hold a lamp,
> They observe the position pointed by the needle,

❶ VON KLAPROTH H J. Lettre M. le Baron Alexander de Humboldt sur l Invention de la Boussole[M]. Paris: Dondey Dupré, 1834.

❷ JAL A. Archéologie Navale[M]. Paris: Arthus Bertrand, 1840.

Try to escape from getting lost on sea,
So reliable is this skill,
That it surpasses the bright polestar,
It should be honored as we respect the pope.

Although in 1190 Neckam spoke of its application in European voyages, he did not give detailed information of its formation and construction. However, when we read off these lines written by Guiot de Provins, some of the questions could be answered. Obviously, the early navigational compasses in Europe during the 12th and 13th centuries were just the wet compasses which Chinese had used long before. Their production method was the same as Chinese: first thread the iron needle magnetized by a lodestone into a smooth stalk; then float it on the round slot in the middle of the compass engraved with bearings (*tianchi* or Heavenly Pond called by ancient Chinese); when the needle stops turning, its two ends will point north and south respectively. The method described by French Guiot de Provins was essentially the same as what was said by Zeng Gongliang (1044), Shen Kuo (1088) and Kou Zongshi (1116), who all lived in the Northern Song Dynasty. Small differences lay in the number of bearing scales. In addition, Europeans focused on the direction of the north, while Chinese put extra emphasis on the south. People of the northern Song Dynasty inserted the needle into Juncus effusus (one kind of rushes) so as to increase surface buoyancy. Europeans used the same principle and inserted the needle into the straw rod. This proves that the early European compass was made just by applying the Chinese technology.

Another person worth our attention is Jacques de Vitry (1178~1240), a French chronicler, whose Latin name was Jacobus de Vitriaco. He was born in Vitry le Franois near Versailles (in central France, perhaps Reims) and studied at the University of Paris. In 1216, Jacques was elected Bishop of Acre (now in Israel). In 1219 he began to write *The Historia Hierosolymitana* (also called *Historia Orientalis*), a history of the Holy Land from the advent of Islam until the crusades of his own day, but he only completed two parts. In his book, he mentioned the application of compass in real life. ❶ Scottish astrologer Michael Scot (1175~1234), his contemporary, also talked about two kinds of lodestone in the world in his book *Liber Particularis* (which mainly concerns the

---

❶ SARTON G. Introduction to the History of Science[M]. Baltimore: Williams & Wilking Co., 1931.

foundations of the earth), one pointing to the south and the other to the north. ❶ As stated in the previous chapters, as early as over 200 years ago, Chinese scientist Shen Kuo published similar view about the lodestone. In fact, the needle markings of the early European wet compasses also pointed south, conforming to the traditional Chinese ones. E. G. R. Taylor pointed out that as late as 1670, western astronomers used the south pointing compasses, not the north-pointing compass used by sailors. ❷ During the time from 1228 to 1244, Flemish Thomas de Cantimpre (1204 ~ 1275), who was a Roman Catholic medieval writer, preacher and theologian, completed his famous encyclopaedia *De Natura Rerum*. The book was written around the years 1230~1245. In this enormous encyclopedia, Thomas compiled the natural history knowledge of 20 categories, including what would now be called anthropology, zoology, botany, mineralogy, astronomy, astrology, and meteorology, etc. Of these categories, the part discussing the valuable stones (gems) introduced the wet compasses for navigation.

When talking about the early compasses during the period from the 12th century to the 13th century, Abaraham Wolf, contemporary English historian of science, also said: "This kind of early device is mainly the wet compass. People make magnetized iron needle float on the water in the wooden water bowl, and then keep an eye out for its pointing direction. Sometimes they adopt magnetized iron floater." But he did not interpret what is the iron floater. We speculate that it would be something like the south-pointing fish used by the people in the Song Dynasty in China, and readers could refer to the related content mentioned above. Abaraham Wolf admitted: "Chinese people have long known the nature that the lodestone can point south when in a free state. However, not until the 12th century did the European literature begin to mention this new navigational device. It is obvious that western people did not know its important value before." However, he did not know whether this kind of device was imported by Arabians or European sailors from the orient or was found independently by European. ❸

❶ HASKINS C H. Studies in the History of Mediaeval Science[M]. Cambridge: Harvard University Press, 1927.

❷ TAYLOR E G R. The south pointing needle[M]. Imago Mundi: Yearbook of Early Cartography, 1951.

❸ WOLF A. A History of Science, Technology and Philosophy in the 16th and 17th Centuries[M]. London: Allen & Unwin Ltd. , 1935.

Was the navigational compass an independent invention by European or was it made under the influence of Chinese technology? This is the question posed at the beginning of this chapter of the book, so it is necessary for us to discuss it. To this end, we need to make a comparison between Chinese compasses and European compasses. As we all know, the precondition of making the magnetic compass is that people should understand the lodestone's directivity and polarity. In the 3rd century B. C. , Chinese people found its polarity and made natural lodestone into south-pointing devices, which were called *sinan*. During the $9^{th} \sim 10^{th}$ centuries, people used man-made magnetic iron to make geomantic compass and found the secret of magnetic declination. In the 11th century, the wet compass was widely used in navigation, and at that time, Chinese ships even sailed as far as the Persian Gulf, red sea and East Africa. They enjoyed close economic cooperation and frequent personnel exchanges with the Arabian world. As a result, Arab became the medium of mutual communication between China and Europe. The compasses used by Arabian sailors were made strictly by Chinese technique. Even the implications of the device names were the same as Chinese. Though the related records about the compasses in Arabian literature were slightly later than those of Europe, but the time of Arabians' grasping compass technology practically might be earlier.

But if one turned his eyes to Europe, he would find that European did not know the polarity of the lodestone until the 12th century. When they understood the property at the end of the century, it was over 1000 years later than Chinese. They worked with artificial magnet to make compasses 300 years later than Chinese. Their navigation by magnetic compasses was roughly 100 years later than Chinese. What is noteworthy is that the earliest navigational compasses used in Europe were wet compasses, just like China, and their shapes and structures were the same as Chinese compasses. Their time difference and similarities can only be interpreted by technology transfer theory in the maritime navigation era of marine transportation between China and the western world. According to the communication technology theory developed by H. Chaltler[1], R. E. M. Wheeler[2], A. L. Kroeber[3] and Joseph Needham,

[1] CHATLEY H. The Origin and Diffusion of Chinese Culture[M]. London: China Society, 1947.
[2] WHEELER R E M. Archaeology and the Transmission of Ideas[J]. Antiquity, 1952 (26):180.
[3] KROEBER A L. Stimulus Diffusion[J]. American Anthropologist, 1940(42):1.

one kind of civilization could accept the ideological system or pattern of another civilization entirely, even by accepting one slightest hint or illumination of certain idea from another civilization. Sometimes only one hint or inspiration from another civilization would give rise to a series of reforms. The technological maturity in a faraway place in the world would encourage another nation to figure out the similar problems by using the same technology. Kroeber called this kind of phenomenon stimulus diffusion.

Examples of Chinese papermaking, printing and gunpowder spreading to western countries are just typical cases of stimulus diffusion, which are discussed respectively in previous chapters of the book, so is the spreading of compasses to the west. From the discovery of the lodestone's polarity to the compasses made by artificial magnetic needles, Chinese explored it for over 1000 years. They experienced several stages, such as the stage of *sinanyi*, at which people used natural lodestone to make a magnetic spoon turn on the copper board, and the stage of the man-made magnetic needle floating on the water. All these manifest hardships of original technology in its evolvement. However, for Europeans, from utter ignorance of the lodestone's directivity and polarity to the sudden jumping to the wet compasses for navigation, they had not any traces about its original technology and its evolvement, so it is obvious that they made them totally by accepting foreign existing skills and experiences. Moreover, this foreign experiences could only come from Chinese or Arabian sailors, with the latter a greater possibility. When Friedrich Engels (1820 ~ 1895) began to list the great invention chronology from ancient times to the middle Ages, he found that these inventions, such as compass, printing, movable type and hemp paper from China were never thought of or mentioned before by ancient European. Moreover, he asserted that the magnetic needle was spread to Europe just through Arabians around the year 1180. ❶ In other words, Chinese compasses were spread to the west through Arabians in the year of about 1180.

During the period of 1180 to 1250, these 70 years was the early stage of of development for European compasses. Then came the trial stage by copying Chinese compasses made in the Song Dynasty, which was marked by using wet

❶   恩格斯. 自然辩证法[M]. 北京：人民出版社，1984.
      ENGELS F. Dialectics of Nature[M]. Moscow: Foreign Languages Publishing House, 1954.

compasses for navigation. However, besides using a compass, sailors, in fact, often adopted celestial navigation as additional technology aids when they sailed on the sea. Neckam and Guiot de Provins said that when the weather became dark or at night and people could not see the moon or stars in the sky, they had to resort to a compass. Moreover, at this time, French people were deeply involved in it and left a lot of records on compasses, ranking before the Englishmen. From the relevant records, European, at that time, seemed to accomplish little in the practice of compass application and they did not give any new knowledge beyond the ken of Chinese people in the Song Dynasty.

4.5.3 Dry Compass and Navigation Diagrams in the 13th Century Europe

However, since 1250, namely, from the latter part of the 13th century, the situation began to change because of a series of research work done by Pierre de Maricourt (1224 ~ 1279), a French experimental physicist. He conducted several experiments on magnetism and wrote the first extant treatise describing the properties of magnets, which made the compasses in Europe enter upon a new phase and inspired European to study magnetic phenomenon scientifically. He also made some technological improvements for magnetic compass localization.

Petrus Peregrinus de Maricourt was born in Maricourt in the north of France. His Latin name was Petrus Peregrinus. Here, Petrus in Latin was equivalent as Pierre in French or Peter in English, while Peregrinus in Latin means the foreign people living in Roman. It might be a nickname. Some people thought his family name was Peregrinus or Maricourt, which was partly due to misapprehension. In the middle Ages, European sometimes connected his name with his native place, not using his family name, so Peregrinus de Maricourt means Pierre from Maricourt. Almost nothing is known about Peregrinus' life, but people only knew that he joined the crusades and participated in the battle to lay siege to Lucera City of Italy in 1269. He was one of Roger Bacon's contemporaries, respected as a tutor and a perfect mathematician by Bacon, both of whom were the most knowledgeable scholars of Europe in the 13th century. In his letter *Epistola Petri Peregrini de Maricourt ad Sygerum de Foucaucourt, militem, de magnete* (Letter of Peter Peregrinus of Maricourt to Sygerus of Foucaucourt, Soldier, on the Magnet) of 1269❶,

---

❶ SARTON G. Introduction to the History of Science[M]. Baltimore: Williams & Wilkins Company, 1931.

Peregrinus explained how to identify the poles of the compasses and introduced his rescarch achievement. In fact the letter was just a booklet. In addition, the version of the letter published by Italian Tiberius Cavallo in 1800 was easier for use. ❶

Pierre de Maricourt's letter was more commonly known by its short title, *Epistola de Magnete (Letter on the Magnet)*. It can be divided into two parts:

The first part has 10 chapters: (1) The purpose of This Work; (2) On Experimental Methodology; (3) Methods to Identify a Lodestone; (4) Two Ways to Determine the Poles of a Lodestone; (5) Methods to Distinguish Magnetic Pole from Terrestrial Pole of Meridian Line; (6) How one Lodestone Attracts the Other; (7) Methods to Rub an Iron with a Lodestone to Make It Magnetized; (8) How a Lodestone Attracts Iron; (9) Why One Pole of a Lodestone Attracts the Other Pole; (10) An Inquiry into the Cause of the Natural Energy of a Lodestone.

The second part has three chapters, focusing on compasses' application, which describes three devices that utilize the properties of magnets: (1) Portable Compasses and Sundials for Direct Measurement of Directions of Sun, Moon and Stars; (2) Construction of a Better Instrument for the Same Purpose Above; (3) Art of Making a Perpetual Motion Machine. However, there were not any illustrations offered in these chapters.

In the first part, the author Pierre de Maricourt of *Epistola de Magnete* pointed out that when doing research work, one should be diligent in verifying or correcting theoretical viewpoints. He performed a series of experiments on a sphere made by a natural lodestone, determined the poles of compass, and proved the tendency of the two poles to point to the due north and the due south, with the two opposite ends the strongest in magnetism. He also proved that like poles repel each other and unlike poles attract each other. He described that even though a lodestone was broken into two pieces, each one still had magnetic property and two poles. He also interpreted cosmos motion by referring to the spherical magnetic substance moving with the heaven. In the first chapter of the second part, Pierre de Maricourt introduced an improved wet compass, which had directrix and a graduated circle with 360° in the circumference. The better instrument mentioned in the second chapter was a

---

❶　CAVALLO T. A Treatise on Magnetism in Theory and Practice with Original Experience[M]. 3rd ed. London, 1800.

dry compass，which was made by putting pivoting magnetic needle on the metallic pivot and the graduated circle in the capsule and covering it with a glass cover. This is the earliest record about a dry compass in Europe. In a word，the booklet *Epistola de Magnete* not only summarized the laws of magnetic attraction and repulsion，but also added new contents，such as a perpetual motion machine. Sutton thought that it was a rare example of practical experiments on magnetics for European.

The works of Pierre de Maricourt inherited the experimental spirit in studying magnetics from *Mengxi Bitan* written by Shen Kuo，and carried it forward. He also started similar cause embodied in his work *De Magnete* (论磁石) written by English author William Gilbert (1544~1603). William Gilbert acknowledged his debt to Pierre and incorporated this thirteenth-century scientist's experiments on magnetism into his own treatise. As for dry compasses，they were，just like wet compasses，invented by Chinese. As mentioned before，the material object of a dry compass buried in 1197 was unearthed from the tombs of the Song Dynasty in Linchuan of Jiangxi Province in 1985. ❶ *Shilin Guangji* published around 1135 also recorded some anecdotes that artists often used dry compasses when performing conjuring tricks. This illustrates that dry compasses with the magnetic needles pivoting on the azimuth plate had already appeared in China in the early 12th century. Later this kind of compasses was introduced to the west. In 1269，French Pierre de Maricourt improved them，and put them in a capsule covered with a glass cover，thus making them into portable instruments. Afterwards，this kind of dry compasses was widely used by sailors.

Just like Chinese or Arabian navigators，European sailors went on voyage by referring to some navigational diagrams. These navigational diagrams are called *haibu* (海簿，the book of navigation) or marine charts by ancient Chinese. The word "portolani" in Latin shares the same meaning. A navigational diagram is a graphic representation of a maritime area and adjacent coastal regions，which is generally marked with safe routes and courses，coasts and harbors which ships often pass through，islands and some natural landmark or cultural scenery. All these diagrams are drawn by previous navigation practices and experiences and they possess great practical value in navigation. Courses or directions were determined only by observing stars before the

❶ 陈定荣，徐建昌. 江西临川宋墓[J]. 考古，1988，(4)：329-334.

invention of compass. People marked the heights of the lodestars above ground level (A. G. L.). Since the invention of compasses, people only marked the needle positions or both. Sometimes the chart would also mark the mileage from one place to another place. Navigators can sail easily and safely in boundless ocean based on the routes, directions and mileages marked in the marine charts. *Zheng He's Navigation Map* in the Ming Dynasty embodied the most typical traditional drawing methods of the navigational diagram, but it was not the earliest map in China. As a matter of fact, navigators in the Song and Yuan Dynasties employed the maps of the same type. European navigational diagrams were later than Chinese diagrams, but they also contained similar contents for navigation.

According to the record, King Louis Ⅸ of France (1214~1270) sailed from the harbor city of France Aigues Mortes on the vessel of Genoa of Italian in 1270, and he crossed Mediterranean and reached Tunisia in the north of Africa. After the vessel sailed on the sea along the west coast of Italy for six days and passengers on board still could not see the Sardinian coast, the king began to worry. At this moment, the officials on board showed him the map, and told him the position where the vessel was, and asserted that they were close to the port of Cagliari in the south of Italy. This is the Europeans' first mention of a navigational diagram used for navigation. ❶Afterwards, Ramon Llull (1235~1315), a Franciscan tertiary and Majorcan writer in Catalan Christian Church of Spain, recorded the same anecdote in his book *Arbor Scientiae* or *Arbre de Sciencia* (*The Tree of Knowledge*). *Arbor Scientiae*, like an encyclopedia, was written by Ramon Llull in Rome between 1295 and 1296. When explaining how the sailors *found* their way on the sea, Ramon wrote: "ad hoc instrumentum habent chartam, compassum, acum et stellam maris", which means "because the vessels had the navigational diagram and the compass to determine directions." The diagram marked the needle positions of places and data from star observations, so the diagram was called compass charts. Roger of Hoveden (1174~1201), an English chronicler, also suggested in his historical work *Chronica* 732~1201 that there were some kind of navigational maps or nautical books. ❷

---

❶ SARTON G. Introduction to the History of Science[M]. Baltimore: Williams & Wilkins Co., 1931.

❷ HUNT W. Dictionary of National Biography[M]. London: Smith, Elder & Co., 1891.

Europeans, who were good at sailing on the sea, could trade by sea safely after they brought in compasses, axial rudders and watertight compartments from China. They embarked on exploratory voyages of discovery, not only completing Great Geographical Discoveries, but also colonizing other weak countries and new commodity markets with firearms. European capitalism enjoyed great political and economic interests by employing these inventions. Christopher Columbus (1451 ~ 1506), an Italian explorer, navigator and colonizer, completed four voyages across the Atlantic Ocean by means of compass under the auspices of the Catholic Monarchs of Spain. Christopher Columbus discovered magnetic declination on his first voyage to the New World in 1492. However, this was not new to Chinese, because Chinese noticed it several hundred years ago, earlier than Europe. For European, it was new, not because there was something wrong with compass, but because the magnetic poles of the earth were not coincident with the north and south poles of the meridian line. The explanation of this phenomenon boosted the knowledge of magnetics in Europe.

Chapter 5　Influence of Paper and Printing on Development of World Civilization

5.1　Promotive Role of Paper in Development of Culture in China and Foreign Countries

5.1.1　Appearance of Paper: Revolution in History of Development of Scripts Carriers

The creation of *wenzi* （文字, scripts) is one of the important milestones in the history of human civilization development. The earliest scripts were produced in the form of abstracted and simplified pictures, which later evolved into *xiangxing wenzi* （象形文字, hieroglyphs). With the further simplification, transformation and phonetic transcription, hieroglyphs finally took the appearances of *biaoyi wenzi* （表意文字, ideographs) and *pingyin wenzi* （拼音文字, alphabetic writings). So far, it has been a history of four or five thousand years. The invention of scripts and its constant improvement see the expansion of space and time of human beings' idea exchange and information transmission because, with the scripts, human beings can sufficiently express what they think and what they want to say, record them on the writing materials and transmit them to the relatively distant places, even to the later generations. In addition, the scripts endow human beings with the possibility to write their activities into the *xinshi* （信使, authentic history books), create their classics and thus develop their own spiritual civilization. It therefore could be said that a nation could be regarded as a civilized one only if it has its own scripts. Moreover, the development of human civilization relies on the inheritance of classics recorded in different writing materials. Classics take on the different forms with the different writing materials and change with the times.

Before the invention of the writing material of paper, the scripts carriers used in each civilization area of ancient times took on different forms with different materials and changed with times. For example, in the Yin and Shang Period of China, our Chinese ancestors carved scripts on tortoise shells and animal bones, and then stringed them together, while in the Spring and Autumn and Warring States Periods (770 B. C. ～221 B. C.) as well as the later long period of time, it was prevailing for Chinese people to typecast scripts on bronze objects, carve scripts on stones and write on *jiandu* (简牍, the bamboo and wood slips) and *jianbo* (缣帛, silk); Assyriarr and Chaldeans carved scripts on *niantupei* (黏土坯, clay tablets) and then had them burned into clay bricks; Ancient Egyptians and Arabians used as script carrier the *shacaopian* (莎草片, papyrus) abundant in the Nile Valley; People of India and other countries in the southeast Asia utilized the palm leaves called *beiye* (贝叶, pattra leaves) by ancient Chinese people, having them pricked after writing and stringed together; ancient Europeans casted scripts on plate metals or wrote them on *yangpipian* (羊皮片, parchment); some other nations employed tree leaves as writing material. In a word, various materials from the fields of mineral, plant and animal had been used. Obviously, all of them had their own advantages and disadvantages. Metals, stones, shells, bones and bricks were hard and durable, but the relatively big size of scripts on them led to the small number of scripts they can carry. Meanwhile, their bulk and weight made it inconvenient to write on, store and take. In other words, only those light materials which can be directly written on, conveniently stored and taken could be used as the writing materials of classics such as bamboo slips, wooden slips, silk, papyrus, palm leaves and parchment. Moreover, the classics of these materials could be preserved and used for a long time.

Among these convenient writing materials, just as mentioned above, the bamboo slips, wooden slips and silk were mainly adopted by ancient Chinese people. The rectangle bamboo and wooden slips, after being processed, can hold thirty scripts each. They were stringed together slip by slip according to the order of scripts on them, usually up to the length of about two meters, and then rolled up. When it came to be used, they were rolled out and laid in row. As for the silk, after being written on, a wood axis was usually added to one end for the convenience of rolling. As a result, the classics of ancient China mainly took the shape of pipe and were denominated in *juan* (卷, roll). However, the other light writing materials of papyrus and palm leaves, because of their fragility, could not be rolled up, and thus could only be stringed together in the form of modern books. It is true of the writing materials of parchment. When it came to writing and reading, Chinese ideographs were usually written only on the front of writing materials along the vertical direction from the up to the down and read along the horizontal direction from the right to the left, while the alphabetic writings were written on the front and back of writing materials and read both from the left to the right, which were some of the differences in the forms of classics between China and foreign countries. But in general, compared with the natures of bamboo slips, wooden slips, silk and parchment, papyrus and palm leaves were poor in their flexibility, insect resistance and humidity proof, so they were difficult to be preserved for a long time. What is more, they were not smooth on their surfaces and could be only available in some areas. Consequently, they were impossible to be widely-used writing materials.

Seemingly, bamboo slips, wooden slips, silk and parchment were better writing materials. However, being expensive, the latter two could be afforded only by a few people and thus could not pervade among the folk, while because of the small number of scripts each bamboo and wooden slip holds, a large number of bamboo and wooden slips were bound to be needed when it was to write a book. Obviously, with the increase in the length and number of books, the disadvantages of these three materials became more and more salient. For example, the writing of the Latin version of *Vulgata* (《圣经》, *The Bible*) needed 100 pieces of parchment; the writing of *Shi Ji* needed 36 thousand slips of wood or a piece of silk whose value was equivalent to that of 360 kilograms of rice. Therefore, it was an inevitable tendency for a new type of writing material to come to replace the heavy bamboo and wooden slips and the expensive silk. This material was paper. In the early Western Han Dynasty (206 B. C. ~ 24 A. D.) of the second century B. C. , China invented the technique to make *mazhi* (麻纸, hemp paper) from *pomabu* (破麻布, ragged hemp cloth). In the Eastern Han Dynasty (25 ~ 220), the *pizhi* (皮纸, libriform paper) made from *muben renpi* (木本韧皮, woody phloem) came into being. In the late Tang Dynasty (618~907) of the ninth century, the *zhuzhi* (竹纸, bamboo paper) appeared. In fact, it was as early as the fourth century that the bamboo and wooden slips had been completely phased out and that paper had become the main writing material.

Compared with all the previous ancient writing materials, paper had its own advantages as follows: (1) It was smooth and white on its surface, good at its *shoumo* (受墨, ink absorbability), large in its width and thus could hold many scripts; (2) Being light and pliable, it was able to be rolled up and out, easy to be pasted and convenient for taking and storing; (3) It was inexpensive because the raw materials used to make it pervade worldwide and it could be made anywhere; (4) Being durable and multifunctional, it could be further pressed into supplies for the industrial, agricultural, military and daily uses. In view of these points, it could be said that paper was an almighty material unparalleled by any previous writing material. The appearance of paper was a revolution in the history of the development of human beings' scripts carriers because it, as a widely-used material in the world over the past more than two thousand years, had played a great role in the promotion of human civilization development, and would continue to do so within a long period of time of the 21st century.

The reason why paper is superior to the other writing materials is that the latter are obtained only from the simple and mechanical processing of raw materials and still keep the same as their raw materials in components, forms and natures. On the contrary, the process of papermaking is much more complicated. Briefly, the first step is to extract the paper pulp without any other impurities from the effective ingredient (the plant fibers) of raw materials by using chemical methods and the second step involves a series of mechanical processing of the pulp. In this process, raw materials experience the physical changes in their outer appearances and the chemical changes in their components, so it could be said that paper is the product from the deep processing of raw materials.

After China invented paper, she did not monopolize it, but shared it with the people of the world. During the Wei and Jin Periods (220~420), China first transmitted the papermaking technique to her neighboring countries such as Gaogoli (37 B.C.~668 A.D.), Baekje (18 B.C.~660 A.D.), Silla (57 B.C.~935 A.D.), Vietnam and Japan. During the Tang Dynasty (618~907), the transmission extended to the Indian subcontinent, the central Asia, the western Asia and the Arabian countries in the northern Africa. In the 12th century, it reached Europe from Arab. During the period from the 16th century to the 17th century, it arrived in America from Europe. This was an extremely long journey spanning across a thousand of years. In all the places where paper arrived, it became the competitive rival of local writing materials and finally replaced them.

Evidently, people of the world, whether they believed in Buddhism, or in Islam or in Christianism, liked using paper to write. This led to the result that those classics written on the other writing materials were copied on the paper so as to be kept eternal in the world. Gradually, those classics of the other writing materials, with the time going by, disappeared, but their paper copies were passed on from generation to generation. Paper made it possible to have ancient culture continued, developed and shined with great brilliance and thus made immortal historical contributions in the preservation and inheritance of human cultural heritages. For example, the science and cultural classics of China during the Pre-Qin Period (21st B. C. ~ 221 B. C. , the long period before Emperor Qinshihuang's unification of ancient China) had been preserved mainly in *zhiben shujuan* (纸本书卷, paper books and scrolls) since the Han and Jin Period (202 B.C. ~420 A.D.); the works of scholars of ancient Greece and Rome, the Sanskrit classics of India and the Arabian works were also kept due to their *zhichaoben* (纸抄本, paper copies). These preserved civilization works enlightened and inspired people of the world and became the earliest academic materials for the latter generations to absorb. Therefore, it could be concluded that paper is the holy flame to transmit human civilization.

5.1.2   Promotive Role of Paper in Development of Chinese Culture

Compared with writing on the bamboo and wood slips, writing on the paper was faster and more energy-saving because it made the time spent in writing a book greatly shortened and reading more convenient. Usually, paper was manufactured in large scale. The huge production of paper made it be afforded by the general public with the low price. With paper affordable, people could copy books or write books. Due to the potential that books on paper were easier to be popularized and transmitted than those on bamboo slips, wooden slips and silk, paper had prevailed during the period from the Han Dynasties (206 B. C. ~ 220 A. D.) to the Wei, Jin, Northern and Southern Dynasties. In the society of that time, copying books on paper became very popular, which contributed to the fast development of calligraphy art. Meanwhile, because writing on paper was much faster, the forms of Chinese character developed from the *xiaozhuan* (小篆, seal script) and the *lishu* (隶书, official script) to the *kaishu* (楷书, regular script), with the *kaili* (楷隶, official-regular script) and the *xingshu* (行书, running script) being popular. Calligraphers represented by Wang Xizhi and his son appeared. Their calligraphy had been imitated over thousands of years.

The swift increase of paper and books on paper first contributed to the fast development of education, science and culture. In the Western Han Dynasty (206 B. C. ~24 A. D.), the capital city of Chang'an (in the northeast of today's

Xi'an in Shaanxi Province) had seen the founding of the highest educational institution, *Taixue* (太学, Imperial College), in which *Wu Jing Boshi* (五经博士, The Master of Confucian Five Classics) explained to students the Confucian classics of *Yi* (《易》, *The Book of Changes*), *Shu* (《书》, *The Book of History*), *Shi* (《诗》, *The Book of Songs*), *Li* (《礼》, *The Book of Rites*) and *Chunqiu* (《春秋》, *The Spring and Autumn Annals*), which were finished during the period of time from the Early Western Zhou Dynasty (11th Century B. C. ~771 B. C.) to the middle of the Spring and Autumn Periods (770 B. C. ~476 B. C.). Those students who passed the examinations on the Confucian classics would be granted official positions. In the Eastern Han Dynasty, *Taixue* in the capital of Luoyang saw that the number of its students had increased to more than 30 thousands and became the largest higher education institution in the world at that time. ❶ In addition to the central *Taixue*, the governments of the Han Dynasties (206 B. C. ~220 A. D.) set up the public schools in prefectures and counties to enroll students from various places. At the same time, the private schools prevailed. The total number of students in schools all over the country could be amounted to several ten thousands, even to one million. Most of the textbooks they used were *zhiben jingjuan* (纸本经卷, paper texts and books). Compositions were also written on paper, much more convenient, time-saving and energy-saving than on bamboo and wooden slips. The development of education cause provided a large number of intellectuals for the society. They undertook the research of science and technology and the creation of literature arts in addition to the research of Confucian classics. Various classics of the Pre-Qin Period (21st century B. C. ~221 B. C.) were annotated in detail and a plenty of new works came into being. The imperial government also sent the relevant officials to look for books in the folk many times and even organized experts to conduct a systematic check to the secret books kept in the *neifu* (内府, the imperial storehouse) so as to offer reliable books.

During the Han Dynasties (206 B. C. ~220 A. D.), the number of works was much more than that of the Pre-Qin Period (21st Century B. C. ~221 B. C.). The contents in those books had laid a solid foundation for the academic research of the latter dynasties. For example, *Han Shu · Yiwen Zhi* (《汉书 · 艺文志》, *Treatise on Literature* in *History of the Western Han Dynasty*), China's earliest bibliography documentary edited by Ban Gu in the

---

❶ 孟宪承.中国古代教育史资料[M].北京:人民教育出版社,1961.

Eastern Han Dynasty, listed the works of 678 different schools of thoughts, amounting to 14994 books, including the classics of Confucianism, *Xiaoxue* (小学, Chinese traditional linguistics), Taoism, the School of Yin and Yang, the School of Names (Logicians), Mohism, *Zongheng* (纵横, the School of Thoughts on Chinese Traditional Diplomacy and Political Strategies), the Eclectics (a school of thought flourishing at the end of the Warring States Period and the beginning of the Western Han Dynasty), the School of Agriculturists (a school of thought in the Spring and Autumn and Warring States Periods, 770~221 B. C.), the School of Military Strategist, the School of Physicians' Thoughts, Astronomy, Criminal Law, Machinery, Literature, Arts and so on. At the end of the Han Dynasties (206 B. C. ~220 A. D.), Buddhism was introduced and thus the list was added by the works of Buddhism. It was also in the Han Dynasties (206 B. C. ~ 220 A. D.) that ancient Chinese traditional science system was formed, embracing the excellent works of *Zhoubi Suanjing* (《周髀算经》, *The Book of Ancient China's Astronomy and Mathematics*), *Jiuzhang Suanshu* (《九章算术》, *The Nine Chapters of the Mathematical Art*), *Ling Xian* (《灵宪》, *Book of Astronomy*), *Bencao Jing* (《本草经》, *The Canon of Materia Medica*), *Shanghan Lun* (《伤寒论》, *Book of Exogenous Febrile and Miscellaneous Diseases*) and *Fansheng zhi Shu* (《氾胜之书》, *Book of Ancient China's Agriculture*). Among numerous works of the Han Dynasties (206 B. C. ~220 A. D.), *Shi Ji* written by Si Maqian was China's first biographical history book, covering incisive commentaries, rich historical materials and many progressive views on historical events, while *Han Shu* (《汉书》, *The History of the Western Han Dynasty*) written by Ban Gu was the first book of dynastic histories of China. In addition, the masterpieces of *Fu* (赋, Prose-Poetry, a literary form of part prose and part poetry) and poetry of the Han Dynasties (206 B. C. ~ 220 A. D.), mainly adopting rhyme to describe things and people's emotions and to reflect the social reality, had a long-term influence on the literature of the later generations.

During the Wei, Jin, Northern and Southern Dynasties, the thriving of papermaking industry all over the country promoted the sharp increase of books and the further development of science and culture. During the period of Liu Song Dynasty, the first dynasty of Southern Dynasties, the *Mishujian* (秘书监, an administration for the collection, compiling and proofreading of books), Xie Lingyun, produced *Sibu Mulu* (《四部目录》, *Four Book Directories*), recording 64,582 books. This work was 338 years later than *Han Shu · Yiwen*

Zhi (《汉书·艺文志》, *Treatise on Literature* in *History of the Western Han Dynasty*) and embraced books 4.3 times more than the later. During the Liang Dynasty (502~557), Sheng Yue loved collecting books in his later life. He himself alone had 20,000 books, more than the total number of books recorded in *Han Shu · Yiwen Zhi*. This fully reflected the prosperous book-collecting situation of "Each family had collected books all over the country" during the period of Emperor Liangwu, which was the very golden period of the development of *zhixieben* (纸写本, paper books).

Usually, each time when the number of classics sharply rose was always accompanied by the approaching of the climax development of science and culture. This was directly related to the development of papermaking industry. The period of the Six Dynasties was the one in which the Chinese traditional science system was enriched and promoted, with the types and number of books exceeding those of the Han Dynasties (206 B. C. ~ 220 A. D.) and many innovations emerging in the field of knowledge. The writers during this period of time obtained many major achievements in their research fields and exploited many new research directions. Many of research findings were located at the leading position in the world at that time and even up to today have still maintained their academic value.

The agriculturist of Jia Sixie wrote *Qimin Yaoshu* by the generalization of agricultural techniques and experiences along the middle and lower reaches of Yellow River. It was one of classic works on Chinese agriculture, expounding the techniques and tools for the cultivation of crops, oil plants and fruit trees. It also involved the information on animal husbandry, veterinary, agricultural products processing and sideline production as well as paper and ink manufacturing. In addition, it stressed that agriculture should be done according to the specific time and condition as well as the idea of artificial selection and introduced the methods for the fruit tree grafting, beast castration and microorganism fermentation. The British biologist Charles Darwin (1809~ 1882) gave high praise to Jia Sixie's book when he read it in 19th century and regarded it as "ancient China's encyclopedia"[1]. This book later spread widely in Japan and North Korea, producing good effects.

The mathematician of Liu Hui (220 ~ 270) wrote the masterpiece of *Jiuzhang Suanshu Zhu* (《九章算术注》, *Annotation of the Nine Chapters of the*

---

❶ 杜石然. 中国古代科学家传记: 上集[M]. 北京: 科学出版社, 1990.

Mathematical Art) in which one of his biggest achievements was to use the *geyuanshu*（割圆术，the Cutting Circle Method）to calculate the *yuanzhoulü* （圆周率，the ratio of the circumference of a circle）. He calculated successfully the area of *yuannei jie zheng* 192 *bianxing*（圆内接正 192 边形，the inscribed 192-regular polygon of a given circle）and the *Pi* value of 159/50, i. e., 3. 14. According to the record of *Suishu · Lüli Zhi*（《隋书·律历志》，*Treatise on Rhythm and Calendar* in *the History of the Sui Dynasty*），Zu Chongzhi（429～ 500），the scientist of the first dynasty（Liu Song Dynasty，420～479）of the Southern Dynasties（420～589），calculated the *pi* value of 355/113, i. e., the value between 3. 1415926 and 3. 1415927 on the basis of the achievement of Liu Hui who lived in the period of Three Kingdoms. His success in the accurate calculation of *pi* value to seven decimal places made this value the most accurate one in the world at that time and thus be called by historians as "*Zulü*（祖率， Zu Ratio / The *Pi* Value of Zu Chongzhi)". It was much earlier than that of the West where it was not until the 16th century that mathematicians came up with the calculation of the *pi* value of 355/113. In addition，Zu Chongzhi also discussed the solutions to the *ercifangcheng*（二次方程，the quadratic equation） and *sancifangcheng*（三次方程，cubic equation）in his work of *Zhui Shu*（《缀 术》，*Mathematics*）. Yet in his *Daming Li*（《大明历》，*Daming Calendar*）， he pioneered the introduction of *suicha*（岁差，the precession of the equinoxes） into the calendar calculation，using the *xin runzhou*（新闰周，the new leap weeks）of 391 year plus 144 leap month and putting forward the new *muxing zhouqi*（木星周期，Jupiter cycle）. At the same time，the length of *huiguinian* （回归年，the tropical year）of 365. 2428 days he used was only 46 seconds away from today's value and the length of *shuowangyue*（朔望月，the lunar month）of 29. 5309 days he used was less than 1 second away from today's value. However， this advanced work of calendar，for the opposition of the conservatives，was not formally used until the year of 510，10 years after his death. ❶

The cartographer of Pei Xiu（224～271）hosted the mapping of 18 chapters of *Yugong Diyu Tu*（《禹贡地域图》，*Yugong Regional Altas*）during the period from 268 to 271 under the reign of Taishi（265～274）of the West Jin Dynasty （265～316）. According to *Jinshu · Pei Xiu Zhuan*（《晋书·裴秀传》，*Biography of Pei Xiu* in *History of the Jin Dynasty*），*Yugong Diyu Tu*（《禹贡地域图》， *Yugong Regional Altas*）was a giant atlas. In its preface，Pei Xiu proposed the

---

❶   杜石然. 中国古代科学家传记：上集[M]. 北京：科学出版社，1992.

theory of "*Zhitu Liuti*（制图六体, the six principles of mapping)". Meanwhile, in this book of maps, he described in detail the methods for the representation of scale, orientation, distance and terrain, which basically conformed to the scientific principles of modern cartography, thereby producing a far-reaching influence on China's drawing of later generations and occupying an important position in the drawing history of the world. ❶

The geologist of Li Daoyuan（469～527）wrote *Shuijing Zhu*（《水经注》, *Notes on Book of Waterways*）through his literature consulting and field surveys. In the book, Li Daoyuan retraced 1,252 rivers in China, detailed the natural landscape and human landscape of places which he experienced. This book was the systematic and comprehensive geography book at that time.

During the period of the Six Dynasties（222～589）, medicine also saw its great development. The famous doctor Wang Xi（180～265）summed up the achievements of ancient *maixue*（脉学, sphygmology）and wrote the systematic monograph of *Maijing*（《脉经》, *The Classic of Pulse Research*）in 242, in which Wang Xi put forward 24 kinds of *maixiang*（脉象, pulse conditions）. This work laid the theoretical foundation for *zhenmai*（诊脉, the pulse-taking）and was considered the criterion by the later generations. It also had an influence on the medicine circles of Japan, Korea, Arab and Europe. Huang Fumi's（215～282）*Zhenjiu Jiayi Jing*（《针灸甲乙经》, *A-B Classic of Acupuncture and Moxibustion*）was the earliest existing monograph on acupuncture, making great contribution to the development of meridian and acupuncture theories of Traditional Chinese Medicine. Tao Hongjing（456～536）, in his work of *Bencao Jingji Zhu* finished in 500, reclassified 730 kinds of medicine in total on the basis of the summing-up of the achievements of herbalism after the Han Dynasty（206 B.C.～220 A.D.）and of the adding of 365 kinds of new drugs. Giving up the traditional classification of the upper-middle-lower qualities, Tao took the natures of medicinal materials as the criterion and reclassified them into the parts of jade, grass, insects, fish, fruit, wood and rice. In addition, he added a lot of new knowledge in his work. Therefore, his work laid the foundation for the development of later herbalism. Ge Hong（about 281～341）finished the compilation of his work *Zhouhou Beijifang*（《肘后备急方》, *Handbook of Prescriptions for Emergencies*）in 341. As a book on prescriptions, this work was an earlier record of infectious diseases such as smallpox,

---

❶　唐锡仁. 中国古代地理学史[M].北京:科学出版社, 1984.

glanders, phthisis and chiggers diseases. Besides, he was an alchemist. In his work of *Baopuzi · Neipian* (《抱朴子·内篇》, *The Inner Chapter in Book of the Master Who Embraces Simplicity*), the chapters of *Jindan* (《金丹》, *Methods of Using and Making of Golden Elixir*) and *Huangbai* (《黄白》, *Techniques of Refining Gold and Silver*) were alchemy works well-known at home and abroad, containing abundant knowledge of chemistry. This kept the science of China leading the world's science for a long time after the Han Dynasty (206 B. C. ~ 220 A. D.).

As for the literature, a new development climax emerged during the period between the Han Dynasty and the Wei Dynasty. Cao Cao and his sons were the representative poets in this period. They created Jian'an Literature which was characterized by "The Vigorous Style of Jian'an Poems", ambitious, voluminous and vigorous. This kind of poems often reflected the social turmoil and the suffering of the homeless people in wandering and took five-characters poetry as its main verse structure. Yet in the Jin Dynasty (265~420), Zuo Si and Tao Qian were the most outstanding representatives in the literature creation. In the Southern Dynasties (420~589), Liu Xie was regarded as the founder of the earliest literature theory system. His views on literature creation has produced far-reaching influence on the later literature. In his work of *Wenxin Diaolong* (《文心雕龙》, *The Literary Mind and the Carving of Dragons*), a monograph on ancient China's literature theories, he made a systemic review of writers of various dynasties and their works, put forward the views that literature should be beneficial to society and value both form and content, content in particular, and attacked the practice of only pursuing the flowery form. The widespread use of paper also provided a new world for calligraphy and painting creation. The famous calligrapher and painter of the Jin Dynasty (265 ~ 420), Wang Xizhi and Gu Kaizhi (345~406), pioneered the use of paper to write and paint and became the masters of calligraphy and painting. Their work contributed to the result that Chinese painting and calligraphy had extensive international influence abroad.

With regard to the Buddhism, it also got great development during this period of time. The monks from western regions and India came to ancient China to do missionary work, translate scriptures and recruit disciples. The monks of ancient China also went to India to learn their Buddhism and introduced to Chinese people the Buddhist sutra they brought back there. Buddhism advocated of karma, reincarnation as well as the division of heaven and hell after death, offering the people a spiritual ballast. In addition, Buddhism could arouse the interests of scholars for its containing of deep philosophical teachings. Consequently, the number of its believers in China rose at that time. It was also supported by the rulers. According to the record of *Bianzheng Lun* (《辨正论》, *The Theory of Justification*) written by Fa Lin (512~640), in the Eastern Jin Dynasty (317~420), there were 1768 Buddhist temples and 24,000 Buddhist monks and nuns. When it came into the Liang Dynasty (502~557) of the Northern and Southern Dynasties (420~589), the numbers rose to 2845 and 82,700 respectively, with an increase of 1000 in the number of temples and of 3 times more than the number of monks and nuns of the Eastern Jin Dynasty (317~420. Yet according to *Wei Shu · Shi Lao Zhi* (《魏书 · 释老志》, *Treatise on Buddhism and Taoism* in *History of the Wei Dynasty*), in 1477, the first year of Tai He of the Northern Wei Dynasty (386~534), there were 6,478 Buddhist temples and 77,258 Buddhist monks and nuns; during the Yan Chang period (512~515), there were 13,727 temples; at the end of the Eastern Wei Dynasty (534~550), there were more than 2 million monks and nuns, and more than 30,000 temples. The increasing number of Buddhist sutras in society paralleled with that of monasteries, monks and nuns, which had been verified in the Buddhist sutras of this period discovered in Dun Huang Chambers and Xinjiang of China. Papermaking accelerated the circulation of Buddhist sutras. With its sinicization, Buddhism, together with Confucianism and Taoism, was listed as three religions of ancient China, becoming a part of Chinese traditional culture and enriching the content of the latter. The circulation of classics of Buddhism also had positive influences on the fields like languages, literature and art, architecture, philosophy and technology. Moreover, Buddhism had been the social factor to promote the development of papermaking and printing industries. At the same time, it spread from China to Korea and Japan during the Wei, Jin, Northern and Southern Dynasties (222~589).

The emergence of the paper also caused changes in the fields of politics, economy, military, daily life and customs. In the field of politics, the differently-colored paper was employed to write official documents, law, notices, certificates, houschold register, credentials, meeting records and archives, more convenient than other writing materials, thereby greatly improving the work efficiency of all-level governments from the central to the local and promoting their power construction. In the field of economy, papermaking was an industry of low cost and high income and led to the construction of papermaking mills in different regions of China, which promoted the development of local economy, industry and commerce as well as transportation, and thus increased the tax revenue and export trade of China at that time. What's more, the written-on-paper public or private contract, account books and bills ensured the normal and effective operation of social and economic orders. Paper was also an ideal material for the package of many commodities, so it was very convenient for both customers and promotion of products. In the field of military, the application of information-transmitting paper kites, signal-sending paper lanterns, waterproof paper maps, self-defense paper armors, paper umbrellas and paper powder flask improved the function of military marching and combat. As for the field of daily life, paper napkin, toilet paper and menstruation paper had great significance in health care for their use signified a revolution in human living habits. With regard to the field of social customs, from the Northern and Southern Dynasties (420~589) on, paper had been used to make burial articles to replace their corresponding material ones to be buried with dead body, thereby making funerals more and more frugal for it could reduce the waste of materials. However, during the Qin and Han Dynasties (221 B. C. ~220 A. D.) and the period before them, it was the custom to bury the material articles such as copper crash used by people before their death.

5.1.3   Role of Paper in Development of Medieval European Culture

Europe entered the Dark Ages during the Middle Ages. With the division of ancient Rome (395) and the destruction of the Western Roman Empire (476), the ancient Greek civilization gradually faded and the European culture sank. After the recovery of Christian countries in the 10th century, the feudal society obtained a certain development, but the frequent wars brought about the ignorance of developing cultural and educational undertakings, so knowledge progressed slowly. Rudeness and ignorance existed universally among the upper class of society. Knights were usually illiterate, even not able to write her own names. Many kings and Emperors could not read. As a consequence, Henry Ⅳ (Henry Ⅳ, 1050~1106) was praised for his ability of reading. Let alone farmers and handicraftsmen. They all were illiteracies and nobody cared their education. ❶ The

---

❶   KOSMINSKII E A. 中世世界史:第 2 部[M]. 王易今,译. 上海:开明书店, 1947.

general public was of poverty and ignorance. Knowledge was only in the hands of a few church clergies. They used Latin which the general public did not understand as the written language, aiming to cultivate the missionary schools where children were taught to do rote learning of the Latin version of *Vulgata* (《圣经》, *The Bible*). *Vulgata* of Latin version revised by Saint Jerome (Saint Jerome, 342~420) in 405 and *De Civitate Dei* (《上帝之城》, *City of God*) written by Augustinus (Aurelius Augustinus, 354 ~ 430) were employed to function as the official philosophy to control the ideology of people and to lecture theology, and also as the textbooks in the church schools. However, the people who had chance to receive this kind of education were a few ones who enjoyed the privileges. The masses were excluded.

The schools of Europe of the Middle Ages were strictly controlled by the churches and served the latter. Many of them were set in monasteries and the priests they trained were supposed to learn "seven liberal arts" — Latin grammar, rhetoric, logic, arithmetic, geometry, astronomy and music. Rhetoric was taught to train the eloquence on mission; logic, develop learners' ability to prove theological proposition; music, train singing hymns; astronomy, calculate religious holidays; geometry, to be exact, impart knowledge of geography, plants and animals, which was used to give comments in *Vulgata*; arithmetic, only train simple operation. Evidently, these subjects were tools and accessories to promote religion. The teaching method was dogmatic cramming and the independent thinking was not allowed. *Vulgata* and the annotation of godfathers enjoyed the absolute authority, only to be believed. The purpose of learners' acquiring knowledge was only to deepen their religious beliefs. Christian proclaimed afterlife, asceticism and obscurantism, and strengthened its ideological autocracy, thereby stifling creative thinking and blocking the way of intellectual progress. As a consequence, compared with the culture of ancient Greece, it was a kind of cultural retrogress. Although rulers sometimes established the palace schools in order to cultivate the talented people for the management of empire, only a few nobles' children had chances to go to school there. Meanwhile, the subjects they learned were limited to the content like law. During the Middle Ages, the literature was just religious literature, such as hymns, prayers, Christ stories and so forth, and drama was full of superstition, absurd and boring. The folk literature was oral ones on the dialect stories. Therefore, it could be concluded that the spiritual life of the whole society was very poor at that time.

However, during the period from the 11th century to the 14th century, the situation changed. The Crusade (1196~1291) launched by the Pope and the western European feudal lords offered an opportunity for the Europeans to know that the Arabic culture of Islamist were more advanced than that of the Christian world. Therefore, the Europeans introduced the advanced Arabic culture into Europe. In addition, through the Arabs, they also introduced some sciences, technologies and inventions from China, such as papermaking, gunpowder and firearms, compass and magnetism knowledge and alchemy. On the other hand, due to the development of the industry and commerce and overseas trade in European cities, the citizens engaged in handicraft industry and business or the bourgeois class came into being, becoming an emerging class of feudal urban economic development and a new social force to promote the development of science and culture. Moreover, the establishment of the European paper mills laid a material basis for the development of science and culture. For the responses to these changes, many secular schools appeared in succession and evolved into the universities teaching students many subjects. From the intellectuals of the church flew out a batch of unorthodox scholars who took research and academy as their essential work. Breaking through the thinking prison which the church authorities set, they took the advantage of rational knowledge to awaken the masses and lighted a brilliant light for them to go forward in the dark so as to welcome the coming of the new era. All of these new situations occurred after Europe had its papermaking mills because paper made these changes possible and accelerated their development. The reason why China's sciences and inventions which played an important role in the social development of Europe were introduced through the Arabian region was that in the 13th century, the western expedition of the Mongol army cleared the obstacles on the way connecting the East and the West, providing conveniences for the people's exchanges, the goods freights and the cultural contacts.

In 1085, when European crusaders captured Toledo City ruled by Muslim in Spain (in the south of Madrid), they discovered a large number of papered Arabic writings including the translation version of Greek works, which immediately aroused the relevant attention. The archbishop, Raimundode Penafort (1176~1275), established the translation institute and invited people who understood Arabic to translate those papered writings. The period from 1125 to 1280 saw the climax of translation in Europe. Italian Gerar da Cremona

(1116~1187) alone translated eighty Arabic works, containing the translations of Aristotle and Ptolemy's works as well as the works of Arabic scholars like Ibn Sina. Through the cooperation, Spanish, Italian Sicilian, Arabic and Jew even translated Euclidean geometry and Arabic books of algebra, astronomy, alchemy and medicine into Latin. Later, the Greek manuscripts left by Byzantine like *The Complete Works of Aristotle* (《亚里士多德全集》) were searched for and collected and directly translated into Latin. These new Latin translations amazed European, presenting before them long-forgotten but fresh Old Greek spiritual civilization, recently-blossoming Arabic cultural treasure as well as looming Chinese and Indian sciences and civilizations. The transmission and copy of these works gave birth to the explosive growth of people's knowledge and made people find a new research field to suck new ideas and inspirations, and finally to promote the academic renaissance. This was a typical example of learning from the past for the better today.

Since the 11th century, in order to adapt to the needs of the emerging bourgeois, secular city schools and universities had emerged in succession. Although Latin was the lecturing language in these schools, the lectured content was no longer limited to theology. It also involved other subjects such as law, medicine, literature and art. Meanwhile, the teaching contents on astronomy and mathematics were richer than those of the past. Some of teachers were also secular scholars. This type of universities was public ones. Their principals were elected by students and responsible for school management, such as Italian Padua University and Bologna University. The other type of universities was founded by the church, such as Paris University, Oxford University and Cambridge University. In the 14th century, there were over forty universities in Europe, which had outputted many scholars. Each university had its own library. In the reading of books in libraries, both teachers and students acquired much knowledge on philosophy and nature from the newly-translated Greek and Arabic works, which enabled them to naturally realized the absurdity of the Catholic theology and to doubt the existence of God. This made the church authorities very shocked. As a consequence, Albertus Magnus (1200~1280), the theologian born in Germany and the professor of Paris University, attempted to make use of newly-introduced scientific knowledge and Aristotle's theory to serve the theology, claiming that science was only the preparation for the faith. His student Thomas Aquinas (1225 ~ 1274) expounded his ideas systematically in *De Summa Theologica*. Through the employment of philosophic methods to prove the theological propositions, Aquinas argued that the truth existed firstly in the reason, secondly in things, and that God was the reason and the highest truth, so the faith in God was above the faith in the reason. He further employed the fact that the movement of the celestial sphere needs the First Cause to prove the real existence of God. The theology system of Aquinas and his colleagues obtained further development and spread in the church schools, so it was called scholasticism.

The realism of Scholasticism represented by Aquinas's theory was a reaction to the Greek philosophy and the Arab science introduced in the Arabic literature in the late European Medieval Age, thereby being a thought retrogressing, which reflected the deathbed struggle of the late feudal forces in their ideology. ❶ On the other hand, motivated by the empirical spirit embodied in the Arabic and Chinese sciences the European scholars began to do scientific experiments and to establish their own rational concepts to criticize Scholasticism and the church of that time. What is more, some scholars criticized Aquinas's fallacy from the perspective of philosophy. For example, the professor of Paris University, Pierre Abélard (1079~1142), stated that faith must be based on knowledge, advocating freedom and opposing the supreme authority of the church of that time. He said "Suspicion is the road of research." "Research makes it possible to reach the truth." "Faith must be based on understanding."❷ Because he put forward the rallying cry diametrically opponent to the mainstream views of Scholasticism, he was regarded by the old force as "heresy".

British William de Ockham (1295~1349) launched a campaign to deny the argument of the First Cause about the existence of God. By taking *Chaoju Zuoyong* (超距作用, The Action at a Distance) for example, he said moving objects did not need the continuous physical contact of pusher. The magnet could make the iron bar move without the contact with it. Oxford Roger Bacon (1214~1294), understanding Arabic, had carefully studied the knowledge of various scientific fields including newly-introduced Chinese gunpowder knowledge and given a detailed introduction to them in his three main works. He believed that real scholars should take the way of experiment to understand

---

❶ 洪潜. 哲学史简编[M]. 北京: 人民出版社, 1957.

❷ DANPIER W C. A History of Science and its Relation with Philosophy and Religion[M]. 4th ed. Cambridge: Cambridge University Press, 1958.

the natural science. For practicing what he preached, he even engaged himself in the optical and chemical experiments. He argued that the only way to prove whether the previous statements were correct or not was observation and experimentation, so he voiced loudly "Don't be ruled by dogma and authority any more. Have a careful look at the world!" At the same time, French Pierre or Peter (Pierre de Maricourt or Peter Peregrinus, 1205 ~ 1275) persisted in performing a series of magnetism experiments. In *Epistola de Magnete* (《论磁石信》, *Letters on Magnet*), he said those who studied the magnetism must be diligent to act, only by which could they rectify their mistakes in their understandings. Roger Bacon thought that he knew natural science and got the wisdom and peace from the experiments.

The new change occurring in the European literature circle in the late European Medieval Age was the birth of *Divina Commedia* (《神曲》, *The Divine Comedy*) written by the Italian people Dante Alighieri (1265 ~ 1321). This political and philosophical poem describes a story in which the author was led by Roman poet to roam hell, purgatory and heaven in his dream, which implied that the ideal realization of hope in the social reality required the experiences of sufferings and ordeals. In the poem, the author placed the ideal monarch in heaven and put the Pope in hell. This demonstrated fully that the poem tried to unfold the ideas of thought liberation, the pursuit of knowledge, the absorption of classical culture and the kind treatment of heresy. These ideas marked the budding of humanism thoughts. In addition, the poem was written in the dialect of Tuscany in Midwest Italy, which pioneered the creation of European literature in ethnic languages during the Renaissance Period. Meanwhile, William, Abelard and the other relevant scholars' criticism of the old feudal force seeded for the future religious reform, while Roger Bacon, Pierre and the other relevant scholars' experiment spirit and hard pursuit in the natural science became the omen of the inevitable coming of scientific reform. With the efforts of several generations, European took off the past stagnating situation in science, technology and culture in the 15th century, and began to take on a new atmosphere. Although there were still some social factors stifling progress, the power of promoting social advancement were rapidly aggregating and growing. Dante said in *Divina Commedia* "Segui il tuo corso e lascia dir le genti." ("Walk on your own road regardless of whatever others say.") This line inspired many people to challenge the power of the old forces in spite of suffering the damnation and persecution. It was with this fearless spirit that

European progressive people went to confront the reality and the future.

5.2   Influence of Printing on Development of World Civilization

5.2.1   Influence of Printing on Development of Education and Science in China

Although the writing on paper was superior to the writing on any other writing materials, it shared commonness with them, i. e., every book was copied by hand word by word and every copy could only get one manuscript. When the number of books increased, much time and great efforts were needed in people's copying of books. It would be more difficult for Chinese people because the beautiful ideographic Chinese characters were large in their number and complicated in their strokes. What's more, in the copy of Chinese characters, the wrongly-copied characters were often seen. For example, the Chinese character "鲁" (lu) was often wrongly copied as "鱼" (yu) and the character "亥" (hai), "豕" (shi). It would mislead readers. However, the emergence of printing rescued millions of people from the sufferings of their copying of books because the same *yinban* (印版, printing plate) could print millions of manuscripts with the same content and font, which was convenient for the unified proofreading and thus guaranteed the accuracy of characters. Besides, the printed characters were clear and the printed books were cheaper and faster than written books. Therefore, the printed books flew into every corner of society in a larger scale, becoming a powerful media of spreading ideas, knowledge and information. Their role in the past society was just like television and Internet in today's world. The emergence of printing was an epoch-making revolution in the history of graphic communication. Beginning in the Sui Dynasty (581~618) and blossoming in the Tang Dynasty (618~907), printing promoted the fast development of China's education, culture and academies, making the renaissance arrive ahead of time in the East of the old continent.

The early prints were mainly about Buddhist scriptures and Buddha figures for the folk use, because driven by faith, the Chinese Buddhists were keen on duplicating a lot of Buddhist documents. The *jingzhou zhouwen* (经咒咒文, paternosters) in printed and written mantras had the same mana and function, so they preferred to use printed ones and thus needed not to copy mantras dozens of times or even several hundred times. Only the Buddha asked them to copy the same mantra repeatedly for the *jigongde* (积功德, the accumulation of merits). Yet different from Buddhist, Confucius, the father of Confucianism, asked his disciples to "review what they have learned constantly", so it was enough for disciples to copy each book once. This was why people said that Chinese Buddhists made special contributions to the development of printing. Because of this, Buddhism obtained greater development than that of Confucianism after the Sui and Tang Dynasties (581 ~ 907). However, publishers discovered that the method of publishing Buddhist sutras could be used to publish other common books needed by the general public, such as the Chinese tool books like dictionaries and rhyme books, fortune-telling books and calendar books. These books were also needed in the market. Similarly, the rulers also realized quickly that the publication of nine Confucian classics and tree history books were extremely necessary because they believed that the use of these books as textbooks in schools could help unify the thought system of cultivating future officials at all levels and strengthen the thought ruling of official philosophy all around the country. As a result, during the various dynasties after the Tang Dynasty (618~907), the printing plants were built both by the central authority and the local authorities to publish different kinds of books. In a word, printing had greatly contributed to the development of education of various dynasties after the Tang Dynasty (618 ~ 907) and the establishment and improvement of Imperial Examination System in ancient China.

Take the Song Dynasty (960~1276) for example. The number of schools, campus students and learning contents all exceeded that of the past dynasties. During the period of the Five Dynasties (907~960), after the Central Imperial College began to woodblock-print *Jiu Jing* and issued them across the country, the hand-written versions of textbooks and reference books in schools at all levels gradually decreased. During the Northern Song Dynasty (960 ~ 1127), the woodblock printing saw its further development and became very popular. At the same time, the movable type printing appeared. At that moment, the textbooks used by teachers and students were all printed ones, especially the standard printed version issued by the Imperial College nationwide. It was a revolution in the history of textbooks in China, producing a far-reaching influence on China's education. Teachers and students could spend more energies and time in lecturing and learning rather than copying various books, thereby promoting their work and study efficiency. It was true for the intellectuals. Printed books reduced their physical burden of copying books so that they could focus on their research.

The education system of the Song Dynasty (960~1276) got progresses on the basis of the inheritance of the education system of the Tang Dynasty (618~907). The central government established the *Guozijian* (国子监, the Imperial College) to administrate the national educational affairs. Under the Imperial College, there were higher education institutions of Jingshi Imperial College, Taixue Imperial College, Simen Imperial College, Guangwenguan Imperial College, Medicine Imperial College, Calligraphy Imperial College and Drawing Imperial College. Among them, Jingshi Imperial College enrolled children of ancient Chinese officers over grade seven; Taixue Imperial College enrolled children of the ancient Chinese officers below grade seven and civilians; Simen Imperial College enrolled the same students as Taixue Imperial College. The number of students in the rest other imperial colleges was also large. There had been more than 10 thousands of students in higher education institutions during the Song Dynasty (960 ~ 1276). Relatively speaking, the number of college students during the Song Dynasty (960~1276) was bigger than that of the Tang Dynasty (618~907) and what was worth attention was that there were more civilian students during the Song Dynasty (960 ~ 1276). The local official schools included the provincial schools and county schools, with the total student number of 64,000 nationwide. The primary schools pervading all over the country were *jiashu* (家塾, family schools) established by individuals, in which children of 8 years old could be enrolled. The Imperial College also shouldered the function of publication, providing textbooks and reference books for schools at all levels. According to the incomplete statistics, there were 256 kinds of dictionaries, Confucian classics, history books, medical books and books of various thought schools which had been published by the Imperial School.

The private schools, with the teaching content more advanced and difficult than that of family schools, pervaded the country. Here many teachers were scholars, so the teaching quality sometimes was even better than that of the state-run schools. Besides, the education of the Song Dynasty was characterized by the large-scale emergence of *shuyuan* (书院, colleges). Colleges were originally civilian-run. Later, they were advocated by officials, given by the imperial government the plaques, *xuetian* (学田，the school-owned land) and books. What's more, the instructors here were appointed by the imperial government. Therefore, the private schools become half civilian-run and half state-run gradually. At that time, there were four major colleges, namely Bailudong College in Jiangzhou (now Jiujiang City of Jiangxi Province), Songyang College in Xijing (now Dengfeng City of Henan Province), Yuelu College in Tanzhou (now Changsha, the capital city of Hunan Province) and Yingtian College in Jiangning (now Nanjing, the capital city of Jiangsu Province). Each college had its own regulations and rules, *Shanzhang* or *Dongzhu* (山长/洞主，the designations of teachers in colleges during the period from the Song Dynasty to the Qing Dynasty of China), decades to over one hundred dormitories provided for students, rich collection of books including the books about engraved-on-plank *Jiu Jing* (《九经》，*Nine Classics of Confucianism*) granted by the imperial court. Afterwards, colleges prevailed all over the country, equivalent to state-run prefectural education institution in level. They had cultivated many talented people for the country and had profound influences on the later generations. The colleges of later three dynasties of Yuan, Ming and Qing (1271~1840) were established according to the college model of the Song Dynasty. During the Song Dynasty, each state-run prefectural school and college was also very active in the publication of books, publishing over 300 types of books, even more than that of the Imperial College. They integrated publishing and education and could make the two bring out the best in each other, thereby forming another feature in the education of the Song Dynasty (960~1276).

The highly-developed education of the Song Dynasty (960 ~ 1276) gave birth to the social phenomenon that "even child would feel shameful if he could not discuss poetry". The literate people accounted for a large proportion in the total population of the society. For example, during the 12th century, national average population was 30 million among which there were 200 thousands of *Juren* (举人，people with *Juren* Degree, the provincial graduate in the imperial examination system of ancient China); during the 13th century, although the population decreased half, *Juren* quadrupled (400 thousand). Usually, one of 20 candidates would be admitted as *Juren*, so it could be calculated that 40 thousands of candidates took the examination for the *Juren* Degree during the 12th century, accounting for 13% of the total population. During the 13th century, the percentage rose to 26%. In the world of that time, only China had so many intellectuals.

Therefore, it was an inevitable trend that each academic field saw their flourishing during the Song Dynasty. At that moment, the science and technology, taking the four great inventions as its backbones, entered its new climax. This might be known from works written and issued by people of the Song Dynasty. Zeng Gongliang and his colleague wrote 40-volume *Collection of the Most Important Military Techniques*. It was a great encyclopedia of military science which, with the accompany of illustrations, made a detailed introduction and research on all kinds of cold weapons, fortification technology, warships and chariots, gunpowder and firearms, compass and magnetism and so forth. Shen Kuo's (1031 ~ 1095) *Mengxi Bitan* was an encyclopedic academic monograph on his original ideas of mathematics, astronomy and calender, magnetism and compass, movable-block printing, optics, geology, medicine and others. Needham J. regarded this book as a milestone in the science history of China.

In addition, the other fields such as agriculture and medicine also saw the relevant works. Chen Fu's (1076~about 1154) *Nong Shu* (《农书》, *Book on Agriculture*) was one of the first works to summarize the rice planting technology of the Jiangnan area (the area in the south of the lower reach of the Yangtze River of China). It also discussed the plan of using land and put forward the idea that with the proper tending and timely manuring, the productivity of field would be strong forever. Besides, Chen pioneered writing sericulture technology into his book. Han Yanzhi's (1131~about 1206) *Ju Lu* (《橘录》, *Records on Citrus*) recorded the citrus varieties and the technologies on the plant, insect prevention and storage of citrus in Yongjia (now Wenzhou, Zhejiang Province). As the world's first citrus monograph, it was spoken highly of by western scholars. Song Ci (1186~1247), a former *tixingguan* (提刑官, an officer in charge of province-level judicial body during the Song Dynasty), wrote *Xiyuan Lu* (《洗冤录》, *Witness of a Prosecution*) according to his own test practice and theoretical research. As the world's first systematic work on forensic medicine, it included all the content of forensic medicine and was translated into many foreign languages, having a far-reaching influence on the later generations. The medical official Wang Weiyi (987~1067) was ordered by the Emperor to summarize ancient acupuncture technique, accurately determine the positions of acupuncture points and supervise the work of casting acupuncture copper figure model. He made explanations to the acupuncture points of copper figure model and wrote *Tongren Shuxue Zhenjiu Tujing* (《铜人腧穴针灸图经》, *Illustrated Acupuncture of Acupoints of Copper Figure Model*) in 1027. The life-size copper man model, as an important teaching instrument, was regarded as a national treasure at that time. During the Song Dynasty (960~1276), the herbalism got fast development. In 974, *Taizu* (太祖, the founder of the Northern Song Dynasty) ordered Ma Zhi (about 935~1004) and his colleague to check and annotate the written version of *Tang Bencao* (《唐本草》, *Materia Medica of the Tang Dynasty*) so as to publish it. The published one was named *Kaibao Bencao*. They made their contributions in the preservation of ancient herbal medicine book. The scientists Su Song (1019~1101) and his colleague were also ordered to amend *Kaibao Bencao*. They compiled *Bencao Tujing* (《本草图经》, *Illustrated Materia Medica*) on the basis of national medicine census in 1061 and published it in 1062. The book was added with 103 new medicines and illustrated with 923 medicine pictures. It was the earliest printed edition of illustrated herbal medicine book and the world's most excellent pharmaceutic monograph. Tang Shenwei (1056~1163) made another supplement to the existing herbal medicine books and finished his *Zhenglei Bencao* in 1108. Including all the herbal essence before the Song Dynasty (960~1276), it was also illustrated. Ke Zong's (about 1071~1149) *Bencao Yanyi* was finished in 1116 and published in 1119. It was in the form of notes and replenished medical information not included in the old medica, thereby being very characteristic. In addition, due to the development of commerce and overseas trade during the Song Dynasty (960~1276), many foreign medicinal materials were introduced into the medical books. In a word, the herbal medical books published during the whole period of the Song Dynasty (960~1276) were really dazzling and numerous. The traditional Chinese medicine classics such as *Huangdi Neijing · Suwen* (《黄帝内经·素问》, *Plain Questions in the Yellow Emperor's Cannon of Medicine*), *Zhubing Yuanhou Lun* (《诸病源候论》, *General Treatise on the Causes and Symptoms of Diseases*), *Mai Jing* and *Zhenjiu Jing* (《针灸经》, *The Classic of Acupuncture Research*) only existed in the form of written version. However, they all were printed during the Song Dynasty (960~1276) and entered ordinary families.

The Song Dynasty was the golden period of mathematics. Especially, the achievements in algebra were far ahead in the world. Jia Xian (1005~1065), in his work of *Huangdi Jiuzhang Suanfa Xicao* (《黄帝九章算法细草》, *Nine Chapters of Calculation of Yellow Emperor*) finished in about 1050, proposed using the calculating table to present the binomial coefficients of the whole power, which was called "*Jiaxian Sanjiao* (贾宪三角, Jia Xian Triangle)". He also pioneered using the positive root method to seek any higher power. Qin Jiushao (1202~1261) put forward in his *Shushu Jiuzhang* (《数书九章》, *Nine Chapters of Mathematics*) finished in 1247 the method to calculate the numerical value of the high order equation and the solution to the simultaneous first congruence. ❶ One of reasons for the development of mathematics during the Song Dynasty (960~1276) was the first gathering publication of ten famous math books which were compiled into *Suanjing Shishu* (《算经十书》, *Ten Classics on Mathematics*) at that moment. These ten books embodied the math achievements obtained over one thousand years ranging from the Han Dynasties (206 B. C. ~220 A. D.) to the Tang Dynasty (618~907). Later, they were republished in the Southern Song Dynasty (1127~1279), which had the math knowledge popularized.

As for the astronomy, with five large-scale star position observations performed during the years from 1010 to 1106, the accuracy of star positions was much higher than that of the previous generations. The star map drawn on this basis held 1464 stars. This should be attributed to the advanced astronomic equipment. Su Song completed his work of *Xin Yixiang Fayao* (《新仪象法要》, *Instructions on Water-driven Astronomical Clock Tower*) in 1092. In this book, he recorded the structure of *shuiyun yixiangtai* (水运仪象台, water-driven astronomical clock tower) and its 47 blueprints that Su Song and his engineer Han Gonglian were appointed to construct and draw during the years from 1086 to 1092. The tower was 35 Chi (12 meters) high and had three layers. The top layer was the *hunyi* (浑仪, the armillary sphere) for observing the celestial bodies; the middle one was the *hunxiang* (浑象, the celestial globe) for demonstrating the movement of celestial bodies; the bottom layer was the *baoshi zhuangzhi* (报时装置, the time-keeping mechanism). The three layers were connected with transmission system and wheels. The water flowing from a clepsydra turned the wheels whose movement would further control the movements of the armillary sphere, the celestial globe and the time-keeping mechanism. What was noteworthy was that the linkwork escapement, the most important part in modern mechanical clock, had been installed in the time-keeping mechanism. Therefore, it could be said that the water-driven time-

---

❶   钱宝琮.中国数学史[M].北京:科学出版社,1964.

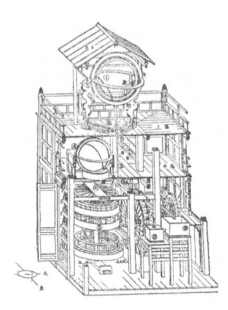

Figure 176　Water-driven Astronomical Clock Tower from Wang Zhenduo（1989）

keeping clock was the direct ancestor of the world's astronomic clock. ❶ Thanks to the circulation of its printed edition, this invention of China could be known by the world（Figure 176）.

The Song Dynasty also achieved various technological and scientific achievements in other fields. For example, the architect Li Jie（about 1060～1110）was ordered to compile *Yingzao Fashi*（《营造法式》, *Rules for Structural Carpentry*）. Published in 1103 and composed of 34 volumes, 357 chapters and 3,555 rules, this book systematically summarized the achievements of architectural sciences and technologies and the management experience of architectural engineering, described in detail the technologies of 13 types of architectural work as well as their operation standards, labors and material quota, and included 193 pictures of architectural engineering. It reflected particularly the new achievements made since the Northern Song Dynasty（960～1127）and the technology experience of architectural craftsmen. The comprehensiveness and systematicness in its content and the science in its narration were rarely seen in the other relevant books in the world. At the same time, the tower constructor Yu Hao of Zhejiang Province wrote *Mu Jing*（《木经》, *Classic on Carpentry*）which was also an architectural monograph consisting of three volumes. Li Xiaomei's（alive during years from 1055 to 1115）*Mo Pu*（《墨谱》, *Recipe of Chinese Ink*）and Chao Guanzhi's（alive during years from 1050 to 1120）*Mo Jing* were two monographs on the research of Chinese ink technology, with the former illustrated. Wang Zhuo's（alive during years from 1115 to 1175）*Tangshuang Pu*（《糖霜谱》, *Book on Making Sugar*）was a monograph on the technology of refining sugar from sugarcane. During the Song Dynasty（960～1276）, the research on the systematic classification of

---

❶　王振铎. 宋代水运仪象台的复原[M]//中国社会科学院考古研究所. 科技考古论丛. 北京:文物出版社,1989.

things was very developed. Many of the relevant works were related to science and technology. In addition to the books mentioned above, there were *Lizhi Pu* (《荔枝谱》, *Research on Classification of Litchi*), *Tong Pu* (《桐谱》, *Research on Classification of Tung Tree*), *Ju Pu* (《菊谱》, *Research on Classification of Chrysanthemum*) and others.

5.2.2   Promotive Role of Printing in Development of Confucianism, History and Literature in China

During the Song Dynasty (960～1276), the printed versions of works on the explanation of Confucianists of the Han, Wei and Six Dynasties (220～589) to the discourses and lines of *Jiu Jing* were widely circulated in society. If the relevant studies continued along this explanation direction, it would be difficult for Confucianism to have a new development. Therefore, when annotating and studying Confucian classics, the people of the Song Dynasty (960～1276) gave up the tradition that Confucianists of the Han Dynasty (206 B.C.～220 A.D.) focused on explanation of words, lines and discourses, but began to focus on the explanation of the world outlooks *Jiu Jing* (《九经》, *Nine Classics of Confucianism*) contained, i.e., explaining Confucianism from the perspective of philosophy. Taking the traditional ethics of Confucianism as the core and absorbing the theoretical thoughts of Buddhism and Taoism, they extended the content of their discussion from the world outlook to the principle of nature development, i.e., the source of the universe and the development of nature. In terms of this, they put forward the concept of *Li* (理, Rational Principle) and claimed that *Li* was not only the supreme principle of human society, but also the root of all the things in nature. Therefore, the people of the Song Dynasty (960～1276) named the developed Confucianism "*Lixue* (理学, Rationalism/The Study of Rational Principle)". The western people called it Neo-Confucianism.

The father of *Lixue*, Zhou Dunyi (1017～1073), proposed concisely in *Taiji Tushuo* (《太极图说》, *Explanations of the Diagram of the Supreme Ultimate*) finished in 1060 the theory of universe creation and the theory of birth of all things, which contained some elements of dialectical thoughts. Zhang Zhai (1020～1077) developed Zhou Dunyi's theory. He put forward the proposition of "The universe is composed of *Qi* (气, Vital Force)" in his work of *Zheng Meng* (《正蒙》, *Correcting Ignorance*) finished in 1076 and used the condensing and dispersing of *Qi* to explain the formation of all things of the world. At the time when *Li* developed into one of the supreme philosophical categories, Cheng Hao (1032～1085) and Cheng Yi (1033～1107) put forward the proposition of "Everything is only a heavenly principle" and that everything in the world derived from *Li*. According to them, the classification of superiors and inferiors was a matter of course and the violation of feudal hierarchy system and moral principles was the violation of heavenly principle and not allowed. So far, the transition of Confucianism into the philosophical and abstract *Lixue* had been completed.

On the basis of the summary, comprehensive understanding and mastery of the doctrines of various schools of Neo-Confucianism, the *Daru* (大儒, the learned and famous scholar) of the Southern Song Dynasty (1127～1279), Zhu Xi (1130 ～ 1200), established a complete and rigorous system of Neo-Confucianism thoughts, which embraced the major achievements of Neo-Confucianism. Zhu Xi read many books on various fields of the history classics, the works on the thoughts of various schools, and the works on astronomy and geology. He was particularly skilled in nature science and had many relevant achievements. The core in his system of Neo-Confucianism thoughts was *Tianli Lun* (天理论, Heavenly Principle Theory) whose key content was *Liqi Shuo* (理气说, Theory of Rational Principle and Vital Force). *Tianli* (天理, Heavenly Principle) or *Li* was *Taiji* (太极, the Supreme Ultimate), the root of all things in the universe. Besides, *Li* and *Qi* exist in a mutual dependence, with *Li* being the root. The reason why there were different things in the universe was that everything in the universe had its own law which was the embodiment of the root *Tianli*. All things in the universe which were derived from the root *Tianli* were different, but they were correlated with each other. Zhu Xi also believed that the moving forms of things included two types of "*hua* (化, process)" and "*bian* (变, change)". The former referred to quantitative change and the latter, qualitative change. In addition, he put forward the views of *gewu zhizhi* (格物致知, the investigation of things), emphasizing that knowledge and action were interdependent with each other. His statements of "For the order of knowledge and action, the former is the first" and "For the importance of them, the latter is more important" illustrated clearly that knowledge should be obtained from practice. Generally speaking, Zhu Xi's philosophy was characterized by Objective Idealism, but it also included many contents of Naive Dialectic. Some of his philosophical ideas called by Joseph Needham as "Natural philosophy of organism" which had a positive effect on the development of nature science. ❶ When Zhu Xi applied his philosophy of *Tianli Lun* into the study of human society, *Tianli* was specified as the feudal order and ethics. Later, his philosophy works were printed and published, which contributed to the development of *Lixue*.

Zhu Xi was also busy with lectures and writings. According to his philosophy system, he not only annotated *Si Shu* (《四书》, *The Four Books*), *Zhou Yi* (《周易》, *The Book of Changes*) and *Shi Jing*, but also compiled and published *Zizhi Tongtian Gangmu* (《资治通鉴纲目》, *The Detailed Outline of History as a Mirror*). Zhu Xi's system of philosophy was regarded by the imperial court as the authentic Confucianism at the end of the Southern Song Dynasty (1127~1279). From then on, *Lixue* of the Song Dynasty (960~1276) had become the official philosophy and governed the mind circle of China for 700 years. The basic principles of *Lixue* soaked various sections of society and became the standards for judging right and wrong as well as good and evil, thereby playing a certain role in the maintaining of stable social order. In addition, the thoughts of *Lixue* were the main content of the school education and the writing of Imperial Examination System of the Yuan Dynasty (1271~1368), the Ming Dynasty (1368~1644) and the Qing Dynasty (1636~1912). It was because of this highly-developed social ideology that China's Feudal System was more powerful, lasted longer and exerted more far-reaching influence than the western one. What's more, it had a long-term influence on the ideas of Chinese-character cultural circle regions and countries like Korea Peninsula, Japan and Vietnam. It should be said that traditional Confucianism including Neo-Confucianism had a profound system composed of both essences and dregs, and thus should not be completely accepted or refused today. Instead, its essences should be carried forward while its dregs should be abandoned. It should also be pointed out that China's Confucianism classics exerted a good influence on the European ideological circles, the thoughts of scholars of the Enlightenment School in the 18ᵗʰ century in particular, after it was introduced into Europe by the western Jesuits in the 17ᵗʰ century. For examples, the French thinker Franois Marie Voltaire (1694~1778) hung the portrait of Confucius in his room, and the German philosopher Gottfried Wilhelm Leibniz (1646~1716) admitted that he was influenced by Zhu Xi's thoughts in the development of his organic natural views. These also proved that *Lixue* of the Song Dynasty

---

❶　潘吉星.李约瑟集[M].天津：天津人民出版社，1998.

(960～1276) had its own valuable idea essence.

The developed printing industry of the Song Dynasty (960～1276) also contributed to the prosperity of the studies of history and literature. First of all, the annalistic and narrative chronicle historic works appeared continuously, such as old and new *Tang Shu* (《唐书》, *The History of Tang Dynasty*), old and new *Wudai Shi* (《五代史》, *History of the Five Dynasties*), *Tongjian Jishi Benmo* (《通鉴纪事本末》, *Narrative Chronicle of Major Events from* 476 B.C. to 979 A.D.), *Sanchao Beimeng Huibian* (《三朝北盟会编》, *Collections of Materials on the Maintaining of Peace Between the Song Dynasty and the Jin Dynasty*) and *Tong Zhi* (《通志》, *General Annals*). The works of epigraphy and archeology included *Jigu Lu* (《集古录》, *A Collection of Ancient Inscriptions*), *Jinshi Lu* (《金石录》, *Records of Ancient Bronzes and Stone Inscriptions*), *Bogu Tu* (《博古图》, *Pictures on Ancient Utensils*) and so forth. Secondly, the publishing of numerous local chorography also started from the Song Dynasty (960～1276). Many historical books of various dynasties before the Song Dynasty (960～1276) were also printed and published during this period of time. Therefore, the various historical works published during this period of time might be more than all the books recorded in *Han Shu • Yiwen Zhi* which listed 14994 books. This led to the result that the tradition of *jiangshi* (讲史, historytelling), a narration of historical anecdotes, prevailed among schools and academies at all levels. Scholars usually paid equal importance to "Confucian classics and history". This further contributed to the result that *huaben* (话本, scripts for historytelling) were very prevalent at that time. These *huaben* were often used by the historytellers in their performance. They were also named *pinghua* (平话, talk show), a demotic and easily-understood dialect. As the primary form of ancient Chinese vernacular novels, its content were the stories on the rising and falling of various dynasties and the wars in historical books, such as *Wudai Shi Pinghua* (《五代史平话》, *Talk Show of History of the Five Dynasties*).

The Song Dynasty (960～1276) also saw that the writing of *Ci Poetry* in the literary world reached its heyday, with its quantity, quality and content exceeding those of the late Tang Dynasty (618～907) and the Five Dynasties (907～960). *Ci Poetry* of the Song Dynasty (960～1276) enjoyed the equal popularity with the poetry of the Tang Dynasty (618～907) in the literary history. There were many outstanding *Ci* poets in this period. Yan Su (991～1055) and Ou Yangxiu (1007～1072) were the early representatives of *Ci Poetry* and their works were characterized by the mild and exquisite language. Liu Yong (985~1053) excelled in the plain description in his *Ci Poems* to reflect the life of civilians in Bianjing City (the present Kaifeng City of Henan Province) by absorbing a large number of spoken words. Su Shi's (1036~1101) *Ci Poems* involved broad range of themes, was full of bold and generous momentum and was not restricted by the traditional rhyme scheme of poetry, thereby opening a new situation for the development of *Ci Poetry* of the Song Dynasty (960~1276). The *Ci* poetess Li Qingzhao (1084~1207), due to the influence of her experiences of the social and political cataclysms of the Northern and Southern Song Dynasties (960~1276), mainly created *Ci Poems* to express her leisurely feelings in her early life, while in her late life, her *Ci Poems* were mainly to express her missing of home and her lamenting of her life. As for Xin Qiji (1140~1207) of the Southern Song Dynasty (1127~1279), his *Ci Poems* mainly expressed his patriotic mood and worries about social and political situation as well as his ambition to take the lost territory back, and his writing style was called "*Jiaxuan Ti* (稼轩体, the writing style of Xin Qiji)" which was featured by boldness, dynamic and coherent meaning as well as long sentences, reaching the highest achievement of *Ci Poetry* of the Southern Song Dynasty (1127~1279); under the influence of Xin Qiji, Chen Liang's (1143~1194) Ci Poems were also full of impassioned patriotic feeling. All the works of these *Ci* poets were collected in Tang Guizhang's (1901~1990) "*Quan Song Ci* (《全宋词》, *The Whole Ci Poems of the Song Dynasty*)" which was composed of 300 volumes, including 1990 *Ci* poems of 1330 *Ci* poets. The master of literary circle Ou Yangxiu advocated the essay style of the poets of Han Yu and Liu Zongyuan of the Tang Dynasty (618~907) and pioneered the new writing style of classic elegance and simpleness instead of the parallel style. He had a thorough understanding of the sufferings of the general public, so he did not use hard-to-understand literary quotations and ornate diction in his works, but rather utilized the simple and elegant language to pour out the pains of people and to uncover the absurdness of the ruling class. What's more, he devoted himself to developing young writers. The poets of Wang Anshi (1021~1086), Su Shi and others had ever received his care. ❶

Because of the efforts of Ou Yangxiu, Mei Yaochen (1002~1060), Wang Anshi, Sushi and others, the literary form of essay style took up the mainstream

---

❶　陆侃如,冯沅君.中国文学史简编:宋代文学[M].北京:作家出版社,1957.

position during the Song Dynasty (960~1276). Among them, Wang Anshi was also an outstanding politician. His achievements in poetry were consistent with his viewpoints of political reform. Su Shi's essays were not characterized by ornate diction but rather by free, bold and generous one. He held that writing should be like the floating clouds and the flowing water. His achievements in literature separated the writing from the principles and thus embodied the further development of the literature of the Northern Song Dynasty (960~1127). In a word, although the poetry of Song Dynasty (960~1276) developed on the basis of that of the Tang Dynasty (618 ~ 907), it had many breakthroughs in both contents and forms.

5.2.3 Role of Printing in Development of Education and Science During the Renaissance Period in Europe

The influence of printing technology on the development of education, science and technology of various dynasties of Korea Peninsula, Japan after the Nailiang Dynasty (710~793) and Vietnam after the Chen Dynasty (1225~1400) was similar to that of China, so here it is not discussed any more. This section attempts to discuss the influence of printing technology on the development of education, science and technology of Europe. As was mentioned above, during the period from the 14th century to the 15th century, Europe had escaped from the standoff situation of culture and science under the influence of Renaissance and came into being with a brand-new appearance. This tendency became more and more salient after the printing technology of China, *huozi yinshuashu* (活字印刷术, the moveable type printing) in particular, was introduced into Europe. *jinshu huozi yinshuashu* (金属活字印刷术, the metal moveable type printing) was very suitable for the alphabetic writing used by European, being able to publish a large number of books without too much investment. In schools, the printed versions of textbooks were used to replace their expensive handwritten copies. The early printings included religious books, school textbooks and language tool books as well as the works on humanities. It was in the 16th century that the printed versions of books on science and technology were sharply increased. These cheap printed books helped more and more people learn to read. The more people could read, the more books they wanted to read, which contributed to the development of education and science. For example, the emerging city bourgeois were eager to send their children to school; scholars may quickly publish their research results; craft workers also felt the necessity of literacy in order to understand the craft methods and illustrations in the printed book; the upper nobles felt keenly that the illiteracy did not match their identities. The development of city industry and commerce, the improvement of technology and the introduction of new technologies (such as gunpowder and compass) continuously caused problems waiting for being solved from the level of theory, which naturally made some cities become academic centers.

During the period from the 11th century to the 15th century, there had been universities in Europe, such as Paris University, Oxford University, Cambridge University, University of Bologna, University of Napoles, University of Padua, University of Prague, University of Krakov, University of Vienna, University of Salamanca, University of Ferrara and University of Saint Andrews. Until 1500, there had been 65 universities all over the Europe. Later, the number rose sharply. In the 16th century, universities had already pervaded the countries of Italy, Germany, France, Belgium, Holland, Swiss, Austria, Poland, England, Span, Czech and Denmark. The countries holding the most universities were Italy, Germany, France and England. In the past, Germany had few universities and thus the Germanic people had to study in Italy and France, but later it nearly caught up with Italy in the number of universities.

What was worth attention was that the cities which had universities were usually the centers of printing and publication.❶ Germany was the first European country to develop the *jinshu huozi yinshuashu*. Its printing center cities of Augsburg, Nuremberg, Cologne, Strasbourg, Frankfurt, Munich and Berlin all had universities. It was true of other countries. For example, the printing centers cities of Rome, Venice and Florence of Italy, the printing center cities of Paris and Lyons of France, the printing center cities of Basel and Zurich of Swiss, the printing center city of Louvain of Belgium and the printing center city of Venice of Austria all had their own universities and a large number of scholars gathered in these cities.

In Europe, the early so-called universities were established according to the madras mode of Arabic empire. They were units subordinate to religion ad mainly taught *The Bible* (《圣经》). The teaching of mathematics and astronomy in these universities served for religion and the degree of teachers' freedom was

---

❶　MARTIN H J. The French Book: Religion, Absolutism and Readership[M]. Baltimore: John Hopkins University Press, 1996.

far behind that of Arabic teachers. Therefore, these universities were completely different from the present universities. Since the 15th century, the situation had had some changes. The teaching content of mathematics, astronomy and medicine had been increased; the research of scholars had been deepened; the quality of published textbooks and reference books had been greatly improved. Meanwhile, due to the continuous increase of urban population, the medical profession became the aim of many people. Therefore, many universities began to set up the major of medicine and studying medicine became a kind of fashion at that time. This stimulated the development of human anatomy and relevant disciplines. The people specializing in the theology also became interested in natural science and engaged personally in the relevant part-time studies. The talents trained in these universities later became active participants in the scientific revolution and the founders of modern science.

The scientific revolution began from astronomy while astronomical studies were stimulated by the ocean navigation and the great geographical discoveries. Since the spreading of the transcripts of *Travels of Marco Polo* finished by the Venetian traveler Marco Polo (1254~1324) in 1299, the horizon of European was broadened. They knew distant China's richness and highly-developed material civilization as well as India's rich products, which in turn provoked their desire to travel to get rich. Yet the introduction of China's compass navigation technology and ocean ship manufacturing technology made it possible for the Europeans to engage in ocean-going voyage in the 15th century. Italian Columbus (Christopher Columbus, 1451 ~ 1506) had read *Travels of Marco Polo* and been addicted to the oriental wealth described in the book. And then Italian astronomer Paolo dal Pozzo Toscanelli (1397~1482) told him that starting from the Atlantic Ocean and sailing westward could bring him to the east. Hence in 1492, Columbus sailed across the Atlantic Ocean with the support of the King of Spain, opened the routes from Europe to the New World, and then ushered in a series of great geographical discoveries. During the process of sailing, one had to tell the direction, to locate the ship in the ocean, to understand the changes in solar terms, the syzygy of the sun and the moon as well as the waxing and waning of the moon, and to draw the precise navigational planet operating table. The observation on the ocean-going ship could provide some new information for these activities. This great development of ocean-going navigation posed new challenges to astronomy, which could not be solved only by the old version of the formerly popular Ptolemaic system of astronomy. Therefore, astronomy needed knowledge innovation.

As early as the first half of the 15th century, the Austrian astronomer in University of Vienna, Georg von Purbach (1423~1461), compiled the table of eclipse on the basis of precise observation and rigorous theoretical calculation. This table was published in 1459 and used for more than 200 years. In 1454, he finished *Theorica Moae Palnetarum* (《新行星理论》, *New Planetary Theory*) which was usually called *Astronomic Handbook*, republished many times and used as textbook. It narrated in detail Ptolemy's system on the mode of planetary motion. His German student Johannes Müller (1436~1476) (another name: Johann Regiomontanus) also studied astronomy in University of Vienna. Both of them were devoted to the translation and annotation of the new translated version of Ptolemy's works. During the years from 1471 to 1475 when Müller was in Nuremberg, with the help of a rich man named Bernard Walther (1430 ~ 1504), he established an observatory for the systematic astronomic observation and a printer specializing in the publication of works of astronomy and mathematics. In 1475, he published his nautical almanacs of the years from 1475 to 1505 which were ever used by navigators like Columbus. What's more, he continued his teacher's work and completed the compiling work of *Epitoma Almagesti Ptolemaei* (托勒密天文学大成节要, *Extracted Gists of Ptolemy's Astronomic Achievements*). However, as an astronomer and a mathematician with running the publishing as his sideline, he put forward the different ideas from the geocentrism and Ptolemy's lunar theory. At the same time, the German philosopher Nicolaus Cusanus (1401 ~ 1464), as an cardinal, also argued in his work *De Docta Ignonamtias* (《愚中之智》, *Wisdom in Ignorance*) that the earth was moving around the sun while it was rotating around its own axis, that all the things in the limitless universe were moving, their complexity exceeding human beings' knowledge, that celestial motion seen on the other celestial bodies was the same as what was seen on the earth, and that people believed that the earth did not move, but in fact it was moving just like the other celestial bodies. What was lack about these hypotheses was the astronomic proving.

The accumulation of observational materials and the relevant thoughts occasionally revealed by the previous scholars laid a foundation for the Heliocentric Theory advocated by Nicolaus Copernicus (1473~1543), the great Polish astronomer. Copernicus studied astronomy in Cracow University. He went to Italy in 1497 and studied in Polonara University, Padua University and

Ferrara University for ten years during which he participated in the observation of celestial bodies and made the records. As early as 1502 when he studied in Italy, he began to discover that the earth and the other planets moved along the track centered by the sun, the core of the Heliocentric Theory and thus provided an alternative theory of the universe to Ptolemy's Geocentric Theory which claimed that the earth was stationary center of the universe and the sun and the planets moved around the earth. After going back to Poland, he not only continued to think over this issue, but also sought the proving from the observational materials at hand and performed the mathematic calculation. During the years from 1510 to 1515, he finished his primary manuscripts of astronomic system titled *De Hypothesibus Moturum Coelestium a se Constituti Commentariolus* (《关于天体运行假说之简论》, *Brief Commentaries on Hypotheses of Celestial Motion*) and circulated its copies only to a few trusty friends. In 1539, the Austrian astronomer Georg Joachim von Lauchen (1514~1576) went to Poland to learn from Copernicus and became one of the devoted followers of Copernicus. Later, he changed his surname into Rheticus. In 1541, Rheticus first offered the publisher the draft of Copernicus in Gdansk of Poland and published the brochure titled as *De Narratio Prima de Libris Revolutionum Copernici* (《哥白尼关于天体运行著作初探》, *Primary Exploration of Copernicus's Works on Celestial Motion*).

Rheticus's former suggestion that Copernicus converted *De Hypothesibus Moturum Coelestium a se Constituti Commentariolus* into a book to publish was adopted. About in 1530, Copernicus finished his manuscript of *De Revolutionibus Orbium Coelestium*, *Libri* Ⅵ (《天体运行论》, *On the Revolutions of the Heavenly Spheres*), but for the sake of rigorous and careful consideration, he constantly revised his manuscript. What he concerned most was that the publishing of his manuscript would bring the persecution from the old force, so he was reluctant to publish his manuscript. Under this situation, Rheticus first published the outline of this manuscript in 1541. It was until his late life in 1543 that his manuscript was published in Nuremberg, Germany. Eventually he saw the print on the sickbed in his dying. Beijing Library had this book which was the second edition published with hemp paper in Basel in 1566. Copernicus' *De Revolutionibus Orbium Coelestium*, *Libri* Ⅵ provided people with a new perspective to observe the universe and used the observational data, mathematical calculation and rigorous logic reasoning to prove the rationality of the Heliocentric Theory, an alternative theory of the universe system to

Ptolemy's Geocentrism which had been widely accepted in the West more than one thousand years and been used by the scholasticism as its religious system. It degraded the earth from the center of the universe to one of planets, thereby bringing the fundamental changes into people's views on the universe. From then on, the natural science was gradually liberated from theology. The spread of scientific works accelerated the overthrowing of the feudalism, but in the anti-feudalism struggle natural science also experienced the baptism of fire.

The Italian astronomer Giordano Bruno (1548～1600) published in Italian *Cena de le Ceneri* (《灰堆上的华宴》, *A Feast over Ashes*) to publicize Copernicus's Heliocentric Theory. In the same year, he published *De l' Infinito Universo e Mondi* (《论无限宇宙和世界》, *On the Infinite Universe and World*). In this book, he argued that the universe was infinite and the celestial bodies scattered among the infinite space, which was different from the views that stars scattered among the crystal sphere centered by the sun. In 1593, he was sentenced to prison by the old feudal force. Because he would rather die than surrender, firmly believing that the sun was the center of the galaxy, he was burnt to death in 1600. However, there were many successors devoting themselves to the scientific revolution. In 1632, the Italian scientist Galileo Galilei (1564～1642) published *Dialogo dei due Massimi Sisemi del Mondo* (《两大世界体系的对话》, *Dialogue Concerning the Two Chief World Systems*) in Italian. This book described the debate between the believers of Aristotle and the believers of Copernicus, but the former was no match for the latter. In 1633, Galileo was tortured to extort a confession and sentenced to life confinement so as to force him to give up his scientific belief. When he was forced to sign on in the judgment, he still said: "but the earth is still in the rotation" ("Eppur si muove"). His book was secretly taken outside Italy and spread abroad. In 1638, he published *Discorsi due Nuove Scienza* (《两种新科学对话集》, *Collection of Dialogues Concerning the Two New Sciences*) which introduced his experiments on the falling bodies from the Leaning Tower of Pisa and his law of falling bodies proposed in the form of mathematics. In addition, he used telescope to observe the celestial bodies and discovered the sunspots. Afterwards, a series of breakthroughs had been achieved in the fields of astronomy, physics, chemistry and mathematics. Until the 17th century when Isaac Newton (1642 ～ 1727) published his *Pholosophiae Naturalis Principia Mathematica* (《自然哲学的数学原理》, *Mathematic Principles of Natural Philosophy*), the scientific revolution was finally accomplished. The modern science first emerged in Europe.

Breakthroughs also occurred to the field of medicine. The Flemish (now Belgian) Andreas Vesalius (1514~1564) was so keen on anatomy that he even went to cemeteries to collect corpses for his research when he learned medicine in Paris University in 1533. In 1537, he was awarded Doctor of Medicine in Padua University and was the professor of anatomy and surgery. For insisting on personal practice of dissection and lectures on the spot, he was very welcome. Jumping out of the restriction from the theories of ancient medical authorities, he made remarkable achievements in the research of organs like bone, blood vessels, stomach and brain on the basis of what he saw in his dissection process. He wrote the book *De Humani Corporis Fabrica* (人体结构, *On the Fabric of Human Body*) which was published in Basel in 1543. The book had fine illustrations and became the representative of medicine and anatomy during the period of Renaissance. He also pointed out that the famous ancient doctor Galen's mechanical application of what he learned in his dissection of animals into the human bodies would led to the medical faults and needed to be rectified. Immediately later, the British William Harvey (1576~1657) who studied medicine in Padua University published his book *Exercitatio Anatomica de Motu Cordis et Sanguinis in Animalibus* (《动物心脏和血液运动的解剖学研究》, *An Anatomical Disquisition on the Motion of the Heart and Blood in Animals*) in 1628. In this illustrated book, Harvey publicized his theory of blood circulation. He showed that the blood flew from the heart to the artery, and then flew to the heart again via vein. This was how the blood circulated in the human bodies. It denied the theory of Galen who believed that the blood flew from the liver to the other parts of human body and did not circulate. Harvey proved that this theory was wrong and proposed that the dissectors "should take the experiments as the basis rather than only rely on the knowledge in books and should learn from the nature rather than philosopher", advocating a new spirit of the age. His research laid a foundation for physiology and made medicine have its own theoretical basis. ❶

During the period of Renaissance, the biggest accomplishments in technology mainly came from mining, metallurgy, casting, chemical industry, machinery manufacturing and other fields. This was reflected in a large number

---

❶　HARVEY W. On Movement of Blood and Heart[M]. Beijing: the Commercial Press, 1962.

of sketches and their expositions designed and drawn by the Italian drawer and engineer Leonardo da Vinci (1452～1519). These sketches involved artillery, crossbows, machine, crane, water pump, spinning wheels, bicycles, aircraft, rolling machine, rotating device and others, all showing the new ideas of Leonardo da Vinci and seen in the notebooks he left, but unfortunately they failed to be timely published. These sketches he designed did neither be put into practice because of insufficient funds. However, the Italian engineer Agostino Ramelli (1530～1590) who had ever learned from da Vinci published *Le Diverse et Artificiose Machine* (《精巧的机械装置》, *The Various and Ingenious Machines*) in Paris in 1588. This book was explained in French and Italian, with 195 illustrations including the device converting the rotary motion into the linear motion, the piston pump, the jackscrew, the windmill-driven vertical grinding mill and others. These machines were all used in the production. The Italian metallurgist Vannoccio Birringuccio (1480～1539) published *Pyrotechnica* (《炉火术》, *Pyrotechnics*) in 1540. Written in Italian and composed of ten volumes, this book was an monograph of metallurgy and casting. Besides, it mentioned gunpowder and firearms. Germany Georg Agricola's (1490～1555) *De re Metallica Libri* Ⅻ (《矿冶全书》, *Encyclopedia of Mining and Metallurgy*) was published in Basel in 1556. It was the representative work of technology during the period of Renaissance, involving the methods for looking for metalliferous vein, the mining technology and equipment, the metal smelting and separation as well as the inspection. It also spoke of the methods for producing aqua valens or inorganic acid, glass and all kinds of inorganic salt as well as the management of mines. The book included 295 illustrations in total and embodied the highest level of technology of Europe at that time.

5. 2. 4　Promotive Role of Printing in Development of Humanist Thought and National Literature in Europe

In Europe, the printing technology promoted the great development of education and the secularization of knowledge, which further contributed to the occurrence of the scientific revolution, and thus led to the great changes in the ideological circle and the literary world. This could be called a chain reaction. During the Medieval Age, Christianity claimed that God was the center and supreme because it was the creator of the universe and man, while man, born to be guilty and insignificant, could enjoy freedom and happiness in the afterlife "heaven" only by offering everything to God and its representative — the

Church. Using the rights of God to suppress the human rights, the Church of the middle ages wanted to make people be tightly dependent on the authority of God and the Church so as to maintain the feudal ruling order. However, during the period from the 14th century to the 15th century, the capitalism of Western Europe got some development. The emerging bourgeoisie required to rid themselves of the shackles of the feudal rule and to search for the reasons for their reasonable existence and development. As a result, they held the banner of reviving the classic culture and launched a new cultural movement which they called Renaissance. Nevertheless, the real purpose of this social movement was not simply to revive the ancient Greece and Roman cultures, but to create a brand-new anti-feudalism culture suitable for the demands of bourgeoisie. Therefore, the word of "Renaissance" did not fully express the features of this age. Accurately speaking, the movement launched in Europe at that time was not restricted only in the scope of culture. Later, it spread to the fields of economy, politics and military. Its real purpose was to throw over the feudal ruling and to establish the capitalist system. ❶

The ideological system of Renaissance Movement was humanism which originated from Latin humanus. It advocated the human-centered idea, eulogized the nobleness of human nature, promoted the authority of human, praised the value and dignity of human, so the protection of human rights was the major concern of humanism. Italy was the birthplace of Renaissance. As early as the 14th century, the humanist Giovanni Boccaccio (1313 ~ 1375) coming from Florence (a city in the middle Italy) revealed the hypocrisy, greed and cruelty of the Church people of the middle ages and the feudal aristocracy in his work of *Decameron* (《十日谈》), and claimed that human beings were born to be equal. He did not agree with the tradition of taking the family background as a standard to determine one's status but advocated personality liberation. Another humanist Francesco Petrach (1304~1374), a poet coming from Florence, first put forward the differences between humanism and theology. His sonnet *Conzoniere* (《抒情诗集》, *Lyrical Ballads*) written in Italian described love, full of the milk of human kindness and covered with the color of anti-feudalism. It also expressed the miss for the glory of ancient Rome and the eager for the unification of Italy.

During the period from the 15th century to the 16th century, the

❶ 朱寰. 世界中古史[M]. 长春: 吉林人民出版社, 1981.

introduction of printing technology fueled the praire fire of humanism and enabled its ideas to spread with a stronger trend in a wider scope. The Italian linguist Lorenzo Valla (1406~1451), in his work *De Falso Credita et Ementito Constantini Donatione* (《论君士坦丁皇帝让权的杜撰和赠地的捏造》, *On the Fabrication of Emperor Constantine's Right-conceding and Land-granting*), exposed through his rigorous research the fact that the history documents was forged in the eighth century. It was not true that the Emperor of Roman Empire Constantine I (280~337) had conceded to the Vatican the jurisdiction rights of four patriarchates except the Roman patriarchate and the secular rule rights of the west region of the empire. In the book, he also stated that during the Middle Ages, the Pope took these "gifts of Constantine" as the supporting evidence to ask the other four patriarchate and the kings of countries in the Western Europe for power, which illustrated that the so-called gifts of Constantine had become the weapon of Pope in the fighting against the secular regimes. Valla's research denied the legitimacy of the Pope's territory and the historical basis of the secular rule rights. Besides, he took use of the linguistic knowledge to uncover the fact that the doctrines worshiped by the churches as the "enlightenment" originally was the errors caused in the translation of *The Bible*. The Italian philosopher Pietro Pomponazzi (1462~1524) refuted the views of immortal soul in his work *De Immortalitate Animi* (《驳灵魂不朽》, *On Soul Immortality*). For he argued that feeling was caused by the things in the outside world and functioned as the basis for the rational knowledge, he suffered persecution and his books were burned.

Valla's research demonstrated that the Vatican played tricks in the translation of *Ta Biblia* (《圣经》, *The Bible*) from Greek into Latin. In order to let people have access to the original appearance of *Ta Biblia* in Greek, Netherlandish (now Holland) humanist Desiderius Erasmus (1467~1536), mastering several languages such as Greek and Latin, published Greek *The New Testament* with his own Latin translation for the first time in Basel in 1516, which broke the monopoly of the Vatican authorities on the translation of *The Bible*. In 1509, he also published *Encomium Mariae* (《愚人自夸》, *The Boast of a Fool*) in Basel to expose the evil of the old feudal forces and its fooling of people by the way that "a fool" preached on the stage to boast themselves. In the book, that old feudal force was portrayed as a group of greedy and lewd guys worshiping stupidly. It played a leading role in the religious reform of Western Europe. Besides, he preached personality freedom and human liberation and opposed the asceticism of the Church of that time in his work of *Diatribe de Libero Arbitrio* (《自由意志论》, *On Libertarianism*). He required the secular rule and advocated the establishment of reasonable Church. Since the middle of the 15ᵗʰ century, the printing offices in Florence, Venice, Basel, Lyon, Paris and Barcelona (Barcelona) published works of many humanists. This showed that the barriers set by the old feudal force could not prevent the spread of new ideas in Europe. Publisher put books in airtight casks and used cart or ship to "smuggle" them around.

During the Medieval Age, Greek and Roman classical secular cultures were actually not welcome, because the tradition that they attached great importance to the real life and the pursuit of happiness was not consistent with the mainstream of that time. At that time, the literature was coated with religious color, dull and boring. During the period of Renaissance, some humanists sorted and published many classical works in the publishing of their own works. They strove to search for the ancient handwritten books to replace those distorted ones in the Middle Ages. Especially those manuscripts which were brought by the Greek people to Italy when they escaped from the Byzantine Empire after its capital Constantinople fell into enemy hands in 1453 were the most precious. They were gradually published after being sorted. For instance, the ancient Roman poet Vergil's (70 B. C. ~ 19 B. C.) idyll *Bucolics* and didactic poem *Georgics* praising the life and loyalty of farmers were published in 1470; the ancient Greek poet Homeros' epic *Iliad* and *Odyssey* were published in 1488 and 1504 respectively; the Latin translations of the whole works of Aristotle were printed in 1469 and their Greek ones, printed during the years from 1495 to 1498; the Latin versions of the whole works of Plato were printed in 1483 and their Greek ones, printed in 1513. In Europe, various university set the course of ancient language. In 1530, the school of three languages (Latin, Greek and Hebrew) was established in Paris, becoming another center of humanism. The Italian humanistic historian Leonardo Bruni (1369~1444), the translator of the works of Plato and Aristotle, discovered in his translation that the major doctrines of scholastic philosophers were built up on the basis of the distorted texts of Aristotle.

On the other hand, the west wind of humanism blew to the literature area of Europe, showering new changes. ❶ French Pierre de Ronsard (1524~1585)

---

❶ Anon. Vozrozhdenie. 文艺复兴[M]. 王以铸,译. 北京:人民出版社,1955.

who had studied medicine and compiled textbooks on medicine was a humanistic lyricist. He organized the seven-member literary group called the Pléiade and firmly claimed the restoring of using French and its application in literary creation, opposing the use of Latin and foreign languages to write. He wrote in French *Odes* (《短歌行》, *Odes*), *Amours de Cassandre* (《卡桑德拉的爱情》, *Love of Cassandre*) and *Hymnes* (《赞歌集》, *Hymns*) to reflect the awakening of national consciousness and the strengthening of patriotic ideas. Fran ois Rabelais (1494~1553), another French writer who also had the background of natural science knowledge, took fairy tales as blueprint in his French novel *La vie de Gargantua et de Pantagruel* (《巨人传》, *The Life of Gargantua and Pantagruel*) to shape images of the ideal monarch, the giant Garguntua and his son Pantagruel, thereby showing the irrationality of feudal system, the darkness of the Church of that time as well as the decay of scholasticism and medieval education. In the novel, he also propagandized humanists' ideas on politics, education and morality, put forward the doctrines like "Fais ce que voudra (Do what you want to do.)" and reflected the requirement of personality liberation.

In Britain of during the Renaissance period, Thomas More (1476~1535) who was born in a aristocratic family and occupied important positions in the cabinet and parliament published *Utopia* in Latin in 1516. In the book, he repudiated the bourgeoisie itself. In the utopia described in the book, just like in the one depicted in *Civitas Solis* (《太阳城》, *Sun City*) of the Italian philosopher Tommaso Campanella (1568~1639), the means of production were owned by the public and everyone was equal and worked together. The utopian socialism was one of the most radical ideas of humanism, so those English and French authors who put forward that idea were persecuted at that time. The British ironist Thomas Nash's (1567~1601) novel *The Unfortunate Traveller* was a realistic work, describing the life of people of various professions and grades under the background of historical events at that time. And William Shakespeare (1564~1616) who came from the lower class of society knew a little of Latin and even less about Greek, but had rich social experience and became one of the greatest dramatists and poets in Britain during the period of Renaissance. He made the the drama genius of British people fully unfolded. This was a literary form without succumbing to the classical tradition but with a strong taste of new era. His drama, different from the drama in Italy, had its own British characteristics. Besides, what he was different from other dramatists was that he was a versatile and productive writer because while he was writing script, in total 37 plays and 154 sonnets, he was an actor, director and theater owner.

Shakespeare's plays which had shaped many distinctive typical characters mainly described the conflict among various social forces during the period of the disintegration of feudal system and the emerging of capitalist, advocated the personality liberation and opposed the feudal shackles and the theocratic rule, thereby having the brilliant color of humanism. His plays, taking the basic material from the daily life and written in easy-understanding English, could be classified into historical drama, comedy and tragedy. Specifically, the historical dramas included *Henry* Ⅳ (1597~1598), *Richard* Ⅲ (1592~1593) and others; the comedies contained *A Midsummer Nights Dream* (1959), *Twelfth Night* (1601) and so forth; the tragedies embraced *Romeo and Juliet* (1595), *Hamlet* (1600~1601), *Othello* (1604), *King Lear* (1605~1606) and others. In his historical plays, Shakespeare praised national unity, objected to the feudal separation and supported kingship, which embodied his political ideals of humanism. In his comedies full of optimistic tone, he praised friendship, love, freedom and equality, which showed his moral views of humanism. In his tragedies, he employed the tragic ending of just and kind characters to accuse feudal forces and to criticize the greed and cruelty of businessmen in the primitive accumulation of capital as well as the money worship in the society.

The Renaissance Movement in Span started from the 16th century. With the establishment of universities and the introduction of new culture of Italy, humanism spread and gave birth to many famous litterateurs. The early representative was Miguel de Saavedra Cervantes (1547 ~ 1616). Young Cervantes went to Italy to make a living and suffered a lot. In 1580 when he returned to Madrid, he started his writing of drama and fiction. His major representative work was the novel of *El Ingenfoso hidalgo Don Quijote DE la Mancha* (《曼查的才智骑士唐·吉诃德》, *The Ingenious Hidalgo Don Quixote of La Mancha*) published in 1605. The novel was written in the dialect of Castillano in central Spain, a dialect which was later spoken nationally and became today's Spanish. It told the stories in which the obsolete knight Don Quixote and his squire Sancho wandered in search of deeds of courage and chivalry, but kept hitting dead ends. Believing there were insurrections from demons, he held his gun to attack and made a lot of ridiculous things. It was when he was dying that he woke up and denied himself. In the novel, Cervantes revealed the peremptoriness, cruelty and hypocrisy of the feudal aristocracy of that time, expressed the miserable plight of people and at the same time denied the medieval knight literature of beautifying the feudal system. Although Don Quixote's behaviors were ridiculous, he had the feudal moral outlooks and at the same time he desired freedom, happiness and social equality. He therefore was the typical representative of personality contradiction and reflected the complex thoughts of people in facing the alternating of old and new eras. Sancho represented the image of farmer and overcome his selfish thoughts during the travelling with his host. Later, he became the governor of the island and implemented the clean and impartial governance, which reflected the author's humanistic political ideals. The ideal Don Quixote failed to achieve was realized by Sancho. From Sancho, people's wisdom and power to defeat darkness and change their social status were seen.

During the period of Renaissance, some scientific and literary works were not written in Latin, but in each country's own national language such as Italian, French, Spanish and English. The publication of these works was of great historical significance. In the medieval Europe, the languages used in the manuscripts were the two ancient languages of Greek and Latin which were the official languages of Greek Byzantine Empire and Roman Empire respectively. Since the Great Schism in Christendom in the 11th century (1054), Latin had been the language promoted by the Roman Catholic Church in western European countries. At that time, few of western Europeans understood Greek and the use of Latin was restricted only to a handful of clergies and nobles, but the ordinary people could not understand these languages. Therefore, the official languages were seriously disconnected with the native languages and grammar spoken by the vast majority of the masses of various countries. After the Western Roman Empire collapsed in 476, Europe was in a state of constant division. For example, the Charlemagne Empire (751 ~ 843) broke up and disintegrated into the prototypes of today's three nations of Germany, France and Italy. After entering the 11th century, Western Europe had already had more than twenty independent kingdoms, each with its own language different from the languages of other kingdoms. Each nation needed to develop its own national culture in the process of its formation because it was believed that there was no future for a nation without its own culture. This required the use of characters suitable for the language of each nation. Nevertheless, at that time, most of Latin literature was the religious works restricted to the churches. They were not understood by the ordinary people of each European nation. In other words, they stopped the ordinary people at the door of culture and killed the possibility for each country to develop their own national cultures.

However, the introduction of printing technology broke this deadlock. The printing technology made it possible to publish a large number of popular books in the native languages of each nation. Definitely, the new ideas and new discoveries published by scholars in their own native languages were easily accepted by the public. Those literary works written by writers in their own national languages were what the public liked to listen to and read. In view of this, it could be said that the real history of European literature started during the time period from the 14th century to the 15th century when the printed books appeared. The literary works each country published not only functioned as the samples written in standard Italian, French, English, Spanish and German, but also reduced the language barrier caused in the use of different dialects within the same country. Later, these written languages gradually became unified literature languages in their respective countries. Refined and standardized by the writers and scholars of their own countries, fixed through the printing, improved in the aspects of their vocabulary, grammar, sentences structure, spelling and pronunciation principle, and constantly revised in the process of their use, these literature languages were finally able to shoulder the function of philosophical discussion just as Greek and Latin did, even better than the latter. This contributed in a large extent to the promotion of state and nation consciousness of each country in Europe and the development of national cultures. Only after possessing its own literature language could a nation be qualified to have its own culture and cultural classics. The printing technology played a midwife role in the formation of literature language of each European nation. At the same time, the emergence of the literature language led to the result that Latin literary works ceased to exist in the 16th century.

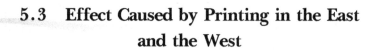

5.3　Effect Caused by Printing in the East and the West

5.3.1　Influence of Printing on Examination System in the East and the West

Science, philosophy, literature and art, education, law and system are all regarded as the ideological reflection of society, politics and economy. At the same time, they, in turn, have an influence on the latter and play a promotive role in the development of the latter. As the product of literature replication technology revolution, the printing technology had the same impact on the politics as well. In China, it was first used by the feudal ruling class to consolidate their feudal rule. During the second century B. C. when Emperor Wu (in office from 156 B. C. ~87 B. C.) of the Han Dynasty (202 B. C. ~220 A. D.) promoted the policy of "rejecting other schools of thought and respect only Confucianism", Confucianism, after being transformed, became the state ideology, which from then on had lasted for over two thousand years until the end of the Qing Dynasty (1644~1912). During the period of the Five Dynasties (907~960), that the imperial court enacted the plank-engraved *Jiu Jing* of Confucianism was precisely to strengthen the rule of official philosophical thoughts on the *chenmin* (臣民, the ruled subjects). Inheriting the tradition of the Five Dynasties (907~960), the *Guozijian* (国子监, the Imperial College) of the Song Dynasty (960~1279) published a large number of Confucian classics and Neo-confucianism works to be used as the textbooks and reference books in the schools at all levels.

For the sake of providing the political officials for central and local government departments at all levels, the Sui Dynasty (581~618) created the system in which the imperial government implemented the public examination to select officials. Without considering the family background of examinees and the recommendation from local governments, the imperial government would award the examinees fame and official positions according to their examination results. This was known as the *Keju Zhidu* (科举制度, the Imperial Examination System). It was a major improvement for the ancient system of selecting officials, having a progressive significance. The Imperial Examination System was developed during the Tang Dynasty (618~907). During the Song Dynasty (960 ~ 1279), it was reformed. The *Dianshi* (殿试, the Palace Examination) was added, i. e., the Emperor would preside over the highest examination in the palace after the examination chaired by the Ministry of Rites in the feudal China. In the Palace Examination, the candidate with the first place was named *Bangshou* (榜首, Number One Scholar) and the ones with the second and third places, *Bangyan* (榜眼, Runner-up Scholar). During the Southern Song Dynasty (1127~1279), the candidate with the first place in the Palace Examination was renamed *Zhuangyuan* (状元, Number One Scholar) and the ones with the second place and the third place were renamed *Bangyan*

(榜眼，Runner-up Scholar) and *Tanhua* (探花，The Third) respectively. The winners in the Palace Examination were directly granted official positions. The regular examination held each year for candidates included three levels of the *Zhoufushi* (州府试，the Provincial Examination)，the *Libushi* (礼部试，the Assembly Examination / the Rite Examination) and the *Dianshi* (殿试，the Palace Examination). The candidates who succeeded in the nation civil service examination in imperial China were granted the degree of *Jinshi* (进士). The imperial government would give them the excellent treatment，so there were many candidates attending *Jinshi* examination. ❶

The advanced development of the printing technology during the Song Dynasty (960~1279) made the printed reference materials and textbooks cheap and affordable to the candidates taking the Imperial Examination，which greatly contributed to the implementation of the Imperial Examination System. This system made it possible for the folk scholars to pursue their scholarly honour through taking the examination and to be selected by the feudal governments at all levels. China's scholar-bureaucrats called by the westerners Mandarinates were advanced intellectuals selected from the examination. They had high academic degrees and were the products of the development of culture.

The Imperial Examination System，as a great invention in the examination system of civil service，was later followed by some other countries. Dr. Joseph Needham who regarded the scholar-bureaucrats as "nonhereditary and rare social elites" believed that China's feudal society ruled by literati official group seemed to be weak，but in fact stronger than European feudal society ruled by military aristocrats and could more effectively prevent the feudal system from the harm from the industry and commerce capitalism. This was because the Imperial Examination System was more reasonable than the Hereditary System of the European feudal aristocracy in the selection of officials. Under the Hereditary System，the offspring of nobles might not be excellent，but they could still occupy the high positions without participating the examination and selection. ❷ The case，however，was different in the Imperial Examination System. It provided opportunities for those talented men who passed the examinations to occupy official positions. For example，the literary giants of

---

❶ 脱脱.宋史：选举志[M].上海：上海古籍出版社，1986.

❷ NEEDHAM J. China and the West[M]//Dyson A，Tower B，eds. China and the West：Mankind Evolving. New York：Humanistics Press，1970.

Ou Yangxiu, Wang Anshi and Su Shi were awarded the top officials in the local and central governments only after they were successfully granted the degree of *Jinshi*（进士）. That they could occupy the high official positions did not rely on the hereditary, but on their own talents which helped them win the examinations. So did their children. However, it was seldom seen in feudal Europe. The Imperial Examination System, therefore, made scholar-bureaucrat feudal system more advanced and powerful than military aristocratic feudal system, be able to absorb and develop the major inventions like printing, gunpowder and compass, and thus had itself enriched and maintained, whereas the western feudal system could not stand the social impact brought about by these inventions.

According to the statistics, there were 40,000 candidates who were granted the degree of *Jinshi*（进士）during the Song Dynasty（960～1279）and 200,000 candidates who were granted the degree of *Juren*（举人）during the 12th century. The number of *Juren* rose to 400,000 during the 13th century. At the same time, there were a large number of candidates who possessed the same education level, but were not granted the titles. They were experts in *Shisan Jin*（《十三经》, *Thirteen Classics*）, literature and history, loved nature science and technology, and had performed the work of observation and research for a long term. Undoubtedly, they became the human resources and forces for the advanced development of academic studies. Those areas where there were the most candidates with the degree of *Jinshi*（进士）were the very places where there were the most printed products and the education was the most developed. For example, during the Song Dynasty（960～1279）, in the five provinces of Liangzhe（now Zhejiang Province）, Fujian, Sichuan, Jiangnan（now Jiangxi Province）and Jiangnan Dong（now Jiangsu Province）, there were 24,172 candidates with the *Jinshi* degree, taking up 84% of the whole country. Yet at the same time, there were 1,168 kinds of printed books in these five provinces, occupying 90% of the whole country（1,303 kinds of books）. Nevertheless, there were only 103 candidates with the candidates with the *Jinshi* degree in Guizhou Province in southeast China where there were the fewest printed books, only two kinds. ❶ Those big printing provinces were also the big papermaking ones. The proportion relationship between the printed books and

---

❶   Needham J. Science and Civilization in China, vol. 5, Paper and Printing Volume by Tsien Tsuen Hsuin. Cambridge: Cambridge University Press, 1985: 379-380.

the Imperial Examination System demonstrated clearly the contribution of printing to the Imperial Examination System. Meanwhile, the implementation of the Imperial Examination System, in turn, promoted the development of printing industry, publishing industry and education. They were in the interactive relationship of mutual promotion.

From the political point of view, compared with the system of choosing officials in the previous generations, the Imperial Examination System had the following characteristics. Firstly, it made the rights to choose officials centralized from the local governments to the central government and thus strengthened the rule of centralism. At the same time, it expanded the social basis of the ruling group by providing the ordinary landlords and civilians with the opportunities to participate in the regime. Secondly, that it connected study, exam with the pursuit of official positions, on one hand, made intellectuals have opportunities to promote their social status and improve their situations, and on the other hand, attracted more and more people to join the team, made them transfer their energies to the reading and testing so as to avoid their going against the superiors and making troubles, thereby contributing to the maintaining of social stability. Thirdly, it overcame the shortcoming in the past exam system which only paid importance to the family background and conduct of candidates but ignored their knowledge and talent. In addition, it selected talented people with the unified national standard to work in the government institutions, thereby improving the quality of officials and strengthening the regime power of local and central governments. In view of these characteristics, the Imperial Examination System lasted for more than one thousand years in China. It was not until the 19th century, the late Qing Dynasty (1644 ~ 1912), that the system was abolished and replaced by the modern examination system.

The Imperial Examination Systems of Tang and Song Dynasties all had a great influence on the West and the East. According to the tenth volume of *Sanguo Shiji* (《三国史记》, *History of the Three Kingdoms*, 1145) written by Jin Fushi (1075~1151), since the unification of the Korean Peninsula during the Xinluo Dynasty (57 B. C. ~935 A. D.), the saint king of the Xinluo Dynasty took for reference the Imperial Examination System of the Tang Dynasty during the fourth year of his reign (788) and set the Three-level Reading System to implement the sate examination to select officials. Candidates who read *Chunqiu Zuoshi Zhuan*, *Li Ji*, *Wen Xuan*, *Lun Yu* and *Xiao Jing* (《孝经》,

Principles of Filial Piety) were leveled as the upper candidates; those who read *Qu Li* (《曲礼》, *Daily Etiquette*), *Lun Yu* and *Xiao Jing*, the middle ones; those who read only *Qu Li* and *Xiao Jing*, the lower ones. Hence, the candidates were classified into the upper level, the middle level and the lower level according to the books they read and the exam programs they took, and then were granted the corresponding official positions. Meanwhile, those who mastered the five Confucian classics, the three historical books and the works on the ideas of various schools of thoughts could bypass their immediate official ranks to be promoted. Meanwhile, some students from the Korea Peninsula during the Xinluo Dynasty (57 B.C. ~935 A.D.) also took the examinations of China and achieved the *Jinshi* degree (进士) when they studied in China. According to *Xuanju Zhi* (《选举志》, *Treatise on Election*), the 77th volume of *Gaoli Shi* (《高丽史》, *History of Korea*, 1454) written by Zheng Linzhi (1395~ 1468), in 958, the ninth year of Guangzong's reign of the Gaoli Dynasty (918~ 1392), the examination system began to be set to mainly test candidates' *Shi* (诗, Poem), *Fu* (赋, Prose-Poetry), *Song* (颂, Ode) and their ideas on the current affairs and policies, which lasted for more than one thousand years. According to the third volume of *Dayue Shiji Quanshu* (《大越史记全书》, *The Comprehensive Chronological History of Great Vietnam*) written by Wu Shilian (1439~1499), in 1075, the forth year of Taining Period of Emperor Renzong in the Li Dynasty (1009~1225), the Imperial Examination System began to be implemented in Vietnam. The later dynasties continued this examination system to select talented people. It was not until the 20th century that it was abolished by the government of Vietnam. During the Nara Dynasty (710~794) of Japan, the *Daxueliao* (大学寮, Institutions of Education and Examination) was established, setting the programs of classics, music, law, writing and calculation. Each program required the learning of Confucian classics, the calculation science of *Jiuzhang Suanshu* (《九章算术》, *The Nine Chapters of the Mathematical Art*) and *Sunzi Suanjing* (《孙子算经》, *The Mathematical Classic of Sun Zi*) and others. The candidates passing the examination would be granted the official position of eight-level rank, similar to the Imperial Examination System.

The government official examination system in modern Europe and America was directly influenced by the Imperial Examination System of China. At first, the French Jesuit, Louis Daniel le Comte (1655~1728), introduced China's Imperial Examination System and pointed out its four advantages.

Firstly, the imperial government's selection of the excellent youth according to their research achievements embodied in the exam rather than their family background could drive them to study hard. Secondly, it could cultivate people's quality of diligent pursuing of knowledge and make the society respect knowledge. Chinese youth's pursuit of the success in the state examinations could improve their cultural qualities. Thirdly, to some extent it prevented the occurrence of the phenomena like greed, spiritual decadence, knowledge emptiness and self-indulgence. Fourthly, it could help the Emperor get the talented people together to select more suitable successors when the officials with bad behaviors were removed from their official positions. In 1735, another French Jesuit, Jean Baptiste du Halde (1674 ～ 1743), compiled the communication of Jesuits from China into a book and published it in Paris. The book also reported the Imperial Examination System of China, pointing out that the Emperor of China set up the Imperial Examination System to select the talents all around the country through the state examinations and to grant them official positions. Exams were held every three years. Those candidates passing the imperial examinations would be granted *Jinshi* degree and enter *Hanlin Yuan* (翰林院, the Imperial Academy). *Hanlin Yuanshi* (翰林院士, member of the Imperial Academy) would be elected by the Emperor to serve as the chief executive, the *zaixiang* (宰相, the prime minister) and the teacher of *taizi* (太子, the crown prince). Besides, they also engaged in the writings. Usually, they were respected by the world. ❶

During the 18th century when the French Enlightenment thinker Voltaire (Francois-Marie Arouet, 1694~1778) read the report above, he praised China's Imperial Examination System in Chapter 195 in his work *Essai sur les Moeurs* (《风俗论》, *On Customs*). He stated that China's administrative organization was one of the best organizations in the world because Chinese government was composed of six ministries of *Li* (吏, Personnel), *Hu* (户, Census Register), *Li* (礼, Regulation), *Xing* (刑, Justice), *Bin* (兵, Military) and *Gong* (工, Architecture), the official rank was divided into nine classes and candidates who were appointed official positions had to pass several strict exams. The French physiocratic economist Francois Quesnay (1694 ～ 1774) published in succession long articles entitled *Despotisme de la Chine* (《中国的专制政体》, *China's Autocratic Regime*) on the four periods of magazine named *Ephéméride*

---

❶　DU HALDE J B. Description de l'Empire de la Chine[M]. Paris: Le Mercier, 1735.

du Citoyen (《公民日志》, *Log of Civilians*) issued during the time from March to April of 1767. The articles stated that, not like in the West, there were no hereditary peers in China and what the officials relied on to achieve their official positions were their talent and achievement. The author praised the Imperial Examination System of China because the examinees, no matter what family background they had, might rely on their own ability to participate the competition and even the children of craftsman could become the governor if they succeeded in the relevant examinations. Therefore, Quesnay believed that what China practised was the enlightened monarchy, worth the learning of the West at that time. George Staunton (1781 ~ 1859), the assistant of George Macartney (1737~1803) who was a British embassy and paid a visit to China in 1793, also introduced and praised China's Imperial Examination System in the twelfth chapter of his work *An Authentic Account of an Embassy from the King of Great Britain to the Emperor of China*.

Under the influence of China's Imperial Examination System, France first set the Civil Official Examination System. Ten years later (1801), it was stopped, but restored in 1840. During the 18th century, some British people wrote articles to praise China's examination system and some of them even advocated that the British government should practice the similar system. Therefore, in 1806, British East India Company first practised the Civil Official Examination System. Afterwards, many British people took use of China's case as an evidence to support the claim that Britain should set the popular civil official system. For example, Thomas Meadows (1815 ~ 1868), the British diplomat in China, introduced China's Imperial Examination System and pointed out that the reason why the Chinese Empire could exist for a long time was that its government officials were composed of those with talents and achievements in his book *Desultory Notes on the Government and People of China* in 1847. Therefore, he advocated that the open civil service examination system should be comprehensively built in Britain so as to improve administrative departments. These calls caused the British government to set up a committee to study this matter and reported to the British parliament in 1853. In 1855, the British government put it into the practice.

Undoubtedly, Britain's Civil Official Examination System set a good example for the United States of America to establish a similar system, but China's influence was still obvious. For example, in 1868, Thomas A. Jenckes, the senator of Rhode Island, first put forward the proposal for the congress to establish the civil official examination system. In his proposal, there was a chapter in which China's Imperial Examination System was introduced. In the same year, Ralph Waldo Emerson (1803~1882), a famous Boston intellectual, delivered an address to welcome two diplomats of the Qing Dynasty (1644~1911), Zhi Gang and Sun Jiayu, in Boston. In his address, he praised China's Imperial Examination System and urged the U.S. congress to approve Jenckes' proposal. His words are paraphrased as follows:

> *There is one point about China's political system in which we are very interested. I believe everyone here remembers the bill that the senator of Rhode Island, Mr. Jenckes, proposed twice for the approval of Congress, i.e., he advocated that civil officials must first achieve their academic qualifications through the success in relevant examinations before they could take office. In terms of correcting bad habits, the Chinese people do walk ahead of us and also ahead of Britain and France. Similarly, Chinese society also walks ahead of us in the aspect of the importance attached to education. China attaches great importance to education, which is the capital worth China' pride.* ❶

Emerson's address in Boston aroused responses, some agreeing with him and some opposing him. Just as in Britain, in the United States, many people who got vested interests from the old system strongly opposed the new ideas. Some of them even protested that using examination to decide the appropriate candidates to be administrative officials was Chinese tradition and "foreign" system, so it was "un-American" system. Due to the obstruction from the conservatives, Jenckes' proposal was delayed and it was not until 1883 that his proposal was finally passed by the Congress of the United States of America. Professor Derk Bodde, the contemporary sinologist of Pennsylvania University, believed that China's Imperial Examination System had two major benefits. Firstly, it was open to all the people in the society, so it was world's most democratic way of the selection of government officials during the times before the modern society. Secondly, it guaranteed that people who became government officials must possess the background of higher education at first. Bodde pointed out that "Today's civil official examination system in fact has

---

❶ Teng Ssu-Yu. Chinese Influence on the Western Examination System[J]. Harvard Journal of Asiatic Studies, 1943(7):267-312.

been accepted by all the democratic countries. More and more people can work in government institutions by their talents rather than political favoritism. As a result, the political corruption pervasive one hundred years ago has disappeared. Undoubtedly, the Civil Official Examination System is the most precious gift China gives to the West."❶

However, the disadvantages of the Imperial Examination System became more and more salient after the Ming Dynasty (1368~1644). The most obvious one was that the exam content was limited to rigid and dogmatic *baguwen* (八股文, eight-legged essays), lack of the practical knowledge of such fields as governing, science and technology. The education of schools was still confined to the Confucian classics, literature and history, without involving the content of science and technology, and thus hindered the development of science and culture, no longer meeting the demands of age. Nevertheless, the western Civil Official Examination System, without completely following the Imperial Examination System of China, only absorbed its reasonable elements and was developed in combination with the western characteristics. As a result, in the West, the Civil Official Examination System made the construction of western capitalist regime improved and strengthened, while in China the feudal system further declined due to the degradation of such a system.

5.3.2 Economic Effect Caused by Printing in the East and the West

The role of printing technology in the field of economy was not less than that in the fields of thoughts, culture, education, science, religion and politics. The case was the same both at home and abroad, so it is necessary to briefly describe it here. Since the Song Dynasty (960~1279), the printing industry had always been one of the prosperous industrial sectors of national economy, holding the stable market. Different from the industries like mining, metallurgy, shipbuilding and machinery, the users of printing products (presswork) were numerous, pervading throughout the country, so the sales of printing products had remained high, without being influenced by other social factors, which made the printing houses and the bookstore businessmen constantly busy. Printing manufacturers often gathered in some certain areas of some provinces and formed the printing centers such as Jianyang of Fujian

---

❶   BODDE D. Chinese Ideas in the West[M]. Washington, D. C.: American Council on Education, 1948.

Province, Hangzhou of Zhejiang Province, Chengdu of Sichuan Province as well as Yangzhou and Nanjing of Jiangsu Province, from which the printing products were often distributed to other cities. Because these printing centers were usually near the papermaking areas, the long-term prosperity of the printing industry naturally stimulated the development of papermaking industry and at the same time promoted the economic prosperity by accelerating the development of commerce and amphibious transportation in these areas. Moreover, the bookstores selling printing products sometimes focused in a certain block of city, forming bookstore street and thus being very convenient for customers to choose and buy the products they wanted. As early as in the Tang Dynasty (618~907), bookstores appeared in Chang'an (Today's Xi'an), Luoyang and Chengdu. During the Song Dynasty (960~1279), the bookstores were also seen in the cities like Kaifeng, Luoyang, Hangzhou, Suzhou, Quanzhou and Guangzhou. This is well proved in the famous *huajuan* (画卷, scroll) reflecting the scene of Kaifeng of the Northern Song Dynasty (960~1127), *Qingming Shanghe Tu* (《清明上河图》, *Riverside Scene on the Pure Brightness Festival*). In this scroll, there is the component of bookstore.

Since the Tang and Song Dynasties, the printed book had been one of the important items of China's export trade and was exported mainly to the Korean Peninsula, Japan, Vietnam and others. They also sent relevant personnel to purchase the printed books in bulk in China. According to the record in the 34th volume of *Gaoli Shi* (《高丽史》, *History of Korea*), the relevant personnel were sent to Jiangnan (South Jiangsu Province) in China to buy books in 1314 and only in the city of Nanjing they bought over 100,000 volumes of books with *baochao*. According to the fifth volume of the same book, the merchant of the Song Dynasty (960~1279), Li Wentong, transported 597 volumes of books to the Korean Peninsula in 1027. These trade activities show that the international market of presswork trade had formed among the countries of Chinese-character cultural circle in East Asia one thousand years ago and lasted until the Qing Dynasty (1644~1911). During this period, merchant shipped books to foreign ports and then brought back the local goods, thereby promoting the development of foreign trade and enlarging the market of book sales. Some coastal booksellers of China had their own ship fleets and shipped with the changes of monsoon between the Chinese and foreign port cities. Meanwhile, merchants from the Korean Peninsula and Vietnam also imported books from China along the land trade routes.

The case above also happened in Europe. During the 15ᵗʰ century, the big printing power, Germany, monopolized the movable type printing technology after mastering this technology. She printed the Latin version of *Vulgata* (《圣经》, *The Bible*), religious paintings, Latin grammar and cards to dump in bulk in other nearby countries and earned a lot of foreign gold coins. The rapid emergence of printing centers across Germany accelerated the development of papermaking industry. The Wittenberg Printing House once used sheepskin to print due to the paper shortage. In order to meet the printing houses' huge demand for paper, papermaking mills were established around country and thus paper replaced other materials to be the major printing material, reducing the production cost and the prices of books. After the 16ᵗʰ century, the printing industry in the countries like Italy, Holland, France, Swiss and Britain also developed fast and became one of their backbone industries. During the period from 1450 to 1500, Europe printed in total 25 thousands of books, 5 millions of volumes. During the 16ᵗʰ century, the production tripled. In some commercial and culture-developed cities, bookstores forested. With the rapid development of printing industry, the bank businesses also expanded because the European publishing industry had been closely connected with the banking industry from the very beginning, i. e. , the funds used to build printing houses were loaned from banks. Besides, the flourishing of printing industry drove the development of industry and commerce. Specifically, the issuing of newspapers around countries in Europe could provide them with important information and make people in this field know the trend of domestic and foreign markets without leaving their homes. They could also promote their products through the newspaper.

In order to promote their products, maintain their credibility and brand, each manufacturer invited designers to design trademarks to spread in society and also printed advertisements or leaflets in various forms, which helped stimulate the development of social industry and commerce. The printing of ads and trademarks started from China. The earliest existing physical object was the advertising-use copper plate used by Liu's Needle Store in Jinan City of the Northern Song Dynasty (960 ~ 1127) and kept in the Museum of Chinese History. The logo used by the store to advertise its sewing-use steel needle was rabbit and the plaque was hung in front of the store. So did the other stores in the cities like Kaifeng and Hangzhou. In the block-printed editions of the Song and Yuan Dynasties, the logos and advertising texts of various stores could sometimes be seen. Yet in Europe, the printed trademarks and advertisements were first seen in the printed books of the 15th century. Later, newspapers and various magazines which could be circulated in a larger number better served advertising. In addition, the product packaging materials of each manufacturer were also printed with the commodity name, logo, function and advertising text, being transported along with the products to various places across country. Their purposes were to differ one's own products from the similar ones of other places and to highlight the features of their own products so as to attract customers' attention. This business tradition had lasted from the Song Dynasty (960~1279) to the Qing Dynasty (1644~1911), bringing the new business to the printing houses. Yet in Europe, that the print served the industry and commerce with the similar ways also started after the 15th century. Today, the print filling the urban and rural areas of each country in the world also originated from it, bringing the huge economic benefit.

In the economic history of the Tang Dynasty (618~907), there were two pioneering moves which had a far-reaching influence on the later generations. One was the use of *yinzhi* (印纸, the taxpaying receipt). In 783, the fourth year of Jianzhong Period under the reign of Emperor Dezong (742~805 on throne) when the governments levied house tax and income tax, they would grant taxpayers a kind of unified printed document on which the taxpayers' name, tax item and taxes were recorded. It functioned as the taxpaying receipt and was also named *yinzhi*. This was the first time for the print to be used in the financial management, improving work efficiency and effectiveness as well as standardizing the financial work. The other was the use of *feiqian* (飞钱, the flying money). During the early Yuanhe Period of Emperor Xianzong (806~820 on throne), the government issued the format-unifying exchange certificate called *feiqian*. Specifically, the merchants who came to trade in Beijing could exchange their coins from the selling of their goods for *feiqian* and got back their coins with *feiqian* when they returned to their own province, which unloaded their burden of carrying a large number of coins on their way home. Therefore, the issuing of *feiqian* was convenient for merchants to do business and thus contributed to the development of the commodity economy. These two systems were inherited and developed during the Northern Song Dynasty (960~1127). In 1048, the eighth year of Qingli Period under the reign of Emperor Renzong (1022~1063 on throne), *yanchao* (盐钞, the certificate for selling salt) was printed and issued. After merchants paid the government with cash for the right of selling salt, merchants could get the certificate for selling salt (*yanchao*) which proved that they had the legitimate right to sell salt. In 1074, the seventh year of Xining Period of Emperor Shenzong (1067~1085), the government ordered merchants to pay the tax of buying tea and then granted them the printed certificate named *chayin* (茶引, the certificate for selling tea) which proved that they had the legitimate right to go to other places to deal in tea. ❶ During the Northern Song Dynasty (960~1127), the government also unified the issuing of the contract of lands and houses which was provided for the folk in their buying and selling of lands and houses. The filled contract paper functioned as the certificate of land trade and tax payment. In order to prevent the local governments to print private contract paper to meet their own private interests, the central government printed the number on the contract paper with the movable type printing. The number on each paper was different and registered in the book. These economic measures led to a huge increase in treasury revenues and also benefited businessmen.

Because of the big development of commodity economy during the Song Dynasty (960~1279), it was no longer convenient for merchants to carry a large number of coins in trade and a kind of currency with small volume, light weight and big face value was needed to replace coins. Inspired by *feiqian* printed during the Tang Dynasty (618~907), a merchant prince in Shichuan Province issued paper-printed exchange certificate called *jiaozi* in 1011, the fourth year of Dazhong Xianfu Period (1008 ~ 1016) under the reign of Emperor Zhenzong (968~1022) in the Northern Song Dynasty. *Jiaozi* in fact was the predecessor of paper currency. In 1023, the first year of Tianshen Period (1023~1032) under the reign of Emperor Renzong (1022~1063), the government established Jiaozi Department to print the currency of *jiaozi* and to administrate the currency affairs in Yizhou City of Sichuan Province, taking the iron coins as reserves. After 1039, the contents on the face of currency were all printed. The government-administrated *jiaozi* issued in 1023 was the origin of paper currency and China was the earliest country to issue paper currency in the world. The use of paper currency was an innovative and revolutionary event in the development of paper currency history, promoting the development of commodity economy and also laying the foundation for the building of capitalist financial system, thereby producing a far-reaching significance.

---

❶　脱脱.宋史:食货志下三[M].上海:上海古籍出版社,1986.

In the process of issuing of *jiaozi*, the government of the Song Dynasty (960～1279) gradually formed a relatively comprehensive bill law or paper currency system including the design and printing of currency faces, issue and circulation, exchange methods, reserve storage and others, which provided the reference and experience for the later various dynasties. For examples, the Jin Dynasty (1115～1234) paralleling with the Song Dynasty (960～1279) followed the latter to issue *jiaochao* (交钞, Jiao paper currency) in 1154 and its circulation period was changed into indefinite circulation. This was a reform in the monetary history; in 1260, the first year of Zhongtong Period under the reign of Emperor Shizu (1215～1294 on throne) of the Yuan Dynasty (1206～1368), *baochao* was issued, with silver as the standard. After the unification of the whole country, the government banned the use of copper and regulated in policies that *baochao* was the only legal currency. The paper currencies used before the Yuan Dynasty (1206～1368) more or less took the role of exchange certificate and was used only in some areas. It was not until the Yuan Dynasty (1206～1368) that *baochao* really threw away its function of exchange. This was a significant event both in China and in the world. Meanwhile, *baochao* of the Yuan Dynasty (1206～1368) was also used in the Korean Peninsula under the reign of the Korean Dynasty (918～1392), the Central Asia under the control of the Arabic Empire (632～1258) and other areas, so it became an international currency similar to today's US dollar.

The currency policies of the Yuan Dynasty (1206～1368) had a more perfect financial system, well-developed institutions, relatively rigorous management and legal system and adequate reserve, so they could reassure the users. At the same time, the stabilization institution was set up to buy or sell gold and silver to maintain money value. All of these measures were later followed by many other countries. The territory of the Yuan Empire (1206～1368) crossed the two continents of Asia and Europe, making the traffic and the cultural exchange between the East and the West enter a new era. In this new age, various inventions of China including paper currency were spread to the West. In the areas west to China, the country which printed the paper currency earliest was Ⅱ-Khanate (1256～1335), one of Mongol Khanates. Its ruler Gaykhatu Khan (1240～1295) ordered to print and issue the paper currency according to the *baochao* system of the Yuan Dynasty (1206～1368) in 1294. On the face of currency, there were three languages of Mongolian, Chinese and Arabic.

In addition, *baochao* issued by the Yuan Dynasty (1206~1368) aroused the attention from the European. In 1253, the King of France Louis Ⅸ sent his Franciscan Rubrouck (Guillaume de Rubrouck) to China. After he returned to France, he recorded information on *baochao* of the Yuan Dynasty (1206~1368) in his work *Itinerarium ad Orientales* (《东游记》, *Notes on Oriental Traveling*) in 1255. Afterwards, the Italian traveler Marco Polo introduced in his travel notes that in the currency-printing work of Baochao Issue Department in Khanbalique, the capital of the Yuan Dynasty (now Beijing), the mulberry paper was used to print rectangular paper currencies with different face values and official seal on them, that those who counterfeited currency would be put to death, that this paper currency circulated all around the country, and that people might use it to buy any items, including gold and silver jewelry and any shop should not refuse the paper money. It also stated that when the foreign trade caravans carried goods to trade in China, *Dahan* (大汗, the great Khan) called 12 experienced and competent professionals to conduct the fair valuation of the goods and then paid them with *baochao* whose face value was equivalent to the value of those goods and a few reasonable profits. If the foreign merchants could not use *baochao* in their own countries, they could use it in China to buy other Chinese goods suitable for the demand of the market in their countries and carry them back. When the paper money suffered damage due to the long-term use, holders could exchange it for new money by paying 3% service charge. Moreover, "All the armies of Emperor were paid with this paper money and they believed that it had the same value as silver and gold. For these reasons, it could be exactly admitted that the command power *Dahan* possessed on treasure was stronger than that of any monarch in the world."❶ Marco Polo's comparatively objective and accurate description of the paper currency system of the Yuan Dynasty (1206~1368) provided a ready-made model for the European countries to implement the paper money system.

During the same period, the merchants of Venice, Genoa and Florence of Italy were keen on trading with China, contributing to the economic prosperity in these areas. In 1340, Florentine Pegolotti (Francesco Balducci Pegolotti, fl. 1305~1365) used three chapters in his work *Practica della Mercantura* (《通商指南》, *Trade Guide*) to introduce how the European merchants trade with China. Specifically, when they set foot on the land of China, they needed to exchange

---

❶　马可·波罗. 马可·波罗游记[M]. 李季，译. 上海：亚东图书馆，1936.

their silver ingots for *baochao* with three types of face values. "*Baochao*, as officially-issued paper money in China, could be used to buy silk goods and other various goods all around China. Its role was the same as that of coin, so it was not necessary to worry that you had to pay more with Bao paper money."❶ Besides, the book also mentioned how many silk brocades could be bought back with the European silver after being exchanged with *baochao*, by which the exchange rate between *baochao* and European coins could be calculated.

During the 14th century, some European merchants, in their use of the paper money in China, discovered that the weight of *baochao* with the face value equivalent to ten Bezant coins was less than that of one Bezant coin. This light and safe new currency could be used to buy any goods in the markets within the territory of China. When the information of paper currency was spread to Europe, it would be followed at the right time. After the 15th century, with the rapid development of commodity economy and trade in Europe, more and more people felt the inconvenience of traditional coins. As a result, the paper currency issued according to the Chinese currency model was officially used in some European countries after the 16th century, exerting a significant influence on the development of banking industry in Europe. ❷ Specifically, in order to meet the demands of the development of business capitalism, the primitive accumulation of capital and the expansion of market, banks were established across Europe, such as Bank of Venice (1580), Bank of Amsterdam (1609), Bank of Hamburg (1619), Bank of England in London (1694), General Bank of Paris (Banque Générale, 1716), each of which became the financial center of their own country. The word "bank" was "banco" in Italian, whose original meaning was "counter" because it originally referred to the counter set by the monetary transactors in the market to receive clients. The English "bank" and the French "banque" all originated from the Italian "banco". Immediately after the establishment of banks, some countries began to print and issue banknotes, such as Sweden (1661), the United States of America (1690), France (1720) and Germany (1806). During the period from the 18th century to the 19th century, the banknotes had pervaded many countries. On the basis of banknotes, the bank and credit systems were

---

❶ 张星烺. 中西交通史料汇编:第 2 册[M]. 北平: 京城印书局,1930.

❷ TEMPLE R. China-Land of Discovery[M]. Wellingborough, UK: Patrick Stephens, 1986.

established, which was an essential step in the development of capitalist economic order. In the process of printing banknotes, the West had absorbed the precious experience which China summarized in her practices of more than six centuries, including face design, anti-fake measures, management, market circulation and other procedures. Meanwhile, the modern national and international financial systems were all established step by step on that basis.

5.3.3 Effect of Printing in Other Fields in the East and the West

In addition to promoting the development of economy, China's printing technology naturally promoted the development of education and the popularization of knowledge in China, the Asian countries surrounding China and the European countries, which further led to the appearance of intellectual generation after generation in these countries. They engaged themselves in research and creative work in the various fields of natural science and humanistic science, published the works of new ideas and achievements or gave new explanations to the ancient works, which in turn caused a series of social effects. However, due to the different social backgrounds and cultural traditions between China and foreign countries, the effects, the manifesting forms and the scope of influence of the printing technology were all different.

In China, the printing technology emerged in the Period of Great Feudal Unity of the Sui and Tang Dynasties (581~907) and got the great development in the Song Dynasty (960~1279). The period from the emergence to the great development of printing technology had seen the mature feudal system, complete cultural relics and regulations, economic prosperity, developed culture and education and a large number of scholars and scientists. In the field of literature, the poems of the Tang Dynasty (618~907) and the Ci Poetry of the Song Dynasty (960~1279) were in the leadership in the literary circle; the prose was approaching to its perfection; the music, fine arts, drama creation were entering the new literary and artistic conception. In fact, feudal China had realized the renaissance ahead in the world during that period. At the same time, the Song Dynasty (960~1279) printed various scientific classics of previous generations and a large number of his new achievements. At that time, China's science and technology with the backbone of Great Four Inventions was in the leading position in the world. Buddhism were also greatly developed, but it was only a religious belief followed by the masses voluntarily, not the ideology the feudal government used to rule the thought of people. Meanwhile, the Buddhist temples were not like the Roman Catholic Church which caused people's grievances because of its peremptory behaviors and greedy money-collecting or heave taxes. China's official philosophy was Confucianism. After the Song Dynasty (960~1279), it began to be philosophized by absorbing the natural scientific achievements and then developed into *Lixue* or Neo-Confucianism, maintaining the feudal rule more effectively. The organic view of nature in Neo-Confucianism had been playing a promotive role in the scientific development. Evidently, the stable and orderly development of science, philosophy, literature and religion in China did not cause any impact on the society, but only led to the academic prosperity of China. China's relatively rich feudal ruling experience equipped the Chinese feudal society with the self-adjustment mechanism and made it possible to absorb those significant inventions to serve the empire. Although the Song Dynasty had the developed commodity economy, its capitalism bud was not enough to become a social force to challenge the feudal system.

On the contrary, the time when the printing technology was spread to Europe was the very time when the feudal system declined and the capitalism developed rapidly. The early bourgeoisie who had accumulated a large amount of wealth was not satisfied with their current social status and thus involved themselves into the vigorous anti-feudal struggle. Eventually, they established their regime. The important pillar of the feudal rule in Europe was the Roman Catholic Church and their churches which placed themselves above the secular regime. In some countries, they even possessed lands and charged exorbitant taxes. The Pope, as the representative of God, had the absolute thought authority and the privilege to explain *The Bible* and doctrines. He ruled people's thought with the medieval Scholasticism and did not tolerate any different voice. Otherwise, he would regard the people with the different voices as "heresy" and imposed persecution on them. The church suppressed the human rights with the divine rights, advocating Obscurantism and Asceticism which asked people to tolerate their sufferings for the sake of their souls to go up to heaven after their death. Evidently, the church and the feudal forces damaged the interests of bourgeoisie and limited its development, so any anti-feudal struggle must first spearhead at the church.

The Renaissance Movement beginning during the 14th century held the ideological banner of humanism which advocated the people-centered idea, praised the value and dignity of human, and required the individuality liberation, freedom and equality. They were the ideological weapon the bourgeoisie used to oppose the feudal shackles and to fight for their own political and economic status. The printing technology made it possible for the humanism works to be widely spread in the society, which led to the formation of the strong public opinion to challenge the authority of the Pope and the church and to shake the foundations of their rule. The anti-feudal struggle, after entering its new active period, was no longer the call of a minority, but the mass movement involving ten millions of people. Humanism ideological trend was reflected in various fields like philosophy, political ideology, literature, art and science. That the authors of each nation insisted on using their own national languages to write was of political significance. The publication of their works helped enhance the national and state awareness of people in each country, develop their national cultures and make their national languages possess the function of philosophical discussion like Greek and Latin. Therefore, it played a promoting role in the formation of national state.

In the anti-feudal and anti-theology struggle, the natural science became a new battlefield. In Europe of the Middle Ages, science was only the dependency to the church. With the development of the printing technology and the increasing of universities during the period from the 14th to the 15th centuries, the medical and scientific knowledges were added to the teaching content. The introduction of ancient Greeks and Arab science books and China's inventions opened the scientific horizon of European and stimulated them to engage in research. Therefore, it was of extreme necessity to break the thought barriers of theology and to oppose it. The publication of Copernicus's work *De Revolutionibus Orbium Coelestium*, *Libri* Ⅵ at Nuremberg in 1543 announced the start of the scientific revolution. This book overthrew with the scientific evidence the Geocentric Theory, which had ruled the European thought circle for 1000 years and been absorbed by the Scholasticism into its religious system. According to the book, the earth was not the center of the universe, but one of common planets rotating around the sun, thereby causing the fundamental change of people's world outlook and shaking the theoretical basis of theology rule. From then on, the natural science had been liberated from theology. The vigorous struggle between science and theology caused the results that Copernicus's book was declared by the Pope as the banned book and that Bruno who propagandized the Heliocentric Theory was burned to death. However, the successors of revolution were limitless. With the efforts of several generations of scientists like Galileo, a series of theoretical breakthroughs had been achieved in the disciplines like astronomy, mathematics, physics, chemistry, anatomy and physiology during the 17th century. The emerging of modern natural science marked by the systematic experiments and the mathematization of natural hypotheses in Europe laid a theoretical basis for the industrial revolution in the 18th century. Besides, under the help of science and technology, the capitalism developed rapidly and the bourgeoisie obtained its regime.

At last, the Religious Reform Movement broke out in Europe in the 16th century, which was the massive revolt movement from the inside of the Catholicism, spearheading directly at the Pope. The peremptory behaviors and heavy taxes of the Vatican provoked the resentment from the secular regimes and the people of each nation. Germany was the nation which was the most tightly controlled by the Vatican, so the persecution it suffered was the heaviest and its resistance was also the most fierce. In 1517, the theology professor Martin Luther nailed his *Ninety-Five Theses* on the gate of the Augustine Church to reveal that the Pope's peddling of a large number of indulgences was in fact the behaviour of deceiving the public and had no legal basis, thereby winding the horn of the reformation. *Ninety-Five Theses* was printed in the form of leaflets and quickly spread throughout Germany and Europe, attracting the support of the public. Luther also published three pamphlets to promote his religious reform principles, to demand Germany to be independent from the control of the Pope, to oppose the Pope's controlling over churches of each country and the church's possessing of land property, to advocate the simplification of the religious rituals and the use of national languages to replace Latin, to deny that the Pope was the representative of God, to promote the believers' direct communication with God without priest's mediation and to refuse the recognition of the church's authority to explain doctrines. Regardless of the persecution from the Pope, Luther established the Protestant church according to his own theory. Later, the Protestant developed in the countries like Germany and France, shrinking the church area controlled by the Vatican. Under the induction of the Religious Reform Movement, the Peasant War broke out in German and gave a heavy blow to the feudal forces and the ruling of the Vatican.

The Renaissance, the scientific revolution and the religious reform were all anti-feudal, anti-theocracy ideological and cultural movements which had laid a foundation for the emerging bourgeoisie to step onto the stage of history, whereas, the printing technology provided for the bourgeoisie means to arouse the response of the public and to cause the powerful public opinion. Once the masses were mobilized, they would take concrete actions. When the bourgeoisie mastered the gunpowder technology, they would have the armed forces to fight against the feudal system and to build the regime that met their own will. Meanwhile, the roar of the cannon sounded the death knell for the European feudal system. During the period from the sixteenth century to the seventeenth century, the Bourgeoisie Revolution in some European countries succeeded in the two fronts of the civil and the military. It was a great revolution with the progressive significance that human beings had never experienced before. Yet the Great Geography Discovery completed with the help of the navigation of compass and the following flourish of maritime business helped the capitalists exploit overseas colonies, world trade and world market, thereby opening a new era of modern world history.

Chapter 6 Influence of Gunpowder and Compass on Development of World Civilization

6.1 The Impact of Gunpowder and Firearms on Development of World Civilization

6.1.1 Gunpowder and Firearms as Revolution in Wars

Throughout human history, the emergence of different weapons was rooted in the social productivity of a certain era. Weapons in the primitive society were mostly made of stones and modest lethality. Progress of social productivity in the slavery society brought into existence bronze weapons like spears, broadswords, halberds, daggers, axes, bows and arrows in the Shang and Zhou Dynasties (商周, 1600 B. C. ～221 B. C.) in China. Sharp in edge and harder than stones, metal weapons were thus more destructive than stone-made ones due to advanced techniques in bronze smelting and casting. Since the Warring States Period (战国时期, 475 B. C. ～221 B. C.) and the Han Dynasty (汉朝, 202 B. C. ～220 A. D.), great progress had been achieved in iron-smelting and steel-making, which resulted in the fact that harder iron and steel replaced bronze, became the major materials for tools and weapons, characterizing the early phase of the feudal society. By the time, China had had many weapons, for example, crossbows. Equipped with an aiming device, crossbows could fire wider in range with less effort to operate compared with the ordinary bows. The trebuchet, also known as *pao* (炮), was a heavy weapon at that time, which could shoot stones on the lever principle. Firearms made of sulfur, oil and other flammable materials were shot together with an arrow at the enemy. The cabin flame-thrower, invented during the Five Dynasties (五代, 907 ～ 979), worked like a pump with the bronze tube filled with oil, unleashing a stronger flame difficult to put out with water.

The feudal European and Arabian regions in the Middle Ages saw an evolution of weapons from stones through bronze to metal ones, so did China. In the eighth and ninth centuries, metal weapons had already become the mainstream ones and heavy trebuchets were widely used. The "Greek fire", which had the flame oil as the main material, was employed by the Byzantine to fight against the Arabs, who later mastered the technology as well. Before the Renaissance, both the West and the East were virtually at the same level in terms of weapons. The knights, whose images are still on display in European museums today, were not only equipped with spears in hand but also heavily armored together with their horses, which would protect them from potential harm, though not quite flexible in moving. Different from the heavily armored knights, the cavalry in China and Arabian regions wore gambesons, easy to move, thus more advantageous on the battlefield.

Before the advent of gunpowder, armies all over the world consisted of the cavalry, infantry and navy. With metal-edged weapons like broadswords and spears, armies engaged in wars in different formations, fighting enemies face to face, in the wild, in cities and on the water. At that time the result of wars hinged on the physical strength and bravery of soldiers together with the military tactics for mechanical damage to enemies. The cabin flame-thrower and the Greek fire were mainly involved in wars on the water by burning enemy warships to incur chaos for soldiers to take advantage of. This tactic was in essence for close-combat and mainly depended on the wind, without which the incendiary agent could not be ignited. To take a fortress with iron fortification, soldiers had to set up aerial ladders before fighting on the city wall, or had to besiege the town for a long time in order to block all of its supplies. In either situation, both the offensive and the defensive in the battle would suffer heavy casualties.

Bows, crossbows or trebuchets were used to shoot the enemy in order to reduce the casualty and avoid face-to-face combats. The maximum shooting range of the crossbow was about $200 \sim 300$ meters on average, which could be extended a bit, but the lethality would be undermined. Trebuchets could shoot stones of $5 \sim 10$ kilograms for 82 meters. [1] If anchored at the foot of the city wall, they were to throw stones on or inside the wall. But several dozens of people were employed to control the rope for this purpose, which was quite strenuous. In addition, it was difficult to transport it because the whole suite, coupled with its frame, weighted considerably with the pole of the trebuchet

---

[1]　吉田光邦. 宋元の军事技术[M]//薮清. 宋元时代の科学技术. 京都：京都大学人科学研究所刊，1967.

being about 3~6 meters long and taking up much space. For an ideal shooting effect, several trebuchets had to work together as a single trebuchet could not hit the target precisely. At that time the Europeans and the Arabs possessed extra-heavy trebuchets which could throw much larger stones, but their cumbersome weight also stood out. In a word, the result of war was determined by the lethality of the military equipment at close distance of both sides, irrespective of the type of battlefield. Both theoretically and practically, the limitations of old-fashioned weapons became increasingly pronounced while the economic cost of and material supplies on them skyrocketed, thus the perennial wars among the medieval countries imposed heavy economic burdens on the people.

The Chinese gunpowder, together with its offspring — firearms, were in essence brand new weapons completely different from what had been used by any other nations. Prior to the birth of gunpowder, metal weapons such as broadswords, spears, halberds or arrows were usually attached to a wooden handle and operated manually. Trebuchets were also operated by hand although they could throw stones mechanically. Therefore, the lethality of the latter depended on manual labor, the only power for these weapons. The manual power being limited, the operation of the trebuchet would require quite a few people in action. Moreover, weapons made of metal, wood and stones could leave nothing more than a physical damage like piercing and colliding.

However, gunpowder as a compound mainly made of saltpeter, sulfur and charcoal, could set off violent chemical reactions when ignited, unleashing considerable amount of gas, heat and chemical energy. The combination of gunpowder with such carriers as metal, wood and stone would become firearms. In addition, gunpowder could be mixed with poison, smoke agent, lime, shards of iron and china, etc., thus firearms were much more powerful as they worked on the chemical energy instead of manual labor. Apart from the mechanical damage and destruction, firearms would wreak chemical havoc, which the previous weapons could not have achieved. The traditional weapons were also termed as "cold weapons" to distinguish from firearms as they could not produce sound, light or heat.

Firearms were advantageous in many ways. Firearms outperformed cold weapons in lethality, destructive effect and the attacking range. For example, a soldier could kill one enemy or one horse with a spear at a time, but he could triple or quadruple the casualty if he threw a grenade. Similarly, the attack of spears or broadswords was limited to humans or horses, unable to get to camps, fortifications or warehouses of the enemy while a bomb or a cannonball could destroy or burn human beings together with buildings. A well armored enemy would be pulverized by a cannon ball one hundred steps away but might evade a crossbow shooting. As far as trebuchets were concerned, they would destroy at most a small part of the city wall while firearms, by contrast, would tear apart a large section of the wall, allowing troops to pour in and occupy the city. In sea battles, the Greek fire or the cabin flame-thrower would work only when they were close to the enemy ships with favorable wind on site. But fire arrows instead would cause incineration to the enemy vessels from a far distance irrespective of the wind, for gunpowder was self-powered, requiring no assistance of the wind.

Grenades, fire arrows, blunderbusses and flame-throwers were generally light and easy to carry. Soldiers could use these firearms either on horseback or on deck. Heavy though, cannons were equipped with wheels and drawn by horses and therefore could be moved to any place, much more convenient than trebuchets. Another advantage of firearms lay in the low cost of production. The labor, money and materials for one suit of armor equaled to the production of grenades that could kill dozens of cavalrymen. In general, firearms generated a military effect stronger than cold weapons at reasonable cost. On the other hand, it was not necessary to abolish all traditional weapons because they could be loaded with gunpowder to become more powerful. For instance, when stones in trebuchets were substituted by gunpowder packs, and the incendiary agent on the arrow with gunpowder, the destructive effect would be enhanced without changing the shape of the weapon.

In view of the above, firearms made from black powder possessed technical advantages unparalleled to previous cold weapons. From 30th century B. C. to 10th century A. D. , development in weapons underwent four stages, namely, stone, bronze, iron and fire and the functions gradually improved, yet the pace was by no means very speedy. Each of the first three-stages lasted one thousand years, with manual labor as power for all weapons. Such developments mirrored the productivity and technology of the primitive society, the slavery society and the early feudal society respectively. The fourth stage witnessed the birth of firearms, and the source of power which was the chemical energy unleashed from gunpowder—a form of natural force generated by the interaction of human labor and work of nature, liberated human beings from the operation of weapons for the first time in history. As a result, anyone, even a child, could quickly set off firearms and discharge immense power by simply igniting the fuse. The fourth stage was fundamentally different from the first three, ushering in a new era for human beings to know more about and harness nature better than ever.

The invention and application of firearms were an epoch-making revolution in the history of man-made weapons as firearms had surpassed previous weapons in lethality, destruction scale and effective range just upon a trigger. Firearms emerged in the tenth century, the heyday of the Chinese feudalism, reflecting the high level of productivity and technology at that time. Ever since gunpowder was invented, new technologies had been applied to it and firearms became the leading arms in battles for nearly one thousand years, gradually replacing old weapons as an irreversible trend in history. The historical significance of gunpowder and firearms lies in that they have completely changed the technology of man-made weapons, starting a new era for chemical-weapons where technological revolutions came up successively.

6.1.2 The Impact of Gunpowder and Firearms on Social Politics and Economy

The historical role of gunpowder and firearms could be further elaborated from the perspective of weapon development as mentioned previously. However, more emphasis ought to be put on the impact of gunpowder on the oriental and occidental cultures, as well as its role in the development of human civilization. Chinese alchemists in the late Tang Dynasty (唐朝, 827~859) already observed the phenomenon of burning and explosion of gunpowder compounds, but unfortunately the Empire Tang collapsed before they were able to harness such natural force and turn it into real arms. The ensuing Five Dynasties and Ten Kingdoms (五代十国, 907~960) saw actually more than ten states on the territory of China then, which hindered the economic development or the enhancement of national power. The economic coordination as a united country then was broken by opposing and warring political regimes, which was definitely retrogression of history. Gunpowder and firearms then played a positive part in the unification of the nation, which was the general trend of history.

During the reign of Emperor Shizong (周世宗, 954~959) in the Later Zhou Period (后周时期, 951~960), some small states were annexed in the hope of a nationwide unification. In charge of a grand army, General Zhao Kuangyin (赵匡胤, 927～976) usurped the Later Zhou regime and established the Northern Song Dynasty (北宋, 960～1127), claiming himself Emperor Taizu (宋太祖, 960～976). It was in his time that firearms began to develop, which secured the Song Dynasty an ever-victorious position in wars against countries that had not yet possessed firearms. From 960～979, the new-born Northern Song Regime, equipped with firearms, swept all of the southern states and unified half of China, leaving the Liao Regime (辽, 916～1125) in the north as the only barrier for a complete reunification. The Liao Regime was founded by an ethnic minority named *Qidan* (契丹), and was then underdeveloped socially and economically. The leaders of the Song Dynasty would have annihilated the Liao Regime if they had adopted a national policy other than "valuing civil administration above military strength" and practiced a strategy of defensive comprise to the Liao Regime with further troubles in the north. Nevertheless, the economy, culture, science and technology began to prosper in the most developed areas of China during the 166 years of the Northern Song Dynasty, which resulted in a full-scale advancement of papermaking, type printing, gunpowder and the compass, exerting a profound influence on the other countries in the coming dynasties.

The 12th and 13th century witnessed the coexistence of three regimes in China, namely, the Song by the Han people, the Jin by the Jurchens and the Yuan by the Mongolians. With the spread of gunpowder, people of these three regimes, i. e. , the Han people, the Jurchens and the Mongolians battled against one other with firearms, all aiming to unify China. Finally the Mongolians established the Yuan Dynasty (元朝, 1271～1368) and since then China has been reunified. Obviously, military power backed by firearms played an important role in terminating separation apart from other factors. It is widely known that there were repeated separatist movements in the Ming (明朝, 1368～1644) and Qing Dynasties (清朝, 1636～1911) and it was because of firearms that none of them had succeeded militarily. While firearms assisted the ruling in cementing their sovereignty, they were also employed by the ruled to shake and even overthrow the feudal governments as well, thus a powerful means for different social classes or groups for political, military and economic aspirations. However, in ancient China firearms had never been used for any religious purposes, which were essentially different from the Western world.

Wars among the Song, the Jin and the Yuan Dynasties were in essence for their political purposes to take control of the entire China. The Song Dynasty, an economically and culturally developed country rich in raw materials, became the target of the other two, which forced the Song Regime to prioritize the advancement of firearms and innovation of technology. Later both the Jin Regime and the Yuan Regime mastered these technologies and applied them to tripartite wars. Consequently, the arms race gave birth to several technological updates on one side or two, which were soon adopted by the others, upgrading the general standard of Chinese firearms to an unrivalled level at that time. With the unification of China, the Yuan Dynasty utilized all previous technologies in firearms to fight beyond the east and west of China, enlarging China as an unprecedented giant empire.

Technology in firearms soared in the Ming Dynasty, which was established after the overthrow of the Yuan Dynasty. Firearms continued to contribute more to the coming changes of political structures of the world. The development of gunpowder and firearms brought about a series of inventions and changes both in China and the world, leaving a long-lasting historical impact.

In the 12th and 13th century, a new sector of industry emerged in China, which specialized in gunpowder and firearms production and was responsible for supply of equipment to a million-size standing army. Due to its involvement in the affairs of national defense, weapon manufacturing was under the direct control of the central or local governments, with no private business being permitted in this regard. A huge number of craftsmen and technicians were hired since various kinds of raw materials and processing procedures were required in the arsenals, powder magazines and weapon depots. Consequently, the mass production of firearms stimulated the whole economy and development of relevant technologies. As a pillar industry, firearms production was highly valued by the government just like shipbuilding, and received strong support in such aspects as human resources, finance and material supplies. Ever since the Song Dynasty, rulers of different times attached great importance to firearms, deeming them the expertise of China, as they knew that firearms were the material guarantee for the fighting capacity and thus the consolidation of their reigns.

Gunpowder in the Northern Song Dynasty, however, contained little nitrate and was made in paste-like form. Wrapped in non-metals, it was intended for incendiary or explosive purposes rather than for launching an arrow. Although more advanced than cold weapons, gunpowder then failed to function fully in these firearms. The dawn of the Southern Song Dynasty (early 12th century) saw the invention of granular gunpowder rich in nitrate, which greatly enhanced its launching power and explosiveness. The new form of gunpowder gave rise to stronger firearms as another technological revolution. Later in the Yuan Dynasty and the Ming Dynasty, more novel technologies followed suit, transforming firearms into a whole new look.

In conclusion, compared with papermaking as well as type-printing, gunpowder had led to very frequent technological revolutions for renewals with ever-emerging product types at shorter intervals for updates, making it the most innovative sector in the Middle Ages. This was true across the globe when gunpowder and firearms were transmitted to the rest of the world. With endless magics, gunpowder brought about changes to the world as well as to itself as an ever-changing item. It had such a trait that had inspired generations of scholars, technicians and craftsmen to innovate and create ceaselessly.

Between the Song and the mid-Ming Dynasties (12th century to 15th century), China pioneered in making and using a series of multifunctional firearms with different types, including bombs, grenades, metal blunderbusses, flamethrowers, counteractive rockets, multiple rocket launchers, multiple-stage rockets, rocket bombs, reciprocating rockets, cannons, landmines, torpedoes, time bombs, signal flare and smoke grenades. Originally such firearms were used in battles together with cold weapons, but the use of firearms increased while that of cold weapons gradually declined as military technology advanced. In the Ming Dynasty, firearms took up 50% of all weapons whereas in the Qing Dynasty, over 60%. As a result, the traditional army formation — the cavalry, infantry and navy, was expanded to include Special Forces responsible for firearms operation. At first these Special Forces were mixed with other troops, fighting as the vanguard (Figure 177), but later it became an independent arm of services with its soldier population growing, known as *Paoshou Jun* (炮手军, Artillery Troops), *Shenji Ying* (神机营, Magic-arms Troops) and *Huoqi Ying* (火器营, Firearm Troops) in the Yuan, Ming and Qing Dynasties respectively, which mainly consisted of the artillery and rocket troops. Then the strategic role of cavalry was gradually replaced by firearms troops, which was a major change in army formations.

As a result of weapon revolution, the increasing use of gunpowder and firearms in wars dramatically strengthened the fighting capacity of armies and

Figure 177　The layout of firearm troops
by Kang Diqian, General of the Early Ming
Dynasty in 17th Century, fighting against
troops of the Qing Dynasty, from
Records of Founders of Qing Dynasty

enhanced the combat effectiveness of war. There were quite a few historical cases in which troops with firearms defeated their enemies who had much more soldiers without firearms. The mode of war was also altered by gunpowder and firearms. At times when cold weapons were the only available fighting appliances, the mode of war was face-to-face battles, i.e., soldiers to soldiers and generals to generals. It was in essence a mode of physically-driven war because the result of battles relied on human strength and skills of weapon maneuvering, in which intensive lineups rushed to the enemies from all directions for a close combat as strategic assistance. But after the advent of firearms there hardly was such a scene. As rockets, blunderbusses and cannons could be launched or shot from a distance with fire or explosion to the enemies, the elite troops and military facilities could be destroyed before any face-to-face contact started, which meant that the offensive might launch an intensive shooting to their enemy to destroy the army and fortification, and then charge in a small scattered formation because an intensive formation in this situation would be an easy target for the other side. Meanwhile assaults were followed by the infantry march into the depth, supported by light firearms such as flamethrowers, bombs, grenades, blunderbusses, guns, together with cold weapons.

On the part of the defensive, the army formation should be scattered and fortification be built according to the landform with the firearms troops arranged at the forefront for blocking actions. Meanwhile cities and battleships should be equipped with heavy firearms as protection. At the time, soldiers were required to improve their skills and gain expertise in the property and usage of gunpowder and firearms. The changes in war modes also put an end to the traditional modes of commanding in military operations. Previously commanders had to combat on site, but with firearms what they had to do was to stay at the rear area and give orders according to the situations.

In the era of cold weapons, military experts regarded it as a bad scheme to take a city by force as it would cause a significant toll. When they really had to, they would like to besiege, dig out a tunnel or employ some other tactics. But the advent of firearms not only made cities vulnerable but facilitated bombing a city, which was seen repeatedly in history. Heavy armors collapsed at the first blow in face of firearms and was transformed to hard helmets and joint gambesons.

Chinese gunpowder and firearms exerted a similar and even greater social influence on Europe than on China when they were spread to other countries by the Mongolian troops. In 1206, risen as a big power in Mobei (漠北, generally referring to the northern part of the Mongolian Plateau in China), the Mongolians immediately moved southward. In 1234 they annihilated the Jin territory and got the vast land to the north of the Yellow River. Then they took over the local arsenals and integrated the firearms makers of the Han ethnicity into their camp and established the Paoshou Army, assimilating the Song's new technologies to fight back. During 1235~1244 and 1253~1258, in order to gain more land, materials and wealth, the Mongolian leaders subjugated all countries in the Central Asia by launching two grand conquests westward, with various kinds of firearms and hundreds of thousand to several millions of army population. In 1237 the troops of Mongke (蒙哥, 1209～1259) arrived at Kiptchak (钦察, now Russia) and conquered Moscow with cannons. In 1238 Batu (拔都, 1209～1256), captured Kiev (基辅, the present capital of Ukrainian) and the army of Haitu (海图, 1235～1301) approached Bohemia (波西米亚, now Czech Republic). In 1241, the Mongolians fought Poland-Germany allies in the grassland near Legnica (莱格尼察, now Poland). For the first time, the head-to-toe armored European cavalries suffered a debacle by rockets, multiple rocket launchers, flamethrowers and metal blunderbusses.

The Mongolian armies swept its enemies all the way through to the central Europe, with the strategic advantage of firearms and the mobility of Fast Riders. The Christian world would have prostrated itself to the Mongol Khan if the military operations did not halt due to the death of Emperor Ogedei (窝阔台, 1186~1241). In 1251, troops were sent westward again when Mongke (蒙

哥，January 10，1209 ～ August 11，1259）succeeded the King of the Yuan Dynasty. Later in 1258 Hülegü Khan（旭列兀，1217～1265）occupied Baghdad，the capital of the Abbasids Dynasty（750～1258），terminating the Arabian Empire at its prime. The Mongolians set up the Golden Horde Khanate（钦察汗国，1243～1480）in the conquered area with Sarai（萨莱，now Astrakhan in Russia）as its capital，the Ilkhanate（伊利汗国，1260～1353）with Tabriz（大布里土，now Iran）as capital，and the Chagatai Khanate（察合台汗国，1225～1242）with Alimali（阿力麻里，now Xinjiang）as the capital.

After the Mongolian Westward Conquests，Kublai（忽必烈，1215～1294）became the Khan（1260～1294），moving its capital to Yanjing（燕京，now Beijing）in 1263，and the country was thus renamed Yuan in 1271. He then annihilated the Southern Song Dynasty and unified China in 1279，taking control of the khanates in the west. The extensive land of the Yuan Empire stretched from the East Asia，through the Middle Asia，the West Asia to the vast land of the East Europe and even close to the West Europe at some point，making it the largest empire in human history.

The Mongolian armies wreaked havoc along its way，but later the khanates were devoted to economic recovery and construction for the consolidation of their reigns. For instance，Hülegü Khan transferred groups of craftsmen，technicians and scholars from inland China to his khanate，where they built water conservancy projects，established cities，temples and academic institutes. Meanwhile，the Mongolians also extended its territory to other directions. Farther to the East it occupied the Korean peninsula，turning it into a vassal state of China. It further launched wars against Japan. Southward it combated Vietnam and Indonesia. The key reason that the Mongolians could implement such large-scale operations lies in that they possessed unique firearms technologies and had a strong economic backup. These crucial incidents had a profound social impact on the world in the late Middle Ages.

The instant consequence of the Mongolian Westward March was the get-through of the Asia-Euro business route，the ancient Silk Road，which once has been blocked. A safe corridor was built for direct exchanges of economy，materials，information and personnel，the Beijing-Europe section of which was under the absolute control of the Yuan Empire. The entire route was dotted with courier stations and guarded by soldiers，protecting travelers and providing convenience，which stimulated trade and economic communication，scientific and cultural exchanges between the East and the West. It was during this period that Chinese gunpowder and firearms, papermaking, compass and other technological inventions were disseminated to Europe via the Silk Road or the Maritime Silk Road.

The Mongolian Westward March led to the dissemination of firearms, which further led to the collapse of feudalism in Europe. Chinese firearms were used in European wars in the 13th century. The anti-feudalism citizens in Europe obtained powerful arms to fight against the feudal lords. It turned out that the castles, heavily-armored cavalrymen and cold weapons in Europe were very fragile in the face of firearms.

Between the 14th and the 16th century, some cities in Italy, France, Germany, the United Kingdom and the Netherlands saw the birth of the capitalist relations of production as a result of the commercial economy. But feudalism stood as a barrier to the further development of the new relation of production. The rising bourgeoisies, such as owners of factories and money houses and merchants, felt that their interests were not guaranteed because they possessed no political status even though they own some financial wealth. They demanded that the domestic markets be opened but were fettered in many aspects by feudalism. As to peasants, they bore even more hatred to the feudal lords who were enslaving and oppressing them. Feudal separations and the division of the country incurred dissatisfaction and a groundswell of opinion for national unification. The new aristocrats, including the Emperors themselves who contended for reform drew a clear line with the old nobles; therefore the political struggles between Emperors and the Pope went on, forming an irresistible trend to abolish the old order of the society. To achieve their goals, the bourgeoisies, with cannons made of metal resources at hand, hit the castles which were regarded as the stronghold of the feudal lords by taking advantages of these social conflicts. The first roaring of cannons in Europe in the 14th century tolled the death knell for the castles together with the feudal power behind.

The internal instability of the European aristocratic feudalism was just like an archipelago — segmented by city-states of different sizes, which made it difficult to form a strong unity, far less steady than the centralized Chinese feudal system. In the times of cold weapons, the feudal lords could well maintain their rules, relying on stone-built castles, heavily armored cavalries and slave-driven Mediterranean battleships. But all these were pulverized in the Firearms Era (Figure 178). Therefore, the application of gunpowder brought revolutionary changes to the European society, and finally led to the collapse of feudalism and the establishment of capitalism in the 16th and 17th century. The social productivity, economic construction and the whole societal presence were reformed later, ushering in a brand-new historic period.

Figure 178　A town — attack with cannons in one of European
wall paintings in 1537 from Needham (1986)

In the second half of the 14th century, firearms technology achieved much progress in Germany, France and Italy. The Chinese front-loading blunderbusses were transformed into breech-loading artillery, also known as the Frankish culverin in the Ming Dynasty. The artillery was anchored on a cannon rack with wheels, ready for field battles and sieges of cities. The 15th century witnessed the invention of the muzzle-loading match-lock and the small-sized version of the breech-loading artillery used for charging. Bombs, grenades and rockets were also created at that time which was used together with cold weapons in wars. Cannons were highly valued at that time because of their augmented firepower and greater range of attack. Same as in China, firearms also led to military revolutions in Europe. Taking France as an example, it set up the first professional artillery in Europe in 1450 and built fortifications, digging trenches at back and placing the covered cannons at the forefront. Charles Ⅶ (1422~1461), the French Emperor then, reclaimed by means of the artillery the northwest Normandy region from the UK, occupying all castles in 1450 with 5 castles per month. ❶ The French strategy of developing firearms was soon imitated and implemented in other countries (Figure 179) as well.

---

❶　OMAN C W C. A History of the Art of War in Middle Ages[M]. New York: Cornell University Press, 1953.

Figure 179   Drawings of trenches dug in the 17th-century European wars with
firearms (Hogg,1980): Gustavus Adopphus (1594~1632),the Swedish King,
inspecting the frontline, with artillery, artillery shells, muskets and powder kegs
in front and soldiers digging trenches to the right

William Shakespeare (1564 ~ 1616), the British dramatist, once complained that one tenth of the armored cavalrymen and crossbowmen had been killed by the loathsome saltpeter on the Lincoln grassland. It is true that battles involving firearms were more brutal than those with cold weapons, yet the feudalism would not be replaced by the capitalism if the cavalry were not wiped out and the castles not destroyed. Such replacement was certainly a progress of society, which was conducive to the enhancement of social productivity, the development of science and technology and the creation of modern civilization. Yet human beings certainly had to pay for it. Some European countries employed firearms for colonization and wealth plunder in Africa, America and Asia, as well as for the establishment of maritime hegemony after they have developed capitalism. They financed the industrial revolution once the capital accumulation was completed, which resulted in a prosperous economy in the capitalist world.

However gunpowder was not always heinous for it was also used for good reasons in addition to application in wars. For example, the Chinese made fireworks from gunpowder for entertainment on festivals. They also applied it to mining, tunneling and road building for economic purposes, which later became a common practice across the world (Figure 180). In Europe it was recorded that the earliest mining with gunpowder took place in 1403 in Italy❶,

---

❶   PARTINGTON J R. A History of Greek Fire and Gunpowder[M]. Cambridge: Heffer& Sons, Ltd. , 1960.

Figure 180   the European embossment
depicting mining with gunpowder
in the 18th century from Needham (1986)

where the local people were renowned for making fireworks. Soon after, gunpowder played a part in such fields as mining of gold, silver and iron, tunneling, canal excavation, railroad construction and traffic road building. Had there not been such innovations, the modern civilization would not have existed. In this sense, gunpowder itself was a form of productivity. Moreover, cannons and rockets were also used for artificial hail suppression, beneficial to agriculture. In China, gunpowder was also used as one of medicine ingredients by herbalists.

6.1.3   The Impact of Gunpowder and Firearms on Modern Science and Technology

Gunpowder and firearms exerted the same influence on science and technology as on society, politics, economy and military affairs. Not only did gunpowder and firearms remove the medieval feudal economy and politics that stood in the way of capitalism, but they erased the rotten medieval ideological system and catalyzed new science and technology. John Mayow, a British chemist in the 17th century once wrote: "Nitre has made as much noise in philosophy as it has in war."❶ For one thing, many researches were required on gunpowder and firearms as they were brand new items devoted by the Chinese, of which the rest of the people in the world had not any knowledge. For another, there were a lot more problems to deal with in the production and applications of firearms, from shooting, flight, target-hitting of cannonballs to explosion. As a result, new subjects and disciplines were founded and new technologies developed in the process of solving such problems.

In fact, gunpowder has been the research interest of the ancient Chinese alchemists, herbalists and technicians. They had accumulated much technological experience in the production, ingredient mixing, and explosion of gunpowder and firearms shooting, which were documented in various military

---

❶   BERNAL J D. Science in History[M]. London: Watts & Co., 1954.

works. Their findings were collected at the cost of many lives and most of them fit into modern scientific theories. For example, the proportion of the saltpeter should be more than that of sulfur and charcoal, and all of the three should be unadulterated. The naming method based on flame reaction and the substance was instructive too, which in fact involved dissolution and crystallization in physics and analysis and testing in chemistry. The casting of cannonballs entailed metallurgy, metal processing and mechanics. In the 14th and 15th century, there had been enough knowledge and scientific documents for further research.

The production and application of gunpowder and firearms were a comprehensive, multi-professional section that involved knowledge of different categories. Either in China or in other countries, it has facilitated the development of technical knowledge and production, enhanced social productivity, economic prosperity and strengthened national defense, which demonstrated that the destructive firearms could also be applied constructively. Apart from technical experience in the field of firearms production, the ancient Chinese also made attempts in theorizing it with the Chinese modes of thinking, the most typical example of which was the theory of combustion and explosion, detailed before in this book. Based on the traditional thinking of *yin-yang* (阴阳) and *wuxing* (五行), the theory proposed explanations for explosion without referring to any superpower. But as time went by a new theory was called for since the ancient theory of combustion and explosion was limited by the scientific standards and turned out to be flawed in some aspects.

Gunpowder and firearms went to Europe during the Renaissance period, and the social and military effect of which was recognized immediately whereas the scientific and technological effect was not realized until sometime later. Humanities, astronomy and mechanics enjoyed a rapid development at that time while chemistry still lagged behind. Though not popular among the public any more, the four-element-theory proposed by Aristotle, which was similar to the Chinese Five-Element thinking, had not yet been replaced by a new one since the European alchemists were not able to to propose new ones. The Europeans were capable of making gunpowder but could not interpret explosion with what they had known about chemistry and physics. They were aware that saltpeter constituted the key part of gunpowder and it was fire that caused the explosion but they were puzzled about the differences between the common fire and that of gunpowder which did not involve air. It has been proved by modern

European chemists that gunpowder is combustible in the vacuum. But why does gunpowder bear such a magical trait and what is the role of saltpeter? It is apparent that the spread of the Chinese gunpowder and firearms to the West posed great theoretical challenges to the European scientific community, to which they had to offer their answers.

It was not an easy task for the European chemists after the Renaissance to answer these questions. They spent four centuries on it, among which nearly 200 years (the 17th century and the 18th century) were on the issue of combustion, which had been the leading research interest in chemistry. Some speculated that saltpeter provided air or air itself contained spiritus nitro-aereus, others assumed that there was phlogiston in gunpowder. The experiments and academic debates continued for a long time, or "the philosophical noise caused by saltpeter and gunpowder" in Mayow's words, ended up with the discovery of oxygen in the second half of the 18th century, which was a revolution in chemistry, bringing it to the modern era thereafter. Since then people got to know that saltpeter provided oxygen for gunpowder as an oxidant, and that it was the high-pressured and heated gas produced abruptly in the limited space that triggered explosion. Thus the misconceptions were eventually dispelled, which had dominated for hundreds of years in the Middle Ages.

The academic commotion caused by gunpowder manifested not only in chemistry but in physics, especially in mechanics. The development of cannons and rockets sparked the interest in the motion of cannonballs or ballistics (Figure 181), thus stimulating the research on kinetics. Science in the ancient

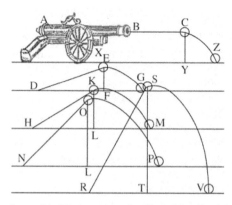

Figure 181　Illustrations of ballistics on projectile orbits fired from different angles of elevation in the European scientific works (1606) (Bernal, 1954)

times mainly dealt with static objects and the stress direction of which tended to be linear on most occasions. Now the situation called for the research on projectile motion as the combination of launching force and gravity was non-linear in most cases, therefore there was a demand for a new, detailed discipline of mechanics based on such motion at the same time. On the other hand, it also required courage to get rid of the shackles of the Aristotelian stereotypes for new science as the Aristotelian theories deified by the scholasticism in the Middle Ages had hampered the development of new mechanics.

Cannonballs "strike" the old society of Europe heavily, and the old scientific thinking even more. According to the Aristotelian opinions, the projectile would rise diagonally and then fall down perpendicularly. Aristotelian scholars never adopted the concept of resultant force in their research on projectile motion, rather they held that there must be a time sequence for forces to work effects[1], and that the motion trajectory would form an acute angle. But in reality, gunners observed a curved trajectory after they ignited the gunpowder and shot the cannonball, which was contradictory to Aristotelian predictions, Niccolo Tartaglia (1500~1557) from Italy clarified it in his works on strategies, gunpowder and shooting tactics that launching force and gravity would exert a joint effect on the entire motion of the cannonball as "there is always a force dragging the cannonball away from its normal route" and the trajectory would be curved. He also proposed an empirical law that the farthest shooting would be reached when the barrel tilted at an angle of 45 degrees, and it would not reach the limit at either more or less than 45 degrees.

In 1638, Galileo Galilei (1564~1642) published *Discorsi e Dimonstrazioni Mate-matiche Intorn a due Nuove Scienze Attenenti alla Mecanica ed i Movimenti Locali*) in Italian, in which he proved after strict experiments and accurate mathematical deduction that the trajectory of the cannonball was parabolic curve, which was determined by the resultant force of the launching force and gravity. He also proved that an elevation angle of 45 degrees would form the greatest resultant force for the farthest shooting range. His theory provided a framework for Tataglia's speculations which were based on his experience as a gunner in China. Galileo elevated experiential speculations to scientific laws, casting away the Aristotelian theories. Developed by the Dutch scholar Christian Huygens (1629 ~ 1695) and integrated by Newton (1642 ~ 1727),

---

[1]   MASON S F. A History of the Sciences[M]. New York: Collier Books, 1962.

mechanics as a discipline became a mature field of science, known as classical mechanics. The major difference between classical mechanics and traditional mechanics lies in that the former is of quantitative nature because it is based on math and rigorous experimentation and is expressed in mathematical terms while the latter stays at the qualitative phase either in ancient China or Europe in Leonardo da Vinci period (1642~1727). The classical mechanics dominated the scientific world until Albert Einstein proposed the Theory of Relativity 300 years later.

Gunpowder and firearms were not only scientific but also technological in nature. Technicians gradually learned that it was the heated air generated inside the cannon that produced energy to throw the cannonball far away. Such principle would help transfer gunpowder from military use to non-military applications, for example, the creation of engines. In this way, gunpowder became the source of energy for civil machines, which inspired generations of technicians in the Renaissance to invent engines, which endowed gunpowder with more peaceful and significant application in addition to engineering blasting. Contemporary French scholar A. Varagnac listed seven categories of energy discovered and exploited by human beings, the first three being fire, agriculture and metal—all ancient forms of energy, while the rest four being gunpowder, steam, electricity and atomic energy — energy in its modern sense. Joseph Needham in *The Gunpowder Epic* elaborated how gunpowder as the fourth energy discovered and harnessed by men played its role in the invention of the heat engine, which was the crucial contribution that gunpowder has made in the revolution of technology.

From the perspective of mechanic engineering, cylinders in steam engines and metal bores are in essence the same while pistons and piston rods could be regarded as cannonballs with handle. It is not technically difficult to transform the latter into the former because both function on the same principle as the cannon. The attempt to invent a gunpowder-driven engine dated back to the da Vinci Period, who himself harbored the plan to make an engine that took gunpowder as its fuel, cylinder as its bore and piston as its cannonball. In 1673 Huygens designed a machine (Figure 182-A) that was fueled by the swelling force of gunpowder to lift weights. ❶ Theoretically feasible as they were,

---

❶ WOLF A. A History of Science, Technology and Philosophy in the 16th and 17th Centuries[M]. London: Allen & Unwin Ltd. , 1935.

gunpowder-fueled engines were difficult to control. Denis Papin (1647~1712) a French technician, adapted the Huygens machine, replacing gunpowder with steam which would produce vacuum after condensation. He succeeded in inventing the first steam engine with pistons (Figure 182-B), which could get water from the mine.

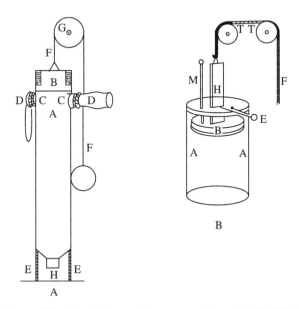

Figure 182   Early forms of power machines (Wolf, 1935)
Figure A was designed by Huygens with gunpowder as power.
A. Cylinder   B. Piston   C. Exhaust pipe   E. Cylinder   F. Rope   G. Pulley   H. Powder chamber
Figure B was designed by Papin with steam as power.
A. Cylinder   B. Piston   C. Latch   F. Rope   H. Piston rod   T. Piston

The steam engine by Papin was further modified in the early 18th century and became a kind of practical power machine in industry. Since then steam engines began to be applied in the textile industry and other sectors, giving rise to steam boats, trains and eventually the Industrial Revolution which has changed Europe and the world. The steam engine was in all sense an international creation that had its historical origin in both the Eastern and Western cultures, so was the internal combustion engine. After the mid-19th century, it occurred to some people that steam as the source of power for the engine could be replaced by Greek fire in Byzantine or the oil flame-thrower cabin in China, thus the internal combustion engine came into being, followed by cars, planes, locomotives, boats, tractors and dynamos that have once again revolutionized the world.

Looking back at the early history of the steam engine and the internal combustion engine, it is apparent that the Chinese metal blunderbuss was actually a type of engine applied in wars, i. e. , the early form of one-cylinder steam engine and internal combustion engine fueled by gunpowder. Cannons are effective in converting thermal energy into mechanic one. The spreading of gunpowder and cannons from China to Europe inspired the Europeans to make civil engines that could work like the cannons. Initially their idea was limited to traditional cannons fueled by gunpowder, yet they found that gunpowder unleashed too much thermal energy, which was difficult to control. Though trial and error, they found that it was possible to make steam engine and internal combustion engine with steam or oil as a substitute for gunpowder. Meanwhile technology in cannon casting and bore spinning ensured the delivery of precision cylinders, which enabled the thermal engine to work smoothly. Quite a few European scholars have admitted that the steam engine was the direct offspring of Chinese cannons and Western pumps, thus a hybrid of the oriental and occidental technology. John Desmond Bernal (1901 ~ 1971), a British scholar, once wrote as follows:

> The steam-engine has a very mixed origin; its material parents might be said to be the cannon and the pump. The awareness of the latent energy of gunpowder persistently suggested that uses other than war might be found for it, and when gunpowder proved intractable, there was a natural tendency to use the less violent agents of fire and steam. ❶

He then continued as follows:

> A new and important connection between science and war appeared at the breakdown of the Middle Ages with the introduction or discovery of gunpowder, itself a product of the half-technical, half-scientific study of salt mixtures ... In their physical aspect the phenomena of explosion led to the study of expansion of gases and thus to the steam-engine, though this was suggested even more directly by the idea of harnessing the terrific force that was seen to drive the ball out of the cannon to the less violent function of doing useful civil work.

Giovanni Vacca (1872~1933), an Italian Sinologist, spoke at a seminar on

---

❶ BERNAL J D. The Social Function of Science[M]. London: Routledge, 1939.

the origin of science in contemporary Europe held in Rome in 1946, to the effect that firearms motivated the invention for machines powered by steam expansion. And the increasing knowledge of the explosion mechanics of machines like firearms led to countless trials of the power it produced, including everything from Papin's earliest efforts to the fabrication of internal combustion engines in recent times. ❶

Lynn White, a British historian, commented in 1962 that cannons were actually the forefather of all modern engines since they were important not only as the weapons in wars but as machines of single cylinder internal combustion. For example, Da Vinci tried to substitute cannonballs with pistons, taking gunpowder as fuels. The patent of Sir Samuel Morland (1625~1695) in 1661, the experimental piston engine by Huygens in 1673 and the air pump by Parisians in 1674 all originated from the cannon on the same principle and gunpowder was not replaced by liquid fuels until the 19th century. He remarked on this issue again in 1977 that the development of internal combustion engines was hampered because inventors assumed that cannons were the earliest type of mono-cylinder internal combustion engines. They did not realize the nature of the chemical explosive mixture from China until the mid-19th century, and found out that cannons were not suitable to provide power for further development like an engine.

The failure of gunpowder-fueled engines gave birth to the steam engine and the internal combustion engine some time later. The following sections will focus on rockets since cannons have been detailed in previous chapters.

Rockets differ from cannons in their fabrication and working principle. Based on the reaction principle, rockets are the aircraft which is launched by the pushing force of high-pressured, high-speed, heated air generated in the combustion capsule. The self-loaded fuels and oxidants make it possible for the rocket to work not only in the air but in the vacuum. The rocket consists of an engine that is loaded with fuels, capable of carrying other stuff and a stabilizer. Rockets today and those in the ancient times share the same components and working principle, though drastically different from each other in performance, fuels and structural complexity. The modern rockets are also equipped with control systems. Undoubtedly, modern rockets have evolved from the ancient ones, the process of which has been elaborated by relevant

---

❶   VACCA G. Perché non si é sviluppata la scienza in Cina[M]. Roma: Partenia, 1946.

studies in history. ❶

Rockets today are powerful strategic weapons in national defense, employed as carriers to launch various kinds of missiles. They are also used for man-made satellites, manned spaceships and space shuttles to go to the outer space, serving as the cutting-edge medium for human beings to conquer the nature and get to know the cosmos. Multi-functional satellites are playing an increasingly important role in military, economic and scientific fields. By virtue of rockets, mankind has entered a new era of interplanetary voyage: extending human activities for the first time to the space outside the Earth atmosphere, setting foot on other celestial bodies, and making the long-drawn scientific fantasy come true to fly to the cosmos. Space voyages own business potentials to transport more people to space stations and other planets and rockets are indispensable if human beings would have to migrate to other planets provided the Sun condenses some day and the Earth is no longer suitable to support life. The rocket science has also facilitated the development of emerging subjects and novel technologies, leading to tremendous changes.

The technology of rocket was originated from China, which has dramatically changed human life and science. In the early 12th century, people in the Southern Song Dynasty made fireworks with solid gunpowder on the principle of rocket. Later in 1161, with military rockets the army of the Song Dynasty defeated the Jin troops, which had much more soldiers in a battle on the waters. The early development of rocket technology was achieved in China too. For example, multiple rockets launchers were invented in the 13th century in order to enhance the combat effectiveness of individual rocket soldiers with intensive fire. The multiple rocket launchers, also called Huolong rockets (火龙箭) at that time, were held together with many rockets connected with a fuse. Operated by a soldier, the launcher would release many rockets simultaneously upon ignition. The Mongolian armies used such rockets in their westward military operations and later the rocket technology was transmitted to Europe in the 14th century. What is now known as the Huolong rockets or the Chinese dragon belching-fire (中国龙喷火箭) in Polish should be the prototype of the Katyusha Rocket by the Soviet army against the Germans in World War Ⅱ.

In the 14th century, technicians in the Ming Dynasty invented the multiple-

❶ VON BRAUN W, ORDWAY F I. History of Rocketry and Space Travel[M]. London-New York: Crowell Co. , 1966.

stage rocket and applied it to battles on the water, trying to extend the shooting range of rockets. Taking the two-stage rocket for example, it was designed for the first and second stage of rockets to be bound together with a fuse. Upon ignition of the first stage the rocket was sent into the air flying for a certain distance before the second stage was ignited automatically to continue the flight. There being enough gunpowder inside, the rockets would fly $1 \sim 1.5$ kilometers over the water. ❶Joseph once observed that "This was a cardinal invention, foreshadowing the Apolla space-craft, and the exploration of the extraterrestrial universe."❷In addition, Wan Hoo (万虎, $1450 \sim 1500$) in the Ming Dynasty made an aircraft with 47 big rockets in 1500, implementing the first manned flying test with it. ❸ His heroic trial has been revered in the international aerospace field.

Both having been introduced to Europe, rockets lagged behind cannons in terms of its progress. Rockets were first used in a battle in Italy in 1380. Although two-stage and three-stage rockets were noted in European books in the 16th and 17th centuries, application of them was not recorded in real warfare. Before the 18th century, the popularity of cannons left rockets unknown. The British and the French did not attach importance to the development of the rocket until their joint army was attacked severely by it when they invaded India. Some European countries set up the rocket camp in the 19th century, independent from the artillery. But the practical development of the rocket began in the 20th century, when the flight of the rocket was demonstrated from the perspective of physics and mathematics. In 1926, Americans successfully launched a new kind of rocket with liquid fuels instead of gunpowder. In the 1950s and 1960s, the Soviet Union and the United States managed to launch manmade satellites with multiple-stage rockets, turning space flight into reality for the space era. In the 1960s, China, the hometown of rockets, also succeeded in launching modern rockets and realized the Chinese manned space project in the early 21st century.

❶   潘吉星. 中国火箭技术史稿[M].北京:科学出版社,1987.

❷   NEEDHAM J, et al. Science and Civilization in China[M]. Cambridge: Cambridge University Press, 1986.

❸   ZIM H S. Rockets and Jets[M]. New York: Harcourt Brace & Co. , 1945.

6.2　The Impact of Compass on Development of World Civilization

6.2.1　Revolutions in Nautical Technology and Great Geographical Discovery

It was a preliminary need in ancient times for people around the world to identify directions in such activities as city construction, house building, district planning, mining, combating, travelling and mapping etc., which could not be achieved without a clear sense of direction, especially for scientific research. Different nations have unanimously defined four primal directions at first, namely the east, west, south and north before the other directions in between were oriented. The method for measuring directions varied with the level of social productivity and scientific and technological development, reflecting the knowledge of the nature. The improved measurements of direction exerted an influence on social economy, military affairs, politics, science and culture for greater convenience. That is why special government officials were assigned in ancient China at the national level for measuring directions which was equally important to observing astronomical phenomena.

It was recorded that the kings of ancient China sometimes presided in measuring directions themselves. For example, *Shi Jing* (《诗经》, *The Book of Poetry*) recorded that Duke Dan of the Western Zhou Dynasty erected a wooden pole on the hills of *Bin* (豳, now Xunyi in Shaanxi Province) to set the standard for four primal directions during his reign (17th century B.C. ~16th century B. C.). *Shu Jing* (《书经》, *The Book of History*) written in the Warring States Period also documented that "Emperor Yu divided China into nine provinces by wading across mountains and cutting trees to indicate directions, and dredged rivers and stipulated tributes according to regional yields."❶ Only by clarifying boundaries could economic and political disputes be avoided, and directions of boundaries must be figured out for these purposes. That is why it was one of the major state events to measure directions, which was true for any other countries as well. Boundaries defined, markers were to be erected and maps made to indicate ownership. City planning and house building would not be implemented

---

❶　尚书正义:禹贡[M]//十三经注疏:上册.上海:世界书局,1935.

until sites were chosen and directions gauged. Archaeological evidences show that the relics of the royal palace of the Shang Dynasty in *Anyang* (安阳, a city in Henan province of today) were constructed on an orderly foundation, clearly based on a plan after direction measurements.

Before the compass came into being, directions were oriented by the sunlight shadow with a *guibiao* (圭表, gnomon, an ancient Chinese sundial) which was improved in the Qin Dynasty (秦朝, 221 ~ 206 B. C) and Han Dynasty (汉朝, 206 ~ 220 B. C.), making direction orientation much more convenient. Meanwhile people managed to find directions by observing the position of the Pole Star. These activities were documented as "referring to the sunlight shadow in daytime and observing the Pole Star at night". In the Song Dynasty, *Qianxingshu* (牵星术, the Astronomy Navigation Technology) was invented which could figure out directions with the navigational board. Others like Europeans, the Arabs, Africans and Latin Americans had similar methods as well. Directions could be oriented in such a way because the solar movements continue regularly, easy to be observed. Likewise, the Pole Star is the only static one in the North Constellation in a diurnal motion, thus easy to observe with human eyes. Even though the position of the Pole Star did move a bit, it was virtually unnoticeable.

Different nations oriented directions by observing the same natural phenomenon with similar principle, though their tools varied. The ancient method for direction orientation was still in use long after the birth of the compass even in the 19th and 20th centuries. For instance, the gnomon is still in use by the secluded aboriginals on an Indonesian island[1], which was the common method in the ancient world for quite a long time before it was discarded. Therefore, it turns out that such methods were effective in precise orientation, especially after they were modified. It also explains why these methods were still in use for a while after the compass was invented, same as the fact that bamboo slips coexisted with paper for a period. But the ancient method for location was eventually abolished because of its inconvenience and limitations which were magnified by the compass.

The obvious limitation of the ancient locating technology lies in its dependence upon remote, extraterrestrial celestial bodies, namely the Sun and

---

[1]   NEEDHAM J, WANG L. Science and Civilization in China[M]. Cambridge: Cambridge University Press, 1959.

the Pole Star, yet the observability of such celestial bodies was affected by the self-rotation of the Earth and the atmospheric changes. The sunlight shadow was only observable in daytime and the Pole Star at night, which meant that either method was applicable for each half of 24 hours at most. People felt quite helpless with the only option and had to wait for a proper opportunity to come when the weather was bad without the Sun or stars in sight. Events would be suspended if it happened on land but when sailors came across bad weather at night, there were potential risks for them to deviate from the normal course, involving unexpected perils.

In view of the reasons above, ships of the Eastern and the Western hemispheres in the ancient times had to sail near the coast with limited days and distance. They had to pull into the shore in the evening and in bad weather, or slow down under the guidance of the beacon. Ancient sailing areas therefore were restricted to the waters close to the Asian, African and the European continents, which meant that all human beings were confined to one half of the land area on the Earth. Although the other half was also inhabited by human beings, communication with its counterparts was rare due to the lack of navigation technology. Astronomical navigation then could not ensure far-reaching voyages, neither could the shipbuilding technology guarantee long sailings. Much of the continents remained undiscovered due to lack of geographical knowledge in ancient times, and in addition, maps at that time could not provide accurate information, leaving blanks in some areas.

The compass invented by the Chinese was completely different from those navigational devices that had been in use. It was designed with a magnet-made gadget with fixed directions. The needle would point to the north and the south at all times with free rotation in the magnetic field, the advantage of which lies in that it works with the force of the geomagnetic field instead of the celestial bodies. Free from the influence of self-rotation and atmospheric changes of the Earth itself, the compass could work for 24 hours in the geomagnetic field irrespective of the weather conditions. In addition to its sensibility to the geomagnetic field, another advantage of the compass was its portability for it could be used on land, on the water or in the air. It was easy to read the direction with help of the dial on the face of the compass. For example, sailors only had to turn on the light for proper orientation at night.

The birth of the compass was an epoch-making revolution in the technology of orientation, dignifying mankind as the real master of directions because it liberated humans from the reliance on celestial bodies and weather conditions for orient directions.

The earlier compass consisted of a manmade magnetic needle and a dial. The needle was placed floating in the water at the heart of the dial, and it was also called "the floating needle" or "the water compass" in some slangs. The water would be drained and the needle put in a magnetic box when temporarily out of use. It was modified in the Southern Song Dynasty, i. e. , the needle was placed on a pivot at the heart of the dial, thus named "dry needle" or "dry compass" as there was no water any more. The ancient geomancers had much research on magnetic for a long time and began to use it originally in the geomancy of tombs and houses before the magnetic traits were applied to navigation. Some other nations in ancient times also noticed the magnetic property in certain metals but they failed to invent the compass. It was then invented in China because the Chinese people not only noticed the magnetic property to attract iron, but also the polarity property to the poles, which was utilized for direction orientation by means of the magnetic needle that resulted in the birth of the compass.

Navigation by compass is the prerequisite for an ocean-going voyage and it is indispensable to have vessels with gigantic sailing endurance as well. China had some fundamental inventions in the design and building of ships that were introduced to other countries later, making critical contributions therein. First, a flexible manipulating device was invented to operate a huge hull and support immense weight and change directions according to the situation at any time. In the Han Dynasty (the $1^{st} \sim 2^{nd}$ centuries), an axial rudder as high as several meters was installed at the stern, capable of swinging left-and-right, and of elevating and descending. In the Song Dynasty (the 11th century) the balanced rudder was invented which made it easier to manipulate the ship and quicker to change directions. Moreover, forceful sculling oars were installed at the prow and on the flanks of Chinese ships as propellers, which were modified as paddle wheels in the Song Dynasty. From the third century on, the Chinese have set up multiple masts on their ships to prop up the rectangular sail made of bamboo slivers, which could be raised and lowered like a shutter to the wind. Finally, the bottom space of the ship was separated by watertight compartments. One compartment being broken, others remained intact, preventing the ship from sinking at one time.

It was a revolution in the history of seamanship to have a separate compass room in the advanced Chinese ships in addition to the above-mentioned inventions in the sailing. The revolution took place in the Song Dynasty in China when the second millennium just began, which brought the giant Song fleet to the east and central Africa and Mozambique in southern Africa more than three hundred years earlier than the Portuguese explorer Vasco da Gama (1469~1524). The Song's voyage marked the curtain-up of the geographical discoveries though on a reverse route compared with the Europeans. The continent of Australia was first discovered by a Chinese, whose landing point was Darwin Harbor today. China had taken the lead in the world for a long time as far as the ocean-going technology was concerned, as Alan Villiers from the United States once commented that "the Chinese are the best among all Asian sailors, and their boats the most fabulous as the technologies they had achieved hundreds of years ago were not seen in European ships until contemporary times, including the watertight compartments to separate the broken hull, the balanced rudder easy for manipulation and the rectangular extendable sail."

It was recorded in both Chinese and Arabian history that the huge vessels of the Song Dynasty were 360 tons in weight, 99 meters in length, 30 meters in height and 35 meters in width. Apart from the cargo, the ship could carry 1,000 people, among whom 400 were convoy soldiers with firearms and 600 sailors. Ships were loaded with food enough for all crew members for one year. The propeller consisted of 8~20 sculls, each being operated by 4~30 people. The ship had a stern rudder of 16.5 meters high and there were many masts, some of which were as high as 33 meters in addition to as many as 50 pieces of bamboo sails. The iron anchor was several hundred kilograms in weight and each of the nails could be as long as 15 centimeters. On board there were public and private rooms, bathrooms and toilets, and even the space for rearing pigs, growing vegetables and brewing. More than a dozen of watertight compartments were installed at the bottom of the hull. At the stern there were a rudder room and a compass room where the helmsmen performed their duties. At that time, the huge vessel had to anchor on the nearby waters and sent three smaller boats to unload the cargo on the wharf because it was impossible to lie at anchor in the harbors of the Persian Gulf. Zheng He (郑和,1371~1433) in the Ming dynasty led seven great ocean-going voyages with massive fleets from 1405 to 1433, the size and scale of which far surpassed those of the Song Dynasty. At its peak, the fleet by Zheng He was composed of 62 boats with 27,000 people and immense commercial goods. The biggest ship was 151.8 meters long and 61.6 meters wide, the deck of which was larger than a football square. Being the largest ship worldwide at that time, it was equipped with 9 masts and 12 sails, weighing 1,500~2,000 tons, only 40 meters shorter but 39 meters wider than Minnesota, an American business ship in 1905. ❶ A five-mast freighter of the Ming Dynasty is rendered below (Figure 183-B) for the imagination of Zheng's ships.

Figure 183   the ocean fleet led by Zheng He during 1405~1433

Figure A: the ocean fleet led by Zheng He (Jin Qiupeng, 1985)

Figure B: the five-mast cargo ship in the North *Zhili* District of the Ming Dynasty, (drawn by Landstorm), the same type as the integrated treasure ship but smaller, (Needham, 1971)

As Villiers once said, the giant ocean-going ships with great endurance and advanced equipments did not appeared in Europe until contemporary times. Joseph echoed this opinion and even corrected Villiers's words "among all Asian sailors" to "among sailors of the world", for he regarded the medieval China as the most developed country in ocean-going technology in the world. In medieval Europe, the multiple-paddle boats propelled by men could only sail in the Mediterranean Sea, and would be overthrown in the Atlantic Ocean. The

---

❶   范中义，王振华. 郑和下西洋[M]. 北京：海洋出版社，1982.

rudderless mono-mast sailing boat would completely lose control of itself in the ocean, thus unable to sail a long distance. Only after the introduction and application of the stern axial rudder, the multiple masts of rectangular sails, the watertight compartment together with the compass did Europeans master the technology to explore the heart of the ocean.

6.2.2 The Impact of Compass and Nautical Technologies on Social Politics and Economy

After the Song Dynasty, Chinese ships equipped with compasses went to the Japanese Sea, the East China Sea, the South China Sea, the Bay of Bengal, the Arabian Sea, the Red Sea and the Gulf of Mozambique visiting over thirty countries in Asia, Africa, Europe and Oceania. Trade relations were established between China and these countries, facilitating the export of silk, china, paper, lacquerwares, bronzewares, ironwares, medicinal plants, patterned cloth, drums, sugar, cinnabar and other commodities as well as the import of spices, ivory, gold and silver, cotton, sappan wood, sulfur, tortoise shells, beeswax, jewels, parrots and fruits, which not only expanded business but enhanced the exchanges of goods and enriched geographical knowledge. Lots of Chinese merchants went to other countries to set up stores, settling down as the first generation of overseas Chinese and contributing to the development of local industry and business. A trade market dominated by the Asians and Africans was set up between the western Pacific Ocean and the Indian Ocean and fostered the civil communication between Asian and African countries, taking up half of the total business volume in the world.

The development of foreign trade increased the percentage of maritime income in government revenues as a new source of fiscal income. Yet the massive outflow of copper coins resulted in the devaluation of the paper currency as a negative impact, but extensively fostered the development of industries like porcelain manufacturing, silk, shipbuilding, sugar-making, bronze-and-iron etc., both in size and in techniques. The bountiful import of some resources like sulfur, beeswax and sappan wood replenished the raw materials required for industries of gunpowder, firearms, dyeing and sewing at home. The commercial economy was stimulated and progressed remarkably in some coastal and interior cities, as evidenced by the rising of wealthy businessmen who sold foreign goods or were engaged in the import and export trade. The development of industry and business, coupled with the prosperity of cities, made it possible for factory owners to employ more workers, and capitalism in embryo began to emerge in China, which could be dated back at least to the Song Dynasty. ❶

To foster foreign trade and expand markets, the Song government assigned ambassadors and trade delegations to Southeast Asia, India, Arab, Africa, Japan and Korea one after another, cementing friendly ties between governments and extending political influence. Meanwhile, tributary ambassadors were well treated in China, and foreign businessmen provided with favorable trade conditions. In such cities as Guangzhou and Quanzhou there were *fanfang* (藩坊, residential areas tailored for foreign merchants). With many Chinese businessmen having settled down in other countries, opportunities were created for sino-foreign exchanges of science and technology. It was against this backdrop that the compass and shipbuilding technology were disseminated to Arab and further to Europe while medicinal plants and spices from Arab and Africa became popular commodities in China too, enriching Chinese herbaceous studies, thus positively affecting the incense-burning traditions of Chinese people in religious or sacrificial ceremonies. However, the overseas trades and the market expansion did not lead to fundamental changes to the society as the Chinese feudalism in the Song Dynasty still remained powerful even though it was at its sunset.

European vessels could only sail on the nearby waters of the coast before the 13th century as the compass and technologies in shipbuilding and navigation were not yet introduced from China. The installation of the stern rudder, multiple masts, paddles and the compass facilitated them to set sail for ocean-going voyages, which triggered dramatic changes in Europe completely different from those in China, and affected the course of the world history. Since the 15th century capitalism has further developed in some European countries. Consequently, factory owners and businessmen not only contended for abolishing the feudal system for domestic markets, but demanded that foreign markets be opened and new channels of resources be established. Yet the trade between Europe and the East was monopolized by the Arabs and the Turkish who founded the Ottoman Empire, selling Chinese and Indian goods to Europeans at a price several times higher than their purchasing prices. The Europeans therefore were eager to rid the Arabian merchants by finding a direct

---

❶　陈高华,吴泰.宋元时期的海外贸易[M].天津:天津人民出版社,1981.

route to China.

After the 15th century gold became the only means of payment either within Europe or between Europe and Asia. But Europe did not possess much gold and silver yet witnessed a large outflow. The Italian traveler Marco Polo (1254 ~ 1324) depicted the prosperity and wealth of China, India and other Asian countries in his travel journals (1299), which casted an impression that there seems to be gold and jewelry everywhere in the East. Inspired, the greedy upper-class Europeans were eager to go eastward seeking gold. In the meantime, with a firm belief in the theory on the spherical earth of the Greek geographer Ptolemy (85~165), Paolo dal Pozzo Toscanelli (1397~1482), an Italian astronomer, inferred that sailing westward from Europe could make it to Asia and he even drew a world map for the voyage, on which he put India at the opposite coast of the Atlantic Ocean. Apparently he did not know that there was the American continent between Europe and Asia. All of these inspired explorers for adventures. Some coastal countries like Portugal and Spain which were then disadvantaged economically, were especially enthusiastic about such adventures, deeming it a good chance for a windfall.

The Portuguese were the first to implement ocean-going voyages. Bartholomeu Diaz (1450~1500), for example, departed from Lisbon in 1486, sailing southward along the west coast of Africa to the southernmost point of the continent, i. e. , the Cape of Good Hope of today, covering a journey of 10,000 kilometers. He intended to proceed, believing that the discovery of the Cape could bring him hope to get to India. Unfortunately, the exploration terminated because of the fatigue of the crew members, but they took back the maps along the voyage. In 1498 Vasco da Gama, another Portuguese, completed the trip based on Diaz's experience to India via the Cape of Good Hope. Since then, Portugal has got control of the Golden Coast of Africa and established colonies therein, plundering expensive spices, gold sand, ivory, jewelry and silk from Indian, and expelling Arabian merchants. What is noticeable is that the ships of the European "explorers" then were not only equipped with compasses, but they carried with them cannons and light firearms learnt from the Chinese. They barbarically massacred and looted the local people who had no firearms.

Almost at the same time when Diaz found the new marine route to Asia, Christopher Columbus (1451~1506), an Italian, influenced by the theories of Toscanelli believed that sailing westward all the way he could arrive in China taking a shortcut. He failed to persuade the kings of Portugal, the United Kingdom and France to finance his voyage but succeeded in Spain. In 1492, he arrived at the Bahama Islands in North America with three clippers, and discovered Cuba and Hatti. Later he reached Jamaica and the north shore of South America. The new continent Columbus found then became the territory of Spain. At his deathbed Columbus still believed these were the eastern part of Asia, regarding the local inhabitants as Indians. It was Amerigo Vespucci (1454 ~ 1512) from Florence who, by investigating the northeast coast of South America, confirmed that the land was not India but a new continent and he drew new maps of it. The new continent was thus named America after him.

After that, Ferdinand Magellan (1480 ~ 1521), a Portuguese adventurer, believed that he would reach Asia with the compass by sailing southwest from Europe and then bypassing the southernmost point of South America. He went to Spain after failure to win endorsement in his own country. In 1519 Magellan started his first voyage around the world with 5 sailing boats and 265 crew members. They first arrived at the east coast of South America and then went further south. In 1520 they crossed the southernmost strait of South America, which was the channel connecting the two oceans, i.e., the Strait of Magellan of today. They continued sailing westward before his sailors saw a calm and tranquil area of waters soon, and named it the Pacific Ocean. Traversing the Pacific Ocean, they then reached the Philippine Islands in 1521. Magellan attempted to colonize these islands, which resulted in confrontations with the local tribes that claimed his life. His fellows, crossing the Indian Ocean and bypassing the Cape of Good Hope, continued the voyage and finally returned to Spain in 1522 with only one boat and 18 crew members. It turned out to be the first successful voyage around the earth by human beings. ❶

The great geographical discovery helped open routes to get to new continents. European bourgeoisies, merchants and declining aristocrats, backed by Kings of their countries, poured into Africa, America and Asia, colonizing these places and looting valuables atrociously for their primitive accumulation of capital. As pioneers in this regard, Portugal and Spain claimed some of American regions as their territories, which is noted as the first colonial

---

❶　See Magidovich J P. 世界探险史:第三部分[M]. 屈端,云海,译. 北京:世界知识出版社,1988.

partition in human history. ❶ The colonizers plundered gold and silver, enslaved the local people for mining and planting, and shipped gold and other valuables back to their home countries. Portugal and Spain turned themselves from poor countries into mushroom millionaires, which enticed others, such as the Netherlands, the United Kingdom and France, into the war of colony partition, resulting in the defeat of the two trailblazers and in their being surpassed by their successors. Since then, the vast land of Asia, Africa and America was turned into the battlefield for big powers of the Western Europe, into which the massive wealth has flown ceaselessly in turn.

Consequently, the transfer of the business sea lanes in Europe and business center from the Mediterranean coast to the Atlantic coast led to the loss of importance of Italian commercial cities and the rise of some Atlantic coastal cities. The plunder stimulated the industrial and commercial development of some colonial countries, and the economic pivot of the Western Europe shifted from Italy to emerging countries like the United Kingdom, France and the Netherlands. Spain and Portugal, which were originally underdeveloped, possessed large quantities of gold, spices, jewels and agricultural produces from their colonies and sold them to other countries. Gradually Antwerp in the Netherland was turned into a trade center in this regard in the mid-16th century. Antwerp, with a dense population of businessmen from different countries, saw the founding of exchanges with capitalist features. Banks, financial and credit industries and trade companies sprang up soon, expanding business both in size and scope. After Antwerp, Amsterdam and London witnessed the establishment of larger exchanges with more mature financial systems.

Capitalism had a chance for further development when the European feudalism was about to collapse. The bourgeoisie who had harvested huge profits from colonial trade began to invest in such industries as textile, shipbuilding, cannon making, mining and metallurgy, machine building, papermaking, printing etc. Factories grew increasingly larger with more specific division of labor. As a result, factory owners encouraged scientific research for competitive products while businessmen tried to develop overseas trade. The discovery of the new ocean routes opened up the global market and the world became an organic one with the isolated world then connected by major sea lanes.

---

❶ 朱寰.世界中古史[M].长春:吉林人民出版社,1981.

The geographical discoveries brought about by compasses and new navigation technology (together with firearms) had their political and economic consequences manifested immediately with the founding of overseas colonies. Some small countries along the Atlantic coast provided the rising bourgeoisie with novel stages, extending their political and economic control to regions in Africa, America and Asia, which were many times larger than their own territories. These colonizers accumulated their primitive capital by exploiting resources, wealth and human labors of their colonies while the self-sufficient economy of the colonies was broken by the occidental colonizers and was included into the European commodity economy in a world market, therefore the look of the world was changed and the new order of politics and economy was established in the world.

Geographical discoveries also triggered the so-called "price revolution". In Europe there soon appeared to be a big price drop accompanied with the skyrocketing price of agricultural produces and daily commodities on the contrary due to oversupply of precious metals, such as gold and silver, exploited from enslaved America Indians and Africans. In this way the bourgeoisie benefited most while the feudal lords living on rents of land were harmed, and workers on wages became more attached to factories and farm owners. In all, the discovery of new sea lanes and the American continent accelerated the disintegration of feudalism and the growth of capitalism in the Western Europe, overtaking Asia, Africa and America. It was definitely a progress for capitalism to replace feudalism, and it was also true that the twirl of the compass needle has brought about tremendous changes to human beings and societies.

6.2.3 The Impact of Compass on Development of Modern Science and Technology

The compass has affected the development of modern science and technology in many ways, not only in a short-term fashion demonstrated above but in the following generations. First of all, the compass brought about the revolution in navigation technology as elaborated above, which triggered the geographical discoveries, leaving direct impact on geography and mapping science. People had long been observing celestial bodies but possessed a rudimentary knowledge of the mother Earth in many aspects in the Middle Ages. Inhabitants on the East and West hemisphere hardly knew anything about each other. For example, there was a *Fusang* Country ("扶桑国", mythological place in Chinese legends) recorded in ancient Chinese books, which might refer to the American continent, yet Europeans were totally unaware of it. The ancient people at best knew half of the earth geographically as they were confined by the surrounding oceans. Only by resorting to ocean voyages could mankind broaden their horizon and know more about the world they live in.

Navigation by means of compass made it possible for humans to explore oceans more in the 14th and 15th centuries. The Chinese and the Europeans set sails in different directions and turned new pages for ocean voyages successively, but the Europeans sailed farther. Geographical knowledge and information increased exponentially during this period with more continents, islands and oceans being discovered one after another. Moreover, Magellan's voyage proved that the earth was spherical and for the first time people got to know that oceans in the world were connected, seeing the sketch of the five continents relatively clearly. Knowledge in geography was upgrading rapidly with new world maps modified constantly to replace previous ones. It was estimated that the amount of new geographical knowledge outnumbered the total of the previous millenniums, ready for a new look of the geography discipline, overthrowing the ancient and medieval geographical conceptions.

Maps were urgently needed for repeated voyages so that positions of the ships could be located in relation with signs in the area, which involved the measurement of the longitude and latitude of different places astronomically and the detection of positions on land of the spherical earth on one planar paper. Gemma Frisius (1508 ~ 1555) in the Netherland first published the method of triangulation and Gerhardus Mercator (1512~1594), a student of Gemma Frisius, drew a world map tailored for ocean voyages in 1569 on the projection and statistics of the longitude and latitude. On the map he divided the longitude lines into parallels with equal distance, latitude ones into parallels perpendicular to longitudes. The distance between two latitude lines increases as the latitude leans to the Polar Regions, thus both longitudinal and latitudinal radians grow at the same rate. Mapping of the sea route was greatly simplified by the Mercator map, with which ships would sail along a fixed linear direction instead of going a roundabout way. His method was a significant development of quantitative cartography, solving the problem of drawing a spherical object on a planar map, from which maps of today have evolved.

The discovery of the American continent resulted in the introduction of corns, peanuts, tomatoes, sweet potatoes and tobacco to other continents, enriching the categories of agricultural produces and enhancing the agricultural development therein. Discoveries of rare animals, plants and rock specimen supplied naturalists with new research subjects, for example, the mass collection of specimen of non-European plants contributed to plant taxonomy. Medicinal herbs growing in America fleshed out pharmacy, for instance, the barks of Cinchona succirubra Pav, native to South America, turned out to be effective to cure malaria after processing, thus highly valued around the world.

As far as the compass is concerned, the Chinese philosophy in design of the device exerted a profound influence on the history of science and technology. The compass was made of a round dial with scales of directions and a rotatable needle, which automatically points to a certain scale after movements. When it was introduced to the West, scientists of other countries invented a series of new scientific apparatus, imitating the compass. In fact, some earlier apparatuses were made after the compass in the post-Renaissance era, though working principles may vary. Up until today, moving needles on a round dial are as popular as the compass in laboratories, factory workshops, cabins of planes, ships and cars. Therefore, the compass could be regarded as the ancestor of all apparatuses with round scales and needles.

Contemporary science and technology to a large extent was dependent on the invention and application of scientific apparatuses of various kinds. Different in functions though, these devices are designed for precise quantitative measurements. With these devices scientists could enhance some observations with eyes or detect some objects or phenomenon that could not be sensed by human beings. They could also draw more reliable conclusions by controlling variables and conditions. It is hard to imagine that modern science could have progressed so much without the assistance of scientific apparatuses. As said by Joseph as follows:

> The magnetic compass is the first and the oldest among all devices with scales and needles that played so important a part in contemporary scientific observations.

As a matter of fact, the compass is the earliest scientific apparatus with a dial, an automatic rotating needle and readings. *Rigui* (日晷, the sundial) was one of the earliest scientific apparatuses with a needle and a scale but it was not as delicate as the compass because its needle was fixed. *Huntianyi* (浑天仪, Armillary Sphere) also has a movable sighting tube and a scale, but it was operated manually, thus hardly comparable to the compass either.

Manufacturing industries of compass or other delicate devices were born to meet the increasing demand for compass and other kinds of navigation equipments, the practitioners of the latter had a huge impact on science. Many scientists were actually practitioners in this field, for example, Robert Norman, author of *New Attractive* in 1581, used to be a sailor and compass maker in the United Kingdom.

Finally, the interest in compass risen in philosophy was no less than that in gunpowder. Europe remained silent since Peregrinus de Maricourt, a French, experimented on the magnet around 1269 after the Chinese scientist Shen Kuo (沈括, 1031~1095, a Chinese scientist and statesman in the Northern Song Dynasty), but research on the magnetics became popular again as the compass began to be widely used in the geographical discoveries of the 15th century. For example, in 1492 Columbus noticed during his voyage to America that the compass needle sometimes leaned to the astronomic meridians instead of pointing to the due north, which the Chinese had already discovered some 600 years earlier. After Columbus, sailors across the world also measured and recorded the declination of the compass needle and obtained lots of relevant data. They found out that the declination would change with the geographical position and time, which was also dealt with in *New Attractive* by Robert Norman, marking a fundamental progress in magnetic research.

The efforts of William Gilbert (1544~1603), a British scholar, led the magnetic subject into a new stage at the end of the 15th century. He published his findings in 1600 in *De Magnete, Mageticisque Corporibus, et de Magno Magnete Tellure, Physiologia Nova*, or *De Magnete* in short. Gilbert first made a spherical magnet named terrella (Figure 184-1) as the experimental object determining the magnetic poles. Then he made a bar magnet (Figure 184-2) and cut it into two pieces. He inferred that it was the Earth that kept the needle pointing to a certain direction when he noticed that the polarity of each piece stayed unchanged. The iron needle always stayed parallel with the magnet surface when placed at any position on the magnetic equator but it became perpendicular to the surface (Figure 184-3) when placed over the poles. These experiments convinced Gilbert that the Earth itself was a gigantic spherical magnet, and he concluded that the declination at the North Pole was larger than that in London based on the movement of the needle around the magnetic poles.

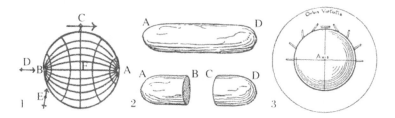

Figure 184   In 1600, Gilbert testing the response of magnetic poles and small magnets
to spherical magnets, (Wolf, 1935)
1. A spherical lodestone with a pointing needle   2. Two types of natural slender magnets
3. Responses of small magnetic needle to the Mini-Earth

When it comes to the cause of the declination, Gilbert thought it was the irregularities of the Earth surface, which, however, was proved wrong by later documents. He proposed that the magnetic power spread from one magnet to all directions, so did the Earth. He further speculated that other celestial bodies including the Sun and the Moon shared the same magnetic features with the Earth. His idea had a big impact on Kepler, who applied it to explain planetary motion. Gilbert's book on magnetic was of great significance, as Bernal wrote as follows:

> De Magnete is a great book in itself, and as an exposition of the new scientific attitude. Gilbert did not confine himself to experiments: he drew from them new general ideas. The one that struck most at the imagination of his time was his idea that it was the magnetic virtue of attraction that held the planets in their courses. It provided the first physically plausible and completely non-mythical explanation of the ordering of the heavens. It certainly made it easier for Newton in his argument against the physically-minded scientists who could conceive force only by the impulsion of material bodies in contact.

The revolution of natural science began with astronomy. The book *De Revolutionibus Orbium Coelestium*, *Libri* VI published by Nicolaus Copernicus (1473~1543), a Polish astronomer in 1543 was the pioneer of this revolution, in which he proposed the heliocentric theory which pitched itself against the dominant geocentric theory in the Middle Ages. The Earth was deprived of the central position of the cosmos, shaking the outlook of the Church and feudalists. Bruno and Galilei were persecuted for promoting the heliocentric theory, bit it did not stop the spreading of truth. Soon Johannes Kepler (1571~ 1630), a German astronomer, published *Prodromus Dissertationum Cosmographicarum Continents Mysterium Cosmographicum* in 1596, in which he defended and developed the theory of Copernicus based on his own observations and documents long accumulated by Tycho Brahe (1546 ~ 1601), a Danish astronomical observer together with another book titled *Harmonics mundi* in 1619. Kepler explained issues not yet clarified by Copernicus and concluded three laws on planetary motion by providing demonstrations in view of astro-mechanics.

Magnetism also played a part in Kepler's explanation of the shape of the planetary orbits. Largely under the inspiration of Gilbert's researches, Kepler supposed that moved like a huge magnet whose axis kept its direction in space unaltered during the revolution of the planet. The two poles were alternately presented to the Sun, which attracted one and repelled the other. Hence the whole planet was alternately attracted to and repelled from the Sun, and the radius vector thus underwent those fluctuations in length which characterize the ellipse orbit. Kepler also speculated on the nature of gravity, which he regarded as "a mutual affection between cognate bodies tending towards union or conjunction, similar in kind to magnetism. That is, gravitation resembles the magnetic force." Therefore Kepler believed that there was a universal force, working effectively between the planets and on the Earth, which might well be the magnetic force.

The contribution by Kepler laid a foundation for the comprehensive theory by Isaac Newton (1642 ~ 1727). In the era of Newton, research on astro-mechanics had been quite fruitful, yet further explanation for planetary motion was called for, especially the elaboration on why the planets orbited the Sun in a closed curve rather than a linear track leading to the outer space. Newton's contribution was the generalization of mechanics on the Earth to the solar system. Same as Kepler's laws, Newton concluded three laws of motion in his *Philosophiae Naturalis Principia Mathematica* in 1687, proving that these three laws also worked in the solar system.

Newton interpreted the oval rotation of the planets around the Sun with the theory of universal gravitation, and clarified that the gravity of the Earth was a form of the universal gravitation. The magnetic force proposed by Kepler became the universal gravitation in Newton's term. Such conversion was no difficulty in terms of philosophy. The system of mechanics by Newton, or the classical mechanics, has become the cornerstone of modern natural science, in which a pedigree from Shen Kuo through Maricourt and Gilbert to Kepler and finally to Newton could be traced clearly. Undoubtedly, the ancestors of the pedigree were the Chinese who invented the compass. In a word, Magnetics has greatly influenced mechanics and astronomy so forth.

Epilogue

It has been over a year and a half since the University of Science and Technology of China Press telephoned me and asked me to take charge of the translation of the book of *The Four Great Inventions of Ancient China*: *Their Origin*, *Development*, *Spread and Influence in the World* by Professor Pan Jixing, which is a national translation project intended to be published abroad and for the overseas readers. I know what it means to me: challenge, opportunity and responsibility.

The translation is really a challenge, a painstaking but rewarding work. The original book was based on nearly 30 years of investigations on the newest archaeological materials and textual research of sino-foreign sources and was a systematic and profound study on the origin, early development and outward spread of China's four great inventions of papermaking, printing, gunpowder and compass. It also discussed the impact of the four great inventions on the development of world civilization. In addition, the book researched the scientific principles and the manufacture processes of these inventions and the new technical restoration of some relevant products. It involves the professional knowledge of various fields of language, literature, history, education, religion, mathematics, physics, astronomy, medicine, economy, military affairs, etc. at home and abroad. Because of the fervent love for the translation and cross-cultural communication, we have overcome the obstacles one by one through the team work.

It is an opportunity, an opportunity to cooperate, to explore, to learn and to gain, though it means hard work. The translation team searched for reference, translating, revising and proofreading. We jokingly called this translation "Star Farming" (its pronunciation is similar to *Sida-faming* "四大发明", the shorted name of the book in Chinese).

Somehow, translating is like farming, taking effort and bearing fruit. For the translation of this book, it is like farming in an alien planet, for the book is imbued with Old Chinese which has grown unfamiliar to us, and also with terminologies of various fields. Therefore, we started to learn how to farm in a new planet. And we made it. And now, we, the farmers, are truly gratified to see the fruit of our endeavor ripen. We smile to ourselves and hope the readers can enjoy the fruit grown by us and smile to us.

Thanks to translation, communication is possible between those who speak different languages in varied cultures. In the past centuries transoceanic ships have taken the four Chinese inventions to every corner of the world, which have long been accepted. Nowadays globalization has brought people closer than ever before with more population learning foreign languages and use of Internet. However translation is still the primary and effective means of intercultural communication for mutual understanding across cultures more empathically. It is hoped that the English version of *The Four Great Inventions of Ancient China: Their Origin, Development, Spread and Influence in the World* will acquaint people with more details about papermaking, printing, gunpowder and compass from the perspective of the Chinese scholar in the multicultural context for better understanding of the Chinese culture, the Chinese people and the Chinese wisdom.

This translation work means responsibility. We deeply know that only responsible translators can achieve functional equivalence of the source text and the target text. Carelessness and irresponsibility will result in the misunderstanding and mistranslation of the original text. As the head of the translation team of this national project, I feel very happy that the translation team are highly responsible, fully dedicated and greatly conscientious, which has guaranteed the good quality of this translation work.

The completion of this translation project is a collaborative effort. Firstly, I want to extend my sincere gratitude to my translation team:

Associate Professor Li Qing (李清) from Anhui Agricultural University for translating 1.1 and 1.2 of Chapter One; Professor Hu Jian (胡健) from Anhui University for translating 1.3 and 1.4 of Chapter One; Lecturer Zhu Linglin (朱玲麟) from Anhui University for translating 1.5.1 of Chapter One, and 3.2, 3.3 and 3.4 of Chapter Three; Lecturer Zhuang Xiaoling (庄晓玲) from Anhui University for translating 1.5.2 of Chapter One, and 3.1 and 3.5 of Chapter Three; Lecturer Bao Man (鲍曼) from Anhui University for translating 1.5.3 of Chapter One, and 2.3 and 2.4 of Chapter Two; Lecturer Ye Lan (叶岚) from Anhui University for translating 1.5.4 of Chapter One, and 2.1, 2.2 and 2.5 of Chapter Two; Lecturer Liu Sheng (刘胜) from Chizhou University for translating Chapter Four; Associate Professor Wu Xiaofang（吴小芳）from Chuzhou University for translating Chapter Five; Professor Xia Beijie（夏蓓洁）from Hefei University for translating Chapter Six; Professor Hao Tugen（郝涂根）from Anqing Normal University and Lecturer Xu Liyue（徐丽月）from Zhejiang College attached to Tongji University are responsible for the proofreading of the whole book; Professor Zhu Yue（朱跃）from Anhui University, head of the translation team, is responsible for the review of the whole book.

Heart-felt thanks also go to editors from University of Science and Technology of China Press who have made the publication of this translation work possible, including Yao Shuo（姚硕）, Li Panfeng（李攀峰）, Yu Xiumei（于秀梅）, Hu Xueyin（胡雪吟）, Han Jiwei（韩继伟）and Huang Chengqun（黄成群）. Without their technical support and their conscientious work, the publication of this work would have been impossible.

Zhu Yue

January 21, 2019

This book is the result of a co-publication agreement between University of Science and Technology of China Press (China) and Paths International Ltd (U.K.)

This book is published with financial support from China Classics International

Title: The Four Great Inventions of Ancient China: Their Origin, Development, Spread and Influence in the World
Author: Pan Jixing
Translators: Zhu Yue, et al.
ISBN: 978-1-84464-542-8
Ebook ISBN: 978-1-84464-543-5
Copyright © 2019 by Paths International Ltd, U.K. and by University of Science and Technology of China Press, China

Paths International Ltd
United Kingdom
www.pathsinternational.com

Published in the United Kingdom

CPSIA information can be obtained
at www.ICGtesting.com
Printed in the USA
LVHW060026170320
650211LV00002B/2